Advances in Solid State Physics
Volume 45

Advances in Solid State Physics

Advances in Solid State Physics is a book series with a history of about 50 years. It contains the invited lectures presented at the Spring Meetings of the "Arbeitskreis Festkörperphysik" of the "Deutsche Physikalische Gesellschaft", held in March of each year. The invited talks are intended to reflect the most recent achievements of researchers working in the field both in Germany and worldwide. Thus the volume of the series represent a continuous documentation of most recent developments in what can be considered as one of the most important and active fields of modern physics. Since the majority of invited talks are usually given by young researchers at the start of their career, the articles can also be considered as indicating important future developments.
The speakers of the invited lectures and of the symposia are asked to contribute to the yearly volumes with the written version of their lecture at the forthcoming Spring Meeting of the Deutsche Physikalische Gesellschaft by the Series Editor early before the meeting. Articles are submitted during the meeting. All articles are refereed.
Advances in Solid State Physics is addressed to all scientists at universities and in industry who wish to obtain an overview and to keep informed on the latest developments in solid state physics. The language of publication is English.

Series Editor

Prof. Dr. Bernhard Kramer

International University Bremen
Campus Ring 1
28757 Bremen
Germany
b.kramer@iu-bremen.de

Bernhard Kramer (Ed.)

Advances in Solid State Physics

45

With 203 Figures

Springer

Prof. Dr. Bernhard Kramer (Ed.)
International University Bremen
Campus Ring 1
28757 Bremen
Germany
b.kramer@iu-bremen.de

Physics and Astronomy Classification Scheme (PACS): 60.00; 70.00; 80.00

ISSN print edition: 1438-4329
ISSN electronic edition: 1617-5034
ISBN-10 3-540-26041-2 Springer Berlin Heidelberg New York
ISBN-13 978-3-540-26041-7 Springer Berlin Heidelberg New York

Springer is a part of Springer Science+Business Media

springeronline.com

Typesetting by the authors using a Springer TEX macro package
Final processing: DA-TEX · Gerd Blumenstein · www.da-tex.de
Production: LE-TEX GbR, Leipzig, www.le-tex.de

Cover concept using a background picture by Dr. Ralf Stannarius, Faculty of Physics and Earth Sciences, Institute of Experimental Physics I, University of Leipzig, Germany
Cover design: *design & production* GmbH, Heidelberg

Printed on acid-free paper 56/3141/YL 5 4 3 2 1 0

Preface

The spring meeting of the Arbeitskreis Festkörperphysik in the World Year of Physics 2005, the Einstein Year, was held from 4–11 March 2005 in Berlin, together with the yearly meetings of the other Arbeitskreise and Fachverbände of the Deutsche Physikalische Gesellschaft. The total number of conference attendees was about 8000, mostly young physicists.

The present volume of *Advances in Solid-State Physics* contains the written versions of selected invited lectures from the Arbeitskreis Festkörperphysik. Many topical talks given at the numerous symposia are included. Most of these were organized collaboratively by several of the divisions of the Arbeitskreis. They cover a number of very interesting and timely subjects.

The book presents, to some extent, the status of the field of solid-state physics in 2005 not only in Germany but also internationally. It is "nanoscience," namely the physics of quantum dots and wires, electrical transport, optical properties, spin transport in nanostructures, and magnetism on the nanoscale, that is of central interest to the physics community. Also, soft matter and biological systems are attracting increasing attention, especially from young physicists.

Hamburg, April 2005 *Bernhard Kramer*

Contents

Part I Disorder, Soft Matter and Biological Systems

Thermodynamics – Past, Present and Future
Werner Ebeling 3
1 Foundation of the Three Fundamental Laws 3
2 The Key Role of Thermodynamics in the 20th Century 8
3 Thermodynamics of Selforganization and Evolution Processes 9
4 Thermodynamics, Non-Linear Dynamics, Information Processing and Life 11
5 Exotic Applications 12
6 Open Problems, Conclusions 13
References 14

Microscopy on Thermal Capillary Waves in Demixed Colloid-Polymer Systems
Dirk G. A. L. Aarts, Matthias Schmidt, Henk N. W. Lekkerkerker, Klaus R. Mecke 15
1 Introduction 15
2 Theoretical Background 16
3 Experimental Methods 20
4 Results and Discussion 22
5 Conclusions 25
References 26

Linear Polymers in Networks: Swelling, Diffusion, and Interdigitation
Mark Geoghegan 29
1 Swelling of Polymer Networks 30
2 Diffusion 35
3 Interdigitation 39
4 Conclusions 43
5 References 43

Biomolecular Layers on Silicon Studied by Optical Spectroscopy
Dietrich R. T. Zahn, Simona D. Silaghi ... 45
1 Introduction ... 45
2 Experimental and Data Analysis ... 46
3 Results ... 48
4 Summary ... 56
References ... 56

Lambert Diffusion in Porous Media in the Knudsen Regime
Stefanie Russ, Stephan Zschiegner, Armin Bunde, Jörg Kärger ... 59
1 Introduction ... 59
2 The Self-Diffusion Problem ... 60
3 The Transport-Diffusion Problem ... 64
4 Summary and Conclusions ... 67
References ... 68

Dynamics of Membranes: From Passive to Active Systems
H.-G. Döbereiner, B.J. Dubin-Thaler, G. Giannone, C.K. Haluska, P.G. Petrov, A. Reinecke, K. Riske, M.P. Sheetz ... 71
1 Classification of Membrane Dynamics ... 71
2 Passive Membranes ... 73
3 Reactive Membranes: Phospholipase Activity ... 77
4 Membranes and Active Gels ... 79
5 Towards Modular Cell Biology ... 81
References ... 82

Disentangling Noise and Fluctuations in Data Sets of Complex Systems
R. Friedrich, D. Kleinhans, J. Peinke ... 83
1 Introduction ... 83
2 Nonlinear Time Series Analysis of Fluctuating Systems ... 84
3 Analyzing Stochastic Systems ... 85
4 Applications ... 86
5 Refined Procedure for Estimating Drift and Diffusion ... 87
6 Summary ... 93
References ... 94

Evolution in Complex Systems: Record Dynamics in Models of Spin Glasses, Superconductors and Evolutionary Ecology
Henrik Jeldtoft Jensen ... 95
1 Introduction ... 95
2 Three Models ... 96
3 Record Dynamics and its Manifestation ... 99
4 Consequences ... 102

5 Summary and Conclusions ... 104
References ... 104

Part II Quantum Transport and Correlations

Flavor Degeneracy and Effects of Disorder in Ultracold Atom Systems
Walter Hofstetter ... 109
1 Introduction and Overview ... 109
2 Optical Lattices and Strong Correlations ... 110
3 Multi-Component Systems ... 112
4 Disorder and Interaction ... 116
5 Summary and Outlook ... 119
References ... 120

Quantum Noise in Mesoscopic Systems
Wolfgang Belzig ... 123
1 Introduction ... 123
2 Probability Distributions and Cumulant Generating Functions ... 124
3 Statistic of Transport in Normal and Superconducting Metal Contacts ... 125
4 Strongly Correlated Statistics ... 129
5 Number Statistics in Fermionic Quantum Gases ... 131
6 Conclusion ... 133
References ... 134

Dynamical Variational Principles for Strongly Correlated Electron Systems
Michael Potthoff ... 135
1 Variational Principles and Approximation Strategies ... 136
2 Various Variational Approaches ... 137
3 New Approximations ... 143
4 Open Problems ... 145
References ... 146

Sol-Gel Approaches to Photonic Crystal Systems
Frank Marlow, Denan Konjhodzic, Helmut Bretinger, Hongliang Li ... 149
1 Introduction ... 149
2 Inverse Opals ... 150
3 Low-n Films ... 154
4 Outlook: Ferroelectric PhC Films ... 158
References ... 160

Theory of Surface Optical Properties
Olivia Pulci, Maurizia Palummo, Margherita Marsili, Rodolfo Del Sole 161
1 Introduction ... 161
2 Reflectance Anisotropy Calculations ... 163
3 DFT Single-Particle Approach ... 164
4 GW: Single Quasi-Particle Approach ... 164
5 Application to InAs(110) ... 166
6 BSE: Excitonic Effects ... 167
7 Application to C(001)2x1 ... 169
8 Conclusion ... 170
References ... 171

Manipulations of a Qubit in a Semiconductor Quantum Dot
Artur Zrenner, Stefan Stufler, Patrick Ester, Max Bichler ... 173
1 Introduction ... 173
2 Single Quantum Dot Photodiodes ... 174
3 Incoherent Spectroscopic Results ... 175
4 Coherent Manipulations ... 178
5 Summary ... 182
References ... 183

The Integrable $sl(2/1)$ Superspin Chain and the Spin Quantum Hall Effect
Holger Frahm, Fabian H. L. Essler, Hubert Saleur ... 185
1 Introduction ... 185
2 The Integrable Vertex Model ... 187
3 Spectrum of the Supersymmetric Spin Chain ... 188
4 Continuum Limit of the Lattice Model ... 193
5 Summary and Conclusions ... 195
References ... 196

Intrinsic Spin Hall Effect
Shuichi Murakami ... 197
1 Introduction ... 197
2 Intrinsic Spin Hall Effect ... 198
3 Disorder Effect and Extrinsic Spin Hall Effect ... 202
4 Discussions ... 203
5 Experiments ... 206
6 Summary ... 206
References ... 207

Part III Superconductivity

Point-contact Spectroscopy on Conventional and Unconventional Superconductors
Gernot Goll . . . 213
1 Introduction . . . 213
2 Experimental . . . 214
3 Conventional Superconductors . . . 215
4 Unconventional Superconductors . . . 219
References . . . 224

Ordering Phenomena in Cuprates
Rudi Hackl, Leonardo Tassini, Francesca Venturini, Christine Hartinger, Andreas Erb, Naoki Kikugawa, Toshitsu Fujita . . . 227
References . . . 237

Two-Gap Superconductivity in Magnesium Diboride
Thomas Dahm . . . 239
1 Introduction . . . 239
2 Two-Gap Superconductivity . . . 240
3 Upper Critical Field Anisotropy . . . 241
4 Microwave Conductivity Peak . . . 243
5 Nonlinear Microwave Response . . . 245
6 Conclusions . . . 247
References . . . 248
References . . . 248

Part IV Semiconductors

Electron Spin Relaxation in Semiconductors
Daniel Hägele, Stefanie Döhrmann, Jörg Rudolph, Michael Oestreich . . 253
1 Introduction . . . 253
2 Anisotropic Dyakonov Perel Spin Relaxation . . . 254
3 Experiments . . . 255
4 Intersubband Spin Relaxation . . . 257
5 Conclusion . . . 260
References . . . 261

Donor Levels in ZnO
Holger von Wenckstern, Swen Weinhold, Gisela Biehne, Rainer Pickenhain, Heidemarie Schmidt, Holger Hochmuth, Marius Grundmann . . . 263
1 Introduction . . . 263
2 Samples . . . 264

3 Transport Properties of ZnO 264
4 ZnO Schottky Diodes 268
5 Summary 272
References 274

Excitonic Properties of ZnO
C. Klingshirn, H. Priller, M. Decker, J. Brückner, H. Kalt, R. Hauschild, J. Zeller, A. Waag, A. Bakin, H. Wehmann K. Thonke, R. Sauer, R. Kling, F. Reuss, Ch. Kirchner 275
1 Introduction and Historic Remarks 275
2 Growth, Doping, Transport, and Deep Centers 276
3 Band Structure, Excitons, and Polaritons 277
4 High Excitation Phenomena 281
5 Conclusion and Outlook 284
References 284

Incorporation of Acceptors in ZnO
B. K. Meyer, J. Sann, A. Zeuner 289
1 Introduction 289
2 Experimental Details 290
3 Experimental Results 290
4 Discussion 296
5 Conclusions 298
References 298

Optical Analysis of Monolayers at Surfaces and Interfaces
Veit Wagner 301
1 Introduction 301
2 Inorganic Surfaces and Interfaces 303
3 Organic Surfaces and Interfaces 309
References 311

Chemistry and Morphological Properties of Metal Interfaces to Organic Semiconductors
D.R.T. Zahn, G. Salvan, G. Gavrila, B.A. Paez 313
1 Introduction 313
2 Experimental 314
3 Interaction of Metals with Perylene Derivatives 315
4 Morphology of the Metal Film 317
5 Structural Properties of the Metal/Organic Interfaces 319
6 Summary 322
References 323

Part V Materials and Applications

The Role of Ion Irradiation Induced Viscous Flow in Thin Film Structuring: Hard Matter Going Soft
S. G. Mayr ... 327
1 Introduction ... 327
2 Phenomenology of Radiation Induced Viscous Flow in Metallic Glasses ... 328
3 Atomic-Scale Mechanism of Radiation Induced Viscous Flow ... 333
4 Generalization to Crystalline Systems ... 335
5 Conclusions ... 337
References ... 338

Atomic-Scale Properties of High-k Dielectrics: Ab Initio Study for Pr-Based Materials
Jarek Dąbrowski, Andrzej Fleszar, Gunther Lippert, Grzegorz Lupina, Anil Mane, Hans-Joachim Müssig, Thomas Schroeder, Roland Sorge, Hans Thieme, Christian Wenger, Peter Zaumseil ... 339
1 Introduction ... 339
2 Theoretical Approach ... 339
3 Fixed Charges ... 341
4 Native Point Defects ... 343
5 Si-related Nitrogen-Coordinated Oxygen ... 345
6 Impurities as Fixed Charges ... 348
7 Summary and Conclusions ... 349
References ... 350

Size Effects in Ba(Pb)TiO_3 Nanopowders by EPR and NMR
Emre Erdem, Rolf Böttcher, Hans Jürgen Gläsel, Eberhard Hartmann, Gert Klotzsche, Dieter Michel ... 351
1 Introduction ... 351
2 Experimental ... 352
3 Results and Discussion ... 353
4 Conclusion ... 358
References ... 360

AlGaN/GaN Electrolyte-Gate Field-Effect Transistors as Transducers for Bioelectronic Devices
Georg Steinhoff, Barbara Baur, Hans-Georg von Ribbeck, Günter Wrobel, Sven Ingebrandt, Andreas Offenhäusser, Martin Stutzmann, Martin Eickhoff ... 363
1 Introduction ... 363
2 Device Structure and Processing ... 364

3 Ion-Sensitivity of GaN Surfaces 365
4 Electronic Characteristics of AlGaN/GaN EGFETs 367
5 Recording of Cell Action Potentials 369
6 Summary 372
References 372

Advanced MOSFET Gate Dielectrics for High-Performance Microprocessors: Materials Selection and Analytical Challenges
E. Zschech, H.J. Engelmann, S. Ohsiek, B. Tracy, E. Adem, S. Robie, J. Bernard, D. Schmeisser 375
1 Introduction 375
2 High-k Materials Selection, Materials Properties, Integration Challenges and Analytical Needs 377
3 Silicon Oxynitride 379
4 Aluminum Oxide 382
5 Metal Oxides 383
6 Summary and Outlook 386
References 387

Neutron Tomography as Tool for Applied Research and Technical Inspection
Eberhard H. Lehmann, Peter Vontobel, René Hassanein 389
1 Introduction 389
2 Neutron Interaction with Matter in Comparison to X-ray 390
3 Set-Up of a Neutron Tomography Device 393
4 Principles and Features of the PSI Reconstruction Tools 395
5 Neutron Scattering Artefacts and Their Correction with the Help of Monte-Carlo Simulations 398
6 Examples for Practical Applications from Science and Industry 401
7 Options for Further Improvements 402
8 Conclusions 404
References 404

Absorption- and Phase-Based Imaging Signals for Neutron Tomography
W. Treimer, N. Kardjilov, U. Feye-Treimer, A. Hilger, I. Manke, M. Strobl 407
1 Introduction 407
2 Theoretical Background 408
3 Phase Contrast 411
4 Refraction Contrast 413
5 Small Angle Contrast 414
6 Some Experiments 415

References 419

Index 421

Part I

Disorder, Soft Matter and Biological Systems

Thermodynamics – Past, Present and Future

Werner Ebeling

Institute of Physics, Humboldt-University Berlin,
Newtonstr. 15, 12489 Berlin, Germany

Summary. We begin with historical remarks on the basic contributions to thermodynamics and statistics with some bias to scientists working in Berlin as Helmholtz, Clausius, Nernst and Einstein. We underline the key role of thermodynamic ideas in the scientific revolutions in the 20th century. Further we discuss several recent applications to natural, evolutionary and informational systems, exotic applications as well as perspectives and open problems.

1 Foundation of the Three Fundamental Laws

Thermodynamics as a branch of science was established in the 19th century by Sadi Carnot (1796-1832), Robert Mayer (1814-1878), Hermann Helmholtz (1821-1894), William Thomson (1824-1907) and Rudolf Clausius (1822-1888). Evidently Mayer was the first who formulated the law of energy conservation. His paper *"Bemerkungen über die Kräfte der unbelebten Natur"* published 1842 in Liebig's Annalen is expressing the equivalence of work and heat. Joule's conclusions on this matter were based on direct measurements of the conversion of work into heat. A great role in the foundation of thermodynamics played physicists working in the middle of the 19th century in Berlin. We will discuss their contribution here in some more detail, just to illustrate the *genius loci*. In particular it was Hermann Helmholtz who determined the direction of thermodynamic research [1, 2]. At 27 years of age Helmholtz – at that time still working as a military surgeon in Potsdam – reported 1847 to the *"Berliner Physikalische Gesellschaft"* about a new principle of conservation of energy. The underlying experimental research which he carried out in the laboratory of his adviser Professor Magnus was primarily devoted to the conversion of matter and heat in such biological processes as rotting, fermentation and muscular activity. From experiments and brilliant generalization emerged the principle of conservation of energy or what is now called the first law of thermodynamics. Neither Mayer nor Joule recognized its fundamental and universal character as clearly as Helmholtz. The work of Mayer and Joule was unknown to Helmholtz at that time. Helmholtz had to fight hard for the recognition of his work. Professor Poggendorf, the editor of the *"Annalen der Physik und Chemie"*, rejected the paper which seemed to him too speculative. Professor Magnus also did not like the work, but at

B. Kramer (Ed.): Adv. in Solid State Phys. **45**, 3–14 (2005)

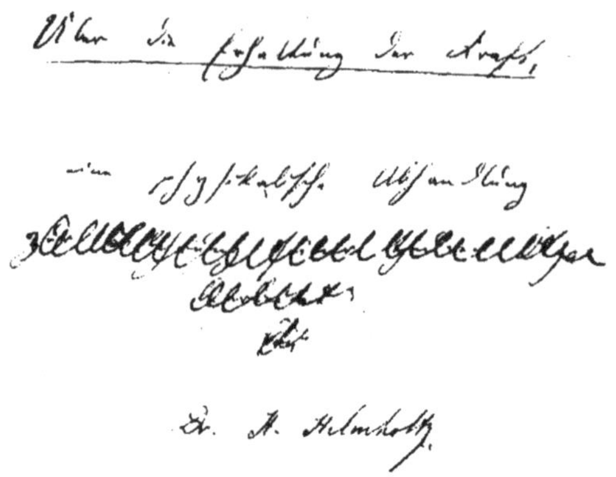

Fig. 1. Title page of the manuscript of Helmholtz's work "Über die Erhaltung der Kraft – eine physikalische Abhandlung"

least he recommended to print it as a brochure, as was quickly managed with the help of Professor Jacobi.

Rudolf Clausius (1822-1888) one of the young colleagues of Helmholtz played an essential role in the further elaboration of the new law [3]. After studying in Berlin, he taught for some years at the Friedrich-Werdersches Gymnasium in Berlin and was a member of the seminar of Professor Magnus at the Berlin University. A report on Helmholtz's work, given to Magnus' colloquium, was the beginning of Clausius' involvement in this matter. Building on the work of Helmholtz and Carnot he published 1850 in Poggendorff's Annalen a first formulation of the second law of thermodynamics. Clausius was fully aware of the impact of his discovery. The title of his paper explicitly mentions *"laws"*. Clausius stated that heat cannot pass spontaneously from a cooler to a hotter body. Unlike Carnot, and following Joule, Clausius interpreted the passage of heat as the transformation of different kinds of energy, in which the total energy is conserved. To generate work, heat must be transferred from a reservoir at a high temperature to one at a lower temperature, and Clausius here introduced the concept of an ideal cycle of a reversible heat engine. In 1851 Thomson formulated independently of Clausius another version of the second law. Thomson stated that it is impossible to create work by cooling down a thermal reservoir. The central idea in the papers of Clausius and Thomson was an exclusion principle: *"Not all processes which are possible according to the law of the conservation of energy can be realized in nature"*. This means, the second law of thermodynamics is a selection principle of nature. Although it took some time before Clausius' and Thomson's work was fully acknowledged, it was fundamental not only for the further development of physics, but also for science in general. In later works Clausius arrived at more general formulations of the second law, in particular he introduced the quotient of the quantity of heat absorbed by a body and the temperature of the body $d'Q/T$ as the change of entropy. In a next step

Clausius was thinking about an atomistic foundation of thermodynamics and published two papers *"Über die Art der Bewegung, die wir Wärme nennen"*, which appeared 1857/1858 in the *Annalen der Physik*. This work is the first comprehensive treatment of the kinetic theory of gases. Clausius developed new terms like the mean free path, cross section etc. and introduced in 1865 the new quantity entropy. Further he derived in 1870 a virial theorem for gases. Parallel to Clausius's work the statistical theory was developed by Maxwell, who derived in 1860-66 the probability distribution for the velocities of molecules in a gas and formulated a first version of a transport theory. In 1867 Maxwell discussed first the statistical nature of the second law of thermodynamics and considered the connection between entropy and information. His *"Gedankenexperiment"* about a demon observing molecules we may consider as the first fundamental contribution to the development of an information theory. In 1878 Maxwell proposed the new term *"statistical mechanics"*.

Ludwig Boltzmann (1844-1906) studied physics at the University of Vienna. He was deeply influenced by Josef Stefan (1835-1903) and Johann Loschmidt (1821-1895). Boltzmann started to work on the kinetic theory of gases. In 1866, he found the energy distribution for gases. In 1871 he formulated the ergodic hypothesis, which is fundamental for the modern version of statistical physics and for the connection to nonlinear dynamics. His work culminated in 1872 with the formulation of a kinetic equation and the H-theorem, which established a connection to the second law. In the year 1872, which was so central for his work, Boltzmann visited Helmholtz in Berlin. In the mean time, after professorships in anatomy and physiology at several German universities, Helmholtz had returned to Berlin to succeed Magnus as director of the physical institute of the university. Then began a very productive period in the history of physical research in Berlin. No burning questions of contemporary physics remained untouched by Helmholtz or his fellow workers, but thermodynamical problems remained central. During Helmholtz's second period in Berlin his work revolved around pure and applied problems of thermodynamics. He developed the concept of free energy and investigated the relationship between the heat of reaction and the electromotive force of a galvanic cell. As president of the Physikalische-Technische Reichsanstalt Helmholtz stimulated studies of the properties and applications of light. The investigations in particular by Wilhelm Wien led later to the development of a thermodynamical theory of heat radiation by Max Planck.

In 1889 Max Planck (1858-1947) succeeded Kirchhoff at the Berlin Chair of Theoretical Physics. He was a pioneer in understanding the fundamental role of entropy and its connection with the probability of microscopic states. Later he improved Helmholtz's chemical thermodynamics and his theory of double layers. He was the first who wrote down explicitely the famous formula

$$S = k \log W \,. \tag{1}$$

Fig. 2. Leading scientists in Berlin: Helmholtz (1821-1894), Clausius (1822-1888), Nernst (1864-1941) Einstein (1879-1955) und Planck (1858-1947)

An independent and more general approach to statistical thermodynamics and the role of entropy was developed by the American physicist Josiah Willard Gibbs (1839-1903). Gibbs developed the ensemble approach, the entropy functional and was the first to understood the role of the maximum entropy method which was later further developed by Jaynes.

The next important contribution to thermodynamics is connected with the work of Walther Nernst (1864-1941) who accepted in 1905 a call on a chair at the Berlin university. In 1905 Nernst detected the "missing stone in thermodynamics", the third law of thermodynamics. Nernst's seminal idea arose from the critical analysis of experimental data on chemical and electrochemical reactions at low temperatures, where there appeared good correspondence between the free energy and the internal energy. Nernst found that the agreement improved at lower temperatures. This led him to the third law. Some years later Planck gave Nernst's new principle the following general and widely known formulation: *"The entropy of all bodies which are in internal equilibrium vanishes at the zero point of temperature"*. After postulating his new theorem Nernst and his collaborators took great efforts to prove this new law of nature. The specific heat, being of special importance, was determined for several substances at low temperatures. This was a very difficult scientific problem which called for the construction of equipment and instruments from scratch and finally led to a new and very fruitful branch of investigations – low temperature physics and technology.

Let us devote the final part of this section to the question *How to present the three fundamental laws today?*. The experience in teaching of physics shows that the three basic laws of physics are difficult to understand. Students reproduce quite often just several formulae without reaching a deeper understanding. Therefore we want to discuss here the problem: How to present the three laws today in a most clear version?

The Zeroth Law: Thermodynamic systems possess a special state – thermodynamic equilibrium. A system in this particular state shows no changes after isolation. Systems in thermodynamic equilibrium are characterized by a scalar, transitive variable T, the temperature.

The First Law: Thermodynamical and other macroscopic real systems are characterized by an extensive quantity energy E. Energy can neither be created nor destroyed. Energy can be exchanged with other systems and appears in such processes in different forms, as e.g. heat, work, chemical energy. Energy can be converted from one form to other forms and moved to other systems,

$$dE = d_i E + d_e E, \quad d_i E = 0, \quad d_e E = d'Q + d'A + \sum \mu_k dN_k \,. \tag{2}$$

In isolated systems we find $dE = 0$ and consequently, energy is conserved. A precise definition of energy is not known. The question: *"What is energy?"* is commented by Poincare in the following way: *"In every instance it is clear what energy is and we can give at least a provisional definition of it; it is impossible however, to give a general definition ... . One sees it dissolve before one's eyes, leaving only the words: There is something, that remains constant (in isolated systems)".*

The Second Law: Thermodynamical and other macroscopic real systems are characterized by an extensive quantity entropy S. Entropy can be created but never be destroyed. Entropy can be exchanged with other systems and in particular by exchange of heat,

$$dS = d_i S + d_e S, \qquad d_i S \geq 0, \qquad d_e S = \frac{d'Q}{T} + \ldots \,. \tag{3}$$

In isolated systems we find $dS \geq 0$, i.e. that the entropy will always increase or remain constant (in thermodynamic equilibrium). The expression for the exchanged entropy is not unique, since several definitions of heat exist.

The Gibbs-Helmholtz Fundamental Relation: In thermodynamic equilibrium, energy, entropy, the extensive volume type variables L_k and the particle numbers N_k are depending on each other. This is expressed by the differential relation (Pfaffian form)

$$dE = TdS + \sum l_j dL_j + \sum \mu_k dN_k \,. \tag{4}$$

The Third Law: Energy and entropy are finite for finite systems and bounded from below $E > 0$, $S > 0$. In the limit $T \to 0$, the entropy as well as its derivatives with respect to extensive variables disappear asymptotically,

$$S \to 0, \qquad \frac{dS}{dL_k} \to 0 . \tag{5}$$

The first and the second fundamental laws are valid for any macroscopic process in nature and society. May be, these are the only laws which have a universal range of validity? Quantum theory and general relativity theory modified our understanding of the energy and entropy concepts, however their fundamental role for all macroscopic processes remained untouched. The third law is less fundamental, it is a law of thermal systems only. However it has deep implications for physical systems. Low temperature physics is of increasing importance.

2 The Key Role of Thermodynamics in the 20th Century

The three fundamental laws of thermodynamics had a deep influence on the physics of the development of physics in the 20th century. In particular we mention applications to

- Quantum Theory
- Low Temperature Physics
- Large-Scale Physics (the universe, stellar objects, black holes)
- Small-Scale Physics (nuclei, elementary particles)
- Biological, Ecological and Social Systems,
- Informational Systems.

The pioneers of the first revolutionary applications to physical problems were Planck, Nernst and Einstein. Planck applied thermodynamic methods to radiative processes and searched for relations between energy and entropy. In order to get agreement with experimental findings, he could not avoid the introduction of a new elementary quantum of action h. This was the first revolution in physics raised by thermodynamics. The second one is connected to the work of Nernst who worked since 1905/06 with a group of talented physicists on the experimental verification of his heat theorem. This led to the development of low temperature physics and stimulated the work of Einstein. Einstein started his work on statistical physics in 1902/3 with two very interesting papers on *"'The kinetic theory of thermal equilibrium and the second law of thermodynamics"*, published in the *"Annalen der Physik"*. Here independently of Gibbs, Einstein developed the basic ideas of ensemble theory and the statistics of interacting systems. In his dissertation, presented in 1905 to the Zürich University, he developed a first correct theoretical interpretation of Brownian motion. This work was published in volume 17 of the

"Annalen der Physik". Einstein was at that time only 26 years old. As well known, he published in the same volume of the "Annalen", also two other fundamental papers devoted to the theory of relativity and the theory of the photo effect.

In 1907, Einstein turned to problems of low temperatures connected with the third law. He proposed that quantum effects lead to the vanishing of the specific heat at zero temperature. His theory led to a deeper understanding of the low temperature thermodynamics and may be considered as the origin of quantum statistics. Einstein's work attracted the attention of Nernst and his collaborators and by 1910 they succeeded in confirming this prediction. In this way the third law of thermodynamics as well as the young and still controversial quantum theory found one of its first experimental verifications. In 1913, Nernst together with Planck, was able to bring the *"new Copernicus"* Einstein to Berlin, they could offer the unconventional genius excellent working and living conditions. As a *"paid genius"* in Berlin, Einstein could complete his general theory of relativity, and make further important contributions to thermodynamics and statistical physics. In 1924, he generalized the Bose theory of photon gases, developed a new quantum statistics, the so-called Bose-Einstein statistics. In addition to the Bose-Einstein condensation his ideas about the interaction between radiation and matter should be emphasized. In 1916 his discussion of spontaneous emission of light and induced emission and adsorption forms the theoretical basis of the nonlinear dynamics and stochastic theory of the modern lasers. Concerning the many other fundamental contributions to thermodynamics and statistical physics in the last century we must restrict ourselves to brief remarks. The German-Greek mathematician Constantin Caratheodory formulated thermodynamics on an axiomatic basis. His analyses of such fundamental concepts as temperature and entropy in terms of the mathematical theory of Pfaffian differential forms were not appreciated by most of his contemporaries, although Planck was an early supporter of what has become one of the important branches of modern thermodynamics. Walter Schottky (1886-1976) developed industrial applications of thermodynamics and wrote a famous textbook "Thermodynamik" (1929).

3 Thermodynamics of Selforganization and Evolution Processes

First applications of thermodynamics to the evolution of the universe go back to Helmholtz, Clausius and Boltzmann and are connected with the idea of the *"Wärmetod"*. A completely new approach was based on a cosmological model presented 1922 by the mathematician Alexander Friedmann in Petersburg based on Einstein's general relativity. Friedmann derived the model of an expanding matter-filled universe from Einstein's field equations. The first who applied thermodynamics to this model was George Gamov, a former student of Friedmann. Together with Alpher, Bethe and Hermann he developed in the 40th the thermodynamic model of the Big Bang. The Big Bang theory of the history of the universe is essentially a thermodynamical theory based on thermodynamical relations applied to the very exotic early stages of the expansion. The assumption of adiabatic expansion leads to the following law of temperature decrease in time,

$$T \simeq \frac{const}{\sqrt{t}} \,. \tag{6}$$

In the last stages of evolution, matter is self-structuring. It forms stars and planets and the temperature gradient between sun and earth - the photon mill - gives rise to selforganization on earth [4]. The earth is an open system which exports entropy in the amount of about $1W/m^2K$. This is the driving force of evolution on earth. Essential contributions to our understanding of the thermodynamic basis of life were given by Mayer, Boltzmann, Schrödinger and Prigogine. The main idea of these pioneers is that the exchange with surrounding is relevant. In open systems with entropy export - the formation of structures does not contradict the 2nd law.

This research lead to the development of a thermodynamics of open systems and a theory of selforganization [4]. Well-known examples of selforganization in nature are the Belousov-Zhabotinsky waves, the Liesegang-rings and Benard's hydrodynamic cells.

Another closely related line of the development of thermodynamics is the foundation of irreversible thermodynamics. We mention only the early work of Thomson, Rayleigh, Duhem, Natanson, Jaumann and Lohr. The final formulation of the basic relations of irreversible thermodynamics we owe to the work of Onsager (1931), Eckart (1940), Meixner (1941), Casimir (1945), Prigogine (1947) and De Groot (1951). Irreversible thermodynamics is essentially a nonlinear science, which needs for its development the mathematics of non-linear processes, the so-called nonlinear dynamics.

Let us discuss now in brief the important question of evolution principles: The most general evolution principle results from the second law which leads to the following requirement for the entropy production

$$P = \frac{d_i S}{dt} \geq 0 \,. \tag{7}$$

For irreversible processes the entropy production is positive, the inverse process would destroy entropy, what is forbidden by the 2nd law.

An independent principle was found by Prigogine,

$$\frac{dP}{dt} \leq 0 \,. \tag{8}$$

Entropy production decreases in the realm of linear processes. A more general principle formulated by Glansdorff and Prigogine states that the change of the force-determined part $d_x P$ is non-positive for all processes. Landauer and several other workers have shown that this statement is not correct for all processes and is not a general evolution criterion.

There exist several more special evolution criteria. For example for all Markov processes with the time-dependent probability $P(x,t)$ there exists a functional (Kullback-Leibler entropy),

$$K = \int dx\, P(x,t) \log \left[\frac{P(x,t)}{P_0(x)} \right] . \tag{9}$$

($P_0(x)$ stationary distribution) which is positive and never increasing (Bergmann, Lebowitz, van Kampen, Schlögl et al.)

$$K \geq 0, \qquad \frac{dK}{dt} \leq 0 . \tag{10}$$

This very general and interesting statement contains the second law and other evolution criteria. We mention that there exist several other statements [2] as e.g. the Jarzynski theorem which states that equilibrium information (on free energy) can be extracted from an ensemble of non-equilibrium measurements.

4 Thermodynamics, Non-Linear Dynamics, Information Processing and Life

The pioneers of this direction of thermodynamics were Mayer, Maxwell, Boltzmann, von Neumann, Szilard, Schrödinger, Brillouin and Wolkenstein. In the 19th century a close relation between statistical thermodynamics and nonlinear science was not known. Henri Poincare, the father of nonlinear science, was the strongest opponent of Ludwig Boltzmann. In recent times it became clear that Poincare's work contains the keys for the foundation of Boltzmann's theory. In particular this refers to the concept of instability of trajectories developed by Poincare. Today nonlinear science and thermodynamics are closely connected, e.g. the thermodynamic formalism plays an important role in nonlinear dynamics as well as the Kolmogorov-Sinai entropy.

A significant progress was made through the investigations of G. Birkhoff and J. von Neumann. The Hungarian Johann von Neumann (1903-1957) came in the 1920s to Berlin attracted by the sphere of action of Planck and Einstein in physics and von Mises in mathematics. Von Neumann made important contributions to the statistical and quantum-theoretical foundations of thermodynamics. Von Neumann belonged to the group of "surprisingly intelligent Hungarians" (D. Gabor, L. Szilard, E. Wigner), who studied and worked in Berlin around this time. Von Neumann formulated a general quantum-statistical theory of the measurement process, including the interaction between observer, measuring apparatus and the object of observation. This brings us back to Maxwell. In fact information-theoretical considerations in statistical physics start with Maxwells speculations about a demon observing the molecules in a gas. Maxwell was interested in the flow of information between the observer, the measuring apparatus and the gas. In fact this was the first investigation about the relation between observer and object, information and entropy. This line of investigation was continued by Leo Szilard, prominent assistant and lecturer at the University of Berlin and a personal friend of von Neumann. His thesis (1927) *"Über die Entropieverminderung in einem thermodynamischen System bei Eingriffen intelligenter Wesen"* investigated the connection between entropy and information. This now classic work is probably the first comprehensive thermodynamical approach to a theory of information processes. The first consequent approach to connect the foundations of statistical physics with information theory is due to Jaynes in the years (1957-1975). The information-theoretical method is based on the maximum entropy principle. In highest generality this approach was developed by Rouslan Stratonovich.

We have to mention also of the important contribution of Erwin Schrödinger to the foundation of statistical and biological thermodynamics. In 1927, Schrödinger succeeded Planck in the chair of theoretical physics. In the fall of 1933 he resigned from this post and after some years of traveling he found his final refuge in Dublin. Here in 1944 he published two little but very influential books *"Statistical Thermodynamics"* and *"What is Life?"*, which considerably influenced the development of science and especially statistical thermodynamics and its applications to life sciences.

We consider life is a high (the highest?) form of selforganization, it is connected with export of entropy to the surrounding, and information processing. Information processing is a "conditio sine qua non" for life. Living systems are "by definition" information processing systems originated from natural evolution (they not based on design and this takes time!). Thermodynamic models are important for the understanding of living systems (studies of the balance of matter, energy, entropy–export and production) [4]. Thermodynamics also plays a key role for modeling ecosystems.

5 Exotic Applications

We consider here a new application of thermodynamics to hydrogen and deuterium plasmas at Mbar pressures [5, 6]. This is a problem of much interest for astrophysical applications since hydrogen is the most abundant element in the universe. Recently several new experimental devices reach Mbar pressures as gas guns, explosive shocks, wire explosions, laser shocks. Already Wigner and Abrikosov suggested for $T = 0\,\mathrm{K}$ the existence of a phase transition to a highly conducting state in the Mbar region. New theories of dense plasma describe this second phase transition for all temperatures [7]. At low temperatures and pressures, hydrogen is a molecular solid or fluid. At high pressures above 100 GPa, hydrogen is supposed to undergo a transition to a highly conducting state which has been verified experimentally for the first

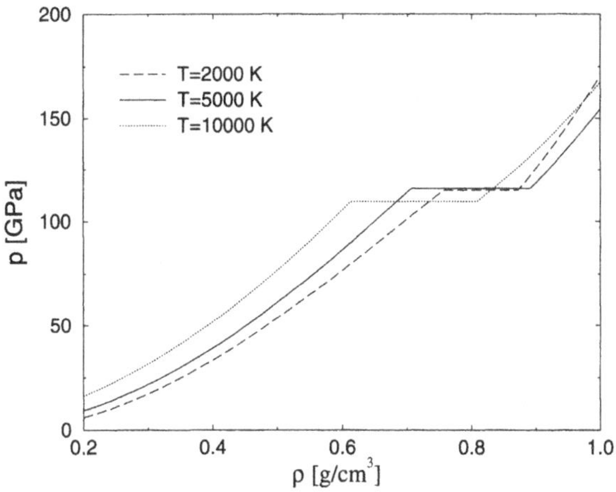

Fig. 3. Pressure as function of the density for various temperatures. A Maxwell construction was performed in the instability region leading to constant pressure in the coexistence region

time in the shock-compressed fluid around 140 GPa and 3000 K [8]. Similar conductivity data have been reported recently for that high-pressure fluid domain [9]. The physical nature of this transition at extreme conditions is not fully explored. The interesting question, whether or not this transition is accompanied by a first-order phase transition with a corresponding instability region, a coexistence line, and a critical point has been treated in our work within advanced many-particle methods adopting a chemical picture. There, the different components in a dense, partially ionized plasma such as molecules H_2, atoms H, molecular ions H_2^+ or H^-, electrons e and protons p interact via effective pair potentials [5, 6]. Several results for hydrogen plasmas are demonstrated in Figs. 3,4. Several estimates for the critical point of the phase transition which is first-order phase transition have also been obtained [7] which are around

$$T_{cr} \simeq 16000\,\mathrm{K}\,. \tag{11}$$

These problems attract in recent times much experimental and theoretical interest, however many problems still remain unsolved, on the experimental as well on the theoretical side.

6 Open Problems, Conclusions

Thermodynamics contributed to the big discoveries of the 20th century and to the theoretical understanding of our world (Weltbild) and survived. We have now good models for many special processes/mechanisms of selforganisation and evolution, and also for many exotic processes. The great open

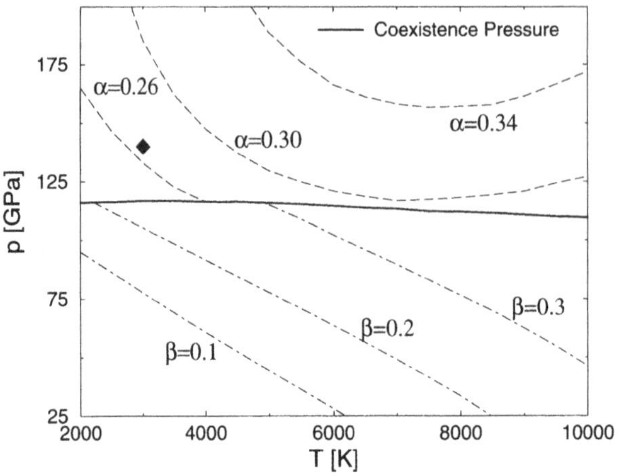

Fig. 4. Coexistence pressure and lines of constant degree of dissociation β and ionization α, respectively, as function of the temperature. The conditions where Weir *et al.* [8] observed metallic conductivity is indicated by a diamond

problems are connected with the theory of far from equilibrium processes and information-processing. In this field most questions are still open. Open problems are in particular connected with evolutionary principles and the the evolution of information processing in nature.

References

1. W. Ebeling, D. Hoffmann, Eur. J. Phys. **12**, 1 (1991).
2. W. Ebeling, I. Sokolov, *Statistical Thermodynamics and Stochastic Theory of Non-equilibrium Systems*, (World Scientific Publ., Singapore 2005).
3. W. Ebeling, J. Orphal, Wiss. Z. Humboldt-Univ. Berlin **39**, 210 (1990).
4. W. Ebeling, A. Engel, R. Feistel, *Physik der Evolutionsprozesse.* (Akademie-Verlag, Berlin 1990).
5. D. Beule, W. Ebeling, A. Förster, H. Juranek, S. Nagel, R. Redmer, G. Röpke, Phys. Rev. B **59**, 14177 (1999).
6. D. Beule, W. Ebeling, A. Förster, H. Juranek, R. Redmer, G. Röpke, Phys. Rev. E **63**, 060202(R) (2001).
7. W. Ebeling, G. E. Norman, J. Stat. Phys. **110**, 861 (2003).
8. W. J. Nellis, S. T. Weir, A. C. Mitchell, W. J. Nellis, Phys. Rev. B **59**, 3434 (1999).
9. V. Ya. Ternovoi et al., Physica B **265**, 6 (1999).

Microscopy on Thermal Capillary Waves in Demixed Colloid-Polymer Systems

Dirk G. A. L. Aarts[1], Matthias Schmidt[2], Henk N. W. Lekkerkerker[1], and Klaus R. Mecke[3]

[1] Van't Hoff Laboratory, Debye Institute, Utrecht University, Padualaan 8, 3584 Utrecht, The Netherlands
[2] Institut für Theoretische Physik II, Heinrich-Heine-Universität Düsseldorf, Universitätsstraße 1, 40225 Düsseldorf, Germany
[3] Institut für Theoretische Physik, Universität Erlangen, Staudtstrasse 7, 91058 Erlangen, Germany

Summary. Recently we have shown how to tune length- and timescales in demixed colloid-polymer dispersions in such a way that thermal capillary waves at the free interface between demixed fluid phases can be studied directly by means of laser scanning confocal microscopy [Aarts, Schmidt and Lekkerkerker, Science **304**, 847 (2004)]. Here, we focus on several static properties of the interface. We show that the probability of fluctuations of the local interface position around its equilibrium value is Gaussian. By comparing two-point correlations of these fluctuations as a function of distance with predictions from capillary wave theory, we obtain results for the interfacial tension and the capillary length. The presented technique enables us to measure also the probability distribution of the tilt angle of the local interface normal and the vertical direction.

1 Introduction

A few years after Einstein's famous 1905 theory of Brownian motion [1] Smoluchowski used molecular kinetic theory to give the first correct explanation of critical opalescence [2]. In the same paper Smoluchowski indicated that due to the thermal motion of the molecules the free interface between two fluids, like that between a liquid and a vapor, inevitably becomes rough. Using the thermodynamic fluctuation theory developed by Einstein in 1910 [3] Mandelstam developed the ideas of Smoluchowski in a quantitative theory in terms of thermal capillary waves [4]. The Smoluchowski-Mandelstam theory was rediscovered by Buff, Lovett and Stillinger [5]. As remarked by Mandelstam it is rare that the circumstances are so favorable that the expected thermal fluctuations indeed can be observed. One such case, in which nowadays direct observations are relatively straightforward, is the Brownian motion. In the period 1908-1915 Perrin was able to make observations in simple, brilliant experiments that confirmed the theory of Einstein and led to the determination of Avogadro's number [6]. The direct observation of critical fluctuations had to wait till 1966 and was the last preoccupation by Debye [7]. Beysens and co-workers did further work on this topic [8]. Clearly, direct observations

B. Kramer (Ed.): Adv. in Solid State Phys. **45**, 15–27 (2005)

Fig. 1. (a) Schematic depiction of a curved interface. A point at the interface $\mathbf{r}$ is written in terms of $(x, y, h(x, y))$. (b) A 2D slice out of (a). The projected angle θ' is related to the interface derivative h_x via $h_x = \tan\theta$

may help to answer basic questions and the direct visual observation of thermal capillary waves that will be reported in this contribution proved to be very rich indeed [9]. Here, we show that in demixed colloid-polymer dispersions the thermal fluctuations of the fluid-fluid interface can be seen directly in real space and we will discuss the static properties of these fluctuations.

The article is organized as follows; We start with a brief derivation of capillary wave theory in Section 2 deriving the relevant expressions that we use to analyze our data, followed by a description of the experimental system and the experimental methods in Section 3. Results are given in Section 4 and conclusions and an outlook are presented in Section 5.

2 Theoretical Background

We here follow the statistical analysis of interface corrugations, first developed by Mandelstam [4], as reviewed by Vrij [10]. The work ΔF to create a corrugation at constant temperature can be written as

$$\Delta F = \Delta F_g + \Delta F_c, \tag{1}$$

where we only consider gravity (ΔF_g) and capillary (ΔF_c) forces, resulting from the displacement of matter against gravity and the creation of extra interface area, respectively. Other contributions, for example arising from bending of the interface, are ignored, the validity of which will be discussed in Section 5. Clearly, this is a mesoscopic approach, which is justified if one looks at distortions much larger than the particle size as is done in light scattering studies on molecular interfaces. The local interface position with respect to the mean interface position has coordinates $\mathbf{r} = (x, y, h(x, y))$, see Fig. 1. This Monge parameterization neglects overhang of the interface as well as bubbles of one phase in the other. To create a corrugation of an area $dxdy$ over a distance h gravity contributes

$$\delta F_g = \int_0^h h' \, g\Delta\rho \, dxdy \, dh' = \frac{1}{2} g\Delta\rho \, h^2 \, dxdy \tag{2}$$

with g Earth's acceleration, and $\Delta\rho$ the mass density difference. The interfacial tension γ contributes

$$\delta F_c = \gamma \Delta A = \gamma dxdy \left(\sqrt{1 + h_x^2 + h_y^2} - 1\right) \approx \frac{1}{2}\gamma \; dxdy \; (h_x^2 + h_y^2), \tag{3}$$

with $h_i \equiv \partial h / \partial i$ and $i = x, y$. Integrating over the total interface area $L \times L$ gives

$$\Delta F_g = \frac{1}{2} g \Delta \rho \int\int dxdy \; h^2, \tag{4}$$

and

$$\Delta F_c = \frac{1}{2}\gamma \int\int dxdy \; (h_x^2 + h_y^2). \tag{5}$$

The height h can be expanded in a Fourier series in a square with length L,

$$h = \sum_{\mathbf{k}} h_{\mathbf{k}} \; \mathrm{e}^{i(k_x x + k_y y)}, \tag{6}$$

with $h_{\mathbf{k}}$ the Fourier coefficients and $\mathbf{k} = (k_x, k_y)$. The summation runs over all Fourier modes k_x and k_y. From Parseval's identity one readily obtains

$$\Delta F = \frac{1}{2} L^2 \sum_{\mathbf{k}} |h_{\mathbf{k}}|^2 \left(g \Delta \rho + \gamma k^2\right), \tag{7}$$

with $k^2 = k_x^2 + k_y^2$ and $k = |\mathbf{k}|$. Mandelstam made use of the equipartition theorem which states that the work necessary to create each mode is equal to $k_B T/2$, with k_B Boltzmann constant and T the absolute temperature. Thus, in the capillary wave spectrum each Fourier component $h_{\mathbf{k}}$ of the interface displacement contributes

$$\langle |h_{\mathbf{k}}|^2 \rangle = \frac{k_B T}{\gamma L^2} \frac{1}{k^2 + L_{\|}^{-2}}, \tag{8}$$

where the brackets on the left hand side denote a thermal average and $L_{\|}$ is the lateral correlation length. It is given through

$$L_{\|} = \sqrt{\gamma / (g \Delta \rho)}. \tag{9}$$

When an external wall is additionally taken into account, $L_{\|}$ is the characteristic (capillary) length of the meniscus [11, 12, 13]. From (8) and again applying Parseval's theorem the mean square interfacial roughness is found to be

$$\langle h^2 \rangle = \frac{k_B T}{4\pi\gamma} \ln \left[\frac{k_{max}^2 + L_{\|}^{-2}}{k_{min}^2 + L_{\|}^{-2}} \right], \tag{10}$$

with $k_{min} = 2\pi/L$ and $k_{max} = 2\pi/l_m$. L denotes the physical system size and l_m is a microscopic length [5]. Thus, the interfacial roughness $L_{\perp} \equiv \sqrt{\langle h^2 \rangle}$ is proportional to $\sqrt{k_B T/\gamma}$.

Since the interface is constantly subject to random forces the distribution $P(h)$ of the interface heights is described by a Gaussian with (10) its variance σ^2. This means the distribution is given by

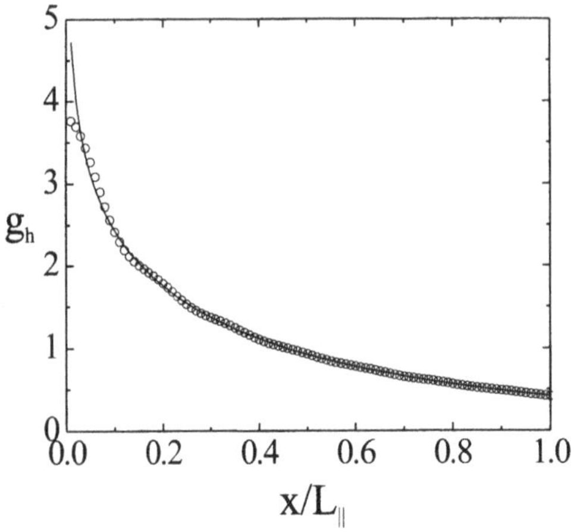

Fig. 2. Static height correlation functions g_h as a function of $x/L_{\|}$ in units of $k_BT/2\pi\gamma$. The symbols denote calculations in which $\bar{k}_{max} = 44$, the full curve is for $\bar{k}_{max} = \infty$

$$P(h) = \frac{1}{\sqrt{2\pi\sigma^2}} e^{-h^2/2\sigma^2}. \tag{11}$$

Further information on the physical properties of the interface can be obtained by considering correlation functions. The static height-height correlation function at a given time t is constructed as

$$g_h(x) = \langle h(x', t')\ h(x' + x, t')\rangle. \tag{12}$$

The angular brackets denote averages over primed quantities. Although we have experimental data only for the x- and not for the y-dependence, as will be shown below, this contains the full information due to the cylindrical symmetry of the interface around the vertical z-direction. It is therefore possible to obtain all information just by performing the analysis along one dimension. The correlation function is obtained by Fourier transforming (8). We can write

$$g_h(x) = \sum_{\mathbf{k}} \frac{k_BT}{\gamma L^2} \frac{1}{k^2 + L_{\|}^{-2}} e^{i\mathbf{k}\cdot\mathbf{s}} = \frac{k_BT}{\gamma L^2} \frac{L^2}{(2\pi)^2} \int d\mathbf{k} \frac{1}{k^2 + L_{\|}^{-2}} e^{i\mathbf{k}\cdot\mathbf{s}}, \tag{13}$$

where we have switched from a summation to an integration and $\mathbf{s} = (x, y)$. Next, we change to cylindrical coordinates and perform the integration over ϕ and obtain

$$\begin{aligned} g_h &= \frac{k_BT}{\gamma} \frac{1}{(2\pi)^2} \int dk \frac{k}{k^2 + L_{\|}^{-2}} \int_0^{2\pi} d\phi\ e^{ikx\cos\phi} \\ &= \frac{k_BT}{\gamma} \frac{2\pi}{(2\pi)^2} \int dk \frac{k}{k^2 + L_{\|}^{-2}} J_0(kx). \end{aligned} \tag{14}$$

The symbol J_0 denotes the Bessel function of the first kind. The integration over k is performed from k_{min} to k_{max}, see just below (10). We can directly set k_{min} to zero, since $2\pi/L \sim 0$. Setting k_{max} to ∞ allows performing the integration and (14) then becomes

$$g_h(x) = \frac{k_B T}{2\pi\gamma} K_0 \left(\frac{x}{L_\parallel} \right), \tag{15}$$

where K_0 is the modified Bessel function of the second kind as a function of $x/L_\parallel$. We can test the effect of setting k_{max} to infinity. We change to $\bar{k} = kL_\parallel$ in (14) and use $k_{max} = 2\pi/\sigma_c$ as a reasonable cutoff with σ_c the colloid diameter. As will be shown below $L_\parallel$ is at least a couple of microns and here we will fix it at a minimal value of 1 μm, whereas $\sigma_c = 142$ nm, such that $\bar{k}_{max}$ is at least ~ 44. In Fig. 2(a) we plot results from equation (14) with $\bar{k}_{max} = 44$ and equation (15). Only at a distance $x/L_\parallel < 0.05$ clear differences can be observed between the two functions. Furthermore, the curve obtained with finite cut-off shows some fine structure. We thus conclude that (15) holds for distances x larger than a small-distance cutoff of the order of the particle size.

A natural extension of the above framework is to analyze the distribution of the angle between the local interface normal and the vertical direction [14], i.e. the tilt angle distribution. This is related to derivatives of h in the x-direction: h_x. Again, the distribution of h_x is Gaussian. The derivative stands in direct connection to the (projected) angle θ' normal to the interface, $h_x = \tan\theta$ (see Fig. 1(b)). It can be shown that the distribution in one dimension is given by [14]

$$P(\theta') = \frac{2}{\sqrt{2\pi\sigma'^2}} \frac{\mathrm{e}^{-\frac{1}{2}\frac{\tan^2\theta'}{\sigma'^2}}}{\cos^2\theta'} \tag{16}$$

with $\sigma'^2 \equiv \langle \tan^2\theta \rangle$ the variance. The special form of this equation is a result of the Jacobian of the transformation of h_x to θ using $\partial h_x(\theta)/\partial\theta = 1/\cos^2\theta$. The variance is extremely sensitive to the value of k_{max} [14].

In molecular fluids γ is of the order of $10 - 100$ mN/m and $\Delta\rho$ is about $10^2 - 10^3$ kg/m^3. Therefore, the interface roughness $L_\perp \sim 0.3$ nm, whereas the correlation length $L_\parallel \sim 3$ mm resulting in extreme ratios of roughness-to-correlation length of 10^{-7}, only accessible through scattering techniques. Here we exploit the scaling up of lengths when going from molecules to mesoscopic colloidal particles of size 140 nm in order to directly observe capillary waves in real space.

Adding polymer to a colloidal suspension may induce a fluid-fluid demixing transition that is widely accepted to be the mesoscopic analogue of the liquid-gas phase transition in atomic substances [15, 16]. The coexisting phases are a colloidal liquid (rich in colloid and poor in polymer) and a colloidal gas (poor in colloid and rich in polymer). The origin of the phase separation lies in the entropy-driven attraction between the colloids, which is mediated by the polymers [17, 18]. It is known from experiment [19, 20, 21, 11, 9],

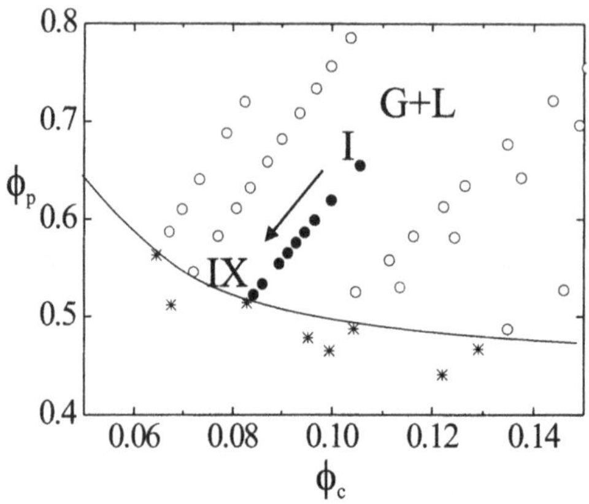

Fig. 3. Phase diagram in (ϕ_p, ϕ_c)-representation. Indicated are state points where gas-liquid phase separation occurs (open and filled circles) and state points in the one-phase region (crosses). The line is an estimate of the binodal and is drawn to guide the eye. State points I-IX (filled circles) are indicated

theory [22, 23, 24, 25, 26] and simulations [27, 28], that in such systems the interfacial tension scales as $\gamma \sim k_B T/\sigma_c^2$ leading to ultra-low values for γ (~ 1 μN/m and below). This, in turn, implies that using colloidal suspensions scales up the interface roughness and simultaneously scales down the correlation length. With the current system (as detailed below) we succeed to bring both the roughness and the correlation length in the μm-regime. In addition, the interplay between ultra-low interfacial tension and relatively large viscosity, η, sets the capillary velocity γ/η (see for example [29]) in the range of μm/s, as opposed to typical velocities of the order of 10 m/s in molecular fluids. The associated characteristic time for the decay of interfacial fluctuations, which we refer to as the capillary time, is given by $\tau = L_{\parallel}\, \eta/\gamma$. In the case of colloids it becomes of the order of seconds. Thus, through the appropriate choice of the colloid diameter we can trace the statics *and* dynamics of the capillary waves at a free interface with optical microscopy [9], but here we will restrict ourselves to the statics.

3 Experimental Methods

We used fluorescently labeled poly(methylmethacrylate) (PMMA) colloidal spheres [30] with radius of $R_c = 71$ nm (obtained from static light scattering) and size polydispersity of less than 10%. The polymer was commercially available polystyrene (Fluka) with molecular weight $M_w = 2 \cdot 10^6$ g mol^{-1} ($M_w/M_n < 1.2$, where M_n is the number average molecular weight) and radius of gyration of $R_g \sim 44$ nm (estimated from data in the literature [31, 32]).

Thus the size ratio, $R_g/R_c = 0.6$, was sufficiently large to obtain stable fluid-fluid demixing [33]. Both species were dispersed in cis/trans-decalin and since all densities were known, mass fractions could be directly converted to packing fractions of colloids, $\phi_c = \frac{4}{3}\pi R_c^3 n_c$, and polymers, $\phi_p = \frac{4}{3}\pi R_g^3 n_p$, where n_c and n_p are the number densities of colloids and polymers, respectively. Samples were prepared by mixing colloid and polymer stock dispersions and diluting with decalin in order to control the overall packing fractions ϕ_c and ϕ_p. The underlying phase diagram is shown in Fig. 3. Large glass cuvettes (of volume ~ 1 cm^3) with extra thin (thickness of 0.17 mm) cover glass walls were used. A confocal scanning laser head (Nikon C1) was mounted on a horizontally placed light microscope (Nikon Eclipse E400). Each measurement was done after one day of equilibration. The microscope was aligned by making use of the interface which serves as a spirit level. We checked that the system was well equilibrated by following the recovery of intensity after bleaching a space region in the gas and/or liquid phase. The recovery appeared to be governed solely through diffusion of particles without any indications of drift (e.g. through convection). Different data sets were acquired at many different state points following several dilution lines. The data sets consisted of approximately 5.1 10^5 interface data points when images were scanned as fast as possible (about 5 frames per second), and about 1.5 10^5 when a delay time between consecutive images was used (of about 10 s) to get rid of some of the time correlation.

Pictures, such as those in Fig. 4, represent an intensity distribution of fluorescent light, $I(x, z, t)$, at a certain time t with x the horizontal (along the interface) and z the vertical (opposite to gravity) components of the space vector. The microscope records the fluorescence of excited dye within the colloids, hence the colloid-rich (liquid) phase appears bright and the colloid-poor (gas) phase appears dark. $I(x, z, t)$ is a direct measure of the local and instantaneous distribution of colloidal particles and provides the starting point for a statistical analysis. Due to the finite resolution [34, 35] we can access length scales $\sim 2R_c$, and we neglect effects induced by the finite time needed to scan each frame, and take I as an instantaneous snapshot (justified by comparing the colloid self-diffusion time with the scanning time). Thus, the real space pictures in Fig. 4 show the structure of a gas-liquid interface practically at the particle scale. We rely on the concept of a local interface between both phases. In the spirit of a Gibbs dividing surface we define an interface position $h_b(x, t)$ (now with respect to the bottom of the image, instead of the mean interface position as in Section 2) such that in one column of vertical length L_z the total intensity can be written as

$$\int_0^{L_z} dz I(x, z, t) = I_{\text{liq}}(x) h_b(x, t) + I_{\text{gas}}(x)(L_z - h_b(x, t)) \tag{17}$$

Here, the values $I_{\text{gas}}(x)$ and $I_{\text{liq}}(x)$ are the average bulk intensities in the gas and liquid phase, respectively, and are taken to be functions of x to account for the microscope objective properties. In practice, integrals in the notation

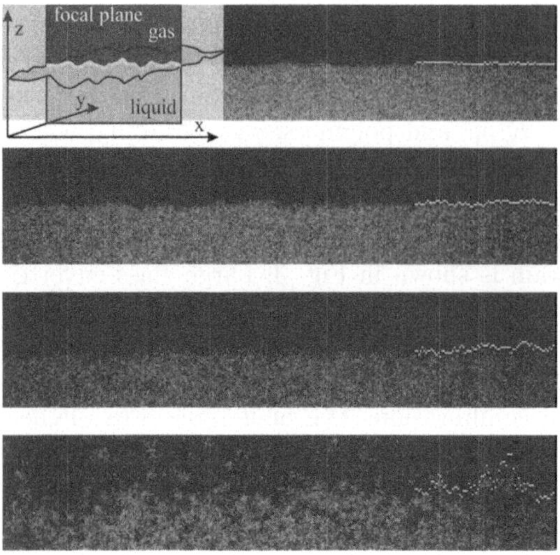

Fig. 4. Capillary waves at the free liquid-gas interface in a phase-separated colloid-polymer mixture imaged with laser scanning confocal microscopy (LSCM) at four different state points approaching the critical point (from top to bottom: state points I, VI, VIII and IX, see Fig. 3). The focal (viewing) plane is perpendicular to the interface and only a very thin slice (of thickness $\sim 0.6\ \mu$m) is imaged (see the inset). Gravity points downwards and the size of each image is $17.5 \times 85\ \mu\text{m}^2$. Thermally excited capillary waves corrugate the interface and their amplitude increases upon approaching the critical point. The bright dots at the right indicate the surface location $h(x)$ obtained with our method

are sums over pixels and we have checked that the results of the subsequent analysis in correlation functions do not depend sensitively on the precise definitions of I_{gas} and I_{liq}. The resulting "height" function $h_b(x)$ (shown as the bright spots in Fig. 4) describes the interface position quite accurately. From top to bottom in Fig. 4 we approach the critical point and both the capillary waves and density fluctuations increase, while the density (intensity) difference between the two phases decreases. For each frame the average interface position is $\bar{h}_b(t) \equiv \langle h_b(x', t) \rangle$, in which the angular brackets denote averages over primed quantities. We now define a new height function that describes the deviations from the mean interface position $h(x, t) = h_b(x, t) - \bar{h}_b(t)$.

4 Results and Discussion

The distribution of heights is shown in Fig. 5 for three different state points. The width of the distribution for state point I is about twice the particle diameter, the order of which is in good agreement with (10). As one approaches the critical point the distributions get broader up to 7 times the

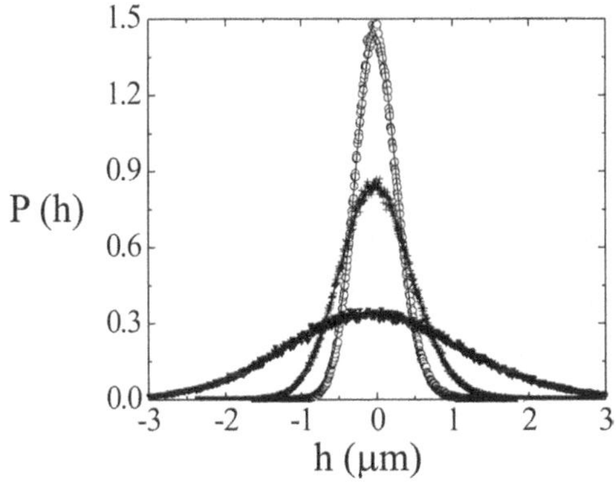

Fig. 5. Distribution of height values, $P(h)$, for three different state points: I (open circles), V (plusses) and VIII (triangles). Full curves are Gaussian fits (11)

particle diameter for state point VIII, but the shape remains that of a normal distribution. However, the physical interpretation of this figure is limited due to the finite resolution of the confocal technique as well as our interface location procedure. Each height $h(x,t)$ appears to consist of the actual height plus a delta correlated "noise" term $\Delta(x,t)$ with properties such that $\langle \Delta(x',t') \rangle = 0$, and $\langle h(x',t')\Delta(x',t') \rangle = 0$ averaged over either x' or t'. Furthermore, $\langle \Delta(x',t')\Delta(x'+x,t'+t) \rangle = \sigma_\Delta^2 \delta(x)\delta(t)$ with δ the delta function. Thus, from Fig. 5 we see that the interface roughness is Gaussian, but the actual physics – although present as can be observed from the trends in the figure – is blurred by the small noise term. To cope with the delta-correlated noise we construct a correlation function, which clearly does not suffer from such noise, except when both $x = 0$ and $t = 0$.

The static correlation function (equation (15)) describes the experimental data points very well, as can be seen in Fig. 6 for various state points with only two physical parameters γ and $L_{\|}$. No bending term in (8) was needed in the analysis. Note that in the original paper [9] the y-axis was given in units of pixel area and not (as indicated) in μm^2, but the analysis was not affected by this. Results for the interfacial tension and the capillary length are displayed in Fig. 7(a) and (b).

To further explore the properties of the interface we examine derivatives of $h(x,t)$ [14]. In Fig. 8 histograms are plotted of the absolute value of θ' for three different state points (as in Fig. 5). One can either use (16) to fit to the data with the variance as fitting parameter or obtain the variance directly from the experiment. In Fig. 8 both methods are plotted. The agreement is yet another confirmation that the interface can be described within a Gaussian interface model. As the critical point is approached the peak in the angle distribution shifts from 0° to 75°, since the interface roughness

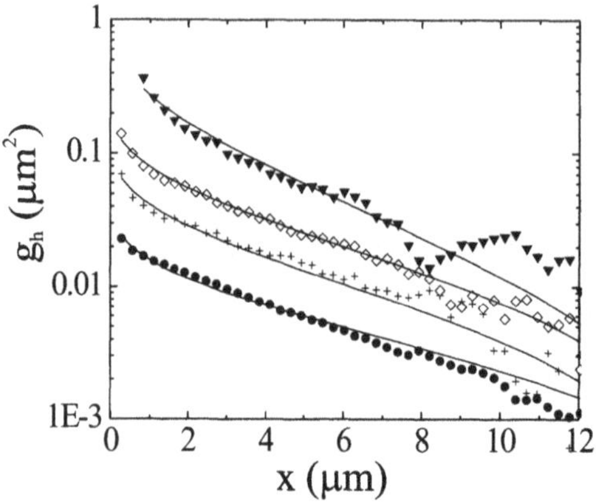

Fig. 6. Static height-height correlation function $g_h(x)$ as obtained from quantitative analysis of LSCM pictures as a function of the (lateral) distance x for state points I, V, VI and VIII (see Fig. 3) approaching the critical point from bottom to top. Experimental results (symbols) are compared with predictions from the capillary wave model (lines)

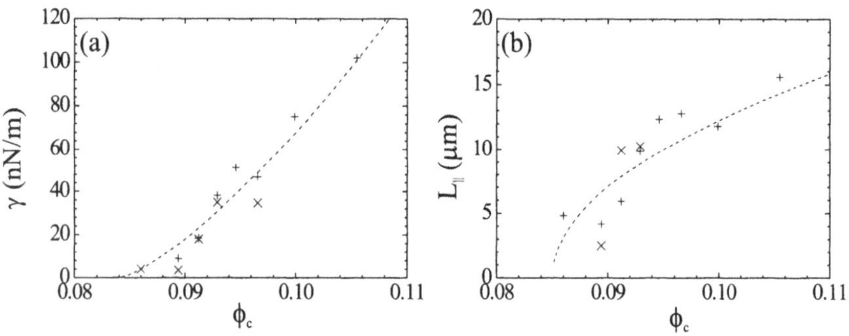

Fig. 7. (a) Interfacial tension γ as a function of the overall colloid packing fraction ϕ_c obtained from $g_h(x)$ (crosses: slow frame rate, pluses: fast frame rate). (b) The capillary length $L_\parallel$ as a function of ϕ_c obtained from $g_h(x)$. Results stem from state points on the same dilution line as state points I-IX. The dashed lines are to guide the eye

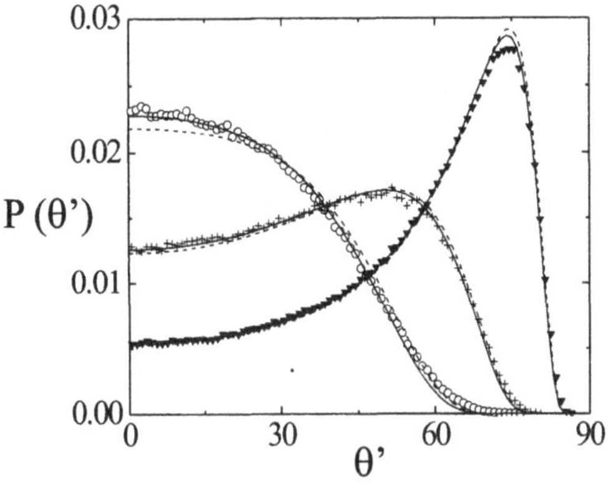

Fig. 8. Angle distributions for three different state points: I (open circles), V (plusses) and VIII (triangles). We either fitted the variance in (16) (full curves) or obtained the variance directly from experiment (dashed curves)

increases and the correlation length decreases. As σ'^2 depends strongly on the molecular interactions, it is in principle possible to obtain the interfacial tension more accurately and to determine the microscopic cut-off k_{max}, i.e. the microscopic length l_m in (10) [14]. It could be interesting if on the particle level a wave-vector dependent surface tension can be detected due to molecular interactions, as predicted for simple liquids [36], and observed in various liquids [37, 38].

5 Conclusions

Using a colloid-polymer mixture allows to carefully tune the interface properties and as a result the thermal capillary waves at a free interface are observed visually by means of LSCM. From the fluorescence intensity difference between the two phases the interface can easily be located and the height functions are constructed. By analyzing the height fluctuations of the interface position as well as the derivatives of these we see that the interface roughness can be described within a Gaussian model. To deal with small noisy contributions to the height function correlation functions are determined. The quality of the fits to describe the correlation functions validate the capillary wave model practically down to the particle level.

The present work opens up a wide range of possibilities, e.g. to study the interface at a particle level by using even larger colloids, explore temperature gradients and mass transport across the interface, and the effects on droplet coalescence [9] and snap-off, on heterogeneous catalysis, freezing of capillary waves at the gel-line, the effects of thermal capillary waves on wetting properties etc.

Acknowledgements

This work was supported by the Stichting voor Fundamenteel Onderzoek der Materie (Foundation for Fundamental Research on Matter), which is part of the Nederlandse Organisatie voor Wetenschappelijk Onderzoek (Netherlands Organisation for Advancement of Research), and we also acknowledge financial support by the DFG SFB TR6 "Colloidal dispersions in external fields".

References

1. A. Einstein, Ann. Phys. **17**, 549 (1905).
2. M. V. von Smoluchowski, Ann. Phys. **25**, 205 (1908).
3. A. Einstein, Ann. Phys. **33**, 1275 (1910).
4. L. Mandelstam, Ann. Phys. **41**, 609 (1913).
5. F. P. Buff, R. A. Lovett, and F. H. Stillinger, Phys. Rev. Lett. **15**, 621 (1965).
6. J. Perrin, Ann. de Chim. Phys. **18**, 5 (1909).
7. P. Debye and R.T. Jacobson, J. Chem. Phys. **48**, 203 (1968).
8. P. Guenon, F. Perrot, and D. Beysens, Phys. Rev. Lett. **63**, 1152 (1989).
9. D. G. A. L. Aarts, M. Schmidt, and H. N. W. Lekkerkerker, Science **304**, 847 (2004).
10. A. Vrij, Adv. Coll. Interf. Sci. **2**, 39 (1968).
11. D. G. A. L. Aarts, J. H. van der Wiel, and H. N. W. Lekkerkerker, J. Phys.: Condens. Matter **15**, S245 (2003).
12. D. G. A. L. Aarts and H. N. W. Lekkerkerker, J. Phys.: Condens. Matter **16**, S4231 (2004).
13. D. G. A. L. Aarts, J. Phys. Chem. B **109**, 7407 (2005).
14. K.R. Mecke and S. Dietrich, preprint, cond-mat/0505294 (2005).
15. W.C.K. Poon, J. Phys.: Condens. Matter **14**, R859 (2002).
16. R. Tuinier, J. Rieger, and C.G. de Kruif, Adv. Coll. Interf. Sci. **103**, 1 (2003).
17. S. Asakura and F. Oosawa, J. Chem. Phys. **22**, 1255 (1954).
18. A. Vrij, Pure Appl. Chem. **48**, 471 (1976).
19. E. H. A. de Hoog and H. N. W. Lekkerkerker, J. Phys. Chem. B **103**, 5274 (1999).
20. B. H. Chen, B. Payandeh, and M. Robert, Phys. Rev. E **62**, 2369 (2000).
21. E. H. A. de Hoog and H. N. W. Lekkerkerker, J. Phys. Chem. B **105**, 11636 (2001).
22. A. Vrij, Physica A **235**, 120 (1997).
23. J. M. Brader and R. Evans, Europhys. Lett. **49**, 678 (2000).
24. J. M. Brader, R. Evans, M. Schmidt, and H. Löwen, J. Phys.: Condens. Matter **14**, L1 (2002).
25. A. Moncho-Jorda, B. Rotenberg, and A. A. Louis, J. Chem. Phys. **119**, 12667 (2003).
26. D. G. A. L. Aarts, R. P. A. Dullens, H. N. W. Lekkerkerker, D. Bonn, and R. van Roij, J. Chem. Phys. **120**, 1973 (2004).
27. R. L. C. Vink and J. Horbach, J. Chem. Phys. **121**, 3253 (2004).
28. R. L. C. Vink and J. Horbach, J. Phys.: Condens. Matter **16**, S3807 (2004).

29. R. F. Probstein, *Physicochemical Hydrodynamics*, section 10.2, (John Wiley & Sons, Hoboken 2003).
30. G. Bosma, C. Pathmamanoharan, E. H. A. de Hoog, W. K. Kegel, A. van Blaaderen, and H. N. W. Lekkerkerker, J. Colloid Interface Sci. **245**, 292 (2002).
31. G.C. Berry, J. Chem. Phys. **44**, 4550 (1966).
32. B. Vincent, Colloids Surf. **50**, 241 (1990).
33. H. N. W. Lekkerkerker, W. C. K. Poon, P. N. Pusey, A. Stroobants, and P. B. Warren, Europhys. Lett. **20**, 559 (1992).
34. T. Wilson, *Confocal Microscopy*, (Academic Press Ltd., London 1990).
35. R. H. Webb, Rep. Prog. Phys. **59**, 427 (1996).
36. K. R. Mecke and S. Dietrich, Phys. Rev. E **59**, 6766 (1999).
37. C. Fradin, A. Braslau, D. Luzet, D. Smilgies, M. Alba, N. Boudet, K. R. Mecke, and J. Daillant, Nature **403**, 871 (2000).
38. S. Mora, J. Daillant, K. R. Mecke, D. Luzet, A. Braslau, M. Alba, and B. Struth, Phys. Rev. Lett. **90**, 216101 (2003).

Linear Polymers in Networks: Swelling, Diffusion, and Interdigitation

Mark Geoghegan

Department of Physics and Astronomy, University of Sheffield,
Hicks Building, Hounsfield Road, Sheffield S3 7RH, United Kingdom

Summary. The diffusion of polymers within heterogeneous media is strongly dependent on the structure of the medium. We review neutron reflectometry and ion beam analysis measurements on (crosslinked) polystyrene films. In particular we consider the diffusion of polymers into and within polymer networks, and consider experiments to determine information on the structure of networks by swelling in their own homopolymer. Finally, we discuss the interpenetration of networks by linear polymer and polymer brushes.

Although many polymers can be readily crosslinked to form networks, a deep understanding of network behavior is still missing despite theoretical attempts to understand their structure dating back some 70 years [1]. Most of the theoretical attempts to study polymer networks have been based on mean-field theory, which contains the crucial assumption that the networks are homogeneous (in order to have a mean field). Later theories have involved more complex variants of the mean-field approach; the replica mean-field trick being the prime example [2]. Other examples include simple scaling theory [3, 4], simulations [5], and mixtures of different methods [6]. The reasons why theory is dominant in this area is because it is particularly difficult to perform experiments on networks that can be considered decent models for theory. These experiments are now appearing, and will be discussed in the present article. Nevertheless, despite recent progress it is probably fair to say that a true understanding of polymer networks is still lacking.

The diffusion of polymers is a widely studied field, and is one in which there is much agreement. Early theoretical attempts to describe the motion of polymers in their molten state led to the theory of reptation [7, 8], which has been amply demonstrated experimentally. Corrections to this theory, for example to describe the motion of the surrounding matrix (constraint release) have also been proposed, demonstrated, and accepted. However, the diffusion of polymers within networks is a particularly interesting topic because of the effect of the topological constraints to the diffusion imposed by the network crosslinks.

The diffusion of linear polymers into networks is much more complicated and no simple theory has been presented to explain any of the observed behavior. When the diffusing species is of high molecular weight, and interdiffusion cannot complete, only partial diffusion occurs. In this case there is the

B. Kramer (Ed.): Adv. in Solid State Phys. **45**, 29–44 (2005)

possibility to study interdigitation between linear polymers and networks as a means of providing adhesion. Adhesion between networks and linear polymers is best demonstrated when that linear polymer is grafted to an inorganic substrate since this addresses an issue of great practical importance.

In this article we discuss these issues, linking the structural importance of networks to the diffusion of polymers within these networks and the interdigitation between polymers (tethered and free) and networks.

1 Swelling of Polymer Networks

From the 1940's the theory of polymer mixtures has been developed by a few workers, most notably Flory. The mean-field theory describing the behavior of blends of two polymers or of polymer and solvent (Flory-Huggins theory) has been very successful because it can predict with very good accuracy the phase separation of polymers. This has been achieved by writing a free energy for a mixture of polymers consisting a thermodynamic attraction or repulsion between the two components, as well as an entropic term. Only near the boundary for spontaneous phase separation (spinodal decomposition) does the theory fail due to the inability of mean-field theory to deal with fluctuations. A very active area of theoretical research in this period has been in considering the behavior of polymer networks. If one could consider a linear polymer mixture and treat it as the sum of an entropic term and an enthalpic term then it would be reasonable to treat networks in the same way except that some acknowledgment needs to be given to the fact that networks swell, and that it takes elastic energy to stretch the network. By treating a network as a Hookian spring, elasticity could be accounted for in the free energy.

There are a variety of ways in which we can write down a mean-field theory to describe a polymer network. We shall consider a mixture of a network and a linear polymer, and will follow this with experimental work on the subject. The enthalpy is a simple pairwise interaction and its free energy can be written as

$$\frac{\Delta F_{enthalpy}}{nk_BT} = \chi\phi(1-\phi) \tag{1}$$

where T is the absolute temperature; k_B is Boltzmann's constant; χ is a measure of the enthalpic interaction between the linear polymer and the network and may be concentration or temperature dependent; n is the number of monomer units of both network and linear polymer (strictly, n is the number of lattice points and each lattice point is occupied by a monomer normalized to the lattice size); and ϕ is the volume fraction of one of the components. We choose this to be the linear polymer, and assuming incompressibility, the volume fraction of the network is $1-\phi$. The entropy is given by

$$\frac{\Delta F_{entropy}}{nk_BT} = \frac{\Phi\phi(1-\phi)}{N}\ln(1-\phi) + \frac{\phi}{P}\ln\phi \tag{2}$$

where N is the average number of monomers between crosslinks and P is the number of monomers in an average linear polymer chain. Φ is a constant that, when set to unity, yields the entropy of mixing for two polymer chains of lengths N and P. For a polymer network, $\Phi < 1$ but the debate over its value is contentious. If one considers the polymer network to be as a simple polymer with infinite molecular weight, then one can set $\Phi = 0$, because it then cannot contribute to the entropy of the system. The physical effect of this is to give the crosslinks the same mobility as entanglements have in a polymer melt. However, in a melt entanglements are completely mobile. As polymers diffuse (reptate) the entanglements move as well. Crosslinks are not mobile because they are fixed on the polymer (we are not considering weak physically crosslinked networks in this article). The motion of the crosslinks must allow their movement through one another, which is physically impossible. For this reason, the mean-field theory with $\Phi = 0$ is known as the "phantom theory" [9-11]. An alternative description for the entropy of a polymer network recognizes that crosslinks cannot travel through one another and are fixed in space. The network strands must still have capacity to change their conformations, albeit not to the same degree as linear polymers. By setting $\Phi = 2/f$, where f is the functionality of the network (the number of strands exiting a crosslink) one can consider crosslinks as fixed but recognize the possibility of the strands to contribute to the entropy. This version of the entropy is the affine Flory model [12, 13]; it is affine in that one can consider the network to behave the same on all length scales. The entropy and enthalpy contributions to the free energy are basically unchanged from the Flory-Huggins mean field theory of linear polymers, save for the appearance of Φ, which, if we set $f = 2$ for linear polymers, defaults to Flory-Huggins behavior. We now need to consider the elasticity of a network, which we account for with another term in the free energy,

$$\frac{\Delta F_{elastic}}{nk_BT} = \frac{\Gamma(1-\phi)}{2N}\left({\lambda_x}^2 + {\lambda_y}^2 + {\lambda_z}^2 - 3\right) \tag{3}$$

where Γ is another constant (given by $(f-2)/f$ for the phantom model, or $\Gamma = 1$ for the affine Flory model). If we assume that the network stretches equally in all directions by a factor λ ($= \lambda_x = \lambda_y = \lambda_z$), then we note that $\lambda^3 = 1/(1-\phi)$, and we can rewrite the above equation as

$$\frac{\Delta F_{elastic}}{nk_BT} = \frac{3\Gamma}{2N}\left((1-\phi)^{1/3} - (1-\phi)\right) . \tag{4}$$

To obtain the total free energy of the network, these terms can be added together,

$$\frac{\Delta F}{nk_BT} = \frac{\Phi\phi(1-\phi)}{N}\ln(1-\phi) + \frac{\phi}{P}\ln\phi$$
$$+\frac{3\Gamma}{2N}\left((1-\phi)^{1/3} - (1-\phi)\right) + \chi(1-\phi) . \tag{5}$$

From this (Gibbs) free energy, it is straightforward to work out the equilibrium swelling of a network by a linear polymer. One must differentiate the free energy per monomer (strictly lattice point) with respect to n, and set this equal to zero. Physically we are equating the free energy of the swollen network with that of any linear polymer outside the network. The result of this calculation is

$$\frac{\Gamma}{N}(1-\phi)^{1/3} - \frac{\Phi(1-\phi)}{N} + \frac{1-\phi}{P} + \frac{\ln\phi}{P} + \chi(1-\phi)^2 = 0\,. \qquad (6)$$

Although this is a relatively straightforward calculation, it remains simple even for more complex situations such as when the network is prepared in solution, when it has less capacity to swell. For network samples prepared in a swollen preparation state, a simple correction to account for the volume fraction of solvent is included into the above equation. For the case of a network swollen by a small molecule solvent, we can use the above equations simply by substituting $P = 1$. We shall not discuss such matters here. However, it is interesting to note that if $\chi = 0$, then P/N is a scaling variable. We turn to this below.

Given that such simple equations are useful tools for predicting the swelling of linear polymers in a network, it is possible to use swelling experiments to test this mean-field theory. Indeed, such experiments have been performed for many years demonstrating that the phantom theory is indeed correct, or at least better than the affine Flory model. However, if one completely stretches a network by solvent swelling (or by any other method) one loses important structural information. It is much more instructive to perturb the network by only slightly swelling it, which can be achieved using linear polymers. Negligible load on a network will result in no resistance because no resistance to stretching network strands will have been met; however at small loads smaller network strands will start to bear some of the energy applied in stretching or swelling the network. At progressively larger loads, more strands will become involved in the swelling process. When the network is completely swollen we cannot see the structural information because it will suffice to take an (effective) average strand length. (We refer to an effective average because some parts of the network, for example dangling ends, do not bear load.) The ideal experiment will not test the fully swollen network, but rather only small degrees of swelling. Such experiments were achieved in the 1980's using different methods. The swelling of polydimethylsiloxane (PDMS) networks by PDMS linear polymer [14] is a simple way of testing mean-field theory because the large linear polymer cannot cause large swelling in the network. An alternative approach is to use small molecule swelling, but to restrict the amount swelling by using the solvent vapor to swell the network [15, 16]. The swelling of networks by linear polymers gave essentially good agreement with the simple Flory-Rehner approach, but the solvent sorption measurements gave rise to anomalous swelling—a peak in the dilation modulus—and a temperature dependence that cannot be explained by the simple Flory-Rehner theory. In fact it was concluded that a fundamental premise of the Flory-Rehner approach failed and the free energy should not be composed of separate entropic, enthalpic, and elastic components. The reasonable agreement in the linear PDMS swelling experiments may well arise from these networks being end-linked, which creates a more uniform network.

Our experiments mimicked those of Gent and Tobias but we chose to swell polystyrene networks with polystyrene linear polymer [17]. The major advantage of polystyrene is that kinetics can be followed by annealing the polymer above its glass transition, and then measuring after cooling the sample to room temperature. Polystyrene networks were prepared by a chemical route, which involved an exchange and Friedel-Crafts reaction functionalizing the aromatic group of anionically synthesized polystyrene with an aminomethyl group. This functional group can then be used for crosslinking, a process that can occur during spin casting of a solution of the crosslinker and functionalized polymer in toluene. The result is a network of known crosslinking density, and in the absence of microgels and other such undesirable clusters. Indeed, we also were able to proton irradiate these networks to cause uniform crosslinking. The swelling of both networks (chemically synthesized and irradiated) was the same; in fact the irradiation route is only suitable for lightly crosslinked networks and the chemical route more suitable for networks with a large crosslinking density, but there is an overlap region where swelling could be compared. The chemistry is fully described in the original paper [17]. To measure the swelling we floated a linear polystyrene layer onto water and picked it up on a polystyrene network (previously spin cast onto silicon) from a glass microslide. The sample was then annealed at 458 K in vacuum until equilibrium was reached (known by measuring the swelling at shorter or longer annealing times). The degree of swelling was measured by the ion beam technique of nuclear reaction analysis (NRA) [18].

NRA data are shown in Fig. 1 to illustrate the swelling of polystyrene networks by linear deuterated polystyrene as a function of the molecular weight of the linear chain, and the average network strand. (The deuteration is necessary for NRA measurements and differentiates the linear polymer from the network.) The equilibrium swelling of the network can be taken from the fits shown and plotted against P/N (we are neglecting the small χ-parameter for the isotopic polystyrene system [19] and assume that the crosslinks do not contribute either), as shown in Fig. 2. The equilibrium swelling is clearly greater than any the simple theory discussed above. To account for the discrepancy, a model of polymer networks based upon their heterogeneity has been developed [4], which treats networks as crosslinked clusters connected to other crosslinked clusters by regions more lightly crosslinked. Thus, when the network is swollen, the linear polymer first enters the lightly crosslinked connecting clusters. The swelling process occurs as a kind of unfolding, with (relatively) large scale reorganization occurring as different regions of network in turn start to swell. The polymeric swelling measurements shown in Fig. 1 are ideal for such a study (small molecule solvents cause all regions

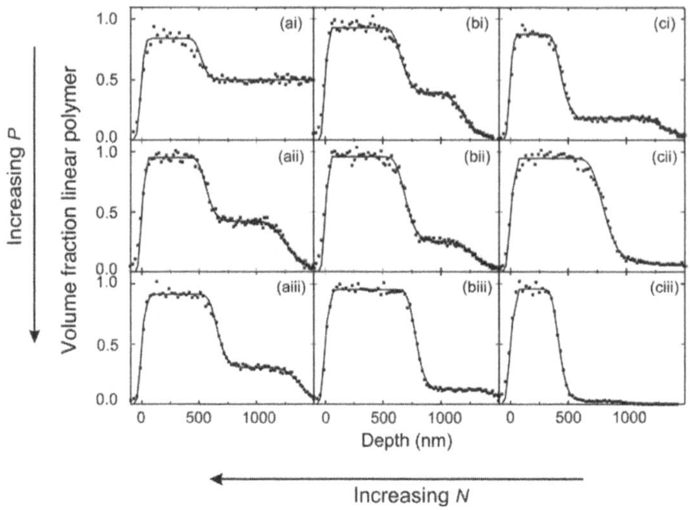

Fig. 1. Equilibrium swelling data and fits as measured by NRA. The left hand side represents the linear polystyrene (since the network cannot penetrate the linear polymer, the volume fraction remains at $\phi = 1$. Spectra a-c are for mesh sizes of $N = 639$, 361, and 110, respectively. Spectra i-iii are for linear chains of $P = 313$, 714 and 1696, respectively. Reproduced with permission from reference [17]. Copyright 2003 American Chemical Society

of the network to be swollen considerably and this can be analyzed with a Flory-Rehner approach). The reason for the large swelling is that even large linear chains are small compared to the length scales of the clusters within the networks. In other words, the linear chains cause the network to swell in the same manner as polystyrene swells in oligostyrene or styrene monomer; the swelling is not so much due to network elasticity but to differences in osmotic pressure between the linear polymer and network. Such logic lends itself to a simple scaling approach [4] with the only unknown parameter coming from computer simulations revealing the internal dimension of the polymer network (using the standard Flory argument, for linear chains this internal dimension is unity, and swelling in a good solvent scales as $P^{3/5}$, but for our network, this is not the case and the scaling parameter must be determined from simulations [3]).

Although the above scaling method has been quite successful at explaining our NRA data, it should be noted that if we do account for a χ-parameter due to crosslinks, there is very good agreement with the data. Supplementary NRA experiments [17] involved changing the chemical potential of the reservoir by adding to the linear polystyrene layer, high molecular weight linear polystyrene that could not swell the network. Only part of this mixture could be used for swelling and consequently, the total swelling was lowered. The crosslink-dependent χ-parameter also was able to account for these data as

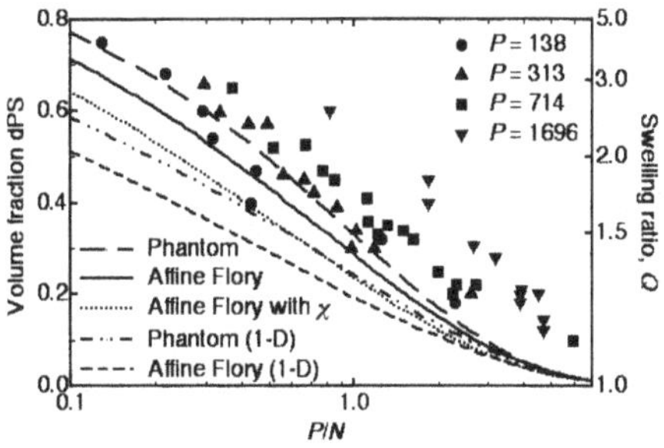

Fig. 2. Scaling plot of the equilibrium swelling ϕ_{eq} and swelling ratio $Q = (1 - \phi_{eq})^{-1}$ as a function of P/N for the affine Flory and phantom models in both one and three dimensions. Also shown is the affine Flory prediction including the effect of the known interaction parameter for polystyrene. Clearly, the use of the interaction parameter predicts smaller swelling ratios, and works in the wrong direction to describe data which demonstrate greater swelling than the theory illustrated here. Reproduced with permission from reference [17]. Copyright 2003 American Chemical Society

well (with unchanged χ-values from the equilibrium swelling experiments). Given the questionable application of mean-field theory to systems with different length scales, this success is surprising, although we should note that crosslink dependent χ-parameters have been applied successfully elsewhere [6, 20-24]. Clearly, the question of swelling of polymer networks will be considered unsolved for some time to come.

2 Diffusion

The diffusion of polymers in a network is a phenomenon of great interest practically and theoretically. Practically it is becoming more important to contain polymer within a network, as this polymer can be released under certain conditions when the network is swollen. Theoretically, there are a variety of models for diffusion of linear polymers in linear polymer melts; the situation for polymer networks is very much worth testing. Similarly, if a network used for a particular purpose contains some uncrosslinked polymer (sol), perhaps as an artefact of the crosslinking process, it is worth understanding how long it would take for that sol to leach out of the network (or indeed if the sol would escape the network at all). We consider here only how linear polymers diffuse within a network and then their diffusion *into* networks.

2.1 Reptation

Like the swelling of polymer networks, the question of how polymers diffuse in a molten network is a very old question. Unlike the swelling problem, the work of de Gennes [7] and Doi and Edwards [8] means that we understand the diffusion of polymers in networks. The idea of the snake-like motion of polymers (reptation) was proposed by de Gennes, but the idea lacked a precise framework until Edwards proposed the "tube" model and gave reptation mathematical rigor in the form of their constitutive equations. Measurements on molten linear polymer blends were sufficient to demonstrate reptation, but there has been some (inconclusive) work on molten polymers. The best of the early experiments were perhaps those performed by Antonietti and Sillescu in 1985 [25], in which a holographic grating technique (Forced Rayleigh Scattering) was used to study the diffusion of deuterated polystyrene in polystyrene networks. The authors were able to identify the expected linear polymer molecular weight dependence of the diffusion coefficient, but not any crosslinking density dependence.

Above a certain chain size, the entanglement molecular weight, molten polymers are entangled. In this way, entanglements can be considered like crosslinks. There are on average N_e monomers between entanglements, where N_e is the entanglement chain length. This means that there are $1/N_e$ entanglements in a given volume. If we add $1/N$ crosslinks, the total number of impediments is given by

$$\frac{1}{N_t} = \frac{1}{N} + \frac{1}{N_e}. \tag{7}$$

As a linear polymer chain diffuses through this molten network, assuming it does not meet any similar linear chains, it experiences a frictional resistance to motion which is proportional to the number of impediments, and also to its own size (i.e. proportional to P/N_t). We can therefore write a tube diffusion coefficient as

$$D_t = \mu T N_t / P, \tag{8}$$

which is the Einstein relation [26] applied along the contour length of the linear polymer (μ is a mobility). We require this tube diffusion coefficient because friction affects diffusion along a length governed by P. The time spent for the linear chain to diffuse along its own length is

$$\tau = \frac{(aP)^2}{D_t}, \tag{9}$$

where a is a monomer length. Of course, the parameter that is measured experimentally is the center-of-mass diffusion coefficient. For a molten polymer the contour length of the polymer corresponds to the (random walk) end-to-end distance $aP^{1/2}$. We then have the center-of-mass diffusion coefficient given by

$$D = \frac{a^2 P}{\tau} = \frac{\mu T N_t}{P^2}. \tag{10}$$

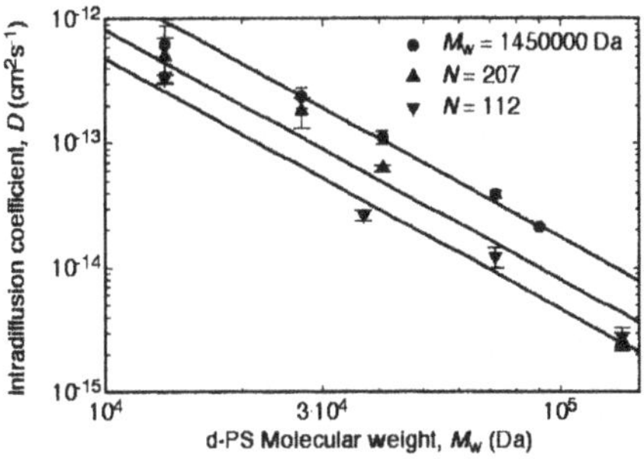

Fig. 3. Diffusion coefficients obtained from ion beam analysis measurements of the intradiffusion coefficient of linear polystyrene in a polystyrene network. The fits are to an inverse square molecular weight dependence. The expected dependence on crosslink density is also in reasonable agreement with these results, with DP^2 from the fits being very close to those calculated with the expected N dependence using the experimental data for the high molecular weight polymer matrix as a standard. The values of DP^2 for the $N = 207$ fits are in very good agreement with the calculated values, whereas those for $N = 112$ are 67% of the calculated value. Reproduced with permission from reference [27]. Copyright 2001 EDP Sciences

To measure such diffusion (intradiffusion), deuterated polystyrene is trapped inside a thin polystyrene network film at a concentration below its overlap concentration. This film was floated onto water and picked up on an identical layer on silicon, which contained normal, instead of deuterated, polystyrene. The diffusion across the boundary between the two layers, neglecting the small isotopic interaction, is purely due to Brownian motion. Such a system was measured by ion beam analysis [27], and in the limited experiments performed, the above theory was supported; the P^{-2} behavior was indeed observed, as was the expected dependence on N, although with only a high molecular weight linear polymer and two values of N used as matrices, such results need further testing. The diffusion coefficients are presented in Fig. 3. There have been measurements on the tracer diffusion of polystyrene into polystyrene networks by forced Rayleigh scattering [25], and these also verified the P^{-2} behavior, but not the crosslinking density dependence. The failure to demonstrate the dependence on N is likely to be due to the highly heterogeneous networks used for these experiments; crosslinking divinylbenzene with styrene monomer creates a crosslinked system but one which is likely to contain microgels.

2.2 Interdiffusion

Interdiffusion is different from intradiffusion in that it is the relaxation of concentration gradients. If a melt of linear polymer is put in contact with an identical polymer network, the linear polymer will gradually diffuse into the network. Here, interdiffusion has a strong concentration dependence of

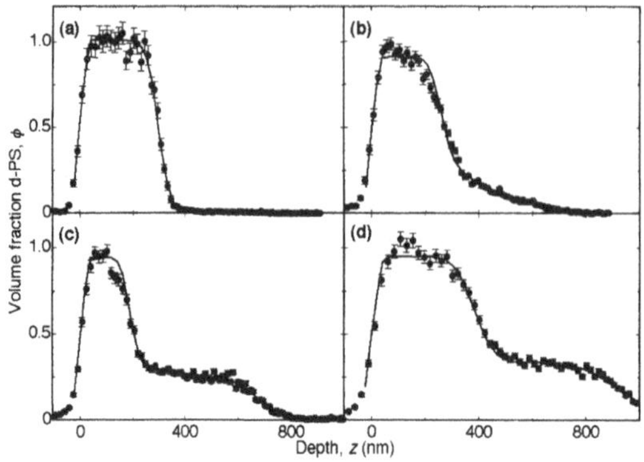

Fig. 4. NRA data showing the interdiffusion of linear polystyrene ($P = 137$, left side of bilayer) into a polystyrene network ($N = 110$). In (a) the bilayer is unannealed; in (b) and (c) it has been annealed for 55 minutes at 413 and 425 K respectively. In (d) the sample is at equilibrium after swelling for 60 minutes at 458 K. Reproduced with permission from reference [27]. Copyright 2001 EDP Sciences

its diffusion coefficient unlike the intradiffusion coefficients described above, which have no concentration dependence.

To analyze this in the simple mean field context is straightforward. There are two competing theories for interdiffusion: the fast [28, 29] and slow [30] theories. The slow theory is perhaps the clearer intuitively: A faster polymer cannot move to a point occupied by a slower polymer until the slower polymer has itself moved. Therefore the interdiffusion coefficient is controlled by the slower moving species. Although there have been demonstrations of the slow theory, it is generally not what is observed, and besides, the center-of-mass diffusion coefficient for a network may be taken as zero, so the slow theory cannot be applicable. The fast theory, by contrast relies on a collective, convective, motion of the fast-moving species to push the slow species out of the way. For a mixture of two linear polymers of polymerization indices P_1 and P_2, the interdiffusion coefficient according to the fast theory is given by

$$D_m = ((1-\phi)P_1D_1 + \phi P_2 D_2)\left(\frac{1-\phi}{P_1} + \frac{\phi}{P_2} - 2\chi\phi(1-\phi)\right). \quad (11)$$

The theory can be readily modified for a polymer network system undergoing swelling [22, 26], giving

$$D_m = (1-\phi)PD\left(\frac{1-\phi}{P} - 2\phi(1-\phi)\chi - \Gamma\phi\left(\frac{2-3\phi(1-\phi)^{2/3}}{6N(1-\phi)^{2/3}}\right)\right). \quad (12)$$

At maximum swelling $D_m = 0$, and the reader may notice that equations (6) and (12) are incompatible, yielding different scaling behaviors for P/N at swelling equilibrium. We show in Fig. 4 NRA data for the interdiffusion of deuterated polystyrene into a polystyrene network. The fast interpenetration of the network at short times is clear but good agreement with equation (12)

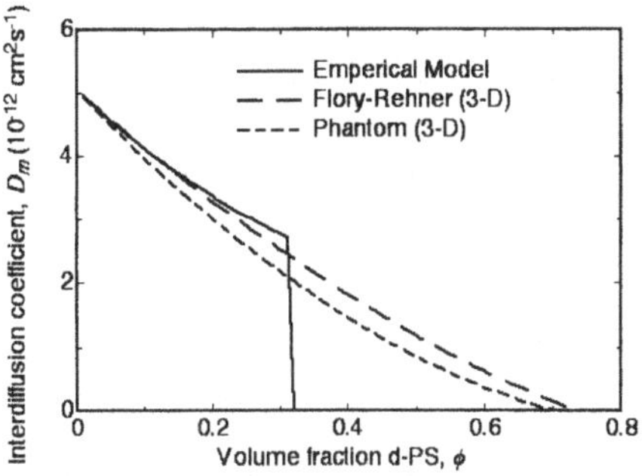

Fig. 5. Empirical interdiffusion coefficient plotted as a function of ϕ for the bilayer data from Fig. 4. Also plotted are the predictions of phantom and Flory-Rehner theory. Reproduced with permission from reference [27]. Copyright 2001 EDP Sciences

theory is not observed; a typical example is shown in Fig. 5, which shows that the interdiffusion coefficient drops off very dramatically close to ϕ_{eq}, but the fast theory shows a much slower decay in interdiffusion coefficient. An interesting point worth commenting on is the behavior of the thick deuterated polystyrene layer in Fig. 4, which does not remain at $\phi = 1$ after annealing. This is not because sol is escaping the gel; we know from measuring film thickness before and after washing in toluene that such networks contain negligible sol. In fact it is likely that heterogeneities in the gel cause deformation of the interface with the linear polymer. Although there is no intermixing, a highly deformed interface would give the impression of this because the NRA experiment averages over regions of 100% network and regions of 100% linear polymer. Such a result shows that in this particular case, the length scale for the heterogeneities is at least 100 nm.

3 Interdigitation

Intermolecular forces encourage adhesion between two materials but such forces are not strong, and are dependent on surrounding media. In the absence of physical and chemical bonds, or electromagnetic forces, polymers represent an ideal tool to permanently adhere two surfaces. In general polymers are sticky; if the energy of adhesion of a particular small molecule is k_BT, then that for a polymer with P monomers in contact with the surface would be P times greater. However, two polymers in contact with each represent an even greater opportunity for adhesion because polymers can entangle. Clearly, the more immiscible the polymers are, the less that they can entangle, and so the choice of polymers is important. The corollary in choosing completely miscible polymers is that, not only will they entangle, but complete intermixing will take place. Therefore, polymer brushes are ideal for interdigitation, with the added advantage that they can be tethered to inorganic surfaces, where adhesion would normally be considered difficult. We first consider experiments designed to measure the interdigitation between a linear polymer melt and a polymer network, before turning to brush-network interdigitation.

3.1 High molecular weight linear polymers

If a network mesh size is too small, or if the linear polymer molecular weight is too large, linear polymer will not swell a network, but rather interdigitate a short amount into the network. The first observation of the correlation between these parameters leading to a cut-off between interdiffusion and interdigitation was made using neutron reflectometry to study polystyrene linear polymers and networks [32]. Another neutron reflectometry study, also involving polystyrene, reported the interdigitation in terms of mean-field theory. The mean-field theory for asymmetric polymer interfaces was generalized by Broseta and co-workers [33]. This theory was incorporated into the one-dimensional swelling theory of Flory and Rehner [34]. Mean-field theory did not successfully describe the behavior of the networks partly on account of its failure to incorporate lateral structure. In Fig. 6 we show the experimental profile for the structure of an interface between linear d-PS ($P = 8214$) and a polystyrene network with $N = 189$ [34], The theoretical prediction (with $N = 200$) agrees well close to the interface, but further into the network, the experimental data predict much more swelling, supporting the swelling measurements described above.

An interesting result obtained in this study is that the amount of interpenetration decreased after some time; it was proposed that after swelling lightly crosslinked regions of the network, a relaxation process occurs, causing some linear polymer to be expelled from the network. In a related experiment, the escape of polymer from a network was studied using neutron reflectometry and novel ion beam techniques [35]. In this case, it was observed that the amount of polymer that segregated to the surface of a network and linear polymer blend film actually decreased after an initial increase. This again points to a second slow relaxation process taking place within the network.

There have been neutron reflectometry measurements on the interface between two polystyrene network layers [34, 36], with detailed experiments showing that the size of the interface between the two layers scales as $N^{1/2}$ [36].

3.2 Polymer brushes

The conformation of polymer brushes interpenetrating into a molten polymer (not necessarily a network) cannot be determined using simple mean-field

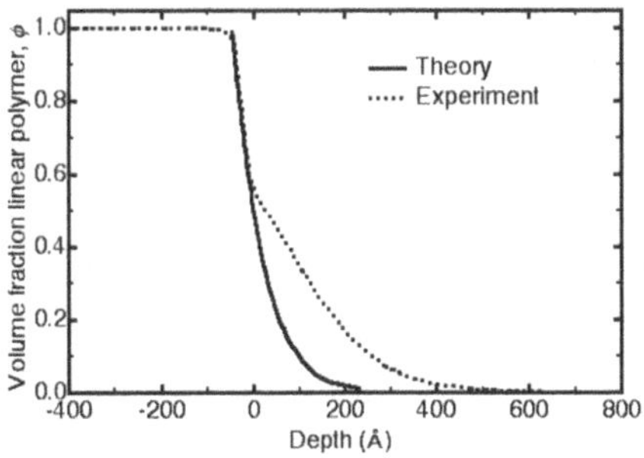

Fig. 6. Volume fraction profile of linear d-PS ($P = 8214$) in contact with a polystyrene network ($N = 189$) after annealing for 48 h at 453 K. The theoretical calculation is assuming one-dimensional swelling using phantom theory (here $N = 200$). In this case the phantom theory and affine Flory theory offer nearly identical profiles

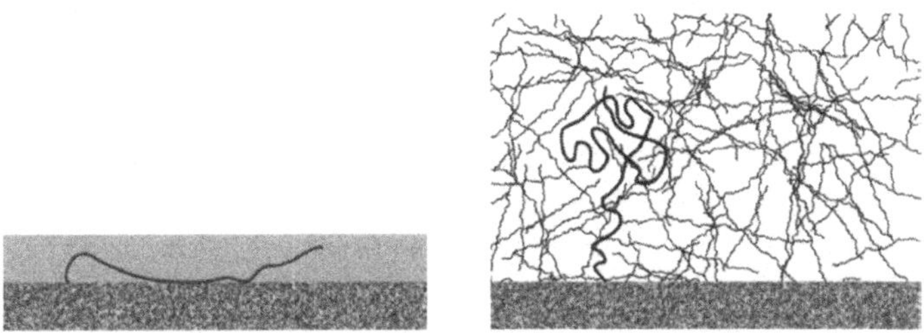

Fig. 7. A molten brush on a substrate will have its free end away from its grafting point (left). In order to reach equilibrium in a network, it will have to enter the network close to its grafting point, a process impeded by crosslinks

theory because there is no excluded volume in the melt. A polymer would like to have a random walk configuration in its melt, but this cannot be applied to brushes because the size of the brush ($\propto P^{1/2}$) would be independent of the grafting density, which is clearly not possible. As a result self-consistent mean-field theory must be used [37]. To this point, self-consistent mean-field theory has not been used to describe a network interpenetrated by a brush. There have been measurements considering the interpenetration of high molecular weight linear polymers by brushes, however [38].

An interesting related problem is the kinetics of brush penetration into a network. A film of (not too densely) grafted polymers will leave the polymers lying along the substrate. However, when placed in contact with a network at a temperature above the glass transition of the brush and network, the tethered polymers will diffuse into the network starting at a point close to where the free end lies, causing the brush to stretch. The ability of the brush to relax (unstretch) will be impeded by the crosslinks, and the route to equilibrium will be due to polymers retracting on the same path that they used

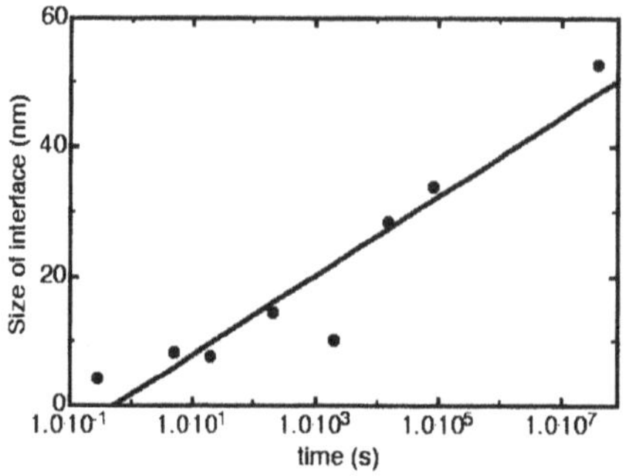

Fig. 8. The interdigitation of polystyrene brushes into polystyrene networks as taken from neutron reflectometry measurements. The time axis is scaled to 423 K using time-temperature superposition; for example, the first datum corresponds to a measurement of a sample annealed in vacuum for 20 minutes at 392 K, whereas the last corresponds to 165 h at 453 K. The fit is to logarithmic interfacial width behavior. Reproduced with permission from reference [41]. Copyright 1999 American Chemical Society

to enter the network. Several attempts need to be made before the polymer finally enters the network close to its grafting point and so maximize its conformational entropy (Fig. 7). This retraction process is very slow, and theoretical simulations [39] and calculations [40] demonstrate that the kinetics of penetration must be logarithmic with time.

Experimentally logarithmic time-dependence is difficult to demonstrate without the use of time-temperature superposition. The relevant parameters for polystyrene are known, so neutron reflection experiments were performed on the interdigitation of polystyrene networks by deuterated polystyrene brushes grafted to a silicon substrate. As expected, the diffusion behavior was indeed logarithmic, as shown in Fig. 8. One additional aspect of this work that is worth noting, is that the polystyrene brush penetrated further into the network than was the case for linear polymers. Although there are several possible explanations for this stretching [41], it is believed to be evidence of heterogeneities. For example, the brush may stretch because it seeks regions of the network where there are fewer crosslinks in order to maximize its configurational entropy (rather like an entropic trap).

An interesting and important challenge in continuing such work is to correlate the adhesion between brush and network as a function of the healing process. Most measurements in the past have been performed using PDMS [42-47], which is liquid at room temperature and so equilibrium is achieved quickly, and the time-temperature superposition approach of the polystyrene networks is inappropriate. However, to perform adhesion experiments at the elevated temperatures of polystyrene is not a trivial task.

4 Conclusions

In all aspects of the work presented above (apart form the reptation measurements), there is evidence for heterogeneities in the networks: Swelling measurements show greater swelling than mean-field theory predicts; interdigitation and segregation measurements show evidence for more than one relaxation process taking place in the networks; brush interdigitation occurs much further into the network than would be the case for a high molecular weight linear polymer matrix. Future experiments and theory aimed at improving our understanding of polymer networks should not be limited to only one length scale. Similarly, future theory should take into account that experimentally, different preparation methods of networks will result in different properties.

5 References

1. P. J. Flory, Proc. R. Soc. London A **351**, 351 (1976).
2. R. C. Ball, S. F. Edwards, Macromolecules **13**, 748 (1980).
3. J.-U. Sommer: Macromol. Symp. **81**, 139 (1994).
4. J. U. Sommer, T. Russ, R. Brenn, M. Geoghegan, Europhys. Lett. **57**, 32 (2002).
5. M. Pütz, K. Kremer, R. Everaers: Phys. Rev. Lett. **84**, 298 (2000).
6. E. Geissler, F. Horkay, A. M. Hecht, M. Zrinyi, J. Chem. Phys. **90**, 1924 (1989).
7. P.-G. de Gennes: *Scaling Concepts in Polymer Physics*, (Cornell University Press, Ithaca 1979)
8. M. Doi, S. F. Edwards, *The Theory of Polymer Dynamics*, (Oxford University Press, Oxford 1988).
9. H. M. James, E. Guth, J. Chem. Phys. **11**, 455 (1943).
10. H. M. James, E. Guth, J. Chem. Phys. **15**, 669 (1947).
11. H. M. James, E. Guth, J. Chem. Phys. **21**, 1039 (1953).
12. P. J. Flory, J. Rehner, Jr., J. Chem. Phys. **11**, 512 (1943).
13. P. J. Flory, J. Rehner, Jr., J. Chem. Phys. **11**, 521 (1943).
14. A. N. Gent, R. H. Tobias, J. Polym. Sci.: Polym. Phys. Ed. **20**, 2317 (1982).
15. G. Gee, J. B. M. Herbert, R. C. Roberts, Polymer **6**, 541 (1965).
16. N. A. Neuburger, B. E. Eichinger, Macromolecules **21**, 3060 (1988).
17. T. Russ, R. Brenn, M. Geoghegan, Macromolecules **36**, 127 (2003).
18. M. Geoghegan, MeV ion beam profiling of polymer surfaces and interfaces in: *Polymer Surfaces and Interfaces III* Eds. R. W. Richards and S. K. Peace (Wiley, Chichester 1999) pp. 43-73
19. F. S. Bates, G. D. Wignall, Macromolecules **19**, 932 (1986).
20. E. Geissler, F. Horkay, A.-M. Hecht, Phys. Rev. Lett. **71**, 645 (1993).
21. G. B. McKenna, K. M. Flynn, Y. Chen, Polym. Comm. **29**, 272 (1988).

22. G. B. McKenna, K. M. Flynn, Y. Chen, Macromolecules **22**, 4507 (1989).
23. G. B. McKenna, K. M. Flynn, Y. Chen, Polymer **31**, 1937 (1990).
24. G. B. McKenna, F. Horkay, Polymer **35**, 5737 (1994).
25. M. Antonietti, H. Sillescu, Macromolecules **18**, 1162 (1985).
26. A. Einstein, Ann. Phys. **17**, 549 (1905).
27. T. Russ, R. Brenn, F. Abel, F. Boué, M. Geoghegan, Eur. Phys. J. E **4**, 419 (2001).
28. E. J. Kramer, P. Green, C. J. Palmstrøm, Polymer **25**, 473 (1984).
29. H. Sillescu: Makromol. Chem., Rapid Commun. **5**, 519 (1984).
30. F. Brochard, J. Jouffroy, P. Levinson, Macromolecules **16**, 1638 (1983).
31. X. Zheng, M. H. Rafailovich, J. Sokolov, X. Zhao, R. M. Briber, S. A. Schwarz, Macromolecules **26**, 6431 (1993).
32. W. Wu, W. E. Wallace, J. H. van Zanten, B. J. Bauer, D. Liu, A. Wong, Polymer **38**, 2583 (1997).
33. D. Broseta, G. H. Fredrickson, E. Helfand, L. Leibler, Macromolecules **23**, 132 (1990).
34. M. Geoghegan, F. Boué, G. Bacri, A. Menelle, D. G. Bucknall, Eur. Phys. J. B **3**, 83 (1998).
35. M. Geoghegan, F. Boué, A. Menelle, F. Abel, T. Russ, H. Ermer, R. Brenn, D. G. Bucknall, J. Phys.: Condens. Matter **12**, 5129 (2000).
36. U. Perez-Salas, R. M. Briber, W. A. Hamilton, M. H. Rafailovich, J. Sokolov, L. Nasser, Macromolecules **35**, 6638 (2002).
37. C. J. Clarke, R. A. L. Jones, J. L. Edwards, K. R. Shull, J. Penfold, Macromolecules **28**, 2042 (1995).
38. C. J. Clarke, Polymer **37**, 4747 (1996).
39. J. M. Deutsch, H. Yoon, Macromolecules **27**, 5720 (1994).
40. K. P. O'Connor, T. C. B. McLeish, Macromolecules **26**, 7322 (1993).
41. M. Geoghegan, C. J. Clarke, F. Boué, A. Menelle, T. Russ, D. G. Bucknall, Macromolecules **32**, 5106 (1999).
42. F. Brochard-Wyart, P. G. de Gennes, L. Léger, Y. Marciano, E. Raphael, J. Phys. Chem. **98**, 9405 (1994).
43. D. Bureau, L. Léger, Langmuir **20**, 4523 (2004).
44. M. Deruelle, L. Léger, M. Tirrell, Macromolecules **28**, 7419 (1995).
45. M. Deruelle, M. Tirrell, Y. Marciano, H. Hervet, L. Léger, Faraday Discuss. **98**, 55 (1994).
46. L. Léger, E. Raphaël, H. Hervet, Adv. Polym. Sci. **138**, 185 (1999).
47. T. Vilmin, C. Tardivat, L. Léger, H. Brown, E. Raphael, Europhys. Lett. **68**, 543 (2004).

Biomolecular Layers on Silicon Studied by Optical Spectroscopy

Dietrich R. T. Zahn and Simona D. Silaghi

Institut für Physik, TU Chemnitz, 09107 Chemnitz, Germany
zahn@physik.tu-chemnitz.de

Summary. Amongst biomolecules the DNA base molecules adenine, cytosine, guanine, and thymine may also find interesting applications in organic electronics. They have optical gaps in the near ultra-violet and have already been considered as charge transport molecules in organic field effect transistors. Still there is very little knowledge on their electronic and optical properties when deposited as layers on inorganic substrates. Here the optical properties of the DNA bases deposited on flat and vicinal, hydrogen passivated Si(111) substrates are studied using spectroscopic ellipsometry (SE) up to 9.5 eV photon energy employing synchrotron radiation and reflectance anisotropy spectroscopy (RAS). The results for the dielectric function reveal strong optical anisotropy for adenine and guanine while the other two molecules form layers with isotropic properties. The experimentally derived dielectric functions are compared to density functional theory calculations of the optical response. Particularly interesting is the RAS response of the DNA bases as a function of thickness when deposited on vicinal Si surfaces. Ordering in the layers is induced by the step and terrace structure of the vicinal Si substrates. Even though the molecular structure is not dramatically different the RAS response is very distinct and allows an unambiguous identification of the bases.

1 Introduction

The usage of DNA base molecules as charge transport molecules in biomolecular electronic and optoelectronic devices is still a challenge. Consequently, the determination of the dielectric functions of DNA base films by spectroscopic ellipsometry (SE) is crucial for the understanding of both their electronic and optical properties. Recently, field effect transistor studies based on a modified DNA base revealed that the prototype bio-transistor gives rise to a better voltage gain compared to carbon nanotubes (CNTs) [1]. The capability to fabricate structures with characteristic dimensions of a few nanometers is nevertheless a key prerequisite for future applications in nanoelectronics or as functional materials on a nanometer scale. With the reduced size of future molecular devices the effects of dimensionality are becoming more and more important. Because silicon can be patterned in many ways, it is possible to use it as a versatile template and combine biomolecules with silicon electronics. Flat Si(111) has a 3-fold symmetry inducing the growth of three equivalent superstructure domains by symmetry but vicinal Si(111) surfaces can

B. Kramer (Ed.): Adv. in Solid State Phys. **45**, 45–57 (2005)

limit overlayer growth to a single domain. Moreover, vicinal surfaces should favour the nucleation along the step edges, thus being potential substrates in controlling the ordering of molecules in so-called molecular nano-wires. In this sense, *in situ* reflectance anisotropy spectroscopy (RAS) is employed in ultra-high vacuum (UHV) conditions for monitoring the ordering of DNA base molecules on vicinal hydrogen passivated Si(111). Additionally, *ab initio* calculations of electronic transitions corresponding to single DNA base molecules carried out based on time-dependent (TD) density functional theory (DFT) methods employing the B3LYP method in combination with the basis set 6-311++G(d, p) [2] are compared to the dielectric functions of the DNA base films.

2 Experimental and Data Analysis

Flat and vicinal p-type (B-doped) silicon(111) surfaces with resistivity in the range of 1–30 Ωcm were used as substrates for the DNA base films. One-side polished flat silicon substrates were provided by Wacker Siltronic while double-side polished vicinal substrates were supplied by Silchem. Prior to biomolecular deposition, the substrates were wet-chemically hydrogen terminated [3]. The "flat" Si(111) (0.35°-off cut angle) surfaces were employed as substrates for the thick biomolecular films and the investigation of their bulk optical properties via spectroscopic ellipsometry. The vicinal Si(111) surfaces used as substrates for thin biomolecular films were off cut oriented by 3° and 6°, respectively, towards the $[\bar{1}\bar{1}2]$ direction. The surface reconstruction was checked by low energy electron diffraction (LEED). In the case of the vicinal surfaces the LEED pattern shows a double splitting of the diffraction points in the $[\bar{1}\bar{1}2]$ direction typical for the formation of steps and terraces. The source materials of high-purity DNA base powders purchased from Sigma-Aldrich were evaporated under UHV conditions (base pressure $\sim 10^{-8}$ Pa) from Knudsen cells. Molecules of thymine and cytosine were evaporated at temperatures of 365 and 410 K with corresponding evaporation rates of 0.8 and 0.3 nm/min, respectively, while adenine and guanine molecules were evaporated at temperatures of 400 and 510 K with evaporation rates of 1.5 and 2 nm/min, respectively. The evaporation rates were *in situ* monitored via a quartz crystal and then *ex situ* calibrated via both ellipsometry and atomic force microscopy film thickness measurements.

In situ as well as *ex situ* ellipsometry measurements were performed on biomolecular films with thicknesses ranging from 40 to 120 nm. *In situ* VUV-SE measurements were carried out in the energy range between 4 and 9.5 eV under an incidence angle of 67.5° using synchrotron radiation as light source at BESSY II. Afterwards, the biomolecular films were investigated *ex situ* via variable angle spectroscopic ellipsometry (VASE) in the energy range of 0.8–5 eV. Spectroscopic ellipsometry measures the change in polarization of light as a function of incident angle and wavelength [4]. The experimentally determined ellipsometric values are Ψ and Δ, which are related to the ratio of Fresnel reflection coefficients $\tilde{r}_p$ and $\tilde{r}_s$ for p- and s-polarized light, respectively, by $\rho = \frac{\tilde{r}_p}{\tilde{r}_s} = \tan\Psi e^{i\Delta}$. The effective dielectric function is connected with the measured ratio of Fresnel reflection coefficients ρ and incident angle ϕ_0 by the following expression:

$$\langle\epsilon\rangle = \langle\epsilon_1\rangle + i\langle\epsilon_2\rangle = (\langle n\rangle + i\langle k\rangle)^2 = \sin^2\phi_0 \left[1 + \tan^2\phi_0 \left(\frac{1-\rho}{1+\rho}\right)^2\right] \quad (1)$$

The dielectric functions of the materials are determined by first constructing a suitable optical model of the sample with physically meaningful structural and optical parameters, and then by iteratively adjusting these parameters to obtain the best fit to the measured ellipsometric data, i.e., the fitting with the minimum mean square error MSE:

$$MSE = \frac{1}{2N-M}\sum_{i=1}^{N}\left[\left(\frac{\Psi_i^{mod} - \Psi_i^{exp}}{\sigma_{\Psi,i}^{exp}}\right)^2 - \left(\frac{\Delta_i^{mod} - \Delta_i^{exp}}{\sigma_{\Delta,i}^{exp}}\right)^2\right] \quad (2)$$

where N is the number of measured (Ψ, Δ) pairs and M is the total number of real fit parameters. The ellipsometry results of the DNA base films were modelled via mathematical models implemented in the WVASE software in the energy range between 0.8 and 9.5 eV. Knowing the corresponding values for film thickness and surface roughness the experimental data were then fitted using an Gaussian oscillator model which ensures Kramers-Kronig consistency of the dielectric function in the entire range from 0.8 to 9.5 eV. Specifically, the optical properties of thymine and cytosine films were evaluated using an isotropic model meaning that the optical response of the sample is equivalent in all three directions x, y, and z. On the other hand, the adenine and guanine films were described by a uniaxial anisotropic model consisting of two sets of Gaussian oscillators corresponding to ordinary ($\perp$ [111]) and extraordinary ($\|$ [111]) contributions. In order to derive accurate optical constants a multi-sample analysis (MSA) procedure was employed [5], [6]. This procedure assumes that the same set of optical constants describes a set of several films with different thicknesses which are coupled together in the fit.

The molecular ordering of DNA base molecules on both vicinal and "flat" H:Si(111) surfaces was monitored *in situ* during the growth by reflectance anisotropy spectroscopy (RAS) in the energy range of 1.5–5.5 eV. RAS measures the difference in normal incidence in reflection for light linearly polarized along two orthogonal directions in the sample surface. In the case of a vicinal Si(111) surface the RAS signal can be expressed as follows:

$$\frac{\Delta r}{r} = Re\left(\frac{\Delta r}{r}\right) + iIm\left(\frac{\Delta r}{r}\right) = 2\frac{r_{[1\bar{1}0]} - r_{[\bar{1}\bar{1}2]}}{r_{[1\bar{1}0]} + r_{[\bar{1}\bar{1}2]}} \quad (3)$$

Due to the fact that RAS is performed in normal incidence the technique is very sensitive to surface changes and capable of measuring very small optical anisotropies of 10^{-3} or even smaller which are hardly detectable by ellipsometry.

3 Results

3.1 Spectroscopic Ellipsometry

Figure 1 shows an example of ellipsometry data evaluation of two guanine samples having film thicknesses of (61.7 ± 0.1) and (84.8 ± 0.1) nm corresponding to samples denoted as A and B. The optical response of guanine films was described by a uniaxial model. A MSE value of 7.655 was obtained by the fitting procedure of both Ψ and Δ spectra in the energy range of 0.8–9.5 eV.

The dielectric function derived for guanine films is included in Fig. 2a together with the corresponding dielectric functions of adenine (Fig. 2b), cytosine (Fig. 2c) and thymine (Fig 2d) films. Besides guanine, adenine films exhibit also strong optical anisotropy with the ordinary ($\perp$ [111]) component of the dielectric function being larger compared to the extraordinary ($\parallel$ [111]) component. This situation indicates a mainly planar orientation of both molecules with respect to the silicon substrate. From the ratio between the extraordinary and ordinary components of the first band in ϵ_2 the average molecular orientation can be deduced. Thus, the molecular planes make an average tilt angle of $\sim$ 10° for adenine and $\sim$ 16° for guanine molecules with respect to the substrate plane. The preferential orientation of guanine and adenine molecules on H:Si(111) is additionally supported by the X-ray diffraction (XRD) results (not shown here). The interplanar spacing determined as 3.29 Åis quite close to the interplanar distance between two stacked guanines in the guanine monohydrate crystal which is 3.30 Å[7]. The favoured planar orientation of the guanine was in addition observed on other kind of substrates such as Cu(111) where the molecules order in a one-dimensional-like structure [8] or on substrates like MoS2 [9] and SrTiO3(100) [10]. In the case of adenine films, the interplanar spacing derived from XRD matches the distance of 3.30 Åbetween stacked molecules. Generally, the adenine molecules prefer to lie down on various substrates e.g. KBr(100) [17], SrTiO3(100) [10], MoS2(0001) [18], HOPG [19].

In contrast to the strong anisotropy revealed by adenine and guanine films, the small molecules cytosine and thymine seem to form isotropic films. The dielectric functions determined from ellipsometry data are compared with the calculated electronic spectra (circles) of single DNA base molecules in Fig. 2. For a better comparison the initial calculated values of the energy positions indicated by bars in Fig. 2 were red-shifted. In each case, the first theoretical electronic transition having $\pi \rightarrow \pi^*$ character is assigned to the highest occupied molecular orbital (HOMO) → lowest unoccupied molecular orbital (LUMO) band gap. In the case of the guanine molecule this is calculated to

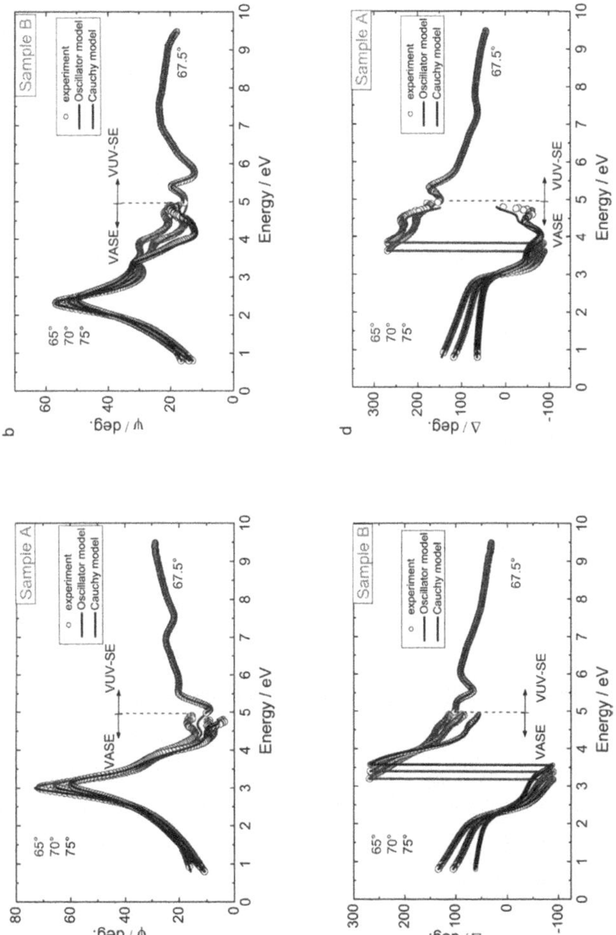

Fig. 1. Ψ and Δ spectra of two guanine films on flat H:Si(111)

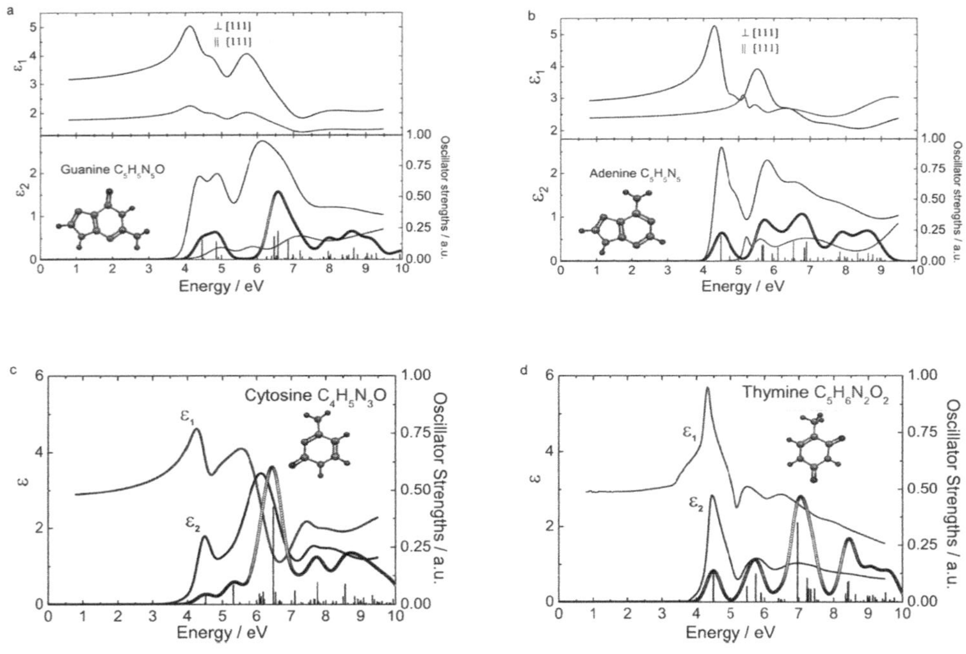

Fig. 2. Dielectric functions of (a) guanine, (b) adenine, (c) cytosine, and (d) thymine films. The calculated electronic spectra (circles) using TD-DFT were simulated using a sum of Gaussian oscillators with a FWHM of 0.2 eV. For a better comparison, the calculated results were red-shifted by (a) 0.65 eV, (b) 0.51 eV, (c) 0.18 eV, and (d) 0.55 eV

be at 4.96 eV. The previous experimental results revealed the first absorption band in the energy range from 4.31 to 4.59 eV [11, 12, 13, 14, 15, 16] while from ellipsometry measurements the HOMO-LUMO band gap appears at 4.31 eV being modelled by a Gaussian oscillator employed perpendicular to [111]. A slightly higher optical band gap of 4.47 eV is measured by ellipsometry in the case of adenine films but still lower than the previous reported results with energy values ranging from 4.51 to 4.63 eV depending on the surrounding environment of molecules [11, 12, 14, 20, 21]. For instance, the highest value of 4.63 eV is observed for the adenine in aqueous solution [11]. The calculated value for the HOMO $\rightarrow$ LUMO band gap of 4.98 eV is fairly high compared to the experimental values. The highest optical absorption onset was observed from ellipsometry at 4 eV in the case of adenine while the lowest at around 3 eV is characteristic for cytosine. Beyond 4 eV the imaginary part of dielectric function ϵ_2 of cytosine exhibits mainly three structures in agreement with those reported in the literature. The first band at 4.46 eV corresponds to the HOMO $\rightarrow$ LUMO band gap with the corresponding theoretical electronic transition at 4.64 eV. The experimental investigations performed on cytosine in various environments revealed a first absorption band in the energy range between 4.40 and 4.70 eV [11, 12, 14, 15, 21, 22, 23, 24, 25, 26]. Above 9 eV one can notice that the real part of the dielectric function ϵ_1 of the cytosine film clearly shows an ascending behaviour. Indeed, the ultraviolet reflectivity measurements performed in the energy range of 4–24 eV using synchrotron radiation on $\sim 1\mu m$ thick cytosine film on a mica substrate revealed additional structures above 9.5 eV [27]. The authors suggested that these additional structures must be due to $\sigma \rightarrow \sigma^*$ transitions.

The imaginary part of the isotropic dielectric function ϵ_2 of thymine corresponding to thymine films reveals also mainly three bands in agreement with the previous experimental data reported in the literature. The asymmetry of the first band in ϵ_2 spectrum was modelled by two Gaussian oscillators having energy positions at 4.44, and 4.64 eV, respectively. It was observed that single crystal absorption measurements generate different energy positions for the first absorption band when the incident electric field is polarized along different crystallographic axis [28]. For instance, the absorption along a axis leads to a band centered at 4.5 eV while the absorption measured along b axis gives rise to another band at 4.38 eV. With the help of these polarization-dependent absorption investigations performed on a single crystal one can conclude that the doublet components observed at 4.44–4.64 eV in the case of thymine are an evidence of the crystalline nature of the film revealed additionally by XRD. Moreover, the doublet band observed in the thymine film seems to be mainly polarized in the ab crystallographic plane which corresponds to the (001) plane [29]. Indeed, Ozeki *et al.* noticed that for instance the thymine anhydrate crystals grow as plates with (100) or (001) planes as prominent faces with the plane of the molecule being nearly parallel to the (001) plane [29]. The calculated HOMO $\rightarrow$ LUMO band gap of 4.99

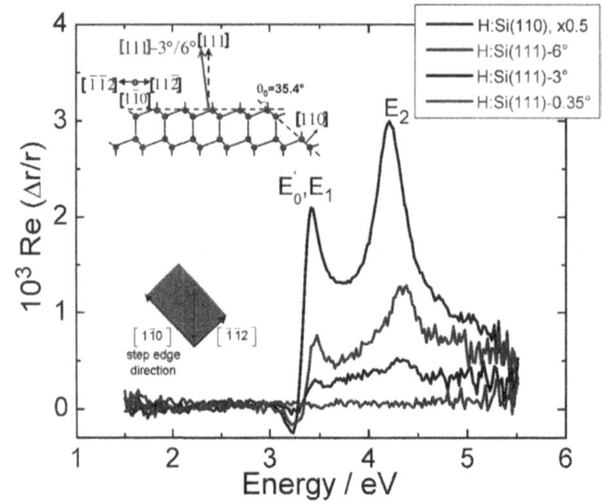

Fig. 3. RAS spectra of various silicon surfaces

eV is quite high compared to previously experimental absorption bands reported at 4.8 eV in gas phase [22, 30], 4.7 eV in various solutions [21, 22, 30] 4.64–4.44 eV in films [12], and 4.5 eV for a single crystal [28, 31]. Generally, the red-shift of the absorption from gas phase to single crystal is proportional with the increase of the intermolecular interaction by decreasing the intermolecular distances. Still, poor agreement in the relative intensities of the bands and energy positions is found when one compares TD-DFT calculations with the ϵ_2 spectra of the DNA bases because the single molecule calculations do not account for the intermolecular interactions so far.

3.2 Reflectance Anisotropy Spectroscopy

The anisotropy of cubic materials, e.g. silicon, arises at the surface due to a broken symmetry as in the case of vicinal Si(111) substrates where the surface anisotropy is induced by the formation of steps and terraces. Such anisotropy is often referred to as surface induced optical anisotropy (SIOA) [32, 33]. The RAS signal of vicinal Si(111) surfaces is similar with the response of Si(110) surface as shown in Fig. 3. The fractional contribution of the [110] direction to the vicinality of (111) surface can be determined from the ratio between the magnitude of the E_2 peak for the vicinal surface and that for the (110) surface [3]. The values determined experimentally from the spectra are about 0.08 and 0.17 corresponding to 3° and 6° off cut angles which are quite close to the expected theoretical values of 0.09 and 0.18.

Figure 4 depicts the *in situ* RAS monitoring of DNA base layers on vicinal H:Si(111)-6° surfaces. During the RAS measurements the substrate geometry was always kept identical with the one sketched in Fig. 3. All four molecules behave optically different when deposited onto the vicinal surface. Very small anisotropies are observed in the case of guanine and cytosine in comparison with the large RAS signals arising from adenine and thymine layers. When increasing the guanine coverage the RAS signal reaches a saturation level for thicknesses above 30 nm. The amplitude of the silicon features increases with thickness. Larger changes can be observed around the silicon E_2 gap which overlaps with the absorption of guanine since the onset is around 3.5 eV. Moreover, the HOMO-LUMO gap of guanine at 4.31 eV is extremely close to the 4.25 eV energy gap of silicon. The lineshape of the signal evolves in a derivative-like lineshape with increasing guanine coverage above 14 nm. For coverage above 20 nm the signal starts to decrease and finally saturates for thicknesses around 30 nm as shown in the inset of Fig. 4a. This can be related to the fact that the molecules form an isotropic layer and consequently the signal vanishes due to equal absorption in orthogonal directions. In order to distinguish between the anisotropy of the substrate and the anisotropy of the guanine films, RAS measurements of guanine deposition on almost flat H:Si(111)-0.35° were carried out (see Fig. 5a). The observed features are attributed to thickness artifacts. The weak anisotropy of guanine films on vicinal substrates seems to be caused by the vicinality which induces ordering of the guanine molecules up to critical thickness of about 20 nm. On the contrary to guanine, adenine exhibits large anisotropy signals. By increasing the adenine coverage the RAS signal evolves in a derivative-like lineshape in the absorption range of adenine while the silicon features gradually vanish. The asymmetric evolution in the lineshape of the derivative-like RAS signals indicates the contribution of at least two oscillators above 4 eV. Following the surface dielectric anisotropy model [34] it is found that the dominant absorption takes place mainly along $[\bar{1}\bar{1}2]$ direction suggesting furthermore strong ordering of adenine molecules with respect to this direction. Weak anisotropy can be observed with a similar derivative-like lineshape also on the almost flat silicon surface (Fig. 5b). Such small anisotropies were not observed by ellipsometry. The RAS monitoring of cytosine deposition on H:Si(111)-6° shows that the lineshape of the signal is strongly thickness-dependent as in the case of guanine. Larger changes can be observed around the E_2 gap of silicon at 4.25 eV which overlaps with the absorption of cytosine with the onset at 3 eV. At certain cytosine coverage a new feature around 4.90 eV appears which is closed to the second electronic transition at 4.95 eV as obtained from ellipsometry results.

For very thin cytosine coverages, the molecules seem to preferentially align on vicinal surfaces with respect to the $[1\bar{1}0]$ direction. Already for higher coverage larger than 12 nm the signal starts to decrease and finally saturates for thicknesses around 20 nm as shown in inset Fig. 4c. There is no anisotropy of the cytosine layers on H:Si(111)-6° meaning that the anisotropy of cytosine on vicinal substrates is caused by the vicinality which induces ordering of the cytosine molecules up to critical thickness of about 10 nm. The strongest

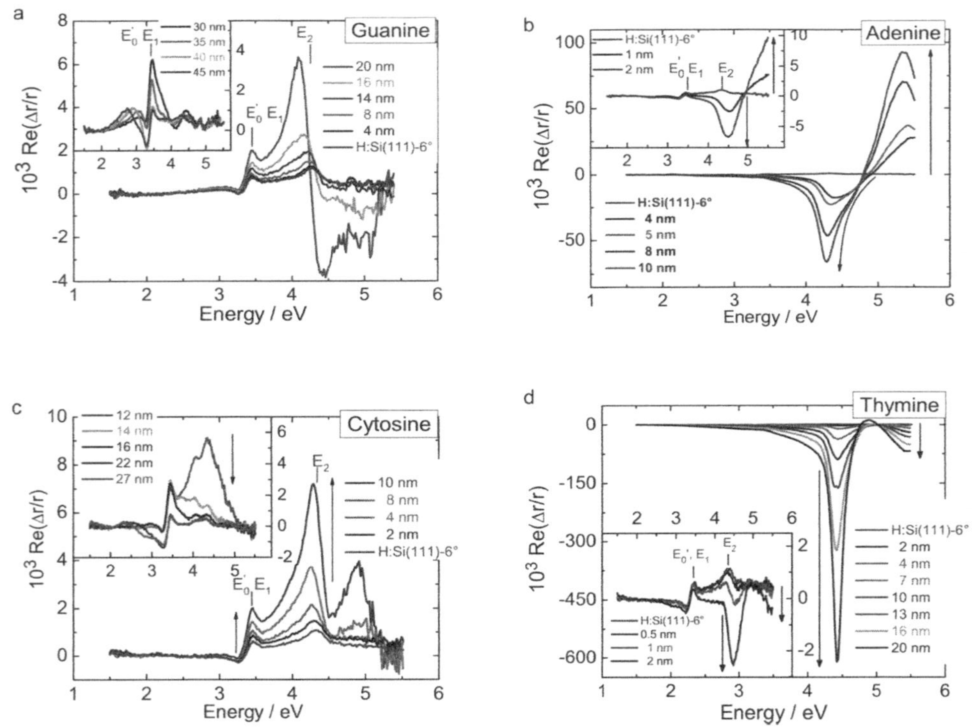

Fig. 4. *In situ* RAS monitoring spectra of: (a) guanine, (b) adenine, (c) cytosine, and (d) thymine layers on H:Si(111)-6° surface

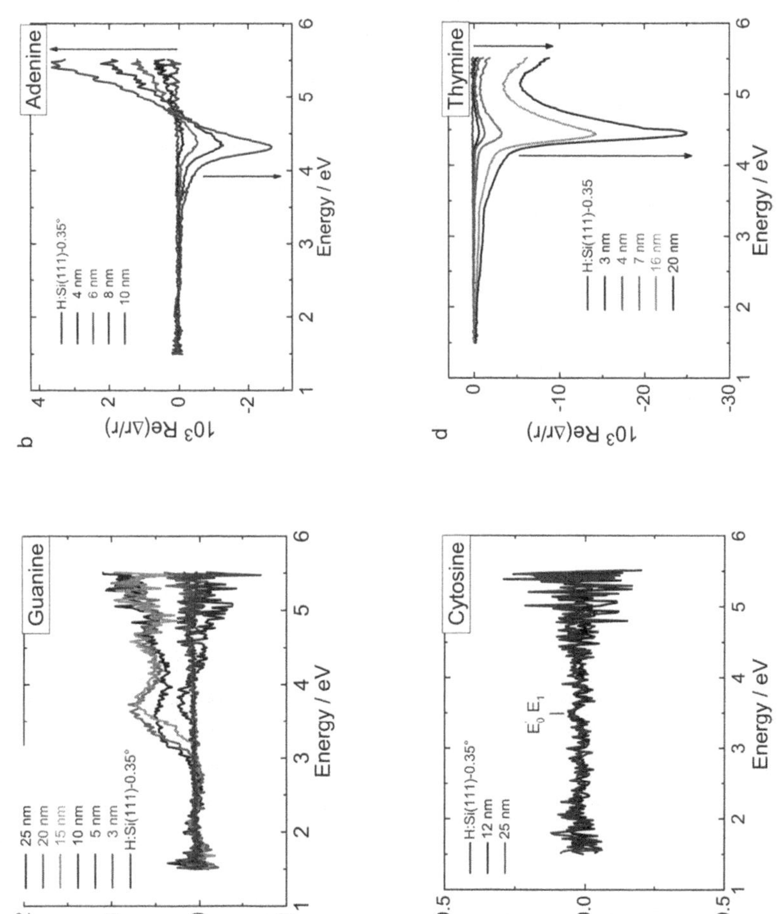

Fig. 5. *In situ* RAS monitoring spectra of: (a) guanine, (b) adenine, (c) cytosine, and (d) thymine layers on "flat" H:Si(111)-0.35° surface

RAS signals were observed in the case of thymine layers on H:Si(111)-6° as reproduced in Fig. 4d. By increasing the thymine coverage the RAS signal of the substrate completely vanishes while large anisotropy signals appear in the absorption range of the molecule. The lineshape of the RAS signal of thymine is very much like ϵ (see Fig. 2d). The negative sign of the signal suggests high absorption along $[\bar{1}\bar{1}2]$ direction and hence a preferential alignment of the transition dipole moments of the molecule with respect to the parallel to the same direction. Smaller anisotropies were also observed on "flat" silicon substrates but still larger signal amplitudes compared to the case of adenine layers on the same type of substrate.

4 Summary

For the first time, the dielectric functions of all four DNA base films were successfully determined in the energy range from near infrared to ultra-violet. Adenine and guanine films exhibit strong optical anisotropy being described by a uniaxial dielectric function with a dominant ordinary component relative to the extraordinary one and hence indicating a mainly planar orientation of both molecules with respect to the silicon substrate. The ellipsometry data of thymine and cytosine films were described by an optically isotropic model. The calculations of electronic transitions gave rise to higher energy values compared to the ellipsometry results due to the fact that the single molecule calculations do not account for the intermolecular interactions which usually lead to a red-shift of energies. Furthermore, the RAS technique was employed for the first time in monitoring the growth of DNA base molecules on both flat and vicinal H:Si(111) surfaces. The RAS measurements revealed information about the molecular ordering of DNA bases induced by the density of steps on silicon surfaces.

Acknowledgements

The financial support by Sächsisches Staatsministerium für Wissenschaft und Kunst (SMWK), Deutsche Forschungsgesellschaft Graduiertenkolleg GRK 829/1 "Accumulation of Single Molecules to Nanostructures" and the BMBF project 05 622 ESA2 is acknowledged. The BESSY facility is also acknowledged.

References

1. Mauricio, P. Visconti, V. Arima, S. D'Amico, A. Biasco, E. D'Amone, R. Cingolani, and R. Rinaldi, Nanoletters **3**, 479 (2003).
2. M. E. Casida, C. Jamorski, K. C. Casida, D. R. Salahub, J. Chem. Phys. **108**, 4439 (1998).

3. Yasuda, D. E. Aspnes, D. R. Lee, C. H. Bjorkman, G. Lucovsky, J. Vac. Sci. Technol. A **12**, 1152 (1994).
4. R. M. A. Azzam, N. M. Bashara: *Ellipsometry and Polarized Light* (Elsevier, North-Holland 1992).
5. M. A. Barret, Z. Borkowska, M. W. Humphreys and R. Parson, Thin Solid Films **28**, 137 (1986).
6. B. M. Ayupov, S. A. Prokhorova, Optics and Spectroscopy **90**, 446 (2001).
7. U. Thewalt, C. E. Bugg, R. E. Marsh, Acta Crsyt. B **27**, 2358 (1971).
8. M Furukawa, H. Tanaka, T. Kawai, J. Chem. Phys. **115**, 3419 (2001).
9. W. M. Heckl, Phys. Bl. **49**, 417 (1993).
10. H. Tanaka, T. Kawai, J. Vac. Sci. Techol. B **13**, 1411 (1995).
11. D. Voet, W. B. Gratzer, R. A. Cox, P. Doty, Biopolymers **1**, 193 (1963).
12. T. Yamada, H. Fukutome, Biopolymers **6**, 43 (1968).
13. L. B. Clark, J. Am. Chem. Soc. **99**, 3934 (1977).
14. Y. Matsuoka, B. Norden, J. Phys. Chem. **86**, 1378 (1982).
15. C. A. Sprecher, W. C. Johnson, Biopolymers **16**, 2243 (1977).
16. J. C. Sutherland, K. Griffin, Biopolymers **23**, 2715 (1984).
17. J. T. Kim, M. Kanai, T. Kawai, and S. Kawai, Jpn. J. Appl. Phys. **33**, 5027 (1993).
18. S. J. Sowerby, M. Edelwirth, M. Reiter, W. M. Heckl, Langmuir **14**, 5195 (1998).
19. T. Uchihashi, T. Okada, Y. Sugawara, K. Yokoyama, S. Morita, Phys. Rev. B **60**, 8309 (1999).
20. L. B. Clark, J. Phys. Chem. **94**, 2873 (1990).
21. W. Voelter, R. Records, E. Bunnenberg, C. Djerassi, J. Am. Chem. Soc. **90**, 6163 (1968).
22. L. B. Clark, I. Tinoco Jr., J. Am. Chem. Soc. **87**, 11 (1965).
23. K. Raksanyi, I. Földvary, J. Fidy, L. Kittler, Biopolymers **17**, 887 (1978).
24. T. P. Lewis, W. A. Eaton, J. Am. Chem. Soc. **93**, 2054 (1971).
25. F. Zaloudek, J. S. Novros, L. B. Clark, J. Am. Chem. Soc. **107**, 7344 (1985).
26. A. F. Fucaloro, L. S. Forster, J. Am. Chem. Soc. **93**, 6443 (1971).
27. D. E. Johnson, and M. Isaacson, Optics Communications **8**, 406 (1973).
28. B. G. Anex, A. F. Fucaloro, A. Dutta-Ahmed, J. Phys. Chem. **79**, 2636 (1975),
29. K. Ozeki, N. Sakabe, J. Tanaka, Acta Cryst. B **25**, 1038 (1969).
30. L. B. Clark, G. G. Peschel, I. Tinoco Jr., J. Phys. Chem. **69**, 3615 (1965).
31. J. S. Novros, L. B. Clark, J. Phys. Chem. **90**, 5666 (1986).
32. D. E. Aspnes, J. Vac. Sci. Technol. B **3**, 1498 (1985).
33. D. E. Aspnes, A. A. Studna, J. Vac. Sci. Technol. A **5**, 546 (1987).
34. R. J. Cole, B. G. Frederick, P. Weightman, J. Vac. Sci. Technol. A **16**, 3088 (1998).

Lambert Diffusion in Porous Media in the Knudsen Regime

Stefanie Russ[1], Stephan Zschiegner[1,2], Armin Bunde[1], and Jörg Kärger[2]

[1] Institut für Theoretische Physik III, Justus-Liebig-Universität Giessen, 35392 Giessen, Germany
[2] Fakultät für Physik u. Geowissenschaften, Universität Leipzig, 04103 Leipzig, Germany

Summary. We discuss molecular diffusion in nanopores with different types of roughness under the exclusion of mutual molecular collisions, i.e., in the so-called Knudsen regime, both, in $d = 2$ and $d = 3$. We show that the diffusion problem can be mapped onto Levy walks and discuss the roughness dependence of the diffusion coefficients D_s and D_t of self- and transport diffusion, respectively. We find that in $d = 2$, the diffusion is anomalous with $D_s \sim \ln t$ and $D_t \sim \ln L$, whereas in $d = 3$, the logarithmic corrections are absent. Both diffusion coefficients decrease significantly when the roughness is enhanced, in remarkable disagreement with earlier findings.

1 Introduction

Diffusion and transport phenomena of gases in disordered and porous media have been the subject of intense research for several decades [1, 2, 3, 4]. Among the experimental and technical applications are heterogeneous catalysis [5], adsorption [6] and separation [7] of molecules. In general, the diffusion of the gas depends on the collisions between the gas molecules as well as on the collisions with the pore boundaries. Of particular interest for many real gases is the so-called Knudsen regime, where the interaction of the molecules with the pore boundaries play the crucial role and the intramolecular collisions can be neglected. Accordingly, for the diffusion properties, the pore geometry is essential. Examples of experimental studies include diffusion in mesoporous materials [8, 9] and the low-temperature range of long-range diffusion in beds of zeolite crystallites [10].

There are essentially two kinds of diffusion experiments: (i) In transport diffusion, the particles diffuse in a non-equilibrium situation from one side of the pore to the opposite side under the influence of a concentration gradient ∇c. The relevant quantity here is the transport diffusion coefficient D_t which describes how the stationary current increases with ∇c,

$$j = -D_t \nabla c. \tag{1}$$

(ii) In self- (or tracer-) diffusion, particles diffuse under equilibrium conditions. The relevant quantity is the self-diffusion coefficient D_s which describes the increase of the mean-square displacement $\langle x^2(t) \rangle$ along the pore with time t,

B. Kramer (Ed.): Adv. in Solid State Phys. **45**, 59–69 (2005)

$$\langle x^2 \rangle = 2\, D_s(t)\, t. \tag{2}$$

In this paper, we use scaling arguments and numerical simulations to describe both types of diffusion in the Knudsen regime. Our main results can be summarized as follows: In 2d-pores, diffusion is anomalous such that D_s depends logarithmically on t and D_t depends logarithmically on the pore length L. In $d = 3$, these logarithmic corrections vanish and diffusion is normal. Both, D_s and D_t are influenced in the same way by the boundary roughness and decrease with increasing roughness. This finding corrects earlier claims that D_t is not influenced by the roughness of the boundary [4].

In the numerical simulations, the particles start at the left side of the pore when transport diffusion is considered or in the middle of the pore when self-diffusion is considered. Each particle performs a random trajectory inside the pore, where it moves with constant velocity v_0. The direction of the first jump in the transport diffusion problem is taken from a homogeneous probability distribution (describing a homogeneous particle flux into the system). When the particle hits the pore boundaries, it is absorbed for a very short while and then emitted into a new random direction. This new direction is chosen according to Lambert's cosine law, i.e., the angle $\vartheta \in [-\pi/2, \pi/2]$ to the normal component of the surface occurs with probability $dP(\vartheta, \varphi) \sim \cos\vartheta\, d\Omega$, with the solid angle $d\Omega = d\vartheta$ in $d = 2$ and $d\Omega = \sin\vartheta d\vartheta d\varphi$ in $d = 3$.

The 2d- and 3d-pores of different roughness that we consider are built by sticking together n units of length and width h, with n up to 5000, which leads to a total length $L = nh$ of the pores. The boundary of each unit is created iteratively by a deterministic generalized Koch curve generator. Figures 1 and 2 show examples of the smooth pore together with the first iterations of pores in $d = 2$ and 3. The highest iteration, $\nu = 3$, yields the highest roughness considered in this work. By construction, the volume of the pores does not depend on ν. For simplicity, we have set the pore width to $h = 1$ in all numerical calculations.

Figure 3 shows some typical trajectories in the 2d-pore with $\nu = 3$. It is easy to see how the roughness influences the motion of the particle. In rough pores, the particle can get "trapped" for some time in certain areas of the pore and this trapping slows the motion considerably down. Each trajectory consists of jumps of lengths $\ell_i = v_0 \Delta t_i$, where v_0 is the constant velocity of the particle along the trajectory and Δt_i the jump time.

2 The Self-Diffusion Problem

The first quantity we are interested in is the distribution $P(|x|)$ of the jump lengths $x_i = h \tan\vartheta_i$ parallel to the channel. Figure 4 shows that asymptotically $P(|x|)$ decays as

$$P(|x|) \sim |x|^{-(1+\beta)}, \tag{3}$$

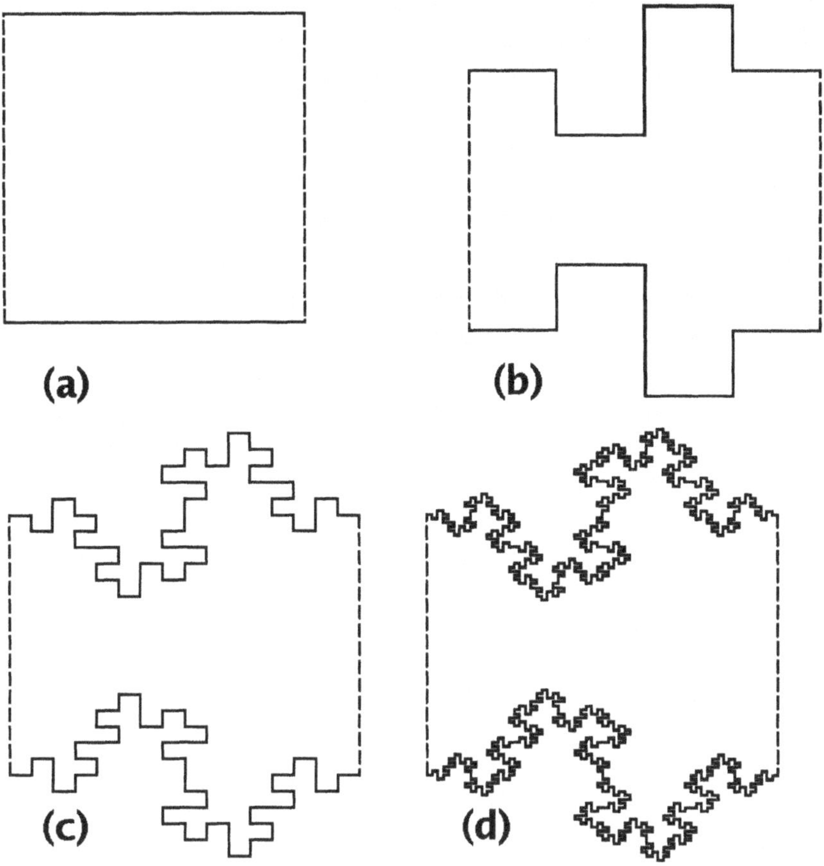

Fig. 1. Units that build up the $2d$-pores with different orders (generations) ν of roughness. (a) The smooth unit. (b-d) The units of generation $\nu = 1$, 2 and 3, respectively. n such units are sticked together to build pores of lengths $L = nh$

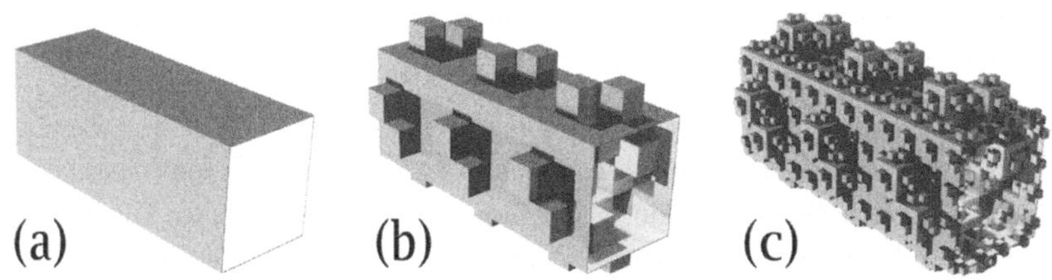

Fig. 2. Geometry of the $3d$-pores with different roughness (generation ν) and $n = 3$. (a) The smooth pore. (b, c) The pores of generation $\nu = 1$ and 2

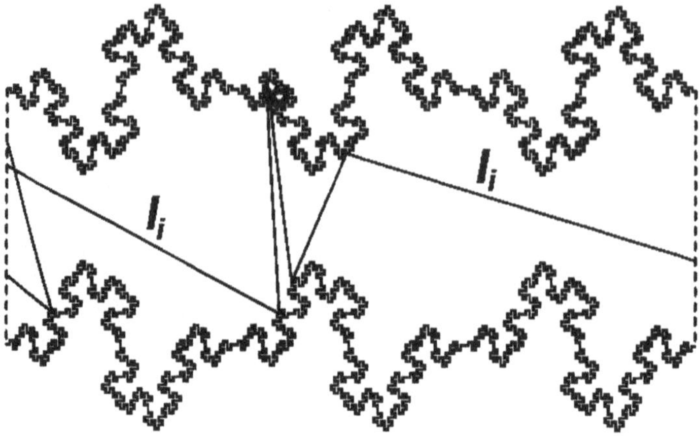

Fig. 3. A two-dimensional pore of length $L = 3h$ and $\nu = 3$ is shown with two typical trajectories that enter at the left side and leave the pore at the two different sides. It can be seen that the particles can be "trapped" for a certain time at some places of the boundary

with $\beta = 2$ for $d = 2$ and $\beta = 3$ for $d = 3$. For the smooth pore in $d = 2$, Eq. (3) can be easily derived analytically.

By definition, the time of each jump is proportional to the jump length l. For very large jumps, we have $x \simeq l$, and the jump time is proportional to the jump length in x-direction, $x \sim t$. Hence, Eq. (3) defines a Levy walk in $d = 1$ [11, 12, 13, 14] and the large jumps determine the diffusion properties. As Fig. 3 shows, in $d = 2$, large jumps occur close to the angles $\vartheta = \pm\pi/2$ (measured to the normal component of the surface) and are thus quite frequent. In $d = 3$, in contrast, the jump length is determined by two angles, ϑ and φ, and only the proper combination of both leads to very large jumps. Naturally, this combined probability of both angles is quite small, leading to very rare large jumps and $\beta = 3$.

It is well known [12, 13, 14] that for a Levy walk, the self-diffusion coefficient D_s scales as

$$D_s(t) = \begin{cases} D_s^0 \ln t & \text{for } \beta = 2 \\ D_s & \text{for } \beta > 2, \end{cases} \tag{4}$$

with the constants D_s and D_s^0. Accordingly, we expect that in $d = 2$, where $\beta = 2$, the diffusion is anomalous with a diffusion coefficient that tends to infinity with increasing time t. Due to the time-dependence of D_s, it is essential to keep t fixed when comparing the different coefficients in different $2d$-pores. In $d = 3$, $\beta = 3$ and we expect normal diffusion since large jumps are rare (see above).

We would like to note that for smooth two-dimensional pores, the logarithmic time-dependence of D_s can be directly obtained from Lambert's cosine law. After N jumps, the elapsed time is $t = \sum_{i=1}^{N} t_i = hv_0^{-1} \sum_{i=1}^{N} (\cos\vartheta)^{-1} =$

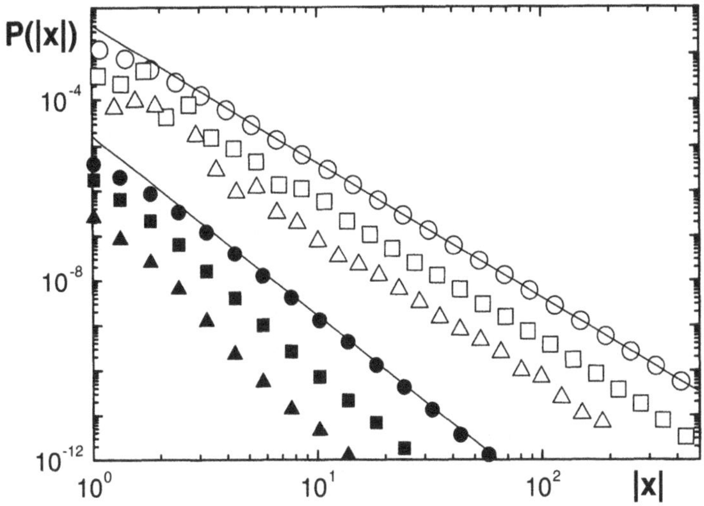

Fig. 4. The distribution $P(|x|)$ is plotted versus $|x|$ for the $2d$-pores (*open symbols*) and for the $3d$-pores (*filled symbols*) of $\nu = 0$ (*circles*), $\nu = 1$ (*squares*) and $\nu = 3$ (*triangles*). For a better overview, the data of $d = 3$ have been shifted down by a factor of 1000. The lines of slopes -3 and -4 are guides to the eye. The average was taken over 10^4 trajectories

$2Nhv_0^{-1}\int_0^{\pi/2}(\cos\vartheta)^{-1}P(\vartheta)d\vartheta = N\pi h/(2v_0)$. For uncorrelated x_i, $\langle x^2\rangle = \langle\left(\sum_{i=1}^N x_i\right)^2\rangle = \langle\sum_{i=1}^N x_i^2\rangle = h^2\langle\sum_{i=1}^N \tan^2\vartheta\rangle$. It is straightforward to show that the maximum angle $\vartheta_{\max}$ that occurs during N steps depends on N via $\vartheta_{\max} = \arcsin(1 - 2/N)$. Replacing the sum by the corresponding integral, we find $D_s \sim \ln t$. It is interesting to note that a homogeneous distribution of ϑ would lead to $D_s \sim t/\ln t$.

To test our predictions, we have simulated the Knudsen diffusion for the different pore geometries. The average was taken over 10^5 random trajectories for each system.

First, we consider two-dimensional pores. Fig. 5(a) shows $\langle x^2(t)\rangle$ versus t. The deviation from a straight line of slope 1 shows that D_s changes with time. To reveal the logarithmic behavior of $D_s(t)$, we plot $\langle x^2(t)\rangle t^{-1}(\ln t)^{-1}$ versus t in Fig. 5(b). For large t, the data reach a plateau, from which we obtain D_s^0 in accordance with Eqs. (2) and (4). The figure also shows that with increasing boundary roughness, the diffusion is considerably slowed down.

Next, we consider three-dimensional pores. Figure 6(a) shows that after a short period of time, $\langle x^2(t)\rangle$ is linear in t. To reveal this normal diffusion behavior, we plot $\langle x^2(t)\rangle/t = 2D_s$ versus t in Fig. 6(b). Again, the data reach a plateau showing that Eqs. (2) and (4) describe the behavior correctly. Also in $d = 3$, D_s decreases strongly with increasing boundary roughness, a behavior which has already been noticed earlier [4, 15, 16].

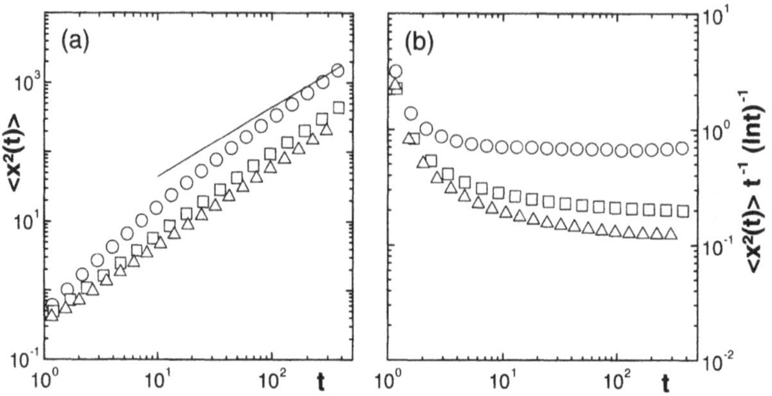

Fig. 5. (a) The mean square displacement $\langle x^2(t)\rangle$ is shown for the $2d$-pores with $\nu = 0$ (*circles*), 1 (*squares*) and 3 (*triangles*) in a double-logarithmic plot. It can be seen that the data deviate from a straight line of slope 1. (b) The scaled mean square displacement $\langle x^2(t)\rangle/(t \ln t)\,(\sim D_s^0)$ is shown for the systems of (a)

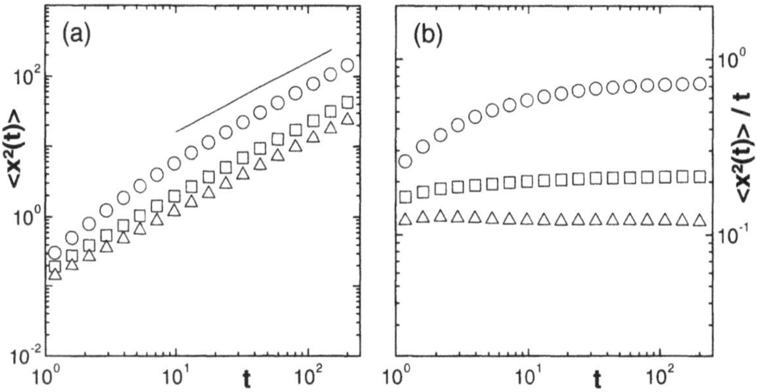

Fig. 6. (a) The mean square displacement $\langle x^2(t)\rangle$ is shown for the $3d$-pores with $\nu = 0$ (*circles*), 1 (*squares*) and 3 (*triangles*) in a double-logarithmic plot. The line of slope 1 is a guide to the eye. (b) The scaled mean square displacement $\langle x^2(t)\rangle/t\,(= 2D_s)$ is shown for the systems of (a)

3 The Transport-Diffusion Problem

Next, we consider the related transport diffusion problem, where a fixed concentration gradient $\nabla c = -e_x\, c_0/L$ is applied between the concentrations $c = c_0$ at the left side and $c = 0$ at the right side of the pore. Particles enter the pore at the left side, perform a random trajectory and are absorbed when they hit either the left or the right side of the pore. This leads, after some relaxation time, to a constant current $\boldsymbol{J}$. According to Fick's law, the current density $\boldsymbol{j} = \boldsymbol{J}/h^{d-1}$ is described by

$$\boldsymbol{j} = -D_t \nabla c = D_t \frac{c_0}{L}\, \boldsymbol{e}_x. \tag{5}$$

We want to understand how the anomalous $\ln t$ dependence of $D_s(t)$ in $d = 2$ is reflected by D_t and if D_t decreases with the boundary roughness in the same way as D_s does. Recently, Malek and Coppens found the surprising result that D_t is not affected by the boundary roughness [4, 15, 16] and thus behaves very different from the self-diffusion coefficient D_s. To see if this result is correct, we have performed extensive numerical calculations of D_t with particular emphasis on the L dependence of D_t. Again, the average was taken over 10^5 random trajectories for each system.

Since the relaxation of the particle flow into a stationary state is very time-consuming, it has become common to derive D_t from the probability f_t, that one single particle that starts at the left side leaves the system at the right side [17]. To calculate f_t, N random trajectories are considered that start at the left side and end when they leave the system (at either of the sides). By definition, f_t is the ratio between the number of particles leaving the pore at the right side and N. Since $|\boldsymbol{j}| = c_0 f_t \langle v_x \rangle$, with the mean velocity $\langle v_x \rangle$ in x-direction, we obtain with Eq. (5)

$$D_t = \langle v_x \rangle f_t L. \tag{6}$$

To estimate the L dependence of f_t, we can use simple scaling arguments. The time t to travel a distance L scales (without logarithmic corrections) as $t \sim L^2$. Assuming that $D_t \sim f_t L$ is equivalent to D_s, we can obtain the L dependence of D_t from the t dependence of D_s and vice versa, yielding

$$D_t \sim \begin{cases} \ln L^2 \sim \ln L, & \text{d=2} \\ \text{const}, & \text{d=3}. \end{cases} \tag{7}$$

The counterintuitive increase of D_t with L in $d = 2$ can be understood in the following way: With increasing L, the number of jumps in the pore is increased and hence, the probability that a very large jump occurs increases with L. On average, the particle is closer to the left side than to the right side and therefore the occurrence of long jumps increases f_t. To put these scaling arguments to a direct test, we have performed extensive computer simulations. In the simulations, we have averaged D_t for each roughness over many different realizations.

First, we discuss the two-dimensional pores. In Fig. 7(a), we plot f_t versus L. We can see that there are strong deviations from a straight line of slope -1, being characteristic for logarithmic corrections. To reveal the logarithmic L-dependence of D_t, we plot $f_t L (\ln L)^{-1}$ versus L in a double-logarithmic plot in Fig. 7(b). For large L, the data reach a plateau, in accordance with Eq. (7).

Next, we consider the three-dimensional pores. Again, we plot f_t versus L in Fig. 8(a). Here, the data are well described by a straight line of slope -1. To see the normal diffusion behavior even clearer, we have plotted $f_t L = D_t / \langle v_x \rangle$ versus L in Fig. 8(b). Again, the data reach a plateau for large values of L, in agreement with Eq. (7).

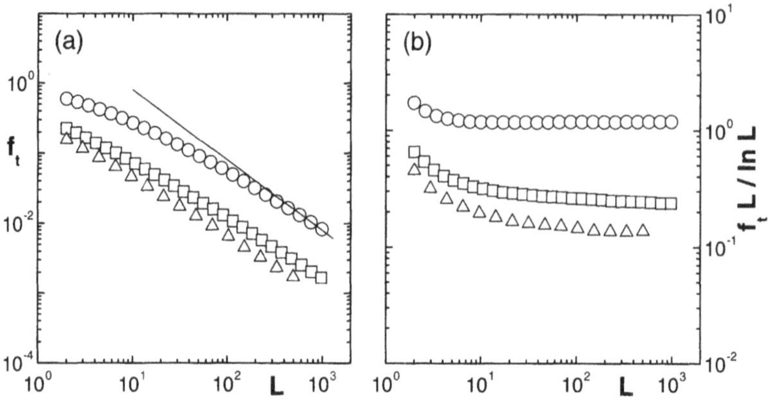

Fig. 7. (a) The probability f_t is shown for the 2d-pores with $\nu = 0$ (*circles*), 1 (*squares*) and 3 (*triangles*) in a double-logarithmic plot. The data deviate from a straight line of slope -1. (b) The scaled probability $f_t L / \ln L\,(\sim D_t)$ is shown for the systems of (a)

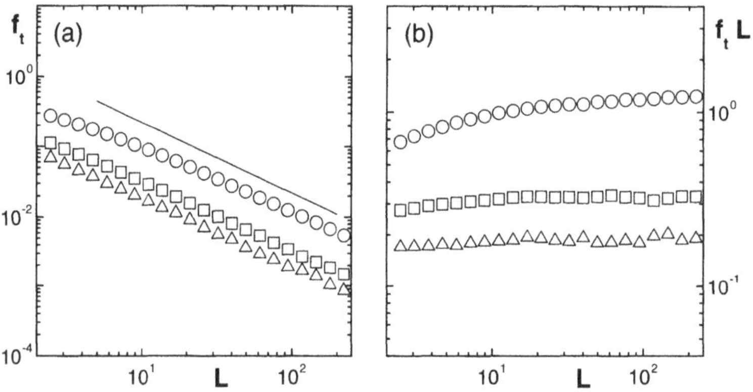

Fig. 8. (a) The probability f_t is shown for the 3d-pores with $\nu = 0$ (*circles*), 1 (*squares*) and 3 (*triangles*) in a double-logarithmic plot. The lines of slope -1 is a guide to the eye. (b) The scaled probability $f_t L\,(= D_t/\langle v_x \rangle)$ is shown for the systems of (a)

From the Figs. 7 and 8, it is obvious that D_t decreases monotonically with the roughness of the pore, in remarkable disagreement with earlier works of Malek et. al. [4, 15, 16]. Comparing Fig. 6(b) with Fig. 8(b), we see that with $\langle v_x \rangle \approx 0.4$, $D_s \approx D_t$ in $d = 3$, as expected. Note that $\langle v_x \rangle$ includes a factor of 1/2 that accounts to the fact that half of the particles starting at the left side leave the system already in the first step.

To find the source for the disagreement with [4, 15, 16], we like to note that our calculation of f_t differs from the treatment of [4], where only those particles were taken into account that penetrated into the system at least to

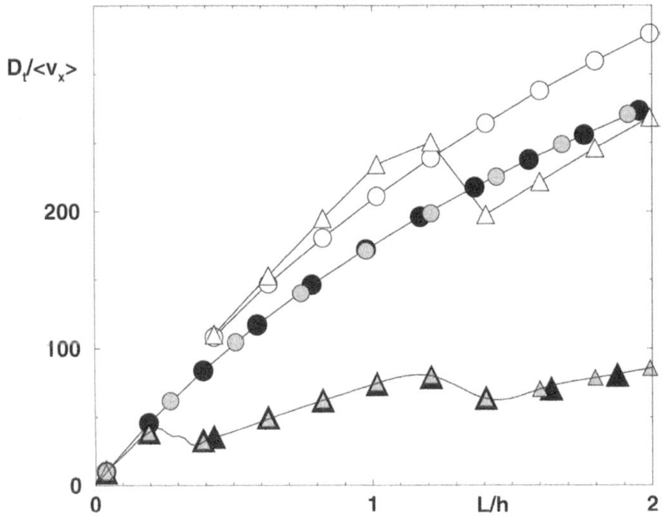

Fig. 9. $D_t/\langle v_x \rangle$ calculated from $\boldsymbol{j}$ (filled symbols) is compared to $D_t/\langle v_x \rangle$ from f_t for $x_{\min} = 0$ (grey symbols) and $x_{\min} = 0.4$ (open symbols) for the $2d$-systems of $\nu = 0$ (circles) and $\nu = 3$ (triangles). The data from $\boldsymbol{j}$ and f_t match perfectly when $x_{\min} = 0$, whereas major deviations occur for $x_{\min} > 0$

some predetermined value $x_{\min} > 0$. To show explicitly that this treatment does not give the correct D_t, we have calculated for small system sizes in $d = 2$ both the stationary current density $\boldsymbol{j}$ and f_t as a function of L for several values of $x_{\min}$. This allows us to compare $D_t/\langle v_x \rangle$ calculated from Eq. (5) with $D_t/\langle v_x \rangle$ calculated from Eq. (6). The results for $x_{\min} = 0.4$, obtained for $\nu = 0$ and one specific realization of $\nu = 3$ respectively, are shown in Fig. 9. Both curves only match perfectly for $x_{\min} = 0$, i.e. when all particles are included in the calculation of f_t. Otherwise one obtains spurious results. Figure 9 also shows that for certain values of $x_{\min}$, the effect of the surface roughness is even canceled by the effect of $x_{\min}$, which possibly leads to the incorrect results of [4, 15, 16]. Accordingly, it is essential to include into the calculations all trajectories, even if they leave the pore already after the first step. Omitting the short trajectories increases the value of f_t in an unpredictable and rather arbitrary way, such that the modified f_t cannot be used to determine the transport diffusion coefficient D_t.

4 Summary and Conclusions

In summary, we have given a comprehensive description of the phenomena of self- and transport diffusion (dependent on time and on pore length, respectively) in the Knudsen regime in $d = 2$ and 3. In real porous systems, the molecular diffusion depends on both, the boundary roughness and the content of molecules in the pore. At finite molecular concentration both, collisions with the boundary and with other molecules determine the diffusion behavior. In the limit of a dilute molecular system, the collisions between different molecules can be disregarded and only the collisions with the boundaries are relevant.

In this paper, we have focused solely on this case, i.e. on diffusion in the Knudsen regime. We considered both, two- and three-dimensional pores. Following the usual assumption that the angle distribution after collisions with the boundaries follows Lambert's cosine law, we have shown that self-diffusion as well as transport-diffusion decreases strongly with increasing roughness of the pore. These results are in marked contrast to earlier investigations [4, 15, 16], where the transport diffusion coefficients were found to be insensitive to the surface roughness. We also have shown that the diffusion coefficients have logarithmic corrections in $d = 2$, i.e. $D_s \sim \ln t$ and $D_t \sim \ln L$, which are absent in $d = 3$, where $D_s \approx D_t = \text{const}$.

Acknowledgements

The authors thankfully acknowledge stimulating discussions within the DFG-NWO sponsored International Research Training Group "Diffusion in Porous Materials", in particular with M.-O. Coppens, M. Kainourgiakis, F. Kapteijn, T. Steriotis and and S. Vasenkov. The $3d$-pores have been rendered with the POV-Ray raytracer.

References

1. F. J. Keil, R. Krishna, M.-O. Coppens, Rev. Chem. Eng. **16**, 71 (2000).
2. J. Kärger, D. M. Ruthven, *Diffusion in Zeolites and Other Microporous Solids*, (Wiley & Sons, New York 1992).
3. N. Y. Chen, T. F. Degnan, C. M. Smith, *Molecular Transport and Reaction in Zeolites*, (VCH, New York, 1994).
4. K. Malek and M.-O. Coppens, Phys. Rev. Lett. **87**, 125505 (2001).
5. G. Ertl, H. Knötzinger, J. Weitkamp, *Handbook of Heterogeneous Catalysis*, (Wiley-VCH, Chichester 1997).
6. F. Schüth, K. S. W. Sing, J. Weitkamp, *Handbook of Porous Solids*, (Wiley-VCH, Weinheim, 2002).
7. R. Krishna, B. Smit, S. Calero, Chem. Soc. Rev. **31**, 185 (2002).
8. I. Ardelean, C. Mattea, G. Farrher, S. Wonorahardjo, R. Kimmich, J. Chem. Phys. **119**, 10358 (2003).
9. R. Valiullin, P. Kortunov, J. Kärger, V. Timoshenko, J. Chem. Phys. **120**, 11804 (2004).
10. O. Geier, S. Vasenkov, J. Kärger, J. Chem. Phys. **117**, 1935 (2002).
11. S. B. Santra, B. Sapoval, Phys. Rev. E **57**, 6888 (1998).
12. D. Ben-Avraham, S. Havlin, *Diffusion and Reactions in Fractals and Disordered Systems*, (Cambridge University Press, Cambridge 2000).

13. G. Zumofen, J. Klafter, Phys. Rev. E **47**, 851 (1993).
14. J. Klafter, M. F. Shlesinger and G. Zumofen, Phys. Today **49**, 33 (1996).
15. K. Malek, M.-O. Coppens, J. Chem. Phys. **119**, 2801 (2003).
16. K. Malek, M.-O. Coppens, Colloid Surf. A-Physicochem. Eng. Asp. **206**, 335 (2002).
17. J. W. Evans, M. H. Abbasi, A. Sarin, J. Chem. Phys. **72**, 2967 (1980).

Dynamics of Membranes: From Passive to Active Systems

H.-G. Döbereiner[1,2,3], B.J. Dubin-Thaler[1], G. Giannone[1], C.K. Haluska[3], P.G. Petrov[3,4], A. Reinecke[3], K. Riske[3], and M.P. Sheetz[1]

[1] Department of Biological Sciences, Columbia University, New York, USA
[2] Department of Physics, Columbia University, New York, USA
[3] Max Planck Institut of Colloids and Interfaces, Potsdam, Germany
[4] School of Physics, University of Exeter, Exeter, UK

Summary. Fluid lipid membranes, which, together with filamentous proteins like actin, form the basic structural component of cells, have been a central topic of biological physics for the last twenty years. Traditionally, soft matter research has mainly focused on equilibrium or near equilibrium properties. In this respect, we review our work on the dynamics of simple model membranes, which are mechanically or chemically perturbed. We consider both lipid as well as diblock-copolymer membranes. Future opportunities for physics are opening up in the study of active biological systems using a modular approach. We summarize our recent work on the spreading dynamics of mouse embryonic fibroblast cells on adhesive substrates. In a somewhat reductionistic approach such an advancing cell can be considered to consist of a few main modules. An active motile gel wrapped by a membrane relaying environmental information is coupled to the cellular signaling network and the reproductive machinery. We monitored the dynamics of the leading membrane edge reflecting dynamic phase transitions of the enclosed active gel. We argue that observing cellular modules and phases will provide a powerful tool for quantitative cell biology. Physics must formulate new general concepts for interacting active modules and protein networks.

1 Classification of Membrane Dynamics

Biological matter typically is far from equilibrium. It is driven by a multitude of biochemical reactions which actively remodel cellular composition and structure. Energy and primary compounds for synthesis are provided by an intricate coupling of catabolic and anabolic pathways. Nevertheless, quite often cellular subsystems are in quasi-equilibrium with respect to the rest of the cell. Then, it makes sense to consider equilibrium properties which depend on slowly changing parameters. Particularly simple model systems for these cases are *passive membranes*, which are composed of amphiphilic lipid molecules organized in a bilayer [1, 2]. The mechanical equilibrium shapes of such lipid bilayer membranes are obtained by minimizing the elastic energy of the membrane [3, 4]. However, generically one does have a non-equilibrium *reactive membrane* in chemical and mechanical contact with the cytoskeleton and the environment via the extracellular matrix. As a subclass, reactive

B. Kramer (Ed.): Adv. in Solid State Phys. **45**, 71–82 (2005)

membranes include so-called *active membranes* [5] which do not necessarily change their composition over time but have active elements, e.g., a proton pump. These active elements typically operate in a steady state consuming chemical energy.

We can classify the dynamics of *passive membranes* according to the perturbation causing membrane motion. We consider two situations. First, in mechanical and thermal equilibrium with the aqueous buffer, a membrane receives constant thermal kicks from the water molecules causing visible membrane shape fluctuations [6, 7]. The thermal energy stored in elastic deformations is then dissipated into the environment. Whether this happens directly via viscous motion of the aqueous buffer or is mediated by viscous flow within the fluid membrane depends on the viscosity of the membrane and the length scale of the particular membrane shape deformation [8]. We will see that dissipation for micron size lipid membranes is governed by hydrodynamics with typical longest time scales of a few seconds [6, 7], whereas polymer membranes have a strongly slowed down dynamics due to their high membrane viscosity [9]. Second, we may quickly change an elastic constant of the membrane via external perturbation. As a result, the mechanical equilibrium is perturbed and the membrane shape relaxes towards a new morphology reestablishing force balance. We will discuss experiments where such perturbations were achieved via a bulk photochemical reaction [10] or sudden temperature quenches [11, 12].

Chemical reactions of a membrane with its environment will in general change its molecular composition. As a result, such *reactive membranes* have modified internal molecular conformations. Consequently, their elastic parameters change and the membrane shape and structure adapts accordingly, provided the reactions proceed sufficiently slow. We will consider an enzymatic reaction producing a binary from a single component membrane [13].

In addition to experiments with simple model membranes, we will discuss observations of living cells, where we monitored the dynamics of the leading membrane edge of spreading mouse embryonic fibroblasts [14]. The dynamics of their plasma membrane reflects the motile phases of an active gel consisting of fibrous actin and the molecular motor myosin along with additional regulatory proteins. Our notion is that one can describe the various motile phases and dynamics phase transitions of an *active gel* with well defined and comparatively simple physical mechanisms. In contrast, the trajectories in phase space are controlled by the biochemical signaling network. Thus, the dynamic behavior of the leading membrane edge of a spreading fibroblast results in a time series of short segments each of which can be mapped to a physical phase of the active gel. The overall sequence is functionally controlled by the cell. It is this control which distinguishes a living cell from a soft polymer blob. The situation is analogous to the movement of a car. There are short and simple basic motile patterns, like acceleration, driving with constant velocity, or left and right turns, which all correspond to config-

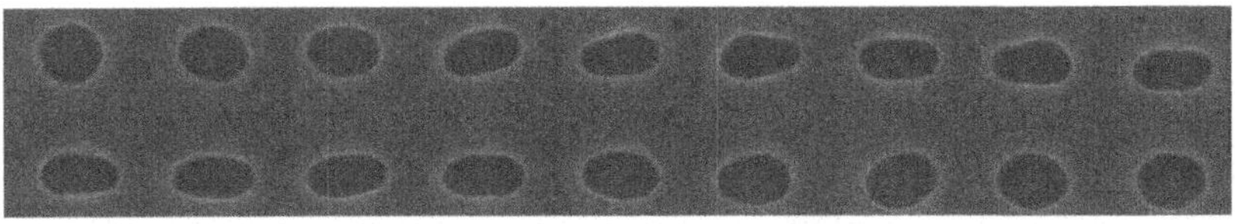

Fig. 1. Shape fluctuations and shape transitions of a zwitterionic phospholipid (SOPC) vesicle swollen in raffinose solution and incubated at room temperature in glucose solution with 50 μM external potassium hexacyanoferrate. All vesicles shown in Figures 1-6 are visualized with a phase contrast microscope (Zeiss 135, 40x). A sequence of shapes is depicted across a light-induced oblate-prolate transition and vice versa [15]. Note the strong pear fluctuations of the prolate shape, indicating metastablility towards the budding transition. Snapshots are taken every 14 seconds

urational states of the various automobile components. However, the overall route, which the car takes, can only be understood by taking into account the decisions and plans of the driver.

2 Passive Membranes

In this section, we juxtapose the relaxation dynamics after a perturbation of giant vesicles made from lipid molecules to those made by diblock copolymers. Giant vesicles form spontaneously in aqueous buffer or are produced by electro swelling with a variation of sizes in the micrometer range. In all cases, we report on inert single component fluid membranes.

2.1 Photochemical Morphology Switch

The morphology of vesicles is determined by the bending elastic energy of their membrane. Within the last 15 years, excellent agreement between experiment [4] and theory [3] has been obtained. As a function of the volume-to-area ratio of the vesicles and the spontaneous curvature of the membrane, on finds several shape classes corresponding to minimal elastic energy shapes. Tuning the spontaneous curvature induces transitions between different vesicle shapes. In Fig. 1 and 2, a time series of such a transition between an oblate and a prolate vesicle is shown [10, 15]. We changed spontaneous curvature using a reversible photochemical reaction of hexacyanoferrate which allowed us to control the pH of the external solution [10]. Variations of average illumination intensity gave an accessible range of about 1-2 units in pH. High pH leads to a weak relative charging of the external membrane monolayer presumably via binding of OH^- ions to the quaternary ammonium of the zwitterionic headgroup of the stearoyl-oleoyl-phosphatidylcholine (SOPC) lipid used for vesicle preparation [16]. Thus, we could quickly increase the spontaneous curvature of the vesicle membrane by varying amounts. A large increase in

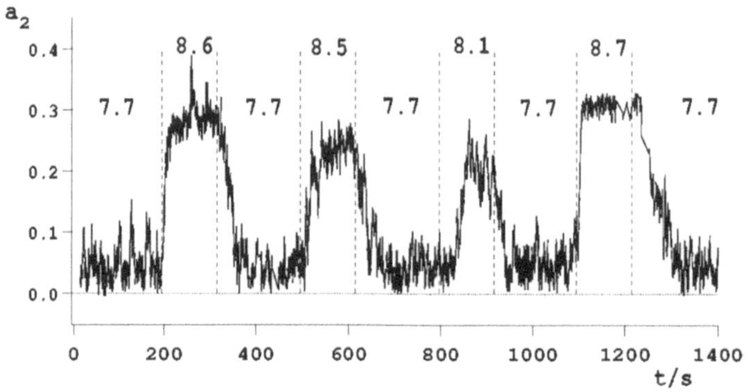

Fig. 2. Time series of vesicle ellipticity as a function of the pH gradient across the membrane. External pH values are indicated. Internal pH stays constant throughout the sequence. Ellipticity is measured as the second Fourier mode a_2 in a polar representation of the equatorial vesicle cross section, $r(\varphi) = r_0 \left[1 + \sum_n a_n \cos(n\varphi)\right]$. Vesicles are oriented along their long axis. The shape sequence shown in Fig. 1 corresponds to the first period of increased external pH. Oblate shapes are characterised by small values of $a_2 \sim 0.05$, whereas for prolate vesicles a_2 is in the range 0.2 to 0.4

pH, corresponding to a large increase in spontaneous curvature, leads to an immediate transition of the oblate vesicle into a prolate shape, see Fig. 2. For small changes in spontaneous curvature, we observe a metastable oblate shape which decays into the prolate only after some time delay. Changes in spontaneous curvature ΔC_0 could be quantified by comparing thermal fluctuations of prolate vesicles to Monte Carlo simulations [17]. We found that $\Delta C_0 = 1.2/\mu m\ \Delta pH$. Note that after spontaneous curvature is reduced again, it takes generally longer for the prolate vesicle to decay into the oblate than vice versa. This is caused by a combination of two factors. First, the elastic driving forces are different for the two shapes, and, second, the accessible phase space for decay is more restricted for the prolate vesicle than for the oblate. Whereas an oblate can become prolate in all directions, a prolate vesicle has to shrink along its long axis first to transition to an oblate via an intermediate non-axial symmetric ellipsoid. In general, the correlation time for vesicle ellipticity within either the prolate or oblate shape is on the order of a few seconds.

2.2 Polymersomes

We now come to shape dynamics of polymersomes. We discuss giant vesicle experiments with membranes made of the diblock-copolymer polybutadiene (32)-b-polyethylenoxide (20) (PB-PEO)[11, 12, 18]. Polymersomes have a number of remarkable properties [19]. First, PB-PEO membranes are capable of forming hexagonal superstructures [11] in addition to the normally abundant vesicles with spherical topology. Second, compared to lipids, synthetic polymer chains offer ample diversity in designing novel artificial membranes with tailor-made mechanical characteristics. The possibility to include polymeric segments that respond specifically to pH, ionic strength, or temperature allows membrane functions to be controlled in different environments. They may be used as drug delivery vehicles or serve as adjustable chemical containers. And third, the surface viscosity of polymer membranes made of PB-PEO, $\eta_s = 1.5 \times 10^{-3}$ $dyn\ s/cm$, is three orders of magnitude higher than for lipids in the fluid phase [9]. This is expressed in strongly increased correlation times for vesicle ellipticity on the order of a minute.

Tubes

Let us first recall, the slow relaxation dynamics of tubular polymersomes after thermal quench [12]. Tubular vesicles of various lengths and diameters were subject to thermal quench from high to low temperatures. We observed that reducing the temperature between 10 and 50 K induced sequential beading of the tubes starting symmetrically from the two ends (see Fig 3 for a typical example). The gradual change of polymersome morphology proceeds at constant temperature and corresponds to the relaxation to a new equilibrium shape. Remarkably, the whole process is very slow. We find a velocity of 35 ± 5 nm/min for neck closure. This is due to the large surface viscosity of PB-PEO membranes. The vesicle end-to-end distance reacts in a characteristic way to temperature reduction. One observes an instantaneous initial reduction of the tube length, reflecting a change in tube area with temperature. Some time after a temperature quench, which varies with the initial conditions of the vesicle, beading starts. We can very well rationalize these observations by noting that (a) tubular vesicles belong to the prolate class in the phase diagram of vesicle shapes [4], (b) beading corresponds to budding transitions into multiply budded configurations, and (c) reducing the temperature induces an increase of the membrane spontaneous curvature in our system as found as well in thermal trajectories of high genus vesicles [11]. Compared to lipid vesicles, the latter effect is temperature reversed.

Non-Spherical Topologies

We already mentioned that in addition to topological spherical shapes, polymersomes form more complex morphologies with low and high genus g, i.e., number of handles of a single vesicle. Lipid-bilayer vesicles with a small number of handles have been investigated before [3, 20, 21]. Here, we show a shape sequence for a $g = 4$ polymersome [18] illustrating again the ultra slow equilibration dynamics of a PB-PEO polymersome after temperature quench, see Fig. 4. Decreasing temperature increases the spontaneous curvature of the

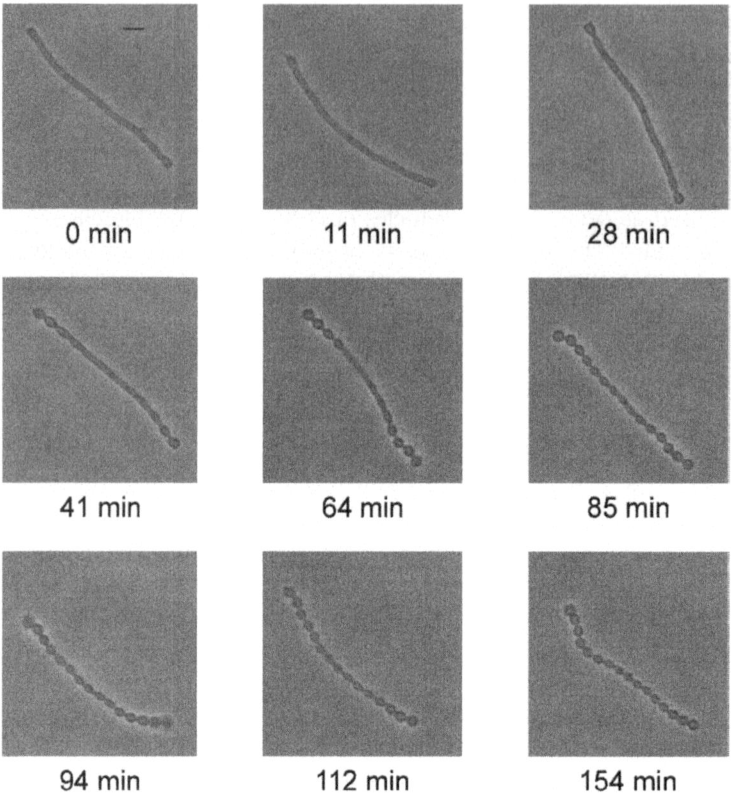

Fig. 3. Sequential beading of a tubular polymersome in sucrose (inside)/glucose (outside) solution after a thermal quench from 38 to 25 °C. The initial diameter of the tube is 2.4 μm. The time elapsed after reducing the temperature is indicated below each picture: 0-28 min, there is a time delay before the vesicle starts to bead; 28 min, beginning of neck closure from the end of the tube; 28-112 min, progression of beading toward the tube center; 112-154 min, completely beaded tube. This configuration is stable in time. The bar corresponds to 10 μm

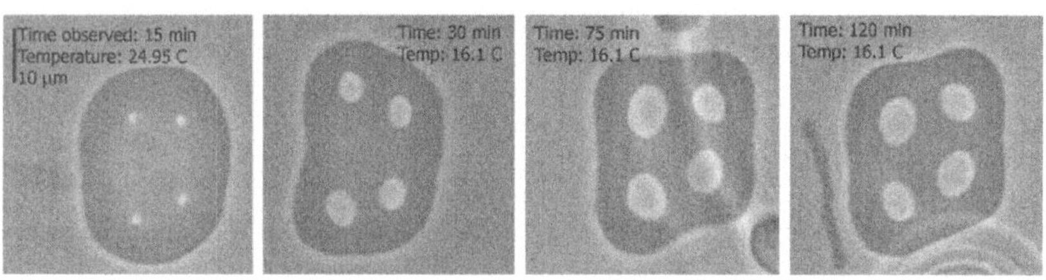

Fig. 4. Low genus button vesicle in sucorse/glucose solution. A $g = 4$ PB-PEO polymersome relaxes slowly after a temperature quench to a new equilibrium shape. Note the coexisting tubular polymersome with spherical topology

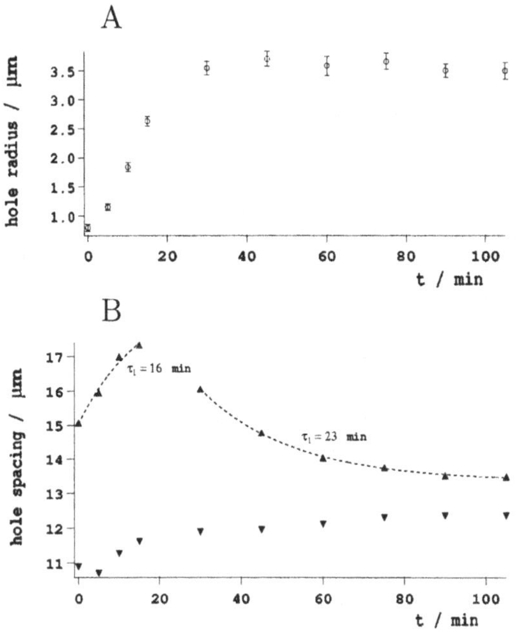

Fig. 5. Dynamics of the vesicle in Fig. 4. The hole radius (A) and the largest and smallest hole spacing (B) are plotted as a function of time after temperature is quenched by $\Delta T = 9$ K

polymer vesicle and the previously stable shape becomes unstable. First, the hole radii adjust. Then, the global shape changes symmetry from a wagon-like configuration to a tilted square. Effectively, these transformations may be parameterized by the average radius and spacing of the holes. In Fig. 5, we show the time dependence of these quantities. Note the overshoot of the hole spacing due to the quicker relaxation of hole radius. The shape changes are characterized by relaxation times on the order of 20 minutes.

3 Reactive Membranes: Phospholipase Activity

The enzyme phospholipase C (PLC) prefers phosphatidyl choline (PC) lipids as substrates, where it catalysis the cleavage of the PC head group to yield a diacylglycerol. PLC from *C. perfringens* is also known as α-toxin. It is responsible for a severe form of food poisoning with about 10.000 cases/year in the US. Single giant unilamellar vesicles made from SOPC were observed under a phase contrast light microscope before and after addition of α-toxin to the external solution [13]. To keep the particular vesicle under investigation in the field of view, the enzyme was introduced by slowly exchanging the external solution with a micropump. Solution exchange was performed under conditions of low or vanishing activity. As the enzymes hydrolyze the SOPC

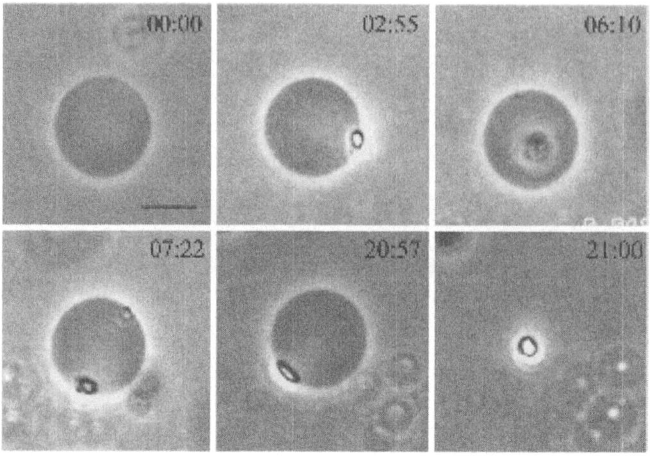

Fig. 6. Sequence of images of a giant SOPC vesicle in 10 mM HEPES, 100 mM sucrose/glucose (inside/outside) after the addition of 0.02 U/ml α-toxin in 0.5 mM $CaCl_2$. The enzyme solution was injected at 25 °C, a temperature at which the activity of α-toxin is negligible. After that the temperature was increased to 37 °C. Enzyme activity resulted in the formation of several oily domains consisting mainly of the reaction product diacylglycerol. The times shown on the snapshots refer to the time after the sample reached 37 °C (min:s). The bar on the first image denotes 10 μm

lipids of the vesicle, the choline headgroups of SOPC are removed and diffuse into the aqueous buffer, whereas SOG, a diacylglycerol, is left in the bilayer. Shown in Fig. 6 is a sequence of pictures of one vesicle after the injection of α-toxin. After a few minutes at 37 °C, a bright domain appeared on the vesicle surface, followed by a second one later on. The vesicle with the two domains remained stable for quite a long time, after which it suddenly collapsed into a single bright drop. We found that vesicles showed this behavior irrespective of detailed conditions (enzyme concentration, $CaCl_2$ concentration, temperature). The domains clearly extend beyond the bilayer plane. At late times, membrane domains appeared to be smoother than right after formation. In fact, the lens-like appearance points to a liquid character of these structures forming a liquid drop within the membrane. Indeed, we have shown with differential scanning calorimetry and fluorescence anisotropy measurements that the reaction product in the membrane corresponds to SOG and is in a fluid phase. In the final product, there is only a broad isotropic peak visible in small angle x-ray spectra at 30 Å. The bilayer peak at 50 Å is lost [13]. Thus, we have fluid bulky SOG-rich domains developping within the bilayer. This corresponds to lateral phase separation as the relative concentration of SOG in the bilayer increases. We could determine a contact angle $\theta = (23 \pm 7)°$

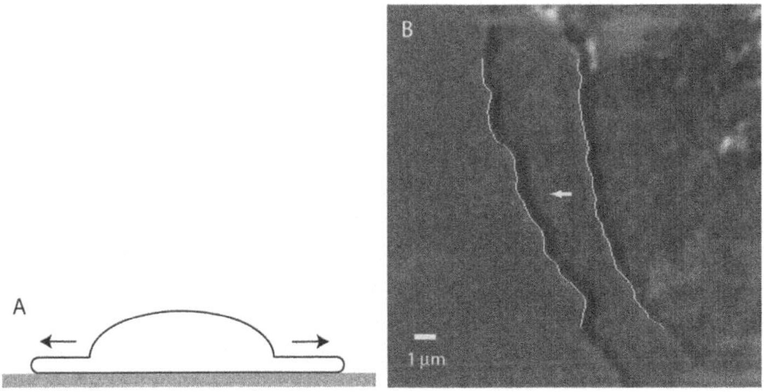

Fig. 7. A: During cell spreading, a thin lamellipodial sheet extends from the cell body onto the substrate. B: Two overlayed snapshots of the leading membrane edge of a lamellipodium moving from right to left are shown in differential interference contrast. The edge position is marked with a white contour overlay

(corrected for membrane curvature) between the fluid lenses and the fluid bilayer membrane.

4 Membranes and Active Gels

Cells need to be mobile in order to perform many critical biological functions. The reorganization of extracellular matrix in wound healing by fibroblasts, the positioning of nerve cells, or the engulfment of bacteria in the immune reaction of white blood cells are particular examples [22]. In the laboratory, fibroblast cells spread onto a suitable coated substrate, extending a 200 nm thick sheet from the cell body called the lamellipodium, see Fig. 7. This process is driven by actin polymerization at the leading membrane edge. The meshwork of actin fibers is crosslinked by various proteins. The molecular motor myosin II enables the meshwork to contract by moving along actin fibers and relative to other cytoskeletal elements. Thus, the lamellipodium is an active gel enclosed in a flat membrane bag adhering to the substrate. The physics of active gels has recently attracted a lot of attention. Rheological experiments of simple mixtures of purified actin and myosin solutions [23] have been carried out. Within a quite general theory, dynamic phase transitions involving extended and contracted fiber density states have been found as a function of fiber-motor coupling strength [24].

A substantial amount of experimental and theoretical work has been done on cells spreading on matrix-coated surfaces [25, 26]. However, only quite recently, quantitative experiments of cell spreading and subsequent migration could be performed with high spatial and temporal resolution. We found that there are well defined and distinct states of spreading [27, 28] which

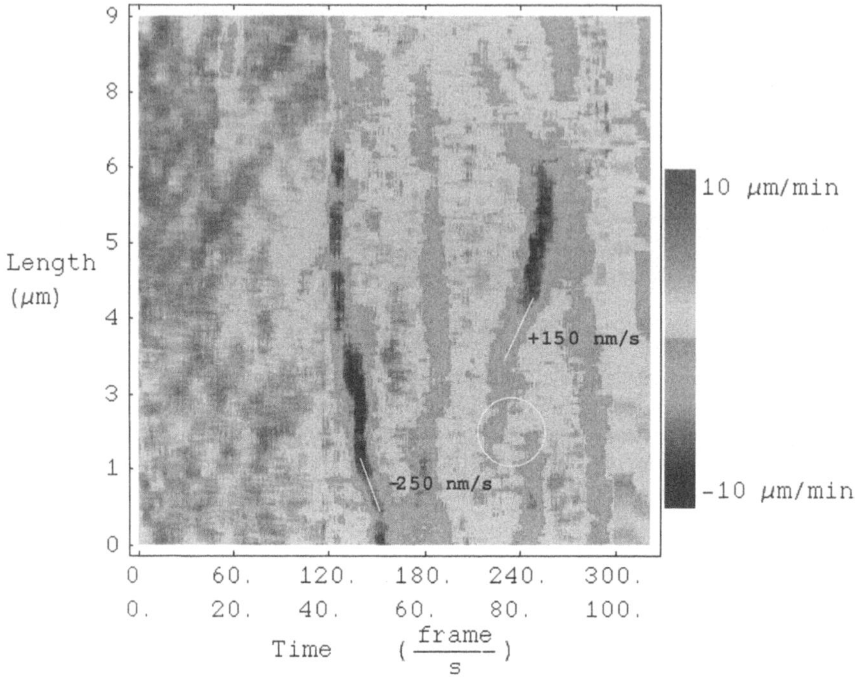

Fig. 8. Normal velocity map of the spreading membrane segment shown in Fig. 7 as a function of time. Movies were digitized at $1/\Delta t = 3Hz$. Note the two qualitatively different sections before and after time $t = 40\ s$ corresponding to a continuous and a periodically contractile spreading phase, respectively. The period of the latter is $T = 17 \pm 4\ s$. White bars indicate lateral waves of maximum contraction velocity running up and down the cell edge. The corresponding speeds are given. The encircled region marks a phase shift of a contraction

can be considered phases of motility due to the existence of dynamic phase transitions between them [14].

We monitored membrane adhesion area A during spreading. We found distinct classes of angular isotropic and anisotropic spreading cells [27]. For the isotropic class, the adhesion area A revealed three subsequent motile phases with distinctly different power law growth [14]. Initially, there is a basal phase where cells test the suitability of the substrate to adhere and area growth is mimimal. Then follows a phase of fast continuous spreading. Finally, the cell slows down again. The latter phase is characterized by periodic local contractions of the cell [28]. The transition from continuous to contractile spreading was monitored for isotropically spreading cells using high resolution differential interference contrast, see Fig.7 B. A typical normal velocity map of an advancing cellular contour over time is shown in Fig. 8. We found that a region of continuous, uninterrupted spreading (red shadows) precedes a sequence of periodic membrane retraction events (blue stripes). These two different states of membrane dynamics correspond to the continuous spreading and contractile phase of the lamellipodium, discussed above. The two phases can be clearly distinguished using discrete Fourier transformations of the velocity map. The continuous spreading phase is characterized by a strong boundary maximum of the spectrum. In contrast, in the contractile phase the spectrum develops a pronounced peak at a finite frequency [14]. The peak position of the power spectrum serves as an excellent phase indicator with a sharp jump at the phase boundary.

5 Towards Modular Cell Biology

The active actomysoin gel within the lamellipodium can be considered a functional module of the cell serving as a tool to test substrate elasticity. The cell employs the period contraction discussed above to pull on the substrate. Other modules, for example, are the microtubuli network responsible for chromsome seperation during cell division, or the plasma membrane relaying external signals and material exchange with the environment [29]. We suggest that the various motile phases discussed above can be mapped to specific states of the actomyosin module, e.g., the continuous spreading phase is characterized by high actin polymerization activity, whereas the periodic contractile phase needs motor activity in addition to continuing polymerization. As a general ordering principle, we have proposed a hierarchical organization of module proteins related to phase structure, phase regulation parameters, and pure signaling components controlling these parameters [14, 30]. Structural proteins together with their associated regulating cofactors form physical entities whose functional states can be understood independently of biochemical signaling. Introducing phases and modules allows to identify the relevant parameters driving phase transitions. Such a high level description of collective motile properties could serve as a paradigm for quantitative biology. Using classical ideas of condensed matter physics and employing new concepts to describe active systems together with the currently emerging methods to analyze protein interaction networks, biological physicist will provide an important contribution to an integrated understanding of living cells.

Acknowledgements

HGD is grateful for enjoyable collaborations and discussions with Stefan Dilling, Rumiana Dimova, Stefan Förster, Gerhard Gompper, Wojcieck Góźdź, Josephine Lee, Udo Seifert, Olaf Selchow, and Harry Xenias. This work was made possible by a Heisenberg fellowship and a grant from the Deutsche Forschungsgemeinschaft to HGD as well as generous support from Reinhard Lipowsky.

References

1. *Physics of Amphiphilic Layers*, Eds. J. Meunier, D. Langevin, and N. Boccara), Springer Proc. Phys. **21** (1987).
2. R. Lipowsky, *Nature*, **349**, 475 (1992).
3. U. Seifert, *Adv. Phys.*, **46**, 13 (1997).
4. *Giant Vesicles, Perspectives in Supramolecular Chemistry* **vol. 6**, Eds., P. Walde & L. Luisi, John Wiley & Sons (2000).
5. J.-B Manneville, P. Bassereau, S. Ramaswamy, J. Prost, Phys. Rev. E **64**, 21908 (2001).
6. S. T. Miler, and S. A. Safran, Phys. Rev. A **36**, 4371 (1987).
7. H. P. Duwe, J. Käs, and E. Sackmann, J. Phys. (France) **51**, 945 (1990).
8. U. Seifert, and S. A. Langer, Europhys. Lett. **23**, 71 (1993).
9. R. Dimova, U. Seifert, B. Pouligny, S. Förster, and H.-G. Döbereiner, Eur. Phys. J. E **7**, 241 (2002).
10. P. G. Petrov, J. Lee, and H.-G. Döbereiner, Europhys. Lett. **48**, 435 (1999).
11. C. K. Haluska, W. T. Góźdź, H.-G. Döbereiner, S. Förster, and Gerhard Gompper, Phys. Rev. Lett. **89**, 238302 (2002).
12. A. A. Reinecke and H.-G. Döbereiner, Langmuir **19**, 605 (2003).
13. K. A. Riske and H.-G. Döbereiner, Biophys. J. **85**, 2352 (2003).
14. H.-G. Döbereiner, B. Dubin-Thaler, G. Giannone, H. S. Xenias, and M. P. Sheetz, Phys. Rev. Lett. **93**, 108105 (2004).
15. H.-G. Döbereiner and U. Seifert, Europhys. Lett. **36**, 325 (1996).
16. J. Lee, P. G. Petrov, and H.-G. Döbereiner, Langmuir **15**, 8543 (1999).
17. H.-G. Döbereiner, G. Gompper, C. K. Haluska, D.M. Kroll, P.G. Petrov, and K. Riske, Phys. Rev. Lett. **91**, 048301 (2003).
18. C. K. Haluska, Master Thesis, Potsdam University 1999.
19. B.M. Discher et al., Science 284, 1143 (1999).
20. F. Jülicher, U. Seifert, & R. Lipowsky, *Phys. Rev. Lett.* **71**, 452 (1993).
21. X. Michalet & D. Bensimon, *Science* **269**, 666 (1995).
22. D. Bray, Cell Movements: From Molecules to Motility (Garland Publishing, New York, 2001), 2nd ed.
23. D. Humphrey, C. Duggan, D. Saha, D. Smith and J. Käs, Nature **416**, 413 - 416 (2002).
24. K. Kruse, J. F. Joanny, F. Jülicher, J. Prost, K. Sekimoto, Phys. Rev. Lett. **92**, 078101 (2004).
25. G. Jiang, G. Giannone, D. R. Critchley, E. Fukumoto, M. P. Sheetz, Nature **424**, 334 (2003).
26. I. B. Bischofs, S. A. Safran, U. S. Schwarz, Phys. Rev. E **69**, 021911 (2004).
27. B. Dubin-Thaler, G. Giannone, H.-G. Döbereiner, and M. P. Sheetz, Biophys. J. **86**, 1794 (2004).
28. G. Giannone, B. Dubin-Thaler, H.-G. Döbereiner, N. Kiefer, A. N. Bresnick, and M. P. Sheetz, Cell **116**, 431 (2004).
29. L. H. Hartwell, J. J Hopfield, S. Leibler, A. W. Murray, Nature **402**, C47 (1999).
30. H.-G. Döbereiner, B. J. Dubin-Thaler, G. Giannone, and M. P. Sheetz, J. Appl. Physiol. **98**, 1542-1546 (2005).

Disentangling Noise and Fluctuations in Data Sets of Complex Systems

R. Friedrich[1], D. Kleinhans[1], and J. Peinke[2]

[1] Institute of Theoretical Physics, Westfälische Wilhelms-Universität Münster, Wilhelm-Klemm-Str. 9, 48149 Münster, Germany
fiddir@uni-muenster.de

[2] Institute of Theoretical Physics, Carl von Ossietzky University Oldenburg, Germany

Summary. An overview on a new operational method is given which disentangles deterministic dynamical evolution laws from the impact of fluctuating forces for nonlinear Langevin equations. An important extension is also reported. The extension overcomes possible problems which may arise for data sets sampled at low frequencies, and data sets which do not contain sufficient amount of data.

1 Introduction

Time series analysis has always been one of the major lines of research in the field of nonlinear dynamics. It has enormously contributed to the development of the theory of complex systems - (for an overview see the monograph of H. Kantz and T. Schreiber [1]). The ultimate goal of nonlinear data analysis is certainly the determination of a dynamical system for a suitably defined state vector from a treatment of experimental time series. This dynamical system contains all the information which is accessible by an inspection of the dynamical behavior of the system.

A major shortcoming of nonlinear time series analysis is its focus on deterministic dynamical systems. The impact of noise, usually, has been treated as a random variable which is additively superimposed on a trajectory generated by a deterministic dynamical system. The problem of dynamical noise, i.e. fluctuations which interfere with the dynamical evolution, apparently has not been addressed in great details.

However, as has become evident starting already from the treatment of Brownian motion by A. Einstein in the year 1905 [2], the dynamics of macroscopic variables of a system composed of a large number of microscopic subsystems, involves the impact of fluctuations, which can be traced back to the dynamics of the microscopic degrees of freedom. This fact has enormous consequences especially for the description of nonlinear complex systems. Such systems can exhibit the phenomenon of self organization [3]: By varying control parameters, qualitative changes of behavior can arise and transitions among different states can be induced. Close to these transitions the behavior can be treated by order parameters. A complete description has to include

B. Kramer (Ed.): Adv. in Solid State Phys. **45**, 83–94 (2005)

fluctuations and the equations determining the dynamics of order parameters take the form of Langevin equations [3], [4]. This scheme has been highly successful in understanding complex systems ranging from physics to biology.

Analyzing data sets of complex systems, therefore, means dealing with dynamical noise. The main task consists in disentangling trends, i.e. deterministic dynamical evolution laws, from the impact of the fluctuating forces. In the last years, an operational method has been developed, which is able to cope with this task for the class of nonlinear Langevin equations [5], [6]. The purpose of the present article is to give an overview on this method and to report on an important extension. This extension overcomes possible problems which may arise for data sets sampled at low frequencies and data sets which do not contain sufficient amount of data.

2 Nonlinear Time Series Analysis of Fluctuating Systems

It is certainly not possible, to specify a general method dealing with data of all kinds of fluctuating systems. One has to restrict oneselves to a certain subclass in order to develop the accurate tools for data analysis. A large class of systems relevant for complex systems [3] are stochastic processes generated by Langevin equations [7], [8].

2.1 Systems Considered

We consider the class of systems which are described by a stochastic variable $\mathbf{X}(t)$ contained in a d-dimensional state space $\{\mathbf{x}\}$ which evolves in time according to a Langevin equation of the form

$$\frac{d}{dt}\mathbf{X}(t) = \mathbf{f}\big(\mathbf{X}(t),t\big) + \mathbf{F}\big(\mathbf{X}(t),t\big) \,. \tag{1}$$

Here, $\mathbf{f}$ is a nonlinear function depending on the stochastic variable $\mathbf{X}(t)$ and may be also explicitly depending on time t. The d components F_i of the nonlinear stochastic part $\mathbf{F}\big(\mathbf{X}(t),t\big)$ are assumed to be of the form

$$F_i\big(\mathbf{X}(t),t\big) = \sum_{j=1}^{d} g_{ij}\big(\mathbf{X}(t),t\big)\Gamma_j(t) \,, \tag{2}$$

where the terms Γ_j are considered to be (δ-correlated) Gaussian white noise forces with vanishing mean,

$$\langle\Gamma_j(t)\rangle = 0, \qquad \langle\Gamma_i(t)\Gamma_j(t')\rangle = \delta_{ij}\delta(t-t') \,. \tag{3}$$

Because of a possible nonlinearity of the function $\mathbf{f}$, also systems exhibiting chaotic time evolution in the deterministic case are included in (1).

2.2 Fokker-Planck Equation

The dynamic behavior of stochastic systems formulated by a Langevin equation (1) can also be described by an evolution equation for the conditional probability density $p(\mathbf{x}, t; \mathbf{x}', t')$ in state space $\{\mathbf{x}\}$. This is achieved by formulating the Fokker-Planck equation

$$\frac{\partial}{\partial t} p(\mathbf{x}, t \,|\, \mathbf{x}', t') = -\sum_{i=1}^{d} \frac{\partial}{\partial x_i} D_i^{(1)}(\mathbf{x}, t)\, p(\mathbf{x}, t \,|\, \mathbf{x}', t') + \frac{1}{2} \sum_{i,j=1}^{d} \frac{\partial^2}{\partial x_i \partial x_j} D_{ij}^{(2)}(\mathbf{x}, t)\, p(\mathbf{x}, t \,|\, \mathbf{x}', t') \,. \tag{4}$$

The quantities $\mathbf{D}^{(1)}$ and $\mathbf{D}^{(2)}$ are called drift and diffusion coefficient.

Under consideration of Itô's definitions of stochastic integrals [8] the coefficients $\mathbf{D}^{(1)}$, $\mathbf{D}^{(2)}$ of the Fokker-Planck equation (4) and the functions $\mathbf{f}$, $\mathbf{g}$ of the Langevin equation (1), (2) are related by

$$D_i^{(1)}(\mathbf{x}, t) = f_i(\mathbf{x}, t), \tag{5}$$

$$D_{ij}^{(2)}(\mathbf{x}, t) = \sum_{l=1}^{d} g_{il}(\mathbf{x}, t) g_{jl}(\mathbf{x}, t) \,. \tag{6}$$

3 Analyzing Stochastic Systems

The task of disentangling trends and fluctuations for systems described by Langevin equations is equivalent to finding drift and diffusion terms from experimental data. The procedure consists of the following steps:

- The underlying process must be a Markov process, i.e. the dynamical noise involved must be short correlated in time.
- The dynamical noise involved must follow a Gaussian distribution.
- The drift and diffusion term have to be reconstructed.

An operational approach is based on the hypothesis that the first two properties are fulfilled and to reconstruct the process. The hypothesis may then be checked selfconsistently by comparing reconstructed and measured data. However, one may also try to fix these properties by an inspection of the data. The Markovian property may, e.g. investigated by verifying the Chapman-Kolmogorov equation:

$$p(\mathbf{x}_2, t_2 | \mathbf{x}_1, t_1) = \int d\mathbf{x}' p(\mathbf{x}_2, t_2 | \mathbf{x}', t') p(\mathbf{x}', t' | \mathbf{x}_1, t_1) \,. \tag{7}$$

Thereby, $t_2 > t' > t_1$. Furthermore, one may also directly consider the conditional PDF $p(\mathbf{x}_3, t_3 | \mathbf{x}_2, t_2; \mathbf{x}_1, t_1)$. If the process is Markovian, this quantity should not depend on t_1. Both criteria can be investigated by an inspection of the data.

Let us now briefly touch a possible approach for the estimation of drift and diffusion terms. To this end we consider the approximation to the Langevin equation valid for small time intervals τ,

$$\mathbf{X}(t+\tau) = \mathbf{X}(t) + \mathbf{D}^{(1)}(\mathbf{X}(t))\tau + g(\mathbf{X}(t))\sqrt{\tau}\eta_t \,. \tag{8}$$

Here, η is a random, normally distributed quantity with zero mean. Averaging the expressions $\mathbf{X}(t+\tau)-\mathbf{X}(t)$, $[\mathbf{X}(t+\tau)-\mathbf{X}(t)][\mathbf{X}(t+\tau)-\mathbf{X}(t)]$ with respect to η leads to the following expressions, where the brackets indicate conditional averages:

$$D_i^{(1)}(\mathbf{x},t) = \lim_{\tau\to 0} \frac{1}{\tau}\langle X_i(t+\tau) - x_i\rangle\Big|_{\mathbf{X}(t)=\mathbf{x}} , \tag{9}$$

$$D_{ij}^{(2)}(\mathbf{x},t) = \lim_{\tau\to 0} \frac{1}{\tau}\langle (X_i(t+\tau) - x_i)(X_j(t+\tau) - x_j)\rangle\Big|_{\mathbf{X}(t)=\mathbf{x}} . \tag{10}$$

According to Ito's interpretation of the process $\eta(t)$ and $\mathbf{X}(t)$ are taken to be uncorrelated.

The crucial point of Eqs. (9) is the possibility to disentangle drift and diffusion by performing the limit $\lim \tau \to 0$. However, problems may arise if this limit cannot be performed reliably. An estimate of drift and diffusion for finite values of τ will yield a mixture of drift and diffusion. In many cases one may find the correct values by extrapolation. Below, we shall describe a refined algorithm which overcomes all problems related with the limiting process $\tau \to 0$.

4 Applications

The evaluation of drift and diffusion coefficients as conditional moments according to (9) has been performed for a variety of complex systems. Thereby, two different types of problems have been addressed. On the one hand side, the method has been applied to analyze stochastic processes in time as described in [5, 6]. On the other hand, disordered fields, which exhibit a cascade process in scale like turbulent flows, have been investigated [9].

4.1 Processes in Time

The method has been tested by investigating various synthetic data sets. It has been demonstrated that, given sufficient amount of data sampled at a sufficiently high rate, the drift and diffusion terms can be extracted quite accurately. Thereby, the evaluation of the diffusion coefficients usually requires more attention. In [10] data sets of noisy nonlinear oscillators as well as the noisy Lorenz attractor have been investigated.

Data sets of real-world data treated by the above method include noisy electrical circuits [6], stochastic dynamics of metal cutting [11]. Furthermore, the investigation of the wind-driven Southern Ocean variability [15], [16] shows the potential of the algorithm for the assessment of meteorological processes. Applications to biological data involve human tremor data [6] or data of heart rate fluctuations [17]. A quite interesting application has been reported for traffic flows [18]. Furthermore, the procedure could be used to detect drift bifurcations in a gas-discharge system [19]. Recently, the above method has been generalized to systems with feedback delay [12]-[14]. Furthermore, it has been shown that the method of Pope and Ching for extracting stochastic dynamical systems is in accordance with the present approach [20].

4.2 Processes in Scale

The procedure has been applied to characterize systems with fractal and multifractal scaling behavior like turbulent flows [9], [21], passive scalar advection [22], financial time series [23], analysis of rough surfaces [24], [25]. The idea is to characterize disordered systems by stochastic processes with respect to a scale variable exhibiting Markovian properties.

5 Refined Procedure for Estimating Drift and Diffusion

As has been mentioned by Ragwitz and Kantz [27] and discussed in [28], [29] the limiting procedure involved in the estimation of drift and diffusion coefficients may render this approach rather difficult. In fact, drift and diffusion are disentangled by the limiting procedure $\tau \to 0$ in Eq. (9). If the value of τ is finite, the dynamical effects due to trends and fluctuations interfere. These problems immediately show up for data sets generated at low sampling rates or for the case of stochastic forces $\mathbf{F}(\mathbf{q}, t)$ with small but finite temporal correlations. Furthermore, uncorrelated noise sources, so-called measurement noise [30] additionally may limit the accuracy of the estimates (9). Furthermore, if the data set does not contain sufficiently many points, the results for drift and diffusion terms, which are actually defined as mean values, may scatter considerably. Both problems can be overcome by an iterative procedure, which will be described in this section.

The starting point is the first estimate of drift and diffusion coefficients by the expressions (9) evaluated for the smallest reliably possible values of τ which evidently can serve as the first approximation. The second step is an embedding of drift and diffusion coefficients into a family of functions $\mathbf{D}^{(1)}(\mathbf{x}, \sigma)$, $\mathbf{D}^{(2)}(\mathbf{x}, \sigma)$ parameterized by a set of free parameters σ. The values of σ obtained in the first step already yield a crude estimate of the stochastic process. The third step consists in optimizing the free parameters σ by an iteration procedure eventually leading to an improved approximation of the stochastic process.

Optimization of the free parameters has to be done by comparing the probability distribution from experiment with the one generated by the stochastic process determined by $\mathbf{D}^{(1)}(\mathbf{x},\sigma)$ and $\mathbf{D}^2(\mathbf{x},\sigma)$. This means that one has to calculate the estimated PDF by solving either the Fokker-Planck equation or the Langevin equation for a certain parameter set σ. In each case one has to determine the joint probability distribution $f(\mathbf{x},t;\mathbf{x}_0,t_0;\sigma) = p(\mathbf{x},t|\mathbf{x}_0,t_0;\sigma)f(\mathbf{x}_0,t_0)$ and has to compare it with the experimental one, $f_{exp}(\mathbf{x},t;\mathbf{x}_0,t_0)$. A suitable measure for the distance is the Kullback-Leibler information [4] defined according to

$$K(\sigma,t,t_0) = \int d\mathbf{x} \int d\mathbf{x}_0 f_{exp}(\mathbf{x},t;\mathbf{x}_0,t_0) \times \ln \frac{f_{exp}(\mathbf{x},t;\mathbf{q}_0,t_0)}{f(\mathbf{x},t;\mathbf{q}_0,t_0,\sigma)} . \tag{11}$$

The minimum of the Kullback entropy yields the best approximation in the class of stochastic processes characterized by the parameters σ. For the determination of the minimum standard methods like gradient or genetic algorithms can be used. In the following we shall consider cases, where this procedure can be drastically simplified. Such a simplification becomes possible if the stationary probability distribution of the process under consideration is known in terms of drift and diffusion coefficients. This is the case for one dimensional processes as well as for the class of potential systems.

5.1 One Dimensional Systems

We start by investigating one dimensional processes, where the variable is denoted by q. The stationary probability distribution can be expressed according to

$$f(q) = \frac{N}{D^{(2)}(q)} e^{\int dq \frac{D^{(1)}(q)}{D^{(2)}(q)}} . \tag{12}$$

This expression relates drift and diffusion coefficients

$$D^{(1)}(q) = D^{(2)}(q)\frac{d}{dq} \ln f(q) + \frac{d}{dq}D^{(2)}(q) . \tag{13}$$

In case where a reliable estimate of the stationary PDF $f(q)$ can be obtained from data, one only has to make an ansatz for the diffusion term. In fact, one may use the ansatz $D^{(2)}(q) = Q + aq^2 + bq^4 + \ldots$ where the coefficients of the Taylor series serve as free parameters, which have to be optimized. The drift then follows as indicated above. This observation restricts the number of free parameters considerably.

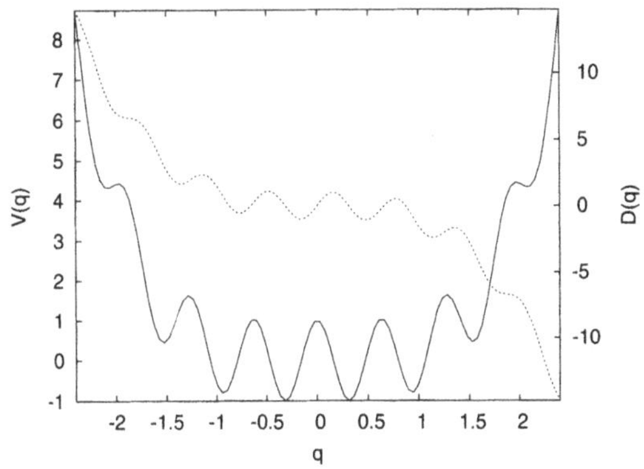

Fig. 1. The drift $D^{(1)}(q)$ and the corresponding potential $V(q)$ for the stochastic process Equation (14)

Let us exemplify the procedure in more detail. We use synthetic data obtained by numerical integration of the Langevin equation [7]. We considered the system

$$D^{(1)}(q) = 10\sin(10q) - 4q^3$$
$$D^{(2)}(q) = 1\,. \tag{14}$$

The drift potential $V(q) = q^4 + \cos(10q)$ is an even function and has four clear minima within the interval $[-1 : 1]$. We considered a time series of 40000 equidistant data points sampled with sampling time $\tau = 0.05$. This series was obtained by numerical integration of the Langevin equation with time increment 10^{-5}. State space has been partitioned into $N = 300$ equidistant bins. The stationary distribution of the process and the two point PDF can be extracted from one time series. In some cases is makes sense to approximate this PDF with an parametric function and differentiate $\ln(f)$ analytically. We differentiated numerically applying local linear least square fits to $\ln(f)$. The advantage of this nonparametric procedure is that it is not based on any prior knowledge on the drift.

The drift $D^{(1)}(q)$ at discrete points can be gained from (13). Then the noise strength Q has to be optimized since it is the only free parameter left. As outlined above we numerically simulate the stochastic process (14) for 300 different initial conditions $p(q, q_0, t)$ taken from the various bins. For a given noise strength Q one ends up with the two point conditional PDF that, together with the PDF extracted from data, are necessary to evaluate the Kullback-Leibler information.

Figure 4 shows the Kullback-Leibler measure as a function of Q and clearly indicates the minimum at the value $Q = 1$. Finally, the drift $D^{(1)}(q)|Q = 1$ can be approximated with parametric functions to get an analytical expression.

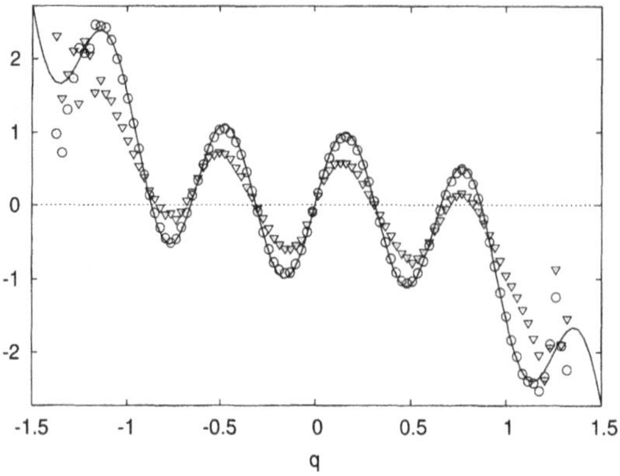

Fig. 2. The drift $D^{(1)}(q)$ of the stochastic process Equation (14) as estimated by the expression 9 and obtained by the refined method. The drift obtained by the iterative procedure is indistinguishable from the one used to generate the time series

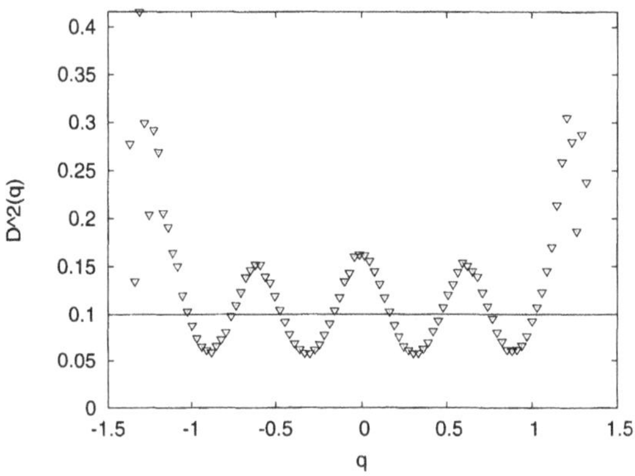

Fig. 3. The diffusion term $D^{(2)}$ of the stochastic process Equation (14) as estimated by the expression 9 shows a dependency on the variable q. The original process was characterized by purely additive noise, $Q = const.$ Drift and diffusion could not be disentangled by the estimate using finite time increment τ. The refined method yields the correct estimate for the noise strength Q

Let us now consider a system with multiplicative noise. We simulated time series for a system with drift- and diffusion coefficients

$$D^{(1)}(q) = q - 4q^3$$
$$D^{(2)}(q) = 1 + q^2 \,. \tag{15}$$

We simulated 10^5 equidistant data points with sampling time $\tau = 0.01$. Time series where generated by a time step 10^{-5}.

The basic idea for the analysis of the data is to make a parametric ansatz for $D^{(2)}(q)$ and gain the corresponding $D^{(1)}(q)$ from the relation (13). We

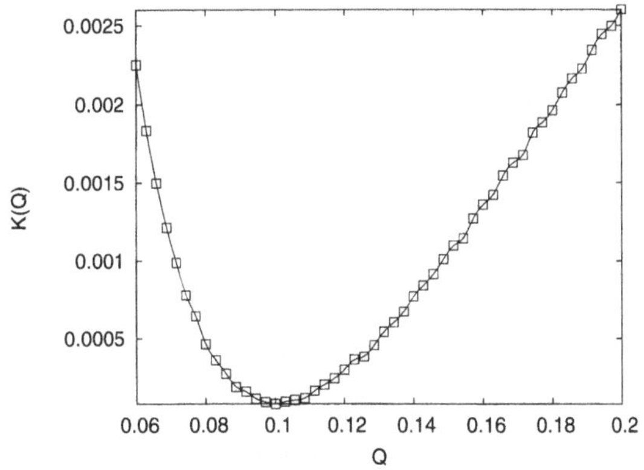

Fig. 4. The Kullback-Leibler information as a function of the noise strength (process of Equation (14)) exhibits a clear minimum at the value of Q used to generate the time series

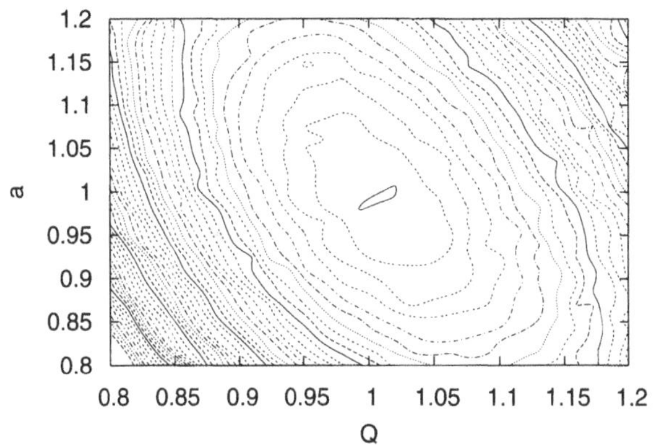

Fig. 5. Contour plot of the Kullback-Leibler information as a function of the parameters a, b for the reconstruction of the stochastic process Equation (15)

shall call this a semi parametric ansatz. Since the first estimation (9) for the drift function gives some idea about its shape $D^{(2)}(q) = a + bq^2$ with $a, b \approx 1$ seems to be a reasonable parameterization. The subsequent procedure is now similar to the one for additive noise.

The Kullback-Leibler information for a fixed set of parameters again is calculated numerically integrating the Fokker-Planck equation for various parameter combinations a,b. In this case the information has to be minimized with respect to the variation of two parameters. In figure 5 one easily finds a minimum at $a = 1$, the minimum with respect to the parameter b is clearly visible. Numerically we find $b = 1.06$ to be the optimal parameter. Knowing the diffusion function one can again extract $D^{(1)}(q)$ from the data.

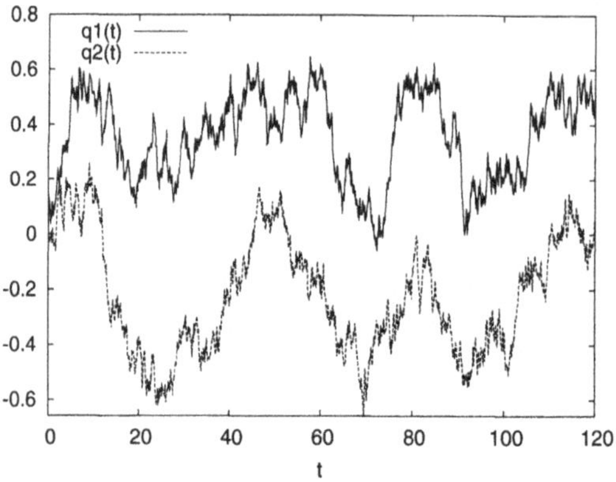

Fig. 6. Segment of the time series for the variables q_1, q_2 of the stochastic process of Equation (19)

5.2 Potential Systems

We want to point out that the procedure for one dimensional systems can be extended to the so-called class of potential systems for which the drift vector $\mathbf{D}^{(1)}(\mathbf{q})$ is obtained from a potential V($\mathbf{q}$) and $g_{ik} = Q\delta_{ik}$:

$$\begin{aligned} \mathbf{D}^{(1)}(\mathbf{q}) &= -\nabla_{\mathbf{q}} V(\mathbf{q}) \\ \mathbf{D}^{(2)}_{ij} &= Q\delta_{ij} \,. \end{aligned} \tag{16}$$

For this class of systems, the stationary PDF can be determined

$$f(\mathbf{q}) = Nexp(-V(\mathbf{q})/Q \quad . \tag{17}$$

This relationship again allows one to estimate the ratio $V(\mathbf{q})/Q$ from the stationary PDF. An ansatz for the drift term, as a consequence, possesses one free parameter Q

$$\mathbf{D}^{(1)}(\mathbf{q}) = -Q\nabla_q lnf(\mathbf{q}) \,. \tag{18}$$

In cases, where the gradient of the logarithm of the stationary probability distribution can reliably be determined, the drift term can be parameterized by a single parameter Q. The Kullback-Leibler information is then a function of one parameter and the optimal value of Q can be found graphically.

We have considered the stochastic process for two variables q_1, q_2 with drift term

$$D^{(1)}_1(\mathbf{q}) = \epsilon q_1 - q_1\left(q_1^2 + Bq_2^2\right) \tag{19}$$

$$D^{(1)}_2(\mathbf{q}) = \epsilon q_2 - q_2\left(Bq_1^1 + q_2^2\right) \,. \tag{20}$$

The vector field of the drift can be derived from the potential

$$V(\mathbf{q}) = -\frac{\epsilon}{2}\mathbf{q}^2 + \frac{1}{4}\mathbf{q}^4 + \frac{B-1}{2}q_1^2 q_2^2 \,. \tag{21}$$

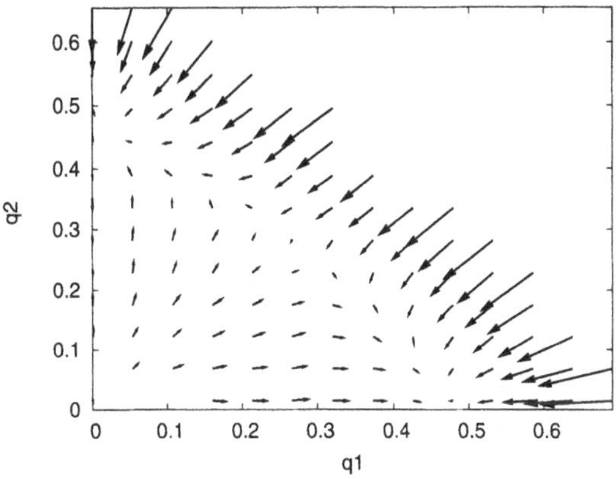

Fig. 7. The drift vector field $\mathbf{D}^{(1)}$ of the stochastic process Equation (19). The drift obtained by the iterative procedure is practically indistinguishable from the one used to generate the time series

These set of equations arise in the theory of pattern formation, where it describes e.g. the selection between square and stripe patterns [3]. A segment of the time series is exhibited in Fig. 6. The obtained drift vector field $\mathbf{D}^{(1)}(\mathbf{q})$ is exhibited in Fig. 7.

6 Summary

Summarizing, we have outlined an operational method for the estimation of drift and diffusion terms from experimental time series of stochastic Langevin processes. We presented a refined procedure, which does not rely on determining drift and diffusion coefficients as conditional moments in the small time increment limit. Although the small time increment limit is considered to be a first approximation an iterative refinement of the estimated stochastic process is performed by minimization of the Kullback-Leibler distance between estimated and measured two time probability distributions. Although the effort evidently increases with increasing dimensions of the stochastic processes under investigation the proposed algorithm seems to be feasible for a modest number of degrees of freedom since the necessary steps involve routine calculations like simulations of stochastic processes, evaluation of Kullback-Leibler entropies and optimization methods for searching the optimal stochastic process.

References

1. H. Kantz, T. Schreiber, *Nonlinear Time series Analysis*, (Cambridge University Press 1997).
2. A. Einstein, Annalen der Physik **17**, 549 (1905).
3. H. Haken, *Synergetics: Introduction and Advanced Topics* (Springer-Verlag, Berlin 2004).
4. H. Haken, *Information and Self-Organization- A macroscopic approach to complex systems* (Springer-Verlag, Berlin 2004).
5. S. Siegert, R. Friedrich, J. Peinke, Physics Letters **A 234**, 275 (1998).
6. R. Friedrich, S. Siegert, J. Peinke, St. Lück, M. Siefert, M. Lindemann, J. Raethjen, G. Deuschl, G. Pfister, Phys. Lett. **A 271**, 217 (2000).
7. H. Risken, *The Fokker-Planck equation*, (Springer-Verlag, Berlin 1983).
8. C. W. Gardiner, *Handbook of Stochastic Methods*, (Springer-Verlag, Berlin 1983).
9. R. Friedrich, J. Peinke, Phys. Rev. Lett. **78**, 863 (1997).
10. J. Gradisek, S. Siegert, R. Friedrich, I. Grabec, Phys. Rev. **E 62**, 3146 (2000).
11. J. Gradisek, I. Grabec, S. Siegert, R. Friedrich, Mechanical Systems and Signal Processing **16** (5), 831 (2002).
12. T. D. Frank, P. J. Beek, R. Friedrich, Phys. Lett. **A 328**, 219 (2004).
13. T. D. Frank, R. Friedrich, P.J. Beek, Stochastics and Dynamics **44**, 9 (2004).
14. T. D. Frank, P. J. Beek, R. Friedrich, Phys. Rev. **E 68**, 021912 (2003).
15. P. Sura, S. T. Gille, Journal of Marine Research **61**, 313 (2003).
16. P. Sura, Journal of the Atmospheric Sciences **60**, 654 (2003).
17. T. Kuusela, T. Shepherd, J. Hietarinta, Phys. Rev. **E 67**, 061904 (2003).
18. S. Kriso, R. Friedrich, J. Peinke, P. Wagner, Physics Letters **A 299**, 287 (2002).
19. H. U. Bödeker, M. Röttger, A. W. Liehr, T. D. Frank, R. Friedrich, H. G. Purwins, Phys. Rev. **E 67**, 056220 (2003).
20. E. Racca, A. Porporato, Phys. Rev. **E 71**, 027101 (2005).
21. Ch. Renner, J. Peinke, R. Friedrich, J. Fluid Mech. **433**, 383 (2001).
22. M. Tutkun, L. Mydlarski, New Journal of Physics **6**, Art. No. 49 (2004).
23. R. Friedrich, J. Peinke, Ch. Renner, Phys. Rev. Lett **84**, 5224 (2000).
24. G. R. Jafari, S. M. Fazeli, F. Ghasemi, S. M. V. Allaei, M. R. R. Tabar, A. I. Zad, G. Kavei, Phys. Rev. Lett. **91**, 226101 (2003).
25. M. Wächter, F. Riess, H. Kantz, J. Peinke, Europhys. Lett. **64**, 579 (2003).
26. P. Sura, j. Barsugli, Phys. Lett. **A 305**, 304 (2002).
27. M. Ragwitz, H. Kantz, Phys. Rev. Lett. **87**, 254501 (2001).
28. R. Friedrich, Ch. Renner, M. Siefert, J. Peinke, Phys. Rev. Lett. **89**, 149401 (2002).
29. M. Ragwitz, H. Kantz, Phys. Rev. Lett. **89**, 149402 (2002).
30. M. Siefert, A. Kittel, R. Friedrich and J. Europhys. Lett. **61**, 466 (2003).

Evolution in Complex Systems: Record Dynamics in Models of Spin Glasses, Superconductors and Evolutionary Ecology

Henrik Jeldtoft Jensen

Department of Mathematics, Imperial College London, South Kensington campus, London SW7 2AZ, UK
h.jensen@imperial.ac.uk
www.imperial.ma.ic.ac/~hjjens

Summary. Recent research on the non-stationary nature of the dynamics of complex systems is reviewed through three specific models. The long time dynamics consists of a slow, decelerating but spasmodic release of generalized intrinsic strain. These events are denoted quakes. Between the quakes weak fluctuations occur but no essential change in properties are induced. The accumulated effect of the quakes, however, is to induce a direct change in the probability density functions characterizing the system. We discuss how the log-Poisson statistics of record dynamics may be an effective description of the long time evolution and describe how an analysis of the times at which the quakes occur enables one to check the applicability of record dynamics.

1 Introduction

Out of equilibrium systems are often treated as being in a stationary state characterized by time independent statistical measures. Although this is probably the case in some situations there are many instances where this is not so and where one may miss essential aspects of the behavior if attempts are made to treat the phenomena as stationary or nearly stationary.

Complex systems often display evolving macroscopic properties. The most important task of a theoretical treatment is then to understand the link between the microscopic fluctuations, which will often exhibit an approximate time reversal symmetry and the macroscopic directed evolution. The description should as well explain the nature of the emergent macroscopic dynamics.

Here we review how the concept of record dynamics, developed by Sibani and Littlewood [1], has successfully served as a paradigm for the description of the evolution of three very different models: the relaxation of a spin glass following an initial temperature quench, the penetration of an external magnetic field into a disordered type II superconductor and a model of evolutionary ecology. In all three cases macroscopic variables, which exhibit a degree of intermittent dynamics, can be identified. Furthermore, the sequence of transitions between metastable configurations can be analyzed in terms of the record statistics.

B. Kramer (Ed.): Adv. in Solid State Phys. **45**, 95–106 (2005)

The work reviewed here is a result of collaboration with Paolo Sibani, Paul Anderson and Luis P. Oliveria. Some details of the specifics have been published in [2, 3, 4, 5]. The concept of record dynamics have been developed by Sibani and his collaborators over a long period, see e.g. [1, 6, 7, 8, 9].

Below we first introduce the models in sufficient self-contained detail. Next we describe how the long time dynamics in each case are manifestations of record dynamics and discuss its consequences.

2 Three Models

Here follows a brief description of the definition of the microscopic dynamics of the three models considered.

2.1 Spin Glass

We consider a three dimensional Edwards-Anderson spin glass

$$H = -\frac{1}{2}\sum_{\langle ij\rangle} J_{ij} S_i S_j . \tag{1}$$

with nearest neighbor Gaussian couplings [10] and Ising spin $S_i = \pm 1$. At time zero the temperature is instantaneously dropped from infinity to a very low value. The subsequent dynamics is realized by use of Monte Carlo dynamics, see [3, 4, 8].

2.2 Magnetic Relaxation

We use Monte Carlo (MC) simulations of a generalized three dimensional layered version of the Restricted Occupancy Model (ROM) model to capture the long time relaxation of interacting vortex matter [11, 12, 13, 14, 15, 16, 5].

The length scales of vortex interactions can be very large compared with the average separation between vortices. At high magnetic induction each vortex interacts with many others suggesting that a simplified coarse grained description in terms of vortex densities may be applicable. For layered superconductors it is natural to introduce two separate length scales: the first is the range of the interaction parallel to the planes, this is the London penetration depth λ. The second length scale is the vortex correlation length, $\xi_{||}$, parallel to the applied field (which we imagine to be perpendicular to the copper oxide planes for high temperature superconductors). The exact identification of this length scale is difficult and is likely to depend on the anisotropy of the material, the nature of the pinning, the strength of the magnetic induction and on the temperature. This length scale may be related to vortex line cutting [17, 18, 19, 20, 21, 22, 23, 24]. These length scales respectively give the horizontal, l_0, and vertical, l_1, coarse-graining length and therefore the lattice spacing of our model. Horizontally we have $l_0 = \lambda$ and perpendicularly $l_1 \sim \xi_{||}$. Smaller length scales are ignored. For our purposes this approximation is acceptable because the length scales smaller than λ seem to have little influence on the long time glassy properties of vortex matter.

The behavior of vortex matter is determined by the competition of four energy scales [25]: intra and interlayer vortex-vortex interaction, vortex-pinning interaction and thermal fluctuations, all of which are schematically included in the ROM model.

The Hamiltonian of the ROM model is thus the following,

$$H = \sum_{ij} A_{ij} n_i n_j - \sum_{i} A_{ii} n_i + \sum_{i} A_i^p n_i + \sum_{\langle ij \rangle_z} A_2 \left(n_i - n_j\right)^2 , \qquad (2)$$

where n_i is the number of vortices on site i of the lattice. In a superconducting sample the number of vortex lines per unit area is restricted by the upper critical field (B_{c2}) [26], so in the model the number of vortices per cell can only assume values smaller than $N_{c2} = B_{c2} l_0^2 / \phi_0$ [27, 14]. Hence the name Restricted Occupancy Model. Moreover, as we are interested in a simulation setup that does not require magnetic field inversion and the vortex-antivortex creation is strongly suppressed, we simply consider $n_i \geq 0$.

The first two terms in Eq. (2) represent the repulsion energy due to vortex-vortex interaction in the same layer, and the vortex self energy respectively. Since the potential that mediates this interaction decays exponentially at distances longer than our coarse-graining length λ, interactions beyond nearest neighbors are neglected. We set $A_{ii} := A_0 = 1$, $A_{ij} := A_1$ if i and j are nearest neighbors on the same layer, and $A_{ij} := 0$ otherwise.

The third term represents the interaction of the vortex pancakes with the pinning centers. A_i^p is a random potential and for simplicity we consider that A_i^p has the following distribution $P\left(A_i^p\right) = (1-p)\,\delta\left(A_i^p\right) - p\delta\left(A_i^p - A_0^p\right)$. The pinning strength $|A_0^p|$ represents the total action of the pinning centers located on a site. In the present work we use $|A_0^p| = 0.3$.

Finally the last term describes the interactions between the vortex sections in different layers. This term is a nearest neighbor quadratic interaction along the z axis, so that the number of vortices in neighboring cells along the z direction tends to be the same.

The parameters of the model are defined in units of A_0. The time is measured in units of full MC sweeps. The relationship between the model parameters and material parameters is discussed in [27, 14]. The model has been demonstrated to reproduce a very broad range of experimental observations including dynamical aspects of magnetic creep and memory and rejuvenation of voltage-current characteristics [28, 11, 12, 13, 14, 15, 16].

Each individual MC update involves the movement to a neighbor site of a single randomly selected vortex. The movement of the vortex is automatically accepted if the energy of the system decreases; if the energy of the system increases, the movement is accepted with probability $\exp(-\Delta E/T)$ [29].

The external magnetic field is modeled by the edge sites on each of the planes. The density at the edge is kept at a controlled value. During a MC sweep vortices may move between the bulk sites and the edge sites. After each MC sweep the density on the edge sites is brought to the desired value. Initially the external field is increased to a desired value ($N_{ext} = 10$ vortices per edge site) by a very rapid increase in the density on the edge sites.

After this fast initial ramping the external field is kept constant, while we study how the vortices move into the sample. The age of the system, t_w, is taken to be the time since the initial ramping.

2.3 Tangled Nature

The Tangled Nature model is an individual based model of evolutionary ecology. We give a brief outline of the model here. Details can be found in [30, 31, 2]. An individual is represented by a vector $\mathbf{S}^\alpha = (S_1^\alpha, S_2^\alpha, ..., S_L^\alpha)$ in the genotype space $\mathcal{S}$, where the "genes" S_i^α may take the values ± 1, i.e. $\mathbf{S}^\alpha$ denotes a corner of the L-dimensional hypercube. In the present paper we take $L = 20$ as this gives space of a reasonable size to explore (over a million genotypes) whilst not being computationally prohibitive. We think of the genotype space $\mathcal{S}$ as containing all possible ways of combining the genes into genome sequences. Many sequences may not correspond to viable organisms. The viability of a genotype is determined by the evolutionary dynamics. All possible sequences are made available for evolution to select from. The number of occupied sites is referred to as the diversity, here analogous to the number of species or species richness [32]. As explained later, genotype, species, site and node are synonymous throughout.

For simplicity, an individual is removed from the system with a constant probability p_{kill} per time step. A time step consists of *one* annihilation attempt followed by *one* reproduction attempt. One generation consists of $N(t)/p_{kill}$ time steps, which is the average time taken to kill all currently living individuals. All references to time will be in units of generational time.

The ability of an individual to reproduce is controlled by a weight function $H(\mathbf{S}^\alpha, t)$,

$$H(\mathbf{S}^\alpha, t) = \frac{1}{cN(t)} \left(\sum_{\mathbf{S} \in \mathcal{S}} J(\mathbf{S}^\alpha, \mathbf{S}) n(\mathbf{S}, t) \right) - \mu N(t) , \tag{3}$$

where c is a control parameter, $N(t)$ is the total number of individuals at time t, the sum is over the 2^L locations in $\mathcal{S}$ and $n(\mathbf{S}, t)$ is the number of individuals (or occupancy) at position $\mathbf{S}$. Two positions $\mathbf{S}^a$ and $\mathbf{S}^b$ in genome space are coupled with the fixed random strength $J^{ab} = J(\mathbf{S}^a, \mathbf{S}^b)$ which can be either positive, negative or zero. This link is non-zero with probability θ, i.e. θ is simply the probability that any two sites are interacting. To study the effects of interactions *between* species, we exclude self-interaction so that $J^{aa} = 0$.

The conditions of the physical environment are simplistically described by the term $\mu N(t)$ in equation (3), where μ determines the average sustainable total population size, i.e. the carrying capacity of the environment. An increase in μ corresponds to harsher physical conditions. Notice that genotypes only adapt to each other and the physical environment represented by μ. We use asexual reproduction consisting of one individual being replaced by two copies mimicking the process of binary fission seen in bacteria. Successful reproduction occurs with a probability per unit time given by

$$p_{off}(\mathbf{S}^{\alpha}, t) = \frac{\exp[H(\mathbf{S}^{\alpha}, t)]}{1 + \exp[H(\mathbf{S}^{\alpha}, t)]} \in [0, 1]. \tag{4}$$

This function is chosen for convenience. We simply need a smoothly varying function that maps $H(\mathbf{S}^{\alpha}, t)$ to the interval $[0, 1]$ and it is otherwise arbitrary. We allow for mutations in the following way: with probability p_{mut} per gene we perform a change of sign $S_i^{\alpha} \rightarrow -S_i^{\alpha}$ during reproduction.

Initially, we place $N(0) = 500$ individuals at randomly chosen positions. Their initial location in genotype space does not affect the nature of the dynamics. A two-phase switching dynamic is seen consisting of long periods of relatively stable configurations (quasi-Evolutionary Stable Strategies or q-ESSs) interrupted by brief spells of reorganization of occupancy which are terminated when a new q-ESS is found, as discussed in [30].

3 Record Dynamics and its Manifestation

In this section we review the macroscopic intermittent dynamics of the three models and show that in all cases record dynamics is an efficient description of the statistical aspects of the temporal evolution. Before that we need to sketch the notion of *record statistics.*

Let $\chi(t)$ denote an uncorrelated stochastic signal distributed according to the probability density function(pdf) $p(\chi)$. By the record of the signal we mean $R(t) = \max\{\chi(t') | t' \leq t\}$. Obviously $R(t)$ is a piecewise constant function which jumps discontinuously as a fluctuation manages to take $\chi(t)$ to a new record value. The times t_k at which this happens are called the record times. It was pointed out be Sibani and Littlewood [33] that the probability that exactly q records occur during a time interval $[t_w, t_w + t]$ is to a good approximation given by

$$p(q) = \frac{\langle q \rangle^q}{q!} \exp\{-\langle q \rangle\}, \tag{5}$$

where $\langle q \rangle = \alpha \log(1 + t/t_w)$. This is a Poisson distribution in the logarithm of time. For a mathematical process the logarithmic rate $\alpha = 1$. Here we include the possibility $\alpha \neq 1$, which may happen for a physical process as an effect of over or undercounting of the true number of records. For example $\alpha > 1$ can occur if the recorded record times are produced by more than a single independent record process. In contrast $\alpha < 1$ can e.g. be an effect of not being able resolve all the record events of record process. The average number of records per time unit decreases inversely proportional with time, namely

$$\frac{d\langle q\rangle}{dt} = \frac{\alpha}{t_w + t}. \tag{6}$$

It is important to note that Eqs. (5) and (6) are independent of $p(\chi)$. Thus the statistical properties of a record signal $R(t)$ are very general and do not depend on the properties of the underlying fluctuating signal $\chi(t)$.

3.1 Spin Glass

After an initial quench from very high temperature it is natural that the dynamics of the spin glass leads to a relaxation towards ever lower energy. The specifics of how this happens was analyzed in great detail by Sibani and collaborators [9, 8] for the Edwards-Anderson spin glass. They followed the temporal evolution of the total energy $E(t)$ given in Eq. (1). They identified the sequence of local minima $E_{min}(k)$ and local maxima $E_{max}(k)$ from which they defined the k-th barrier as $\Delta E_k = E_max(k) - E_min(k)$. The set of barriers turn out to be monotonously increasing $\Delta E_k < \Delta E_{k+1}$, but only marginally so in the sense that $\Delta E_k \simeq \Delta E_{k+1}$. Thus to exit the k-th metastable state visited by the spin glass, a barrier slightly larger than any encountered previously has to be overcome. The set of time instances at which the spin glass manage to move from one metastable configuration, i.e. the quake times t_k, was found to follow the log-Poisson distributed characteristic of record statistics, see E.q. (5).

3.2 Magnetic Relaxation

The magnetic pressure exerted by the external magnetic field in the ROM model introduced above will force the number of vortices in the bulk, $N_v(t)$, of the sample to increase with time. We have found that $N_v(t)$ is essentially a record signal [2, 5], see Fig. 1.

The sequence of record times follow to good approximation the statistics derivable from Eq. (5) expected for an uncorrelated record signal [2].

3.3 Evolutionary Ecology

That directedness of the temporal evolution of the quenched spin glass and also of the superconductor in an external field is to be expected. It is less obvious why the Tangled Nature model of biological evolution exhibit a gradual adaptation towards more stable configurations. The diffusive nature of the dynamics in genotype space may suggest the breaking of time reversal symmetry. But this in itself doesn't point to a reason why the total number of

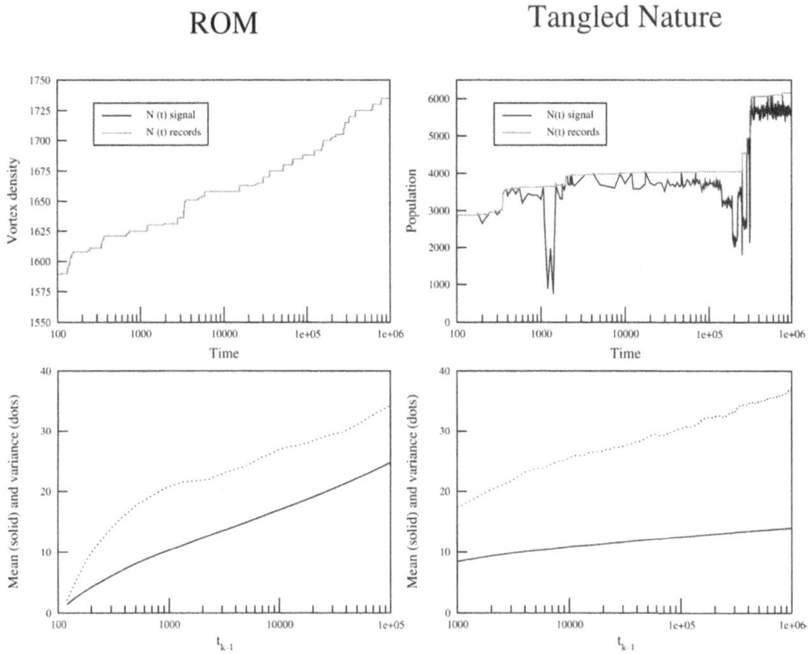

Fig. 1. Time dependence of the ROM and the Tangled Nature model

individuals in the model is increasing, on average, with time. Nor does diffusion imply that the configurations tend to be come more stable with time.

The following analysis suggests an explanation. First assume that no mutations can occur. The population dynamics is controlled by the fixed probability p_{kill} and the offspring probability p_{off}. Fluctuations in the population size $N(t)$ will lead to fluctuations in p_{off} according to Eq. (3). The overall stability of the population is ensured since the fixpoint condition $p_{off} = p_{kill}$ is stable. This follows because an increase in $N(t)$ will lead to a decrease in H (see Eq. (3) and therefore in p_{off}. Similarly a decrease in $N(t)$ leads to a increase in H. Now consider the effect of mutations. The evolutionary dynamics is driven by the mutations which move individuals between positions in genotype space. These mutations are random and lead to symmetric fluctuations in the weight function H as an effect of changing the coupling term H, see Eq. (3). Let us schematically write H as $H = \mathcal{J} - \mu N$. Mutations induce fluctuations of the form $\mathcal{J} \mapsto \mathcal{J} + \delta\mathcal{J}$. Assume $\delta\mathcal{J} > 0$ and that the same fluctuation with opposite sign $-\delta\mathcal{J}$ occurs with equal probability. The mutation leading to $\delta\mathcal{J}$ is, however, more likely to become established since

$$p_{off}(h + \delta\mathcal{J}) > p_{off}(H - \delta\mathcal{J}), \tag{7}$$

for values of H where p_{off} is convex; which is the case for $p_{off} < 1/2$ according to Eq. (4). This is a mathematical way of paraphrasing Darwin's description of how favorable variations become entrenched [34].

It was found from simulations of the Tangled Nature model that $N(t)$ gradually increases. Furthermore, the configurations occupied in genotype space becomes, on average, more stable with time [31, 2]. In Fig. 1 we compare the record signal derive from $N(t)$ with $N(t)$. It is clear that there are much larger deviations between the record and the signal itself, than was the case in the ROM model. Nevertheless, we do believe that the statistics of the record times obtained from the record signal derive from $N(t)$ essentially corresponds to the transition times between the q-ESS epochs.

4 Consequences

We briefly review some of the most prominent consequences, which may be derived from the record dynamics.

4.1 Spin Glasses

A detailed description of how certain aspect of intermittency, aging and memory in spin glasses can be considered a consequence of record dynamics was given in [3, 4]. That the fluctuations clearly separates into 'quake' fluctuations and Gaussian equilibrium like fluctuations show up directly in a study of the heat exchange between the spin glass and its thermal bath. The pdf for the heat exchange consists of a Gaussian part and an exponential tail. The exponential tail is produced by the large releases of energy that occur when a quake takes the spin glass from one metastable state to the next. The exponential tail is only visible when the heat exchange is collected during a time interval, δt that is short compare to the time t_w passed since the initial quench from high temperature. Recall that the quake activity decreases roughly inversely proportional with time. If $\delta t \gg t_w$ there is not enough quake activity during the sampling time δ to make the exponential tail of the pdf visible though peak of the much more frequent Gaussian fluctuations. In this way the one point distribution of the heat exchange is able to probe the aging of the spin glass [4].

Aspects of memory and rejuvenation in a temperature shift experiment can also be related to the record dynamics [3]. Let $b(t_w, T)$ denote the largest energy barrier overcome by the quakes during the time t_w since the initial quench. This barrier will, according to the Arrhenius law, determine the rate of quake activity, r_q at times about t_w. On the other hand, from the view point of record dynamics, we also have $r_q(T) \propto 1/t_w$. A drop in the temperature $T \mapsto T'$ at t_w will not change the barrier $b(t_w, T)$ established by the record dynamics prior to t_w. The Arrhenius activation of the quakes will however drop when the temperature is lowered producing an effective age of the system $t_w^{eff} \propto 1/r_q(T') > t_w$. It is interesting to note that the effect of the temperature drop has two opposite effects concerning the apparent age of the system: (a) the drop in T make the energy barriers look bigger. So in this respect the spin glass appears to be older. (b) The amount of energy delivered to the heat bath during a quake is higher and in this respect the spin glass appears younger. A detailed discussion is given in Ref. [3].

4.2 Magnetic Relaxation

We now explain how record dynamics might explain the observation that the rate of thermally activated creep is found to be essentially temperature independent for broad ranges of the temperature [5]. We described in section 2.2 and 3.2 that, in the ROM model, the temporal evolution of the total number of vortices in the bulk of the sample is a record signal. A prominent feature of the record time of an uncorrelated stochastic process is that the statistics of the records, and in particular the rate with which the records occur, is independent of the properties of the underlying stochastic process. For the specific case of thermally induced fluctuations this implies that the rate of the record will be temperature independent. In Fig. 2 we show that the temperature dependence of the ROM model compare well with experimental creep rates. Both exhibit a broad range of weak temperature dependence. Details can be found in [5]

4.3 Biological Evolution

In figure 1 we showed that the total population size of the Tangled Nature model, despite of large fluctuations, may also be related to record dynamics. The nature of the metastable states between the quakes is in this case difficult to determine. But we believe they are closely related to the quasi-Evolutionary Stable Strategies which have been identified in the model [30]. We have found that the number of extinctions and creation events per time decreases in the Tangled Nature model [31]. Similar behavior is observed in macro-evolution [37]. The dynamics of the Tangled Nature model is intermittent as is to some extend clear from Fig. 1. The intermittency is, however, much more evident from analysis of the time dependence of the configurations in genotype, see space [30, 31]. The fossil record has also been interpreted as exhibiting intermittency, see e.g. [38]. Accordingly by comparing the Tangled Nature model and the dynamics of the fossil record we are able to suggest that the decreasing extinction rate and the intermittency, or punctuated equilibrium, is a result a hitherto unrevealed record dynamic that somehow controls the macro-dynamics of biological evolution.

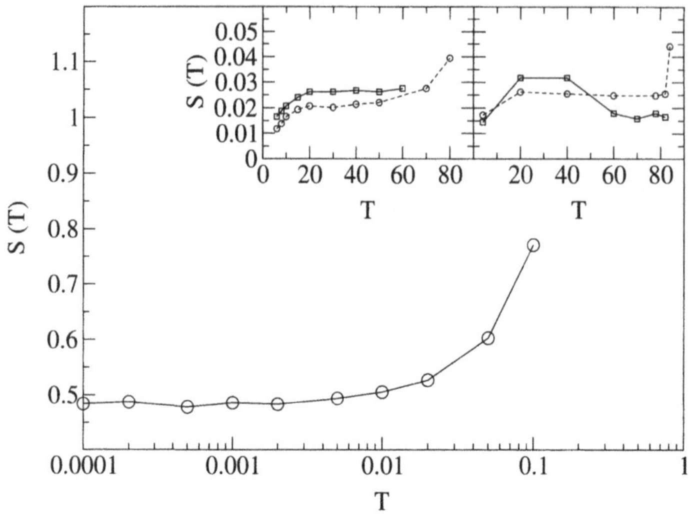

Fig. 2. Main panel: Numerical results for the creep rate versus temperature. Details in [5]. Insets: experimental results for the creep rate versus T. The right inset shows data from Keller *et al.* for melt processed YBCO crystals with the magnetic field applied along the *c* axis (squares) and *ab* plane (circles). The left inset shows data from [35] for unirradiated (squares) and 3 MeV proton-irradiated (circles) YBCO flux grown [36] crystals with a 1 T magnetic Field applied parallel to the *c* axis

5 Summary and Conclusions

The relevance of record dynamics to the long time evolution of complex systems was indicated by reviewing three very different model studies. The most prominent characteristics of record dynamics are log-Poisson distribution of the number of records, or quakes, occurring during a time interval and a rate of events, which decreases inversely proportional with time. It will be very interesting to analyst the dynamics of other complex systems from the view point of record statistics. The analysis requires access to the time instances at which quakes occur. If that is not available analysis of the probability density function of a single fluctuating quantity such as the heat exchange of a spin glass may suffice to reveal the existence of an underlying record dynamics.

References

1. P. Sibani and P. B. Littlewood, Phys. Rev. Lett. **71**,1482 (1993).
2. P. E. Anderson, H. J. Jensen, L. P. Oliveria, and P. Sibani, Complexity **10**, 49 (2004).
3. P. Sibani and H. J. Jensen, JSTAT, P10013 (2004).
4. P. Sibani and H. J. Jensen, Europhys. Lett. **69**, 563 (2005).

5. L. P. Oliveria, H. J. Jensen, and P. Sibani, Phys. Rev. B, in press.
6. P. Sibani and A. Pedersen, Europhys. Lett. **48**, 346 (1999).
7. J. Dall and P. Sibani, Comp. Phys. Comm. **141**, 260 (2001).
8. J. Dall and P. Sibani, Eur. Phys. J. B **34**, 233, (2003).
9. P. Sibani and J. Dall, Europhys. Lett. **64**,8 (2003).
10. S. F. Edwards and P. W. Anderson, J. Phys. F **5**, 965 (1975).
11. M. Nicodemi and H. J. Jensen, Physica C **341-348**, 1065 (2000).
12. M. Nicodemi and H. J. Jensen, J. Phys. A **34**, L11 (2001).
13. M. Nicodemi and H. J. Jensen, Phys. Rev. Lett. **86**, 4378 (2001).
14. H. J Jensen and M. Nicodemi, Europhys. Lett. **54**, 566 (2001).
15. M. Nicodemi, Phys. Rev. E **67**, 041103 (2003).
16. M. Nicodemi, J. Phys.: Cond Matt. **14**, 2403 (2002).
17. T. Puig and X. Obradors, Phys. Rev. Lett. **84**, 1571 (2000).
18. M.F. Goffman et al., Phys. Rev. B **57**, 3663 (1998).
19. H. Safar, E. Rodriguez, F. de la Cruz, P. L. Gammel, L. F. Schneemeyer, and D. J. Bishop, Phys. Rev. B **46**, 14238 (1992).
20. D. Lopez, G. Nieva, F. de la Cruz, H. J. Jensen, and D. O'Kane, Phys. Rev. B **50** 9684 (1994).
21. D. Lopez, E. A. Jagla, E. F. Righi, E. Osquiguil, G. Nieva, E. Morre, F. de la Cruz, and C. A. Balseiro, Physica C **260**, 211 (1996).
22. M. B. Gaifullin, Yuji Matsuda, N. Chikumoto, J. Shimoyama, and K. Kishio, Phys. Rev. Lett. **84**, 2945 (2000).
23. R. Busch, G. Ries, H. Werthner, G. Kreiselmeyer, and G. Saemann-Ischenko, Phys. Rev. Lett. **69**, 522 (1992).
24. D. T. Fuchs, R. A. Doyle, E. Zeldov, D. Majer, W. S. Seow, R. J. Drost, T. Tamegai, S. Ooi, M. Konczykowski, and P. H. Kes. Phys. Rev. B **55**, R6156 (1997).
25. G. Blatter et al., Rev. Mod. Phys. **66**, 1199 (1994).
26. M. Tinkham, *Introduction to Superconductivity.* (McGraw-Hill, New York 1996).
27. M. Nicodemi and H. J. Jensen, Phys. Rev. B **65**, 144517 (2002).
28. D.K. Jacksopn, M. Nicodemi, G.K. Perkins, N.A. Lindop and H.J. Jensen, Europhys. Lett. **52** 210 (2000).
29. K. Binder, Rep. Prog. Phys. **60**, 487 (1997).
30. K. Christensen, M. Hall, A. di Collobiano, and H. J. Jensen, J. Theor. Biol. **216**, 73 (2002).
31. M. Hall, K. Christensen, S.A. di Collabiano, and H.J. Jensen, Phys. Rev. E **66**, 011904 (2002).
32. Charles J. Krebs, *Ecological Methodology*, (Benjamin/Cummings, 2nd edition 1999).
33. P. Sibani, C. Schön, P. Salamon, and J.-O. Andersson, Europhys. Lett. **22**, 479 (1993).
34. C. Darwin, *The Origin of Species by Means of Natural Selection*, (John Murray, more recent edition by Penguin Books 1968, London 1859).
35. L. Civale, A. D. Marwick, M. W. McElfresh, T. K. Worthington, A. P. Malozemoff, F. Holtzberg, J. R. Thompson, and M. A. Kirk, Phys. Rev. Lett. **65**, 1164 (1990).
36. D.L. Kaiser, F. Holtzberg, M. F. Chisholm, and T. K. Worthington, J. Cryst. Growth **85**, 593 (1987).

37. M. E. J. Newman and P. Sibani, Proc. R. Soc. Lond B **266**, 1 (1999).
38. N. Eldredge and S. Gould, Nature **332**, 211 (1988).

Part II

Quantum Transport and Correlations

Flavor Degeneracy and Effects of Disorder in Ultracold Atom Systems

Walter Hofstetter

Institut für Theoretische Physik A, RWTH Aachen,
Templergraben 55, 52056 Aachen, Germany

Summary. Cold atoms in optical lattices offer an exciting new laboratory where quantum many-body phenomena can be realized in a highly controlled way. They can serve as *quantum simulators* for notoriously difficult problems like high-temperature superconductivity. This review is focused on recent developments and new results in multi-component systems. Fermionic atoms with SU(N) symmetry have exotic superfluid and flavor-ordered ground states. We discuss symmetry breaking, collective modes and detection issues. Bosonic multi-flavor ensembles allow for engineering of spin Hamiltonians which are interesting from a quantum computation point of view. Finally, we will address the competition of disorder and interaction in optical lattices. We present a complete phase diagram obtained within dynamical mean-field theory and discuss experimental observability of the Mott and Anderson phases.

1 Introduction and Overview

The achievement of Bose-Einstein condensation (BEC) 10 years ago [1] has opened the new field of interacting quantum gases in the dilute limit. It has become possible to observe quantum phenomena like Bose statistics on a mesoscopic scale, involving a large number of atoms. More recently, fermionic gases have also been cooled to the quantum degenerate regime, using sympathetic cooling of two spin states or boson-fermion mixtures [2, 3, 4, 5]. Although the resulting temperatures $T/T_F \approx 0.1$ are, relatively to the Fermi temperature T_F, much higher than in solids, the Pauli principle has been clearly observed.

In addition to quantum statistics, tunable interactions are another important ingredient in the cold atom "toolbox". The interactions between atoms can be changed by an external magnetic field as a result of Feshbach resonances [6, 7]. In particular, their scattering length can be tuned to positive or negative values, corresponding to repulsive or attractive interactions. This has opened the way to studies of solid-state related phenomena like Cooper pairing and BCS superfluidity of fermions [8, 9]. The resulting BEC-BCS crossover has recently been the subject of intense experimental and theoretical studies [10, 11, 12]

In an independent development, degenerate atomic clouds have been combined with optical lattices, created by standing light waves which generate

B. Kramer (Ed.): Adv. in Solid State Phys. **45**, 109–121 (2005)

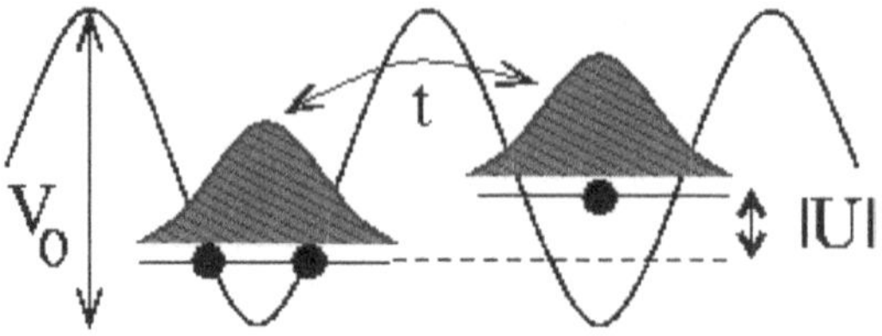

Fig. 1. Cold atoms in an optical lattice of strength V_0, shown here with hopping t and negative onsite interaction U. This situation corresponds to an attractive Hubbard model where multiple occupancy of lattice sites is energetically favorable

an effective periodic potential for the atoms [13, 14, 15]. Interactions can thus be tuned without changing the atomic scattering length. This has been demonstrated in a pathbreaking experiment [14] where interacting bosons were tuned through a quantum phase transition from a superfluid (SF) to a Mott insulating state. Very recently, fermionic K^{40} atoms have been loaded into 3d optical lattices as well [16]. In these new experiments the lowest Bloch band was filled up successively, and the shape of the Fermi surface monitored by time-of-flight measurements. Eventually a completely filled Brillouin zone corresponding to a band insulator was observed.

More generally, fermionic atoms in optical lattices allow for the realization of solid-state type quantum phases like antiferromagnetism or high-temperature superconductivity [17]. Even the spatial dimensionality of the lattice can be tuned. As an example, one-dimensional optical lattices have been realized where the hardcore or Tonks-gas limit of interacting bosons has been observed [18, 19, 20]. Recent progress in numerical methods for simulating 1d quantum systems has lead to interesting predictions about the dynamics of such systems [21, 22].

In the following we will first introduce the basic model describing cold atoms in optical lattices. We will then address systems with multiple flavors, i.e. hyperfine states, which allow realization of new exotic quantum states not accessible in solids. Finally, we will discuss the role of disorder in current and future experiments involving cold atoms.

2 Optical Lattices and Strong Correlations

2.1 Model and Parameters

Atoms can be trapped in standing light waves created by interfering laser beams detuned far from resonance [13, 14, 15]. Due to the AC Stark shift the atoms experience a periodic potential of the form $V(x) = V_0 \sum_{i=1,2,3} \cos^2(kx_i)$ where k is the wave vector of the laser. The natural energy scale for the potential depth V_0 is the recoil energy $E_R = \hbar^2 k^2/2m$. A schematic picture of such an optical lattice is shown in Fig. 1. The eigenstates

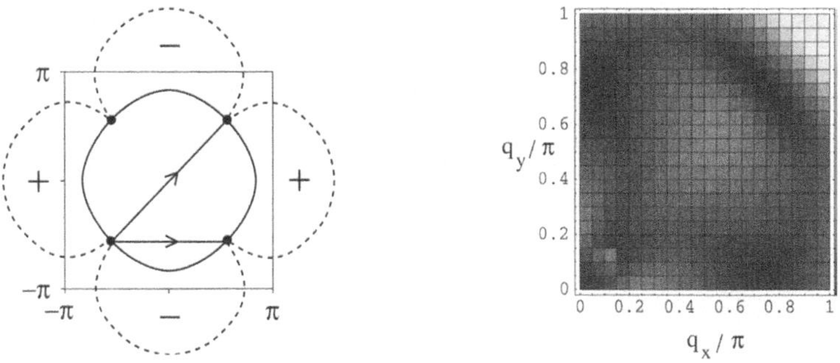

Fig. 2. Probing d-wave pairing in the repulsive 2d Hubbard model via Bragg scattering. Left: schematic diagram of the Fermi surface in 2d (solid line) and the momentum dependence of the gap (dashed line). Right: onset frequency of the quasiparticle continuum in the dynamical structure factor $S(q,\omega)$, plotted as a function of momentum q. At the special wave vectors connecting the nodal points in the left figure, the density response is gapless. Figures taken from [17]

in the periodic lattice potential are given by Bloch bands, and an equivalent representation in terms of Wannier orbitals leads to a tight-binding Hamiltonian. Let us assume for the moment that two different (hyperfine-) spin states are present, which in the following are denoted as $\sigma = \uparrow, \downarrow$. If temperature and filling are sufficiently low, the atoms will be confined exclusively to the lowest Bloch band. In this case the system can be described by a Hubbard Hamiltonian [23, 15]

$$H = -t \sum_{<ij>,\sigma} \left(c^\dagger_{i\sigma} c_{j\sigma} + c^\dagger_{j\sigma} c_{i\sigma} \right) + U \sum_i n_{i\uparrow} n_{i\downarrow} \tag{1}$$

where $\langle ij \rangle$ labels next neighbors, $c_{i\sigma}$ is the fermionic annihilation operator for the Wannier state of spin σ on site i and $n_{i\sigma} = c^\dagger_{i\sigma} c_{i\sigma}$ is the corresponding number density. The parameters for hopping and onsite interaction can then be expressed as $t = E_R(2/\sqrt{\pi})\xi^3 \exp(-2\xi^2)$ and $U = E_R a_s k \sqrt{8/\pi}\,\xi^3$. Here a_s is the atomic scattering length and $\xi = (V_0/E_R)^{1/4}$ is a parameter characterizing the strength of the lattice [15, 17]. It is obvious that by tuning the optical lattice potential V_0 one can achieve arbitrary ratios $|U|/t$ without changing a_s. Optical lattices thus give access to the *strongly correlated* regime without using Feshbach resonances.

2.2 Quantum Simulations

The Hubbard model (1) is of fundamental importance for electronic correlation effects in condensed matter. From this point of view, ultracold atoms can be used to perform *quantum simulations* of solid state physics. Here we illustrate this intriguing idea with the example of high-temperature d-wave superconductivity [17].

Consider the 2d Hubbard Hamiltonian (1) with spin-1/2 fermions and repulsive interaction $U > 0$ resulting from a positive scattering length $a_s > 0$. At half filling $n_i = 1$ this model gives rise to staggered antiferromagnetic order. At lower filling fractions, theoretical arguments suggest a d-wave paired phase, which is a possible candidate for explaining high-temperature superconductivity in the cuprates [24]. However, there is no satisfactory numerical evidence, mainly because quantum Monte Carlo calculations are limited to extremely small systems due to the sign problem.

On the other hand, cold fermions in an optical lattice could be used to *experimentally* probe d-wave pairing in the 2d Hubbard model. The resulting superfluid order can be detected via Bragg scattering, as shown in Fig. 2, which is by now a well-established technique to measure the dynamical density response $S(q, \omega)$ in interacting quantum gases [25]. As already suggested by Feynman [26], such quantum simulations could provide a powerful tool to gain insight into many-body Hamiltonians relevant for solid-state physics.

3 Multi-Component Systems

3.1 Two-Component Bosons with Spin Order

All of the alkali atoms available for trapping and cooling have $2 * (2I + 1)$ low-lying hyperfine states, where I is the nuclear spin. Several of these states can be trapped at the same time: in magnetic traps one is limited by the condition that the states have to be *low-field seekers*, but optical dipole traps, as well as optical lattices, allow confinement of basically any combination of spin states [27], as long as no instability due to three-body collision occurs. Loading a lattice with two hyperfine states of Rb^{87} has been demonstrated experimentally in [28] where also a spin-dependent periodic potential has been implemented. In the following we discuss a proposal, described in detail in [29], how these techniques can be used to engineer quantum spin Hamiltonians which in turn could be relevant for quantum information processing.

Let us consider a system of two bosonic hyperfine states in a lattice, described by the following Bose-Hubbard Hamiltonian

$$
\begin{aligned}
H = & -t_a \sum_{\langle ij \rangle} \left(a_i^\dagger a_j + H.c. \right) - t_b \sum_{\langle ij \rangle} \left(b_i^\dagger b_j + H.c \right) \\
& + U \sum_i (n_{ai} - \frac{1}{2})(n_{bi} - \frac{1}{2}) \\
& + \frac{1}{2} \sum_{i, \alpha = a, b} V_\alpha n_{\alpha i}(n_{\alpha i} - 1) - \sum_{i, \alpha} \mu_\alpha n_{\alpha i} \,. \qquad (2)
\end{aligned}
$$

Here a_i, b_i denote the annihilation operators for two different bosonic pseudospin states, and the number operators are defined as $n_{ai} = a_i^\dagger a_i$,

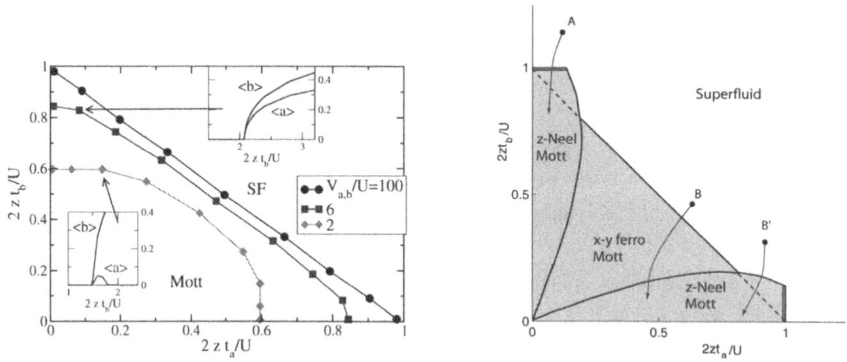

Fig. 3. Left: Phase diagram of the two-component bosonic Hubbard model obtained by mean-field theory. Note that as $V_{a(b)}$ decreases, the Mott domain shrinks. Right: Phase diagram including quantum fluctuations. Figures taken from [29]

$n_{bi} = b_i^\dagger b_i$, with the corresponding chemical potentials $\mu_{a(b)}$. In reality, experiments are performed at a fixed numbers of particles, which in the grand canonical description can be achieved by tuning the chemical potential. The onsite interaction between equal spin states is given by $V_{a(b)}$, and the one between different spins by U. We also assume a spin dependent tunable hopping $t_{a(b)}$ which has already been experimentally realized [28].

We now focus on the case of integer filling $n_a + n_b = 1$, following [29]. We are mainly interested in the nature of the Mott-superfluid transition in this system, and the possibility of spin order in the insulating phase. To address the second issue, it is instructive to first consider parameters $t_{a,b} \ll U, V_{a,b}$ deep inside the Mott phase. States with double occupancy per site are then very unfavorable and can be projected out by a Schrieffer-Wolff transformation. This leads to an effective spin Hamiltonian in the subspace of single occupation [30].

The physics of this XXZ model is well understood and includes an x-y ferromagnetic phase as well as an antiferromagnetic z-Neel ordered state.

The disadvantage of this deep Mott regime is that the critical temperature for magnetic ordering is very low $T_c \sim \max(t_{a(b)}^2/U,\, t_{a(b)}^2/V_{a(b)})$ and therefore experimentally hardly accessible. In order to enhance T_c and study the region close to the Mott-SF transition is necessary to make at least one of the interaction parameters comparable to the hopping. Here we choose $t_{a(b)} \approx U \ll V_{a(b)}$, which means that double occupancy with two different spins is now possible. The main question is whether the spin order discussed above is still visible close to the superfluid. In order to map out the Mott-SF phase boundary, we have used a mean-field approach first proposed in [31], where the kinetic energy is decoupled.

The phase diagram obtained in this way is shown in Fig. 3 (left). Note that as $V_{a(b)}$ decreases, the Mott domain shrinks. In order to resolve different

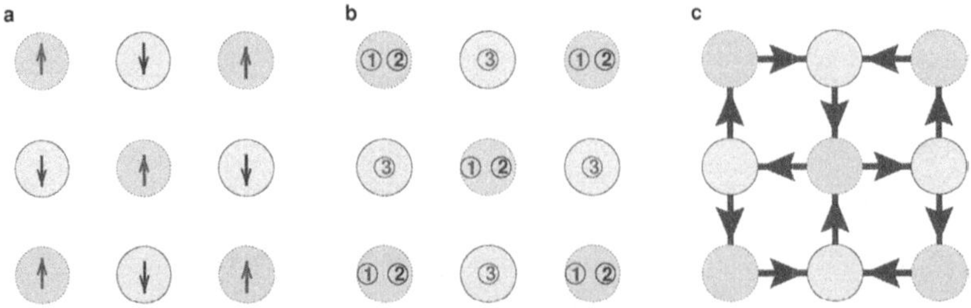

Fig. 4. Types of order in the $U > 0$ fermionic SU(3) Hubbard model. (a) AF spin-density wave for $N = 2$. (b) Flavor-density wave state for N=3. Flavor 1 and 2 prefer one sublattice, flavor 3 the other. (c) Staggered flux state for $N > 6$: particle currents are indicated by arrows. Figures taken from [35]

spin states in the insulator, it is necessary to take into account quantum fluctuations on top of the variational mean-field state and compare the resulting ground state energies. Details of this calculation can be found in [29]. The resulting phase diagram including fluctuations is shown in Fig. 3 (right). Spin ordering persists right up to the SF phase boundary and and can be tuned from xy-ferromagnetic to z-Neel antiferromagnetic by the ratio t_a/t_b. We find hysteresis between the Neel state and the superfluid, while the transition between the xy-state and the SF is continuous. These should be clear signatures for an experimental detection of spin ordered states, using for example Rb^{87} atoms. The spin order can be directly observed by spin-dependent Bragg scattering or via density fluctuations in time-of-flight measurements [32].

3.2 Beyond Solid-State: SU(N) Fermions

We will now show that with the degrees of freedom offered by ultracold atoms it is possible to create new states of matter that have no equivalent in condensed matter. The obvious constraint in solid-state physics is that electrons have only two spin states. Atoms, on the other hand, have large hyperfine multiplets out of which several states can be trapped simultaneously. For fermionic atoms this has been demonstrated with the three states $|F = 9/2, m_F = -5/2, -7/2, -9/2\rangle$ of K^{40} in an optical trap [33]. Alternatively, one could use the three spin polarized $m_s = 1/2$ states of Li^6 which, in a sufficiently large field, have a pairwise equal and anomalously large triplet scattering length $a_s = -2160a_0$ [34].

These systems can be used to realize fermionic Hubbard models with $N > 2$ flavors and approximate SU(N) flavor symmetry. In the following we will discuss the rich physics of these models for finite N, following [35, 36]. The Hamiltonian is given by

$$H = -t \sum_{m,\langle ij\rangle} \left[c^\dagger_{i,m} c_{j,m} + c^\dagger_{j,m} c_{i,m}\right] + \frac{U}{2} \sum_i n_i^2 \tag{3}$$

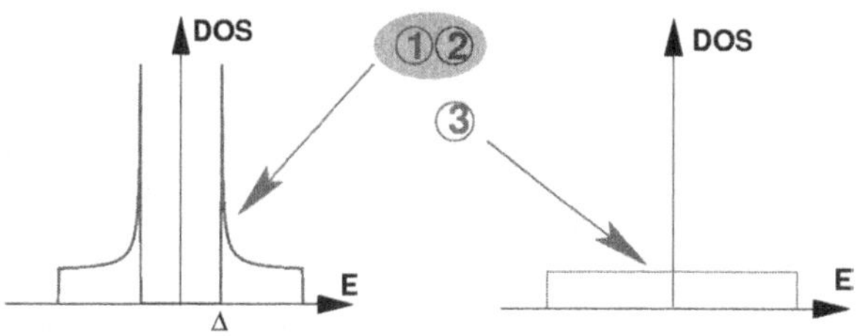

Fig. 5. BCS pairing of three-flavor fermions with SU(3) symmetry. Note that one flavor remains unpaired, with a normal Fermi surface. Figure taken from [36]

where $c^\dagger_{im}$ creates a fermion of flavor $m = 1, \ldots N$ on site i and $n_i = \sum_m n_{i,m}$ is the total number of atoms on site i. Note that the interaction term has local SU(N) invariance while the hopping reduces this to a global one. The values of t and U can be derived from atomic parameters along the lines of section 2.1.

While the large-N limit of this model has been well studied in the context of high-T_c superconductivity [37], few results have been previously obtained for finite N. Consider first the case of repulsive interactions $U > 0$. We have performed a systematic analysis of weak-coupling instabilities using a perturbative functional renormalization group (RG) approach [38]. Although the RG eventually breaks down at strong coupling, it allows to identify the leading instability towards an ordered phase. The analysis performed in [35] focuses on $d = 2$ dimensions.

In Fig. 4 the three relevant types of order at half filling $\langle n_i \rangle = N/2$ are shown. In the spin 1/2 case the system displays staggered antiferromagnetic order, as is well known. For intermediate $N < 6$ the RG yields an instability towards flavor density wave states with ordering wavevector $\mathbf{Q} = (\pi, \pi)$ like in the antiferromagnetic case. This corresponds to a breaking of the SU(N) symmetry, with a degenerate ground state manifold. As N increases, breaking of SU(N) becomes less favorable because the number of Goldstone modes increases. For $N > 6$ the RG indicates a dominant instability of the *staggered flux* type with alternating particle currents around the plaquettes of the 2d lattice (see Fig. 4c).

Let us briefly discuss the temperature scales T_c below which the respective long range orders set in. The critical temperature for flavor density waves at strong coupling scales like t^2/U and can thus be tuned to relatively large values: for $N = 3$ the RG predicts a maximum of $T_c \approx 0.1t$. On the other hand, staggered flux order, like d-wave superconductivity away from half filling, requires significantly lower temperatures, with a typical RG estimate given by $T_c \approx 0.01t$ for $N = 7$. This is about an order of magnitude below the current experimental limit and will require improved cooling techniques.

Next, we focus on the situation with attractive interactions $U < 0$ and $N = 3$ flavors which is relevant for Li^6. A large recent experimental effort has been devoted to the BEC-BCS crossover in spin-1/2 superfluid fermions [10, 11, 12, 39]. A common feature of these experiments with K^{40} and Li^6 is the use of a Feshbach resonance to generate large attractive interaction. These resonances generally occur only between two hyperfine spin states and thus cannot be used to realize an SU(3) symmetric model. However, as pointed out above, Li^6 has a remarkably large and negative background scattering length which in a finite magnetic field is approximately equal for the three spin states with $m_s = 1/2$. In combination with an optical lattice one can thus realize the SU(N) Hubbard model (3) with $U < 0$ and $N = 3$. The possibility of a three-flavor paired state in Li^6, without consideration of the SU(3) symmetry, has already been pointed out by Leggett [40].

Following the analysis in [36] we now discuss how the spin-1/2 BCS state is generalized to three flavors. We introduce a pairing mean-field and Hamiltonian

$$\Delta_{\alpha\beta} = -\frac{U}{N}\sum_{\mathbf{k}}\langle c_{\mathbf{k}\alpha}c_{-\mathbf{k}\beta}\rangle \qquad H_{\mathrm{MF}} = -\frac{1}{2}\sum_{\boldsymbol{k},\alpha,\beta} c^{\dagger}_{\boldsymbol{k}\alpha}c^{\dagger}_{-\boldsymbol{k}\beta}\Delta_{\beta\alpha} + h.c. \quad (4)$$

where $\alpha, \beta = 1, \ldots 3$ are flavor indices and N is the number of lattice sites. We focus on s-wave pairing which is favorable for strong onsite attraction. The Pauli principle then requires antisymmetry $\Delta_{\alpha\beta} = -\Delta_{\beta\alpha}$ in the flavor index. The order parameter can therefore been written as a triplet $\Delta_\alpha = \frac{1}{2}\epsilon_{\alpha\beta\gamma}\langle c_\beta c_\gamma\rangle = (\Delta_{23}, -\Delta_{13}, \Delta_{12})$. From mean-field theory we obtain that all ground states consistent with $\sum_\alpha |\Delta_\alpha|^2 = \Delta_0^2$ are degenerate. This five-dimensional ground-state manifold is consistent with the number of collective modes obtained via Goldstone's theorem.

The remarkable feature of this triplet s-wave state is that superfluid Cooper pairs coexist with a normal Fermi surface (see Fig. 5), i.e. the single-particle spectrum is only partially gapped. This has consequences for the collective modes which we have analyzed within a generalized RPA scheme [36]. They are partially visible in the dynamical structure factor $S(q, \omega)$, which is accessible via Bragg scattering [25]. An example of the calculated density response $\mathrm{Im}\chi^\rho(q, \omega)$ is shown in Fig. 6.

From BCS mean-field theory in two dimensions we find a transition temperature of $T_c = 0.17t$ for typical parameters $n = 3/8$ and $U = -4t$. This amounts to roughly $0.05T_F$ and is within reach of present cooling techniques. Multi-component Fermi systems like Li^6 can thus provide exotic new many-body physics and may even allow quantum simulations of simplified QCD models where only the color degree of freedom is taken into account.

4 Disorder and Interaction

So far in this review we have focused on pure, translationally invariant quantum lattice models. It is indeed one of the main advantages of optical lattices

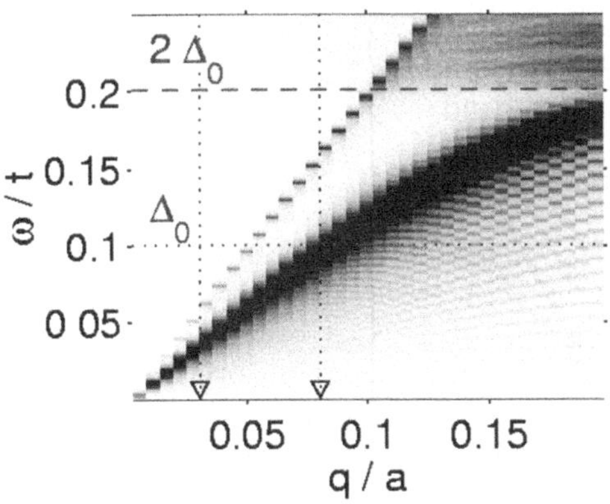

Fig. 6. Density response spectrum $\chi^{\rho}(q,\omega)$ (equivalent to $S(q,\omega)$) of the 2d fermionic SU(3) Hubbard model at $T = 0.01t$, $U = -4t$ and filling $n \approx 0.55$. The Anderson-Bogoliubov mode, the signature of superfluidity, is clearly visible (thick black line) as well as an additional flavor mode indicating the 3-flavor degeneracy. Figure taken from [36]

that perfectly disorder-free systems can be realized. On the other hand, effects of impurities and defects are of central importance in solids, where they often compete with the electron-electron interaction [41, 42]. It is therefore of great interest to realize in a controlled way disordered cold atom systems where localization effects can be studied.

Experimentally, disordered potentials can be created either with speckle lasers [43] or via quasiperiodic optical lattices [44]. Due to the AC stark effect the atoms experience a spatially fluctuating random potential which is stationary in time. Recently, localization effects have been observed in a BEC subject to a speckle laser field [45].

Here we focus on *fermionic* atoms with two spin states in a three-dimensional optical lattice with an additional random potential. A complete presentation of the results discussed here can be found in [46]. The system is modeled by the Anderson-Hubbard Hamiltonian

$$H_{\mathrm{AH}} = -t \sum_{\langle ij\rangle\sigma} c^{\dagger}_{i\sigma} c_{j\sigma} + \sum_{i\sigma} \epsilon_i n_{i\sigma} + U \sum_{i} n_{i\uparrow} n_{i\downarrow} - \mu \sum_{i\sigma} n_{i\sigma}, \tag{5}$$

where ϵ_i is a random onsite potential which we assume to be uniformly distributed in the interval $[-\Delta/2, \Delta/2]$. The parameter Δ is a measure of the disorder strength. We focus on the case of half filling $n = 1$ where on average there is one particle per site. The Hamiltonian (5) describes both the interaction-induced Mott transition into a correlated insulator [47, 48] as well as the Anderson localization transition due to coherent backscattering from random impurities [49].

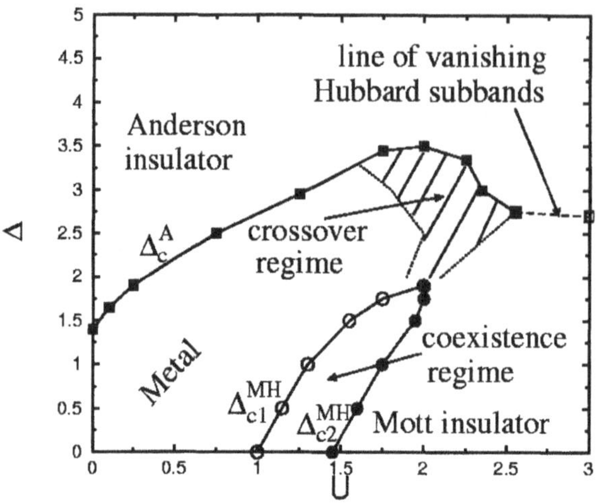

Fig. 7. DMFT ground state phase diagram of the disordered Hubbard model in the nonmagnetic phase at half filling. Figure taken from [46]

Analyzing the model (5) is a challenging problem. Here we present results obtained within the *Dynamical Mean-Field Theory* (DMFT), a non-perturbative technique where local quantum fluctuations are treated exactly [50, 51]. The DMFT has been applied with great success in $d = 3$ spatial dimensions to explain magnetic ordering phenomena and the Mott transition. In the calculation presented here [46] we use a recently developed variant, the *stochastic* DMFT, which is able to describe Anderson localization as well [52, 53].

Within DMFT, the correlated lattice model is mapped onto a self-consistent Anderson impurity Hamiltonian

$$\begin{aligned} H_{\mathrm{SIAM}} &= \sum_{\sigma}(\epsilon - \mu)c_\sigma^\dagger c_\sigma + U n_\uparrow n_\downarrow \\ &+ \sum_{\mathbf{k}\sigma}\left(V_{\mathbf{k}} c_\sigma^\dagger a_{\mathbf{k}\sigma} + V_{\mathbf{k}}^* a_{\mathbf{k}\sigma}^\dagger c_\sigma\right) + \sum_{\mathbf{k}\sigma} \epsilon_{\mathbf{k}} a_{\mathbf{k}\sigma}^\dagger a_{\mathbf{k}\sigma} \end{aligned} \tag{6}$$

where a single correlated lattice site now constitutes the impurity with a random onsite energy ϵ, and the fermions $a_{k\sigma}$ represent a fictitious conduction band with parameters V_k and ϵ_k which have to be determined self-consistently. The chemical potential $\mu = -U/2$ ensures half filling. This effective single-impurity model is solved using Wilson's numerical renormalization group [54, 55, 56, 57]. Within the stochastic DMFT [53] the self-consistency loop involves a geometric disorder average of the local density of states $\rho_{\mathrm{geom}}(\omega) = \exp\left[\langle \ln \rho_i(\omega)\rangle\right]$ which then determines the hybridization function $\eta(\omega) = \sum_{\mathbf{k}} |V_{\mathbf{k}}|^2/\left(\omega - \epsilon_{\mathbf{k}}\right)$ for the next iteration. For more details see [46].

The resulting zero temperature phase diagram as a function of disorder Δ and interaction U is shown in Fig. 7. For weak interaction and disorder the atoms are in a Fermi liquid state ("metal"). There are two different metal-insulator transitions: a Mott-Hubbard transition for increasing interaction U, and an Anderson localization transition as a function of Δ. Our results indicate that the two insulating phases are adiabatically connected. Note, however, that in our DMFT calculation we have so far considered only the paramagnetic phase. For non-frustrated lattices (e.g. simple cubic) it is known that an antiferromagnetic instability occurs in the pure Mott state. We are currently analyzing how far this antiferromagnetic phase extends into the disordered Mott-Anderson insulator [58]. Let us briefly comment on the detection of these different phases. Itinerant versus insulating behavior can be identified by a time-flight measurement as in [16]. In the Fermi liquid state, delocalization of fermions across the lattice leads to an interference pattern which vanishes once the atoms become localized. In order to distinguish the antiferromagnetic Mott insulator from the paramagnetic Anderson insulator one could apply spin-resolved Bragg scattering.

Optical lattices are a promising tool to simulate the above phase diagram experimentally since, in contrast to solids, both parameters U and Δ can be tuned arbitrarily. In particular, measurements can be done both in two and three spatial dimensions, possibly detecting qualitatively new physics in $d = 2$ where DMFT is no longer expected to be a good approximation.

5 Summary and Outlook

In this review we have presented some theoretical aspects of strongly correlated atoms in optical lattices. We have shown that these systems can be used to create analogues of well established solid-state quantum phases, but with much higher tunability of the model parameters. More generally, ultracold atoms can be used to perform quantum simulations of solid-state Hamiltonians like the 2d Hubbard model which may be relevant for high-temperature superconductivity. As another example for such a simulation we have discussed interacting fermions with disorder. Within a DMFT calculation we observe remarkable re-entrance into the itinerant phase due to competing Mott- and Anderson-localization. Finally, we have demonstrated that it is possible to use the highly degenerate internal states of cold atoms to create new exotic quantum states which have no analogue in condensed matter physics. Bosons with multiple spin states can be used to create tunable spin Hamiltonians. Most prominently, we have discussed a new fermionic SU(3) triplet superfluid state which is a toy model for QCD at weak to intermediate interactions. Experimental realization of these quantum phases is within reach and could significantly increase our understanding of the many-body model systems involved.

Acknowledgements

The author would like to thank E. Altman, K. Byczuk, I. Cirac, E. Demler, C. Honerkamp, M.D. Lukin, D. Vollhardt, and P. Zoller for collaborations, and I. Bloch, M. Greiner, M. Zwierlein and W. Ketterle for discussions.

References

1. For a review see special issue of Nature, Vol. **416**, 206 (2002).
2. B. DeMarco and D. S. Jin, Science **285**, 1703 (1999).
3. A. G. Truscott *et al.*, Science **291**, 2570 (2001).
4. F. Schreck *et al.*, Phys. Rev. Lett. **87**, 80403 (2001).
5. Z. Hadzibabic, Phys. Rev. Lett. **88**, 160401 (2002).
6. S. Inouye *et al.*, Nature **392**, 151 (1998).
7. E. Timmermans *et al.*, Phys. Lett. A **285**, 228 (2001).
8. M. Holland *et al.*, Phys. Rev. Lett. **87**, 120406 (2001).
9. Y. Ohashi, A. Griffin, Phys. Rev. A **67**, 033603 (2003).
10. C. A. Regal, M. Greiner, D. S. Jin, Phys. Rev. Lett. **92**, 040403 (2004).
11. M. W. Zwierlein *et al.*, Phys. Rev. Lett. **92**, 120403 (2004).
12. M. Bartenstein *et al.*, Phys. Rev. Lett. **92**, 120401 (2004).
13. C. Orzel *et al.*, Science **291**, 2386 (2001).
14. M. Greiner *et al.*, Nature (London) **415**, 39 (2002).
15. D. Jaksch *et al.*, Phys. Rev. Lett. **81**, 3108 (1998).
16. M. Köhl *et al.*, Phys. Rev. Lett **94** 080403 (2005).
17. W. Hofstetter *et al.*, Phys. Rev. Lett. **89**, 220407 (2002).
18. M. A. Cazalilla, J. Phys. B: At. Mol. Opt. Phys. **37**, 1 (2004).
19. B. Paredes *et al.*, Nature **429**, 277 (2004).
20. Th. Stöferle *et al.*, Phys. Rev. Lett. **92**, 130403 (2004).
21. C. Kollath *et al.*, PRA **69**, R031601 (2004).
22. C. Kollath *et al.*, preprint cond-mat/0411403.
23. J. Hubbard, Proc. Roy. Soc. London A **276**, 238 (1963).
24. D. J. Scalapino, Physics Reports **250**, 329 (1995).
25. D. M. Stamper-Kurn, W. Ketterle, Les Houches Lecture Notes 1999 and cond-mat/0005001.
26. R. Feynman, Int. J. Theor. Phys. **21**, 467 (1982).
27. D. M. Stamper-Kurn *et al.*, Phys. Rev. Lett. **80**, 2027 (1998).
28. O. Mandel *et al.*, Phys. Rev. Lett. **91**, 010407 (2003).
29. E. Altman, W. Hofstetter, E. Demler, M. D. Lukin, New Journal of Physics **5**, 113.1 (2003).
30. L.-M. Duan, E. Demler, and M. D. Lukin, Phys. Rev. Lett. **91**, 090402 (2003).
31. K. Sheshadri *et al.*, Europhys. Lett. **22**, 257 (1993).
32. E. Altman, E. Demler, M. D. Lukin, Phys. Rev. A **70**, 013603 (2004).
33. C. A. Regal, D. S. Jin, Phys. Rev. Lett. **90**, 230404 (2003).
34. E. R. I. Abraham *et al.*, Phys. Rev. A **55**, R3299 (1997).
35. C. Honerkamp, W. Hofstetter, Phys. Rev. Lett. **92**, 170403 (2004).
36. C. Honerkamp, W. Hofstetter, Phys. Rev. B **70**, 094521 (2004).
37. J. B. Marston, I. Affleck, Phys. Rev. B **39**, 11538 (1989).
38. C. Honerkamp, M. Salmhofer, Phys. Rev. Lett. **87**, 187004 (2001).

39. M. Zwierlein *et al.*, Phys. Rev. Lett. **94** 180401 (2005).
40. A. Modawi, A. J. Leggett, Journal of Low Temp. Phys. **109**, 625 (1997).
41. P. A. Lee, T. V. Ramakrishnan, Rev. Mod. Phys. **57**, 287 (1985).
42. D. Belitz, T. R. Kirkpatrick, Rev. Mod. Phys. **66**, 261 (1994).
43. P. Horak, J. Y. Courtois, G. Grynberg, Phys. Rev. A **58**, 3953 (1998).
44. L. Guidoni *et al.*, Phys. Rev. Lett. **79**, 3363 (1997).
45. J. E. Lye *et al.*, preprint cond-mat/0412167.
46. K. Byczuk, W. Hofstetter, D. Vollhardt, Phys. Rev. Lett. **94**, 056404 (2005).
47. N. F. Mott, Proc. Phys. Soc. A **62**, 416 (1949).
48. K. Byczuk, W. Hofstetter, D. Vollhardt, Phys. Rev. B **69**, 045112 (2004).
49. P. W. Anderson, Phys. Rev. **109**, 1492 (1958).
50. W. Metzner, D. Vollhardt, Phys. Rev. Lett. **62**, 324 (1989).
51. A. Georges *et al.*, Rev. Mod. Phys. **68**, 13 (1996).
52. V. Dobrosavljević, G. Kotliar, Phys. Rev. Lett. **78**, 3943 (1997).
53. V. Dobrosavljević, A. A. Pastor, B. K. Nikolić, Europhys. Lett. **62**, 76 (2003).
54. K. G. Wilson, Rev. Mod. Phys. **47**, 773 (1975).
55. T. A. Costi, A. C. Hewson, V. Zlatić, J. Phys.: Cond. Mat. **6**, 2519 (1994).
56. R. Bulla, A. C. Hewson, Th. Pruschke, J. Phys.: Condens. Matter **10**, 8365 (1998).
57. W. Hofstetter, Phys. Rev. Lett. **85**, 1508 (2000).
58. K. Byczuk, W. Hofstetter, D. Vollhardt, in preparation.

Quantum Noise in Mesoscopic Systems

Wolfgang Belzig

Department of Physics and Astronomy, University of Basel,
Klingelbergstr. 82, 4056 Basel, Switzerland
Wolfgang.Belzig@unibas.ch

Summary. Full counting statistics aims at a complete characterization of the distribution of measurement outcomes. In this article I will demonstrate how this concept allows to investigate quantum correlations in a variety of mesoscopic systems. Three examples will be discussed: a) In analogy to Schottky's work on the current fluctuations in a vacuum diode, shot noise in superconducting contacts allows to identify the nature of the elementary charge transfer events. b) The Coulomb interaction in complex quantum dots or molecules leads to a strongly correlated current statistics. c) The density fluctuation statistics in a fermionic quantum gas reflects the crossover from a superfluid state to a molecular Bose-Einstein condensate.

1 Introduction

The probabilistic interpretation is a fundamental ingredient of quantum mechanics. While the wave function determines the full quantum state a system and its evolution in time, observable quantities are related to hermitian operators. Expectation values of these operators determine the average value of a large number of identical measurements. However, different individual measurement yield in general a different result. This is a very general statement and below I will discuss several applications of this idea. For example, in the case of a current measurement in a quantum conductor, this leads directly to the concept of *full counting statistics* (FCS)[1]: during a given time interval a certain number of charges will pass the conductor and the goal is to predict the statistical properties of the number of transfered charges we need a probability distribution. Another systems, I will discuss later is the statistics of particle numbers in the ground state of a Fermi gas with attractive interaction. Such systems with a tunable strength of the interaction has recently been realized in ultra cold gases of trapped atoms and provide an interesting new application of the idea of counting statistics.

In this article I will will first review some concepts of basic statistics, which are relevant for counting statistics. In the rest I demonstrate the use of counting statistics in a number of examples, like two-terminal contacts with normal and superconducting leads or diffusive metals. Next, I will address, whether repulsive interaction, which occur e.g. in the electronic transport through a quantum dot affects the counting statistics. Finally, I address the

B. Kramer (Ed.): Adv. in Solid State Phys. **45**, 123–134 (2005)

particle number statistics in ultra cold quantum gases in the regime of the so-called BEC-BCS crossover, which can be induced by a magnetic field in the vicinity of a Feshbach resonance.

2 Probability Distributions and Cumulant Generating Functions

The fundamental quantity of interest is the probability distribution

$$P_{t_0}(N)\,, \tag{1}$$

which denotes the probability that during a certain period of time t_0 N charges pass through a conductor or some other event has happened N times. The same information is contained in the cumulant generating function (CGF), defined by

$$S(\chi) = \ln\left[\sum\nolimits_N e^{iN\chi}P(N)\right]\,, \tag{2}$$

where we introduced the counting field χ. The normalization condition requires $\sum_N P(N) = 1 \leftrightarrow S(\chi = 0) = 0$. Much of the language used below uses the terminology from electron transport. However, most results are also useful in different contexts, when one replaces e.g. the number of transfered charges with the outcome of any kind of measurement, like e.g. detecting the number of atoms in a part of an atomic cloud by absorbing photons.

2.1 Charge Quantization

We are interested in the long-time limit of the charge counting statistics, which means that no extra charges remain inside the conductor after the counting interval. We will see that the CGF's are in general *periodic* functions of χ, i. e. $S(\chi + 2\pi) = S(\chi)$. This ensures that the total charge transfered is an integer multiple of the electron charge e, which makes sense, since we are talking about electron transport and want to neglect transient effects. However, the interesting question, what the charge of an elementary event is, can be answered by FCS. Suppose the a CGF has the property $S(\chi+2\pi/n) = S(\chi)$. Direct calculation shows that

$$P(Q) = \int \frac{d\chi}{2\pi} e^{-iN\chi+S(\chi)} = \begin{cases} P_n(Q/n)\,, & (Q \bmod n) = 0 \\ 0 & , (Q \bmod n) \neq 0 \end{cases}\,, \tag{3}$$

where $P_n(N)$ is the distribution $S_n(\chi) = S(\chi/n)$. Thus, the probability distribution vanishes for all N which are not multiples of n, and one concludes that the elementary charge transfer is in units of ne, where e is the electron charge. This has interesting consequences in the context of superconductivity, in which multiple charge transfers can occur [2, 3, 4]. The question, whether fractional charge transfers can be detected through FCS is a difficult one (see e.g. [5, 6]).

2.2 Special Distributions

If the elementary events are uncorrelated, the probability distribution is *Poissonian.* With the average number of events is $\bar{N}$ we have

$$P_{\text{Poisson}} = \frac{\bar{N}^N}{(N!)} \exp(-\bar{N}) \leftrightarrow S(\chi) = \bar{N}\left(e^{i\chi} - 1\right) . \tag{4}$$

In the context of electron transport we encounter this distribution mostly for *tunnel junctions* with an almost negligible transmission probability at low temperatures. Here $\bar{N} = G_T V t_0/e$ is simply related to the voltage bias and the tunnel conductance.

As second example we consider the binomial (or Bernoulli) distribution. This is obtained if an event occurs with a probability T and the number of tries is fixed to N_0:

$$P_{\text{binomal}} = \binom{N_0}{N} T^N (1-T)^{N_0 - N} \leftrightarrow S(\chi) = N_0 \ln\left[1 + T\left(e^{i\chi} - 1\right)\right] . \tag{5}$$

In some sense this is the most fundamental distribution in quantum transport: it gives the statistics of a voltage biased single channel quantum conductor if we identify $N_0 = eVt_0/h$.

3 Statistic of Transport in Normal and Superconducting Metal Contacts

3.1 Tunnel Contacts

To illustrate the theoretical method we first calculate the counting statistics of a tunnel junction. As usual the system is described by a tunnel Hamiltonian $H = H_1 + H_2 + H_T$, where $H_{1(2)}$ describe the left(right) terminal and H_T describes the tunneling. The current is calculated in second order in the tunneling amplitudes and we obtain [7] $I(\chi) = (G_T/8e)\int dE \mathrm{Tr}\left(\check{\tau}_K\left[\check{G}_1(\chi), \check{G}_2\right]\right)$, where G_T is the conductance of the tunnel junction. The CGF is

$$S(\chi) = i\frac{t_0}{e}\int_0^\chi d\chi' I(\chi') = \frac{G_T t_0}{4e^2}\int dE \mathrm{Tr}\left\{\check{G}_1(\chi), \check{G}_2\right\} , \tag{6}$$

which is the general expression for the FCS of a tunnel junction. We use the pseudo-unitarity $\check{\tau}_K^2 = \check{1}$ to write

$$S(\chi) = N_{12}(e^{i\chi} - 1) + N_{21}(e^{-i\chi} - 1) , \tag{7}$$

where $N_{ij} = (t_0 G_T/16e^2)\int dE \mathrm{Tr}\left[(1+\check{\tau}_K)\check{G}_i(1-\check{\tau}_K)\check{G}_j\right]$ denotes the average number of charges tunnel from i to j. The statistics is therefore a bidirectional Poisson distribution [5]. It is easy to see that the cumulants are $C_n = N_{12} + (-1)^n N_{21}$. If either $N_{21} = 0$ or $N_{12} = 0$ we obtain the Schottky limit. Furthermore, in equilibrium $N_{12} = N_{21}$ and, remarkably, the FCS is non Gaussian: $(2G_T k_B T t_0/e^2)(\cos(\chi) - 1)$.

3.2 General CGF for Quantum Contacts

Using the Keldysh-Green function method[7], we can find the counting statistics for all conductors, which are characterized by a set of transmission coefficients $\{T_n\}$. Nazarov has shown that the transport properties of such a contact are described by a *matrix current* [8]

$$\check{I}_{12} = -\frac{e^2}{\pi}\sum_n \frac{2T_n\left[\check{G}_1,\check{G}_2\right]}{4+T_n\left(\{\check{G}_1,\check{G}_2\}-2\right)}. \tag{8}$$

Here $\check{G}_{1(2)}$ denote the matrix Green functions on the left and the right of the contact. We should emphasize that the matrix form of (8) is crucial to obtain the FCS, since it is valid for any matrix structure of the Green functions. The *scalar current* is obtained from the matrix current by

$$I_{12} = \frac{1}{4e}\int dE \mathrm{Tr}\check{\tau}_K \check{I}_{12}\,. \tag{9}$$

By integration we find [9]

$$S(\chi) = \frac{t_0}{2\pi}\sum_n \int dE \mathrm{Tr}\ln\left[1+\frac{T_n}{4}\left(\{\check{G}_1(\chi),\check{G}_2\}-2\right)\right]. \tag{10}$$

This is a very important result. It shows that the counting statistics of a large class of constrictions can be cast in a common form, independent of the contact types. Note, that Eq. (10) is just the sum over CGF of all eigenchannels. Thus, we can obtain the CGF of all constrictions from a known transmission eigenvalue density. These are known for a number of generic contacts (see e.g. [10] and Table 1), can be determined numerically, or can be taken from experiment. Below we will discuss several illustrative examples for a single channel contacts.

Normal Contacts

Consider first a single channel with transmission T between two normal reservoirs. They are characterized by occupation factors $f_{1(2)} = [\exp((E - \mu_{1(2)})/k_B T_e) + 1]^{-1}$ (T_e is the temperature). We obtain the result [11, 12] (see Appendix)

$$S(\chi) = \frac{2t_0}{h}\int dE \ln\left[1+T_{12}(E)\left(e^{i\chi}-1\right)+T_{21}(E)\left(e^{-i\chi}-1\right)\right]. \tag{11}$$

Here we introduced the probabilities $T_{12} = Tf_1(E)\left(1-f_2(E)\right)$ for a tunneling event from 1 to 2 and $T_{21}(E)$ for the reverse process. We see that the FCS (for each energy) is a trinomial of an electron going from left to right, from right to left, or no scattering at all. The *counting factors* $e^{\pm i\chi}-1$ thus correspond to single charge transfers from 1 to 2 (2 to 1).

Table 1. Characteristic functions of some generic conductors. The transmission eigenvalue densities $\rho(T)$ are normalized to G/G_Q, where $G_Q = 2e^2/h$ is the quantum conductance. The third column displays the CGF-density, which determines the CGF via $S(\chi) = (t_0 G/4eh) \int dE \, \mathrm{tr}[\check{s}(\{\check{G}_1(\chi), \check{G}_2\}/2)]$

	$\rho(T)[G/G_Q]$	$\check{s}(\Lambda)$
single channel	$\delta(T - T_1)$	$\ln(1 - T_1(\Lambda - 1)/2)$
diffusive connector	$\frac{1}{2}\frac{1}{T\sqrt{1-T}}$	$\frac{1}{4}\mathrm{acosh}^2(\Lambda)$
dirty interface	$\frac{1}{\pi}\frac{1}{T^{3/2}\sqrt{1-T}}$	$\sqrt{2(1+\Lambda)}$
chaotic cavity	$\frac{2}{\pi}\frac{1}{\sqrt{T}\sqrt{1-T}}$	$4\ln\left(2+\sqrt{2(1+\Lambda)}\right)$

At zero temperature and $\mu_1 - \mu_2 = eV \geq 0$ the argument of the energy integral is constant in the interval $\mu_1 < E < \mu_2$ and we obtain the *binomial form* $S(\chi) = (2et_0|V|/h) \ln\left[1 + T\left(e^{i\chi} - 1\right)\right]$. Note that for reverse bias $\mu_2 > \mu_1$ the CGF has the same form, but with a counting factor $e^{-i\chi} - 1$. The prefactor denotes the *number of attempts* $M = et_0 V/h$ to transfer an electron[1]. If the transmission probability is unity the FCS is non-zero only for $N = M$, which therefore constitutes the maximal number of electrons occupying an energy strip eV that can be sent through one (spin-degenerate) channel in a time interval t_0. In equilibrium it follows from Eq. (11) that the counting statistics is [13]

$$S(\chi) = -\frac{2t_0 k_B T_{el}}{h} \mathrm{asin}^2\left(\sqrt{T}\sin\frac{\chi}{2}\right) . \tag{12}$$

When $S(\chi) = -(t_0 k_B T_{el}/h)\chi^2$, the fluctuations are non-Gaussian, except for $T = 1$,

SN-contact

The FCS of a contact between a superconductor and a normal metal also follows from the general expression Eq. (10). Using the Green functions given

[1] The noninteger values of $M(t_0)$ occur due to the quasi classical approximation [12]. A more careful treatment reveals that M itself is described by a probability distribution. For large M the difference is negligible.

in the Appendix we find the result [2]

$$S(\chi) = \frac{t_0}{2\pi} \sum_n \int dE \ln \left[1 + \sum_{q=-2}^{2} A_{nq}(E) \left(e^{iq\chi} - 1 \right) \right] . \tag{13}$$

The coefficients $A_{nq}(E)$ are related to a charge transfer of $q \times e$. For example, a term $\exp(2i\chi) - 1$ corresponds just to an Andreev reflection process, in which two charges are transfered simultaneously. Explicit expressions for the various coefficients are given in Refs. [2, 7].

For low temperatures $k_B T_e \ll eV \ll \Delta$, the CGF becomes

$$S(\chi) = \frac{2et_0|V|}{h} \ln \left[1 + R_A \left(e^{i2\chi} - 1 \right) \right] , \tag{14}$$

where $R_A = T^2/(2-T)^2$ is just the Andreev reflection probability. The CGF is now π-periodic, which according to Sec. 2 reflects that the charge transfer of an elementary event is now $2e$, a consequence of Andreev reflection. Quite remarkably, the statistics is again a simple binomial distribution. In equilibrium, we can adapt the result from Eq. 12 to find

$$S(\chi) = -\frac{2t_0 k_B T_{el}}{h} \text{asin}^2 \left(\sqrt{R_A} \sin \chi \right) \quad (\text{for} \chi \in [-\pi/2, \pi/2]) . \tag{15}$$

The counting statistics is also non-Gaussian, except for $R_A = 1$.

Superconducting Contacts

Now we turn to a slightly more involved problem: a contact between two superconductors biased at a finite voltage V. For $eV < 2\Delta$ the transport is dominated by multiple Andreev reflections (MAR). The microscopic analysis of the average current and the shot noise calculations suggest that the current at sub gap energies proceeds in "giant" shots, with an effective charge $q \sim e(1 + 2\Delta/|eV|)$. However, the question of size of the charge transfered in an elementary event can only be rigorously resolved by the FCS. The answer was given by Cuevas and the author [3] based on a microscopic Green function approach. Independently, Johansson, Samuelsson and Ingerman [4] arrived at the same conclusion using a different method.

We quote here only the final result, which can be straightforwardly obtained using the Keldysh-Green function method. The statistics turns out to be a *multinomial* distribution of *multiple* charge transfers:

$$S(\chi) = \frac{t_0}{h} \int_0^{eV} dE \ln \left[1 + \sum_{n=-\infty}^{\infty} P_n(E,V) \left(e^{in\chi} - 1 \right) \right] . \tag{16}$$

General expressions for the probabilities $P(E, V)$ have been derived in Ref. [3].

It is also possible to obtain simplified expressions in a toy model. Let us neglect Andreev reflections above the gap and replace the quasiparticle density of states by a constant for $|E| > \Delta$. For sub harmonic voltages $2\Delta/(n-1)$ we find

$$S(\chi) = \frac{2\Delta t_0}{(n-1)h} \ln\left[1 + P_n\left(e^{in\chi} - 1\right)\right] , \tag{17}$$

where the probabilities are given by

$$\begin{aligned} P_2 &= \frac{T^2}{(2-T)^2} , \\ P_3 &= \frac{T^3}{(4-3T)^2} , \\ P_4 &= \frac{T^4}{(8-8T+T^2)^2} , \\ P_5 &= \frac{T^5}{(16-20T+5T^2)^2} , \\ P_6 &= \frac{T^6}{(2-T)^2(16-16T+T^2)^2} , \\ P_7 &= \frac{T^7}{(64-112T+56T^2-7T^3)^2} . \end{aligned} \tag{18}$$

Note the limiting cases of these probabilities $P_n \sim T^n/4^{n-1}$ for $T \ll 1$ and $P_n = 1$ for $T \to 1$. We conclude this section by saying that the general results for the CGF [3] allow for a fast and efficient calculation of all dc-transport properties of contacts between superconductors (which may contain magnetic impurities, phonon broadening or other imperfections).

4 Strongly Correlated Statistics

We address the question, whether the Coulomb interaction leads to essential modifications of the counting statistics in the transport through quantum dot. The simplest system, in which this happens, is a two-terminal quantum dot with more than one level. This situation has first been studied in [14]. We assume that the applied bias voltage and the level configuration is such that one level with energy ϵ_- is located below chemical potentials of both leads and the level with energy ϵ_+ in between. The charging energy is so large that only one of the levels can be occupied at the same time.

In this limit the system is described by a Master equation of the form

$$\frac{d}{dt}\begin{pmatrix} P_- \\ P_+ \\ P_0 \end{pmatrix} = \begin{pmatrix} -x\Gamma_L & 0 & \Gamma_R \\ 0 & -\Gamma_L & \Gamma_R \\ x\Gamma_L & \Gamma_L & -2\Gamma_R \end{pmatrix} \begin{pmatrix} P_- \\ P_+ \\ P_0 \end{pmatrix} \equiv \hat{M}\boldsymbol{P} . \tag{19}$$

Here $P_\pm$ describes the occupation of the upper/lower level and P_0 is the probability that the island is empty. The rates are parameterized by the bare tunneling rates $\Gamma_{L/R}$ to tunnel out of the dot to the left or to tunnel onto the dot from the right. The (small) parameter x is determined by the thermal occupations in the left lead: tunneling from the lower state is suppressed by a factor $x = 1 - f(\epsilon_-)$, since the states in the left lead are mostly occupied.

Now we determine the counting statistics using the method of Bagrets and Nazarov [15]. We add the corresponding counting factors $\exp(i\chi)$ to the entries of the Master equation, which involve changes of the dot occupation and refer to tunneling into one of the terminals. This procedure leads to the rate matrix

$$\hat{M}(\chi) = \begin{pmatrix} -x\Gamma_L & 0 & \Gamma_R \\ 0 & -\Gamma_L & \Gamma_R \\ x\Gamma_L e^{i\chi} & \Gamma_L e^{i\chi} & -2\Gamma_R \end{pmatrix}, \tag{20}$$

where we have chosen to count the charges in the left terminal. The counting statistics is determined from the lowest eigenvalue $\lambda_0(\chi)$ of $\hat{M}(\chi)$ according to

$$S(\chi) = -\lambda_0(\chi) t_0 \,. \tag{21}$$

For the present purpose, it is sufficient to determine the lowest eigenvalue perturbatively in $x \ll 1$ and to assume $\Gamma_L \ll \Gamma_R$. The counting statistics is obtained as

$$S(\chi) = -2x\Gamma_L t_0 \frac{e^{i\chi} - 1}{2 - e^{i\chi}} \,. \tag{22}$$

This cumulant generating function reproduces, of course, correctly the first cumulant $C_1 = 2x\Gamma_L t_0$ and the noise $C_2 = 3C_1$, corresponding to a Fano factor $F = C_2/|C_1|$ of 3.[14] The first interesting observation is that the current is *twice* the estimate of the simple thermal tunneling rate $x\Gamma_L$. Furthermore, the noise corresponds to an *effective charge* $q_{\text{eff}} = C_2/C_1 = 3$, which could be interpreted as a Poissonian process of 3 charges. However, we obtain the for the higher cumulants

$$C_3 = 13C_1 \,,\; C_4 = 75C_1 \,,\; C_5 = 541C_1 \,,\; C_6 = 4683C_1 \,. \tag{23}$$

Notably, third (and higher) cumulants obey the relation $C_n > q_{\text{eff}}^{n-1} C_1$ and, consequently, the transport statistics can not be explained by a simple Poissonian process in which multiple charges of size q_{eff} are transfered. Note that this would follow from a counting statistics $-(x\Gamma_L/q_{\text{eff}})(\exp(iq_{\text{eff}}\chi) - 1)$.

A simple physical picture emerges, if we expand the cumulant generating function (22) in terms of Poissonian processes. As a result we find

$$S(\chi) = -x\Gamma_L t_0 \sum_{n=1}^{\infty} \frac{1}{2^n} \left(e^{in\chi} - 1\right) \,. \tag{24}$$

The counting statistics is therefore a sum of independent Poisson processes, in each of which a charge of ne is transfered as signaled by the factor $\exp(in\chi)-1$. Each process is weighted with a probability $(1/2)^n$. This result for the statistics suggests the following interpretation. The lower level is occupied most of the time, since it is well below the chemical potentials of the two leads. Due to Coulomb blockade the other level cannot be occupied at the same time (we assume here that the charging energy is larger than the bias voltage). At a finite temperature, there will be a small rate $\sim x\Gamma_L$ for the electron to hop out to the left lead. Next, the dot will be filled again with an electron from the right lead. However, as we are in the situation $eV_R > \epsilon_+$ both levels will be occupied in the next step with the *same* probabilities of 1/2. In case the lower state is occupied we are back to the initial state and one charge has been transfered in this cycle, which is exactly what is happening in a single level quantum dot in the thermally activated transport regime. However, if the electron has tunneled onto the upper level, it can quickly tunnel out to the left lead and the dot is empty again and another electron can tunnel from the right terminal. Each intermediate cycle occurs with a probability 1/2 (since two states are available) and transfers an additional charge. Therefore each full cycle has a probability $(1/2)^n$ and transfers ne charge.

5 Number Statistics in Fermionic Quantum Gases

We recall some properties of noninteracting Fermi and Bose gases at a given temperature T. Here, individual atoms are independent and we can obtain the statistics as

$$S(\chi) = \mp \sum_k \ln\left[1 \pm f_\pm(e^{i\chi} - 1)\right] , \tag{25}$$

where $f_\pm(k) = \left[\exp((\epsilon_k - \mu)/k_B T) \pm 1\right]^{-1}$ is the Fermi (Bose) occupation function. The chemical potential is determined by the average atom number, i. e. $C_1 = \bar{N} = \sum_k f_\pm$. Obviously, the number statistics of fermions and bosons differs drastically in the degenerate regime. In the nondegenerate regime $f_\pm$ is small and we find the Poissonian statistics of classical particles $S(\chi) = -\bar{N}(\exp(i\chi) - 1)$, both for fermions and bosons. For degenerate fermions $k_B T \ll \epsilon_F$ the statistics is $S(\chi) = -i\tilde{\chi}\bar{N} + \pi^2(k_B T/\epsilon_F)\bar{N}\tilde{\chi}^2$. Here, $\tilde{\chi}/2\pi + 1/2 = [\chi/2\pi + 1/2]$, where [...] denotes the fractional part and ensures the 2π-periodicity of the CGF[16]. Thus, particle number fluctuations are suppressed by T/ϵ_F in comparison to the classical case. Remarkably, the statistics is Gaussian and consequently all cumulants C_n for $n \geq 3$ vanish. This behavior resembles fermions in a 1D wire [16] and can be interpreted as a consequence of anti-bunching. In contrast to that, we obtain quite a different behavior for free bosons. Approaching the degeneracy temperature

$T_C^{\rm BEC} = 2\pi\hbar^2 n^{2/3}/m\zeta(3/2)^{2/3}$ [17] from above, the fluctuations are enhanced due to the large factor $f_-(k)(f_-(k)+1)$. In the condensed regime, the occupation of the ground state becomes macroscopically large. At $T=0$ we obtain $S(\chi) = \ln\left[1-\bar{N}(e^{i\chi}-1)\right]$, which corresponds to a negative binomial distribution and the fluctuations therefore diverge according to $C_n \sim \bar{N}^n$. This can be interpreted as a bunching behavior as is expected for bosons, which however appears here in quite a different context.

We now turn to the FCS for the BCS theory. The BCS wave function is given by [18]

$$|{\rm BCS}\rangle = \prod_k \left(u_k + v_k c^\dagger_{k\uparrow} c^\dagger_{-k\downarrow}\right)|0\rangle\,. \tag{26}$$

The variational procedure yields $v_k^2 = 1 - u_k^2 = \left(1-(\epsilon_k-\mu)/E_k\right)/2$, where $E_k^2 = (\epsilon_k-\mu)^2 + \Delta^2$ is the energy of quasiparticle excitations. The order parameter Δ and the chemical potential are fixed by the self-consistency equations

$$\Delta = -\lambda \sum_k u_k v_k \,,\; \bar{N} = 2\sum_k v_k^2\,, \tag{27}$$

where λ is the BCS coupling constant. After renormalization of the coupling constant λ and considering only the low-energy limit, the gap equation can be related to the two particle scattering amplitude [19, 20, 21].

To calculate the statistics we note that the product form of the BCS wave function greatly simplifies the calculation, since different k states can be treated separately. For a single pair of states $(k\uparrow, -k\downarrow)$ the sum over all possible configurations can be easily performed

$$e^{-S_k(\chi)} = \sum_{n_{k\uparrow},n_{-k\downarrow}=\pm 1} \langle {\rm BCS}|e^{i(\hat{n}_{k\uparrow}+n_{-k\downarrow})\chi}|{\rm BCS}\rangle = u_k^2 + v_k^2 e^{i2\chi}\,. \tag{28}$$

The sum of all states yields the result

$$S(\chi) = -\sum_k \ln\left[1 + v_k^2(e^{2i\chi}-1)\right]\,. \tag{29}$$

We would like to stress that this result holds for a pairing state in any kind of trapping potential (e.g. box or harmonic) and for any dimension. In the following we will concentrate on a box potential in two and three dimensions. The dimension will only enter into the density of states, when transforming the sum over k into an energy integration via the standard expression $N_D = m^{D/2}(2\epsilon)^{D/2-1}/2\pi^{D-1}\hbar^D$ for D=2,3. It should be noted, that in a strictly two-dimensional system the low energy scattering amplitude vanishes $\sim -1/\log\epsilon$ and consequently the gap equation (27) shows a logarithmic divergence for $\epsilon \to 0$. However, for the more realistic situation of a three-dimensional trapped atomic cloud, strongly confined in one dimension, this singularity is eliminated. Results derived for the strictly two-dimensional situation are still valid for the quasi-two-dimensional case, however with the chemical potential μ shifted by the ground-state energy.

Here we only discuss some limiting cases. On the BEC side, $\mu < 0$ and $\Delta \ll |\mu|$ leads to $v_k^2 \ll 1$ for all energies and allows to expand the logarithm in Eq. (29). The result is

$$S(\chi) = -\frac{\bar{N}}{2}\left(e^{i2\chi} - 1\right), \tag{30}$$

which corresponds to a Poissonian number statistics of *pairs of atoms*. This supports the picture of strongly bound pairs, which act like independent particles of a classical gas. Note, that the factor of 2 in the exponent leads to exponentially growing cumulants, viz., $C_n = 2^n \bar{N}/2$. We therefore expect very strong fluctuations. On the BCS side the situation is quite different. Here $\mu = \epsilon_F \gg \Delta$ and we obtain

$$S(\chi) = -i\chi\bar{N} - \pi\bar{N}D\frac{\Delta}{4\epsilon_F}(\cos(\chi) - 1). \tag{31}$$

We observe that the first term is dominant but contributes only to the first cumulant. The fluctuations come from the second term in Eq. (31) which is smaller by a factor Δ/ϵ_F. Furthermore, similar to the degenerate Fermi gas, the odd cumulants C_n for $n \geq 3$ vanish, which is again a consequence of particle-hole symmetry.

6 Conclusion

Full counting statistics aims at a complete characterization of the distribution of measurement outcomes. I have demonstrated how this concept allows to investigate quantum correlations in a variety of mesoscopic systems. Three examples have been discussed: a) In analogy to Schottky's work on the current fluctuations in a vacuum diode, shot noise in superconducting contacts allows to identify the nature of the elementary charge transfer events. b) The Coulomb interaction in complex quantum dots or molecules leads to a strongly correlated current statistics. c) The density fluctuation statistics in a fermionic quantum gas reflects the crossover from a superfluid state to a molecular Bose-Einstein condensate.

Acknowledgements

The ideas presented here are the results of several collaborations. In particular I would like to mention C. Bruder, A. Cottet, J. C. Cuevas, Yu. V. Nazarov, and P. Samuelsson, and C. Schroll. This work was supported by the Swiss NSF, the NCCR Nanoscience, and the RTN Spintronics.

References

1. *Quantum Noise in Mesoscopic Physics*, edited by Yu. V. Nazarov (Kluwer, Dordrecht 2003).
2. B. A. Muzykantskii and D. E. Khmelnitzkii, Phys. Rev. B **50**, 3982 (1994).
3. J. C. Cuevas and W. Belzig, Phys. Rev. Lett. **91**, 187001 (2003).
4. G. Johansson, P. Samuelsson, and A. Ingerman, Phys. Rev. Lett. **91**, 187002 (2003).
5. L. S. Levitov and M. Reznikov, cond-mat/0111057 (unpublished).
6. H. Saleur and U. Weiss, Phys. Rev. B **63**, 201302 (2001).
7. W. Belzig, in [1]
8. Yu. V. Nazarov, Superlattices Microst. **25**, 1221 (1999).
9. W. Belzig and Yu. V. Nazarov, Phys. Rev. Lett. **87**, 197006 (2001).
10. C. W. J. Beenakker, Rev. Mod. Phys. **69**, 731 (1997).
11. L. S. Levitov and G. B. Lesovik, Pis'ma Zh. Eksp. Teor. Fiz. **58**, 225 (1993); [JETP Lett. **58**, 230 (1993)].
12. L. S. Levitov, H. W. Lee, and G. B. Lesovik, J. Math. Phys. **37**, 4845 (1996).
13. L. S. Levitov in [1]
14. A. Cottet, W. Belzig, and C. Bruder, Phys. Rev. B **70**, 115315 (2004).
15. D. A. Bagrets and Yu. V. Nazarov, Phys. Rev. B **67**, 085316 (2003).
16. L. S. Levitov, H. Lee, and G. B. Lesovik, J. Math. Phys. **37**, 4845 (1996).
17. The expression for T_C^{BEC} is calculated for a three-dimensional gas confined by rigid walls. ζ is the Riemann zeta function.
18. J. Bardeen, L. N. Cooper, and J. R. Schrieffer, Phys. Rev. **108**, 1175 (1957).
19. D. M. Eagles, Phys. Rev. **186**, 456 (1969).
20. A. J. Leggett, in *Modern Trends in the Theory of Condensed Matter*, edited by A. Pedalski and J. Przystawa (Springer, Berlin 1980).
21. M. Randeria, J.-M. Duan, and L. Y. Shieh, Phys. Rev. B **41**, 327 (1990).

Dynamical Variational Principles for Strongly Correlated Electron Systems

Michael Potthoff

Institut für Theoretische Physik und Astrophysik, Universität Würzburg, Am Hubland, 97074 Würzburg, Germany

Summary. The self-energy-functional approach (SFA) is discussed in the context of different variational principles for strongly correlated electron systems. Formal analogies between static and dynamical variational approaches, different types of approximation strategies and the relations to density-functional and dynamical mean-field theory are emphasized. The discussion elucidates the strengths of the SFA in the construction of new non-perturbative approximations but also the limitations of the approach and thereby opens up future perspectives.

Variational approaches have a long and successful tradition in the theory of condensed-matter systems as they offer a smart, controlled and systematic way to treat the problem of electron-electron interaction. A well-known variational approach is Hartree-Fock (HF) theory. It is based on the Rayleigh-Ritz principle and provides a practicable and consistent mean-field description of an interacting electron system. As quantum fluctuations are neglected completely, HF theory must be classified as a *static* mean-field approximation. This may be contrasted with dynamical mean-field theory (DMFT) [1, 2] which includes temporal fluctuations in the mean-field picture. The DMFT, however, cannot be derived from the Ritz principle. It must be constructed from some *dynamical* variational principle which involves a dynamical (i.e. time- or frequency-dependent) quantity as the basic object. Dynamical variational principles have already been suggested in the sixties [3, 4] but, compared to the Ritz principle, were employed with rather limited success only. This brings up the following questions: What are the similarities and the differences between different variational principles and approximation strategies? How can the DMFT be considered as an approximation within a variational concept? Can dynamical variational principles be used for constructing practicable and non-perturbative approximations different from the DMFT? An attempt to answer these questions straightforwardly leads to the self-energy-functional approach (SFA) [5] suggested recently. The purpose of this paper is to discuss different variational approaches and to place the SFA into this context with the objective to explore possible future developments.

B. Kramer (Ed.): Adv. in Solid State Phys. **45**, 135–147 (2005)

1 Variational Principles and Approximation Strategies

Consider a many-electron system in the volume V, at temperature T and with chemical potential μ. It is characterized by a Hamiltonian $H_{\boldsymbol{t},\boldsymbol{U}} = H_0(\boldsymbol{t}) + H_1(\boldsymbol{U})$ consisting of a one-particle and an interaction term H_0 and H_1, respectively, which depend on the "model parameters" $\boldsymbol{t}$ and $\boldsymbol{U}$ (a matrix notation is used). In second-quantized form,

$$H_{\boldsymbol{t},\boldsymbol{U}} = \sum_{\alpha\beta} t_{\alpha\beta}\, c^\dagger_\alpha c_\beta + \frac{1}{2} \sum_{\alpha\beta\gamma\delta} U_{\alpha\beta\delta\gamma}\, c^\dagger_\alpha c^\dagger_\beta c_\gamma c_\delta \,, \tag{1}$$

where an index (e.g. α) refers to the states of a one-particle basis.

The characteristic of a variational approach is a certain physical quantity $\boldsymbol{X}$ to be varied, as e.g. the statistical operator, the electron density, the (local) Green function, the self-energy etc. Clearly, at equilibrium this quantity will depend on the model parameters: $\boldsymbol{X}_{\boldsymbol{t},\boldsymbol{U}}$ (and on V, T and μ).

In a variational approach, the quantity is considered to be a variable. The first task is to express a thermodynamical potential Ω (the grand potential, for example) as a function(al) of this variable: $\Omega_{\boldsymbol{t},\boldsymbol{U}}[\boldsymbol{X}]$. As this functional is characteristic for the model system (1), it will depend on the model parameters. At the equilibrium or "physical" value, i.e. at $\boldsymbol{X} = \boldsymbol{X}_{\boldsymbol{t},\boldsymbol{U}}$, one must have $\Omega_{\boldsymbol{t},\boldsymbol{U}}[\boldsymbol{X}_{\boldsymbol{t},\boldsymbol{U}}] = \Omega_{\boldsymbol{t},\boldsymbol{U}}$ where $\Omega_{\boldsymbol{t},\boldsymbol{U}} = -T \ln \mathrm{tr} \exp(-(H_0(\boldsymbol{t}) + H_1(\boldsymbol{U}) - \mu N)/T)$.

Furthermore, the functional $\Omega_{\boldsymbol{t},\boldsymbol{U}}[\boldsymbol{X}]$ should be constructed such that it becomes stationary at the physical value: $\delta\Omega_{\boldsymbol{t},\boldsymbol{U}}[\boldsymbol{X} = \boldsymbol{X}_{\boldsymbol{t},\boldsymbol{U}}] = 0$. This *variational principle* determines $\boldsymbol{X}_{\boldsymbol{t},\boldsymbol{U}}$ once the functional is known. Note that the domain of the functional must be specified in addition since in most cases $\boldsymbol{X}_{\boldsymbol{t},\boldsymbol{U}}$ must satisfy some constraint or sum rule or normalization condition.

Even if the functional is known, however, it is usually impossible to evaluate $\Omega_{\boldsymbol{t},\boldsymbol{U}}[\boldsymbol{X}]$ for a given $\boldsymbol{X}$, and one has to resort to approximations. One may distinguish between three types of approximation strategies:

In a *type-I approximation* one derives the Euler equation $\delta\Omega_{\boldsymbol{t},\boldsymbol{U}}[\boldsymbol{X}]/\delta\boldsymbol{X} = 0$ first and then chooses (a physically motivated) simplification of the equation afterwards to render the determination of $\boldsymbol{X}_{\boldsymbol{t},\boldsymbol{U}}$ possible. This is the most general but worst type, as normally the approximated Euler equation no longer derives from some approximate functional. This may result in thermodynamical inconsistencies.

A *type-II approximation* modifies the form of the functional dependence, $\Omega_{\boldsymbol{t},\boldsymbol{U}}[\cdots] \to \widetilde{\Omega}_{\boldsymbol{t},\boldsymbol{U}}[\cdots]$, to get a simpler one that allows for a solution of the resulting Euler equation $\delta\widetilde{\Omega}_{\boldsymbol{t},\boldsymbol{U}}[\boldsymbol{X}]/\delta\boldsymbol{X} = 0$. This type is more particular but yields a thermodynamical potential consistent with $\boldsymbol{X}_{\boldsymbol{t},\boldsymbol{U}}$. Generally, however, it is not easy to find a sensible approximation of a functional form.

Finally, in a *type-III approximation* one restricts the domain of the functional. The restriction comes in addition to those conditions that are physically necessary anyway (e.g. normalizations) and requires a precise definition of the domain. This type is most specific and, from a conceptual point of view, should be preferred as compared to type-I or type-II approximations as the exact functional form is retained. In addition to conceptual clarity and thermodynamical consistency, type-III approximations are truely systematic since improvements can be obtained by an according extension of the domain. Note that any type-III approximation can always be understood as a type-II one (and type-II approximations as type-I) but not vice versa.

2 Various Variational Approaches

In the following different variational principles and possible approximations shall be discussed according to this scheme, starting with Hartree-Fock and density-functional theory as illuminating examples.

2.1 Ritz Variational Approach

In the Ritz variational approach the ground-state energy is considered as a functional of the quantum state $|\Psi\rangle$. There is a generalization of the Ritz principle to arbitrary temperatures by Mermin [6]. Here the basic variable is the statistical operator ρ characterizing the system's (mixed) state, and the grand potential as a functional of ρ reads:

$$\Omega_{\boldsymbol{t},\boldsymbol{U}}[\rho] = \mathrm{tr}(\rho(H_{\boldsymbol{t},\boldsymbol{U}} - \mu N + T \ln \rho)) \,. \tag{2}$$

Following the classical calculation of Gibbs, it can easily be shown [6] that on the set of normalized and positive definite operators, stationarity of the functional (2) is achieved for $\rho = \rho_{\boldsymbol{t},\boldsymbol{U}} = e^{-(H_{\boldsymbol{t},\boldsymbol{U}}-\mu N)/T}/\mathrm{tr}(e^{-(H_{\boldsymbol{t},\boldsymbol{U}}-\mu N)/T})$. One also has $\Omega_{\boldsymbol{t},\boldsymbol{U}}[\rho_{\boldsymbol{t},\boldsymbol{U}}] = \Omega_{\boldsymbol{t},\boldsymbol{U}}$. An additional feature of the functional (2) consists in the fact that $\Omega_{\boldsymbol{t},\boldsymbol{U}}[\rho] \geq \Omega_{\boldsymbol{t},\boldsymbol{U}}[\rho_{\boldsymbol{t},\boldsymbol{U}}]$ for any ρ. This "upper-bound property" is extremely helpful but specific to the Ritz principle.

For a many-electron system and an arbitrary ρ, the computation of the trace in Eq. (2) is an exponentially hard problem. A nice type-III approximation is the HF approach: Here the variational search is restricted to the subclass of *disentangled statistical operators*, i.e. statistical operators corresponding to independent-electron states. This can be made precise by introducing the important concept of a reference system:

A reference system is a system with a different (microscopic) Hamiltonian $H_{\boldsymbol{t}',\boldsymbol{U}'}$ ($\boldsymbol{t}' \neq \boldsymbol{t}$, $\boldsymbol{U}' \neq \boldsymbol{U}$) but with a macroscopic state characterized by the same values of the thermodynamic state variables as the original system (1): $V' = V$, $T' = T$ and $\mu' = \mu$. The sole purpose of the reference system is to specify the domain of the functional (2): Trial statistical operators are taken from the reference system, $\rho = \rho_{\boldsymbol{t}',\boldsymbol{U}'}$, and are varied by varying the parameters $\boldsymbol{t}'$ and $\boldsymbol{U}'$ within a certain subspace. Hence, the choice of the reference system (and the parameter subspace) defines the approximation.

The HF approximation is given by the choice $\boldsymbol{U}' = 0$ and $\boldsymbol{t}'$ arbitrary, i.e. by trial states $\rho_{\boldsymbol{t}',0} = e^{-(H_{\boldsymbol{t}',0}-\mu N)/T}/Z_{\boldsymbol{t}',0}$. Inserting into Eq. (2) yields

$$\Omega_{\boldsymbol{t},\boldsymbol{U}}[\rho_{\boldsymbol{t}',0}] = \Omega_{\boldsymbol{t}',0} + \mathrm{tr}(\rho_{\boldsymbol{t}',0}(H_0(\boldsymbol{t}) + H_1(\boldsymbol{U}) - H_0(\boldsymbol{t}'))\,. \tag{3}$$

The remaining trace can be computed easily using Wick's theorem as $\rho_{\boldsymbol{t}',0}$ derives from a non-interacting Hamiltonian. The variational parameters $\boldsymbol{t}'$ are fixed by the conditions $\delta\Omega_{\boldsymbol{t},\boldsymbol{U}}[\rho_{\boldsymbol{t}',0}]/\delta\boldsymbol{t}' = 0$. These are exactly the well-known HF equations as can be seen by some straightforward manipulations.

One learns that type-III approximations can be constructed conveniently by the *concept of a reference system.* On the one hand, the reference system should comprise a large space of parameters $\boldsymbol{t}'$ and $\boldsymbol{U}'$ to generate a powerful approximation. On the other hand, the parameter space must be restricted strongly to keep the calculations feasible.

2.2 Density-Functional Approach

For a many-electron system the statistical operator or, at $T = 0$, the ground-state wave function actually is an object that is by far too complex. The *relevant* information is much more efficiently stored in integral quantities, such as the electron density. This is the starting point of density-functional theory (DFT) [7, 8, 9]. The density, i.e. the quantum-statistical *average* of the one-particle density operator $n(\boldsymbol{r}) = \mathrm{tr}(\rho\hat{n}(\boldsymbol{r}))$, represents the basic variable. Normally DFT aims at the inhomogeneous electron gas at $T = 0$ but can also be applied to discrete lattice models [10] and finite temperatures [6].

The grand potential $\Omega_{\boldsymbol{t},\boldsymbol{U}}$ obviously depends on the model parameters. Due to the Hohenberg-Kohn theorem [7], however, it can also be considered as a functional of the density $\boldsymbol{n}$ which is stationary at the physical density: $\delta\Omega_{\boldsymbol{t},\boldsymbol{U}}[\boldsymbol{n}] = 0$ for $\boldsymbol{n} = \boldsymbol{n}_{\boldsymbol{t},\boldsymbol{U}}$. Furthermore, if evaluated at $\boldsymbol{n} = \boldsymbol{n}_{\boldsymbol{t},\boldsymbol{U}}$, it yields the exact grand potential: $\Omega_{\boldsymbol{t},\boldsymbol{U}}[\boldsymbol{n}_{\boldsymbol{t},\boldsymbol{U}}] = \Omega_{\boldsymbol{t},\boldsymbol{U}}$. Keeping the notations introduced above, $\boldsymbol{n}$ is a matrix with $n_{\alpha\beta} = \mathrm{tr}(\rho\, c_\alpha^\dagger c_\beta)$, and the functional reads (cf. Refs. [6, 10]):

$$\Omega_{\boldsymbol{t},\boldsymbol{U}}[\boldsymbol{n}] = \mathrm{tr}(\boldsymbol{t}\,\boldsymbol{n}) + F_{\boldsymbol{U}}[\boldsymbol{n}]\,. \tag{4}$$

Here the trace refers to the one-particle orbitals α, and $F_{\boldsymbol{U}}[\boldsymbol{n}]$ is a *universal* functional, i.e. it depends on the interaction parameters only. Using the Kohn-Sham idea [8, 10], the resulting Euler equation has the form of a one-particle Schrödinger equation.

The variational principle $\delta\Omega_{\boldsymbol{t},\boldsymbol{U}}[\boldsymbol{n}] = 0$ is rigorous but cannot be evaluated as $F_{\boldsymbol{U}}[\boldsymbol{n}]$ is generally unknown (after separating the Hartree and a kinetic-energy term, the remaining exchange-correlation functional is not explicit). Due to the universality of $F_{\boldsymbol{U}}[\boldsymbol{n}]$, however, the density-functional for a reference system with modified one-particle parameters $\boldsymbol{t}'$ reads $\Omega_{\boldsymbol{t}',\boldsymbol{U}}[\boldsymbol{n}] = \mathrm{tr}(\boldsymbol{t}'\,\boldsymbol{n}) + F_{\boldsymbol{U}}[\boldsymbol{n}]$, and thus $\Omega_{\boldsymbol{t},\boldsymbol{U}}[\boldsymbol{n}_{\boldsymbol{t}',\boldsymbol{U}}] = \Omega_{\boldsymbol{t}',\boldsymbol{U}} + \mathrm{tr}((\boldsymbol{t}-\boldsymbol{t}')\boldsymbol{n}_{\boldsymbol{t}',\boldsymbol{U}})$ which can be exploited for a type-III approximation. Choosing as a reference system $H_{\boldsymbol{t}',\boldsymbol{U}}$ the homogeneous electron gas, however, turns out to be too restrictive, as this implies a spatially constant density. The *local* density approximation (LDA) [7, 8], on the other hand, has proven to be very successful [9]. At least for systems with weakly varying density it is well justified. The LDA, however, is no longer a type-III approximation but a type-II one as the *form* of the (exchange-correlation part of the) functional $F_U[\boldsymbol{n}]$ is approximated to have a local dependence on the density only.

As the proof of the Hohenberg-Kohn theorem is based on the Ritz principle [7], the upper-bound property is transferred to the exact functional (4), i.e. $\Omega_{\boldsymbol{t},U}[\boldsymbol{n}] \geq \Omega_{\boldsymbol{t},U}$ for any $\boldsymbol{n}$, but is lost within the LDA due to the type-II character of the approximation.

2.3 Time-Dependent DFT

The weak point of the DFT consists in its inability to describe excitations (see, however, Ref. [11]). This is due to the fact that the Hohenberg-Kohn variational principle is built on the static electron density. Information on excitation properties is contained in dynamic response functions which are accessible in principle via time-dependent density-functional theory (TD-DFT) [12]. In TD-DFT one considers a situation with a time-dependent Hamiltonian and focuses on the time-dependent density $n(\boldsymbol{r}, t) = \langle\Psi(t)|\hat{n}(\boldsymbol{r})|\Psi(t)\rangle$ resulting from a solution $|\Psi(t)\rangle$ of the time-dependent Schrödinger equation as the basic variable. Here the action $A = \int dt \langle\Psi(t)|i\partial/\partial t - H(t)|\Psi(t)\rangle$ can be understood as a functional of $n(\boldsymbol{r}, t)$,

$$A_{\boldsymbol{t},\boldsymbol{U}}[\boldsymbol{n}] = -\int_{t_0}^{t_1} dt \, \mathrm{tr}(\boldsymbol{n}(t)\boldsymbol{t}(t)) + B_{\boldsymbol{U}}[\boldsymbol{n}] \, , \tag{5}$$

where $\boldsymbol{t}(t)$ are explicitly time-dependent one-particle parameters. Contrary to usual DFT, the variational principle $\delta A_{\boldsymbol{t},\boldsymbol{U}}[\boldsymbol{n}] = 0$ does not derive from the Ritz principle, and consequently there is no upper-bound property available. Type-II approximations can be constructed by approximating the universal but unknown part $B_{\boldsymbol{U}}[\boldsymbol{n}]$ of the functional to make it explicit. Far from equilibrium, however, there is no general recipe.

2.4 Dynamical Variational Principle

In the linear-response regime close to equilibrium, excitations are described by Green functions. The one-electron Green function $G_{\alpha\beta}(\omega) = \langle\langle c_\alpha; c_\beta^\dagger \rangle\rangle_\omega$ is the basic quantity in the dynamical variational approach of Luttinger, Ward, Baym and Kadanoff [3, 4]. Employing a coupling-constant integration [3], the grand potential can be understood as a functional of $\boldsymbol{G}$:

$$\Omega_{\boldsymbol{t},\boldsymbol{U}}[\boldsymbol{G}] = \mathrm{Tr} \ln \boldsymbol{G} - \mathrm{Tr}((\boldsymbol{G}_{\boldsymbol{t},0}^{-1} - \boldsymbol{G}^{-1})\boldsymbol{G}) + \Phi_{\boldsymbol{U}}[\boldsymbol{G}] \, , \tag{6}$$

where $\mathrm{Tr} = T\sum_n e^{i\omega_n 0^+} \mathrm{tr}$ and $\omega_n = (2n+1)\pi T$ are fermion Matsubara frequencies. Furthermore, $\boldsymbol{G}_{\boldsymbol{t},0} = (\omega + \mu - \boldsymbol{t})^{-1}$ is the $\boldsymbol{U} = 0$ (free) Green function and $\Phi_{\boldsymbol{U}}[\boldsymbol{G}]$ the (universal) Luttinger-Ward (LW) functional defined as the sum of all dressed closed skeleton diagrams [3]. By construction, $\Omega_{t,U}[G_{t,U}] = \Omega_{t,U}$. In arbitrary order in perturbation theory one has $\delta\Phi_U[G]/\delta G = T\Sigma_U[G]$. Therewith, the Euler equation $\delta\Omega_{t,U}[G]/\delta G = 0$ is given by $G^{-1} - G_{t,0}^{-1} + \Sigma_U[G] = 0$ which is Dyson equation. This shows that $\Omega_{t,U}[G]$ is stationary at the physical Green function $G = G_{t,U}$.

The LW functional is formally given by a diagrammatic sum that cannot be carried out in practice. A self-evident type-II strategy is to sum up a suitable subclass of diagrams to obtain an approximate but explicit expression for $\Phi_U[G]$. In this way the HF approximation can be recovered but there are also new approximations like the fluctuation-exchange approximation [13, 14]. These "conserving approximations", however, are necessarily restricted to the weak-coupling regime.

A type-III approximation, on the other hand, would be non-perturbative by construction. Consider a reference system with modified one-particle parameters: $H_{t',U} = H_0(t') + H_1(U)$. This defines the domain of the functional (6) to consist of Green functions $G_{t',U}$ with arbitrary t'. The interaction is kept fixed ($U' = U$). To evaluate the functional (6) at $G_{t',U}$ requires the evaluation of $\Phi_U[G_{t',U}]$, in particular. Due to the universality of $\Phi_U[\cdots]$ (no t dependence) and due to the choice $U' = U$, one has $\Phi_U[G_{t',U}] = \Omega_{t',U} - \mathrm{Tr}\ln G_{t',U} + \mathrm{Tr}((G_{t',0}^{-1} - G_{t',U}^{-1})G_{t',U})$. Thus,

$$\Omega_{t,U}[G_{t',U}] = \Omega_{t',U} - \mathrm{Tr}(G_{t,0}^{-1}G_{t',U}) + \mathrm{Tr}(G_{t',0}^{-1}G_{t',U}) \, . \tag{7}$$

Hence, on any domain specified by a suitable subspace of one-particle parameters t' which renders the solution of the reference system possible (for fixed U), the functional (6) can be evaluated exactly. A possible (but over-simplified) example is the choice $t' = 0$. It reduces the reference model to the atomic limit where the computation of $G_{t',U}$ and $\Omega_{t',U}$ in (7) is easy. Cluster approximations represent straightforward generalizations of this example.

Unfortunately, this type-III approach for Eq. (6) yields nothing new: Since $\mathrm{Tr}(G_{t,0}^{-1} - G_{t',0}^{-1})G_{t',U} = \mathrm{tr}(t - t')n_{t',U}$ with the one-electron density of the reference system $n_{t',U} = T\sum_n e^{i\omega_n 0^+} G_{t',U}(i\omega_n)$, one gets $\Omega_{t,U}[G_{t',U}] = \Omega_{t,U}[\rho_{t',U}]$ with $\Omega_{t,U}[\rho]$ given by Eq. (2), i.e. the same as in the Ritz variational approach. Interestingly, this implies that upper bounds for the grand potential can be obtained, i.e. $\Omega_{t,U}[G_{t',U}] \geq \Omega_{t,U}$ for arbitrary t'.

2.5 Dynamical Mean-Field Approach

Equipped with these insights, one can address the question of *deriving the DMFT from a variational principle.* Originally, the DMFT was introduced as the exact theory of lattice models with local (Hubbard-type) interactions in infinite spatial dimensions $D = \infty$ [2]. Later on, it was recognized [15, 16] that in $D = \infty$ the lattice model $H_{t,U}$ can be self-consistently mapped onto an impurity model $H_{t',U}$ with the same interaction U. Using this self-consistent mapping procedure as an approximation ("dynamical mean-field approximation"), one can treat lattice models for any finite D.

Instead of considering Dyson equation in the form, $\boldsymbol{G} = (\boldsymbol{G}_{t,0}^{-1} - \boldsymbol{\Sigma}_{\boldsymbol{U}}[\boldsymbol{G}])^{-1}$ (with $\boldsymbol{\Sigma}_{\boldsymbol{U}}[\boldsymbol{G}]) = (1/T)\,\delta\Phi_{\boldsymbol{U}}[\boldsymbol{G}])/\delta\boldsymbol{G}$), which is solved by the exact $\boldsymbol{G}_{t,U}$, the DMFT considers the following simplified equation between *local* quantities at lattice site i:

$$(\boldsymbol{G})_{ii} = (\boldsymbol{G}_{t,0}^{-1} - \widetilde{\boldsymbol{\Sigma}}_U[\boldsymbol{G}])_{ii}^{-1} . \tag{8}$$

Here $\widetilde{\boldsymbol{\Sigma}}_U[\boldsymbol{G}]$ is the derivative of the LW functional but with *local* vertices only as it is the case for an impurity model. Clearly, this is a type-I approximation. Eq. (8) is often called the DMFT self-consistency condition. This is because its solution is achieved by an iterative procedure in practice: Starting with a guess for $\boldsymbol{\Sigma}$, one computes the local lattice Green function as $(\boldsymbol{G})_{ii} = (\boldsymbol{G}_{t,0}^{-1} - \boldsymbol{\Sigma})_{ii}^{-1}$ at first. This is not yet a solution of Eq. (8) since in general $\boldsymbol{\Sigma} \neq \widetilde{\boldsymbol{\Sigma}}_U[\boldsymbol{G}]$ for this $\boldsymbol{G}$. For the necessary update of $\boldsymbol{\Sigma}$ define $(\boldsymbol{G}_{t',0})_{ii} = 1/(1/(\boldsymbol{G})_{ii} + (\boldsymbol{\Sigma})_{ii})$. Assuming that $\boldsymbol{G}_{t',0}$ can be understood as the free impurity Green function of an impurity model $H_{t',U}$ for some $\boldsymbol{t}'$, the (numerical) solution of the impurity problem yields a new $\boldsymbol{\Sigma} = \widetilde{\boldsymbol{\Sigma}}_U[(\boldsymbol{G})_{ii}] = \widetilde{\boldsymbol{\Sigma}}_U[\boldsymbol{G}]$. Iteration of this cycle until self-consistency yields a solution $\boldsymbol{G}$ of Eq. (8). Note that the resulting DMFT self-energy is local.

Here, the question is whether Eq. (8) can be interpreted as an Euler equation of some variational principle. Starting with the functional (6), one can try a type-II approximation by replacing $\Phi_{\boldsymbol{U}}[\boldsymbol{G}]$ with the LW functional of the impurity model $\widetilde{\Phi}_{\boldsymbol{U}}[\boldsymbol{G}]$. This implies $(1/T)\delta\widetilde{\Phi}_{\boldsymbol{U}}[\boldsymbol{G}]/\delta\boldsymbol{G} = \widetilde{\boldsymbol{\Sigma}}_U[\boldsymbol{G}]$, and the resulting Euler equation reads: $\boldsymbol{G}^{-1} = \boldsymbol{G}_{t,0}^{-1} - \widetilde{\boldsymbol{\Sigma}}_U[\boldsymbol{G}]$. This equation is easily seen to be equivalent with Eq. (8) since $\widetilde{\boldsymbol{\Sigma}}_U[(\boldsymbol{G})_{ii}] = \widetilde{\boldsymbol{\Sigma}}_U[\boldsymbol{G}]$ by definition. Hence, DMFT can be understood as a type-II approximation.

Another functional has been suggested recently [17]:

$$\Omega_{t,U}[\boldsymbol{G}] = \operatorname{Tr}\ln\frac{1}{\boldsymbol{G}_{t,0}^{-1} - \boldsymbol{\Sigma}_{\boldsymbol{U}}[\boldsymbol{G}]} - \operatorname{Tr}(\boldsymbol{\Sigma}_{\boldsymbol{U}}[\boldsymbol{G}]\boldsymbol{G}) + \Phi_{\boldsymbol{U}}[\boldsymbol{G}] . \tag{9}$$

Clearly, $\Omega_{t,U}[\boldsymbol{G}_{t,U}] = \Omega_{t,U}$, and furthermore the corresponding Euler equation, $\big((\boldsymbol{G}_{t,0}^{-1} - \boldsymbol{\Sigma}_{\boldsymbol{U}}[\boldsymbol{G}])^{-1} - \boldsymbol{G}\big)\cdot(\delta\boldsymbol{\Sigma}_{\boldsymbol{U}}[\boldsymbol{G}]/\delta\boldsymbol{G}) = 0$, is equivalent with Dyson equation, $\boldsymbol{G} = (\boldsymbol{G}_{t,0}^{-1} - \boldsymbol{\Sigma}_{\boldsymbol{U}}[\boldsymbol{G}])^{-1}$ (assuming local invertibility of the functional $\boldsymbol{\Sigma}_{\boldsymbol{U}}[\boldsymbol{G}]$). The functional (9) therefore yields a valid variational principle. As a type-II approximation, one may replace $\Phi_{\boldsymbol{U}}[\boldsymbol{G}] \to \widetilde{\Phi}_{\boldsymbol{U}}[\boldsymbol{G}]$ and $\boldsymbol{\Sigma}_{\boldsymbol{U}}[\boldsymbol{G}] \to \widetilde{\boldsymbol{\Sigma}}_U[\boldsymbol{G}] = (1/T)\delta\widetilde{\Phi}_{\boldsymbol{U}}[\boldsymbol{G}]/\delta\boldsymbol{G}$ in the functional (9). The resulting Euler equation is equivalent with the DMFT self-consistency equation (8) which implies that DMFT can also be understood as a type-II approximation to the functional (9).

Attempts to prove that a stationary point of the type-II approximated functionals (6) or (9) must be an extremum have failed [17]. Furthermore, while (as shown above) a type-III approximation to the principle based on Eq. (6) with $\boldsymbol{U}$ fixed yields upper bounds to the grand potential, the DMFT cannot be obtained as a type-III approximation starting from Eqs. (6) or (9):

Choosing an impurity model $H_{t',U}$ as a reference system to generate trial Green functions and to define a restricted domain of the functional (6) or (9), respectively, concurrently means that the optimal Green function will be local. This is obviously a very poor approximation for the Green function of a lattice model and differs from the DMFT result. The discussion shows that the question whether or not the DMFT grand potential is an upper bound to the true grand potential is still open.

2.6 Functionals of the Local Green Function

It is also possible [18, 19] to focus on the local Green function $\boldsymbol{G}^{(\mathrm{loc})} = (\boldsymbol{G})_{ii}$ (instead of the full $\boldsymbol{G}$) and to set up a variational principle of the form

$$\delta\Omega_{\boldsymbol{t},\boldsymbol{U}}[\boldsymbol{G}^{(\mathrm{loc})}] = 0\,. \tag{10}$$

A functional which is stationary at the physical $\boldsymbol{G}^{(\mathrm{loc})} = \boldsymbol{G}^{(\mathrm{loc})}_{\boldsymbol{t},\boldsymbol{U}}$ and which yields $\Omega_{\boldsymbol{t},\boldsymbol{U}}[\boldsymbol{G}^{(\mathrm{loc})}_{\boldsymbol{t},\boldsymbol{U}}] = \Omega_{\boldsymbol{t},\boldsymbol{U}}$ can be constructed order by order in the interaction strength [18]. Unfortunately, the diagrammatic formalism is much more cumbersome as compared to the construction of the LW functional. As is shown in Ref. [19], the dynamical mean-field approximation is equivalent with a simple ($\boldsymbol{U}$-independent) approximation to the kinetic-energy part of the functional. So the DMFT appears as a type-II approximation again.

2.7 Self-Energy-Functional Approach

The motivation to characterize the DMFT as a type-III approximation is the following: If it is possible to recover the DMFT merely by restricting the domain of the functional corresponding to an exact variational principle, different choices of the domain will place the DMFT in a systematic series of different and possibly new approximations which, as the DMFT, are all non-perturbative and thermodynamically consistent.

For this purpose it is helpful to focus on the self-energy. Within the self-energy-functional approach (SFA) [5], the self-energy functional

$$\Omega_{\boldsymbol{t},\boldsymbol{U}}[\boldsymbol{\Sigma}] = \mathrm{Tr}\ln\frac{1}{\boldsymbol{G}_{\boldsymbol{t},0}^{-1} - \boldsymbol{\Sigma}} + F_{\boldsymbol{U}}[\boldsymbol{\Sigma}] \tag{11}$$

is considered. Here, $F_{\boldsymbol{U}}[\boldsymbol{\Sigma}] = \Phi_{\boldsymbol{U}}[\boldsymbol{G}[\boldsymbol{\Sigma}]] - \mathrm{Tr}(\boldsymbol{\Sigma}\,\boldsymbol{G}_{\boldsymbol{U}}[\boldsymbol{\Sigma}])$ is the Legendre transform of the Luttinger-Ward functional which is well defined provided that the functional $\boldsymbol{\Sigma}_{\boldsymbol{U}}[\boldsymbol{G}]$ is invertible locally. $F_{\boldsymbol{U}}[\boldsymbol{\Sigma}]$ is universal (independent of $\boldsymbol{t}$) by construction and $-(1/T)\delta F_{\boldsymbol{U}}[\boldsymbol{\Sigma}]/\delta\boldsymbol{\Sigma} = \boldsymbol{G}_{\boldsymbol{U}}[\boldsymbol{\Sigma}]$ which is the inverse of the functional $\boldsymbol{\Sigma}_{\boldsymbol{U}}[\boldsymbol{G}]$. Obviously, $\Omega_{\boldsymbol{t},\boldsymbol{U}}[\boldsymbol{\Sigma}_{\boldsymbol{t},\boldsymbol{U}}] = \Omega_{\boldsymbol{t},\boldsymbol{U}}$. The Euler equation $\delta\Omega_{\boldsymbol{t},\boldsymbol{U}}[\boldsymbol{\Sigma}]/\delta\boldsymbol{\Sigma} = 0$ is given by $(\boldsymbol{G}_{\boldsymbol{t},0}^{-1} - \boldsymbol{\Sigma})^{-1} = \boldsymbol{G}[\boldsymbol{\Sigma}]$ and equivalent with Dyson equation.

To construct a type-III approximation, a reference system $H_{\boldsymbol{t}',\boldsymbol{U}} = H_0(\boldsymbol{t}') + H_1(\boldsymbol{U})$ with unchanged interaction part is considered. The one-particle parameters $\boldsymbol{t}'$ are taken such that the different "correlated" sites (non-zero on-site interaction) are decoupled. Instead, $\boldsymbol{t}'$ shall include an arbitrary hopping to "bath" sites (zero on-site interaction) with arbitrary one-particle energies. In case of the Hubbard model on a lattice with L sites, the corresponding reference system constructed in this way is a set of L decoupled single-impurity Anderson models (which in case of translational symmetry are equivalent). Trial self-energies $\boldsymbol{\Sigma}_{\boldsymbol{t}',\boldsymbol{U}}$ are local by construction. The Euler equation resulting from this type-III approach reads $\partial\Omega_{\boldsymbol{t},\boldsymbol{U}}[\boldsymbol{\Sigma}_{\boldsymbol{t}',\boldsymbol{U}}]/\partial\mathbf{t}' = 0$, i.e.:

$$\left((\boldsymbol{G}_{\boldsymbol{t},0}^{-1} - \boldsymbol{\Sigma}_{\boldsymbol{t}',\boldsymbol{U}})^{-1} - \boldsymbol{G}_{\boldsymbol{U}}[\boldsymbol{\Sigma}_{\boldsymbol{t}',\boldsymbol{U}}]\right) \cdot \frac{\partial \boldsymbol{\Sigma}_{\boldsymbol{t}',\boldsymbol{U}}}{\partial \mathbf{t}'} = 0 \,. \tag{12}$$

Now let $\boldsymbol{t}'$ (the bath parameters) be such that $\boldsymbol{G}_{\boldsymbol{t}',\boldsymbol{U}}$ solves the DMFT self-consistency condition (8). Since $\widetilde{\boldsymbol{\Sigma}}_{\boldsymbol{U}}[\boldsymbol{G}_{\boldsymbol{t}',\boldsymbol{U}}] = \boldsymbol{\Sigma}_{\boldsymbol{t}',\boldsymbol{U}}$, one has $(\boldsymbol{G}_{\boldsymbol{t}',\boldsymbol{U}})_{ii} = (\boldsymbol{G}_{\boldsymbol{t},0}^{-1} - \boldsymbol{\Sigma}_{\boldsymbol{t}',\boldsymbol{U}})_{ii}^{-1}$. Hence, this $\boldsymbol{t}'$ solves Eq. (12). (Note that $\partial\boldsymbol{\Sigma}_{\boldsymbol{t}',\boldsymbol{U}}/\partial\mathbf{t}'$ is local). So *by a restriction of the domain of the self-energy functional* (11) to local self-energies, *the DMFT is characterized as a type-III approximation.*

Interestingly, a type-II approximation does not work: A replacement of the form $F_{\boldsymbol{U}}[\boldsymbol{\Sigma}] \to \widetilde{F}_{\boldsymbol{U}}[\boldsymbol{\Sigma}]$ in Eq. (11) yields the Euler equation $\widetilde{\boldsymbol{G}}_{\boldsymbol{U}}[\boldsymbol{\Sigma}] = (\boldsymbol{G}_{\boldsymbol{t},0}^{-1} - \boldsymbol{\Sigma})^{-1}$ where $\widetilde{\boldsymbol{G}}_{\boldsymbol{U}}[\boldsymbol{\Sigma}] = -(1/T)\delta\widetilde{F}_{\boldsymbol{U}}[\boldsymbol{\Sigma}]/\delta\boldsymbol{\Sigma}$. If this was equivalent with the DMFT self-consistency condition, a local self-energy would be a solution. This would imply, however, that $\widetilde{\boldsymbol{G}}_{\boldsymbol{U}}[\boldsymbol{\Sigma}]$ is non-local for a local $\boldsymbol{\Sigma}$. Hence, $\widetilde{F}_{\boldsymbol{U}}[\boldsymbol{\Sigma}]$ cannot be the Legendre transform of $\widetilde{\Phi}_{\boldsymbol{U}}[\boldsymbol{G}]$ where $\widetilde{\Phi}_{\boldsymbol{U}}[\boldsymbol{G}]$ (as above) is the Luttinger-Ward functional with vertices restricted to a single site. An alternative choice for $\widetilde{F}_{\boldsymbol{U}}[\boldsymbol{\Sigma}]$, however, does not suggest itself.

One can conclude that a functional of the self-energy is necessary and sufficient to obtain the DMFT as a type-III approximation while a functional of the Green function is necessary and sufficient to get the DMFT as a type-II approximation. The decisive point is that rather a local self-energy can be tolerated as an approximation than a local Green function.

3 New Approximations

The immediate return of these considerations is a number of non-perturbative and thermodynamically consistent type-III approximations as shown in Fig. 1. These differ from the DMFT by a different restriction of the domain for the self-energy functional (11), i.e. by a different reference system with a different subspace of variational parameters $\boldsymbol{t}'$ but the same interaction ($\boldsymbol{U}' = \boldsymbol{U}$). The evaluation of a type-III approximation requires the repeated computation of the grand potential and the Green function or self-energy of the reference system to get $\Omega_{\boldsymbol{t},\boldsymbol{U}}[\boldsymbol{\Sigma}_{\boldsymbol{t}',\boldsymbol{U}}]$ which must be optimized with respect to $\boldsymbol{t}'$ subsequently.

The class of possible reference systems is essentially spanned by two parameters, namely $n_{\mathrm{s}} - 1$ which is the number of additional bath sites per correlated site and N_{c} which is the number of correlated sites in a cluster

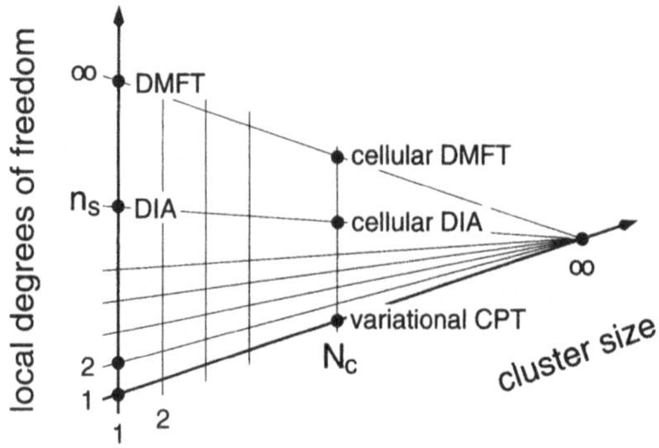

Fig. 1. Different possible approximations within the SFA (see text)

that is decoupled from the rest of the correlated sites (Fig. 1). The DMFT is obtained for $N_c = 1$ and $n_s = \infty$ since a continuous bath ($n_s = \infty$) is necessary to represent an arbitrary local free Green function.

The choice $N_c = 1$ but $n_s < \infty$ yields new approximations ("dynamical impurity approximations", DIA) which are inferior as compared to the full DMFT but allow for much simpler and faster calculations when n_s is small. The most simple but non-trivial approximation ($n_s = 2$-DIA) has been shown [5, 20] to already cover the essence of the DMFT scenario [1] for the Mott metal-insulator transition in the Hubbard model. At the critical point for $T = 0$ the calculations can be done even analytically [20], and with increasing n_s the grand potential, static quantities and the entire phase diagram rapidly converge to the full DMFT results [5, 21]. The DIA is similar but superior as compared to the exact-diagonalization approach [1]. Even for small n_s the approach is thermodynamically consistent and, off half-filling, respects the Luttinger sum rule, for example. The DIA has also been employed successfully to study the influence of phonons on metal-insulator transitions in the Holstein-Hubbard model [22, 23].

Nothing new is obtained for $n_s = \infty$ and $N_c > 1$: Here the SFA recovers the cellular DMFT [24]. (Note that the dynamical cluster approximation [25] is a type-II approximation). More interesting is the case $n_s = 1$ and $N_c > 1$ which turns out [26] to represent a variational generalization of the cluster-perturbation theory [27]. This V-CPT is well suited to describe phases with spontaneously broken symmetry and has been employed to study one-particle excitations and antiferromagnetic order in the $D = 2$ and $D = 1$ Hubbard model at half-filling [28] and charge ordering in the extended Hubbard model [29]. A further application concerns antiferromagnetism in quarter-filled ladder systems [30]. An impressing example of the power of the V-CPT approach has been given recently in a study of the competition between antiferromagnetism and d-wave superconductivity in the hole- and electron-doped Hubbard model [31]. The question of phase separation is addressed in Ref. [32].

Summing up, the SFA is able to unify different cluster theories and local approximations within a single and consistent framework which offers a large flexibility in the use of bath sites, ficticious fields, boundary conditions and particle reservoirs [26]. The formalism provides a controlled compromise between the demands for a non-perturbative and systematic theory working in the thermodynamic limit on the one hand, and the limited computational capabilities to diagonalize finite-size systems on the other.

4 Open Problems

The self-energy-functional approach allows to construct a series of consistent approximations which improve systematically as $N_c \to \infty$. It is by no means clear, however, whether bath sites $n_s > 1$ help to speed up the convergence with respect to N_c and whether a cluster extension of DMFT or the V-CPT is more efficient. This can be clarified only empirically by considering different lattice models in different dimensions. As a few bath sites have turned out to be sufficient for reproducing the essential mean-field ($N_c = 1$) physics in a number of studies of the single-band Hubbard model, further applications of the DIA are worthwhile to explore e.g. the mean-field phase diagrams of more complex (multi-orbital) models. Furthermore, one may also envisage the application of a simplified DMFT where a single (but continuous, $n_s = \infty$) bath is optimized for a multi-orbital model. This might be well justified for not too low temperatures.

On the technical side, there are two main future tasks: The full diagonalization and the Lanczos method which have been used so far, should be supplemented by a "reference system solver" based on stochastic techniques to improve the scaling of the numerical effort with the system size. Secondly, it would be advantageous to have an iterative technique at hand that directly yields a solution of the SFA Euler equation without the need for numerical differentiation. First results using full diagonalization [21] are encouraging.

On the conceptual side, the question for the possibility to give strict upper bounds to the grand potential is still open. Probably, a positive answer requires to establish a link to the Ritz variational principle. On the other hand, no example is known yet where the SFA grand potential at a stationary point is *lower* than the exact one.

There are different directions into which the formalism may be extended. As the coherent-potential approximation for the disorder Anderson model has the same (mean-field) status as the DMFT for the Hubbard model, it suggests itself that a self-energy-functional approach can also be constructed for systems with disorder (and interaction). First applications [33] demonstrate that such a theory [34] is feasible. A challenge consists in the extension of the theory to include two-particle Green functions in a generalized variational principle. Here the recently proposed functional-integral formulation of the SFA [35] can be helpful. Two-particle correlation functions are interesting by themselves and may furthermore facilitate an even greater flexibility in the choice of reference systems. At the same time such an approach could provide a conceptual clear way to treat models with non-local interactions. Currently, this problem is circumvented by a more pragmatic decoupling procedure [29].

Acknowledgements

The author would like to thank M. Aichhorn, E. Arrigoni, F.F. Assaad, M. Balzer, R. Bulla, C. Dahnken, W. Hanke, A. Millis, and W. Nolting for discussions. Support by the Deutsche Forschungsgemeinschaft within the Sonderforschungsbereich 410 and the Forschergruppe 538 is acknowledged.

References

1. A. Georges, G. Kotliar, W. Krauth, M. J. Rozenberg, Rev. Mod. Phys. **68**, 13 (1996).
2. W. Metzner, D. Vollhardt, Phys. Rev. Lett. **62**, 324 (1989).
3. J. M. Luttinger, J. C. Ward, Phys. Rev. **118**, 1417 (1960).
4. G. Baym, L. P. Kadanoff, Phys. Rev. **124**, 287 (1961).
5. M. Potthoff, Euro. Phys. J. B **32**, 429 (2003).
6. N. D. Mermin, Phys. Rev. **137**, A 1441 (1965).
7. P. Hohenberg, W. Kohn, Phys. Rev. **136**, 864 (1964).
8. W. Kohn, L. J. Sham, Phys. Rev. **140**, 1133 (1965).
9. R. M. Dreizler, E. K. U. Gross, *Density Functional Theory*, Springer, Berlin (1990).
10. K. Schönhammer, O. Gunnarsson, R. M. Noack, Phys. Rev. B **52**, 2504 (1995).
11. L. J. Sham, W. Kohn, Phys. Rev. **145**, 561 (1966).
12. E. Runge, E. K. U. Gross, Phys. Rev. Lett. **52**, 997 (1984).
13. N. E. Bickers, D. J. Scalapino, S. R. White, Phys. Rev. Lett. **62**, 961 (1989).
14. N. E. Bickers, D. J. Scalapino, Ann. Phys. (N.Y.) **193**, 206 (1989).
15. A. Georges, G. Kotliar, Phys. Rev. B **45**, 6479 (1992).
16. M. Jarrell, Phys. Rev. Lett. **69**, 168 (1992).
17. R. Chitra, G. Kotliar, Phys. Rev. B **63**, 115110 (2001).
18. R. Chitra, G. Kotliar, Phys. Rev. B **62**, 12715 (2000).
19. A. Georges, cond-mat/0403123.
20. M. Potthoff, Euro. Phys. J. B **36**, 335 (2003).
21. K. Pozgajcic, cond-mat/0407172.
22. W. Koller, D. Meyer, Y. Ono, A. C. Hewson, **66**, 559 (2004).
23. W. Koller, D. Meyer, A. C. Hewson, Y. Ono, cond-mat/0406241.
24. G. Kotliar, S. Y. Savrasov, G. Pálsson, G. Biroli, Phys. Rev. Lett. **87**, 186401 (2001).
25. M. H. Hettler, A. N. Tahvildar-Zadeh, M. Jarrell, T. Pruschke, H. R. Krishnamurthy, Phys. Rev. B **58**, R7475 (1998).
26. M. Potthoff, M. Aichhorn, C. Dahnken, Phys. Rev. Lett. **91**, 206402 (2003).
27. D. Sénéchal, D. Pérez, M. Pioro-Ladrière, Phys. Rev. Lett. **84**, 522 (2000).
28. C. Dahnken, M. Aichhorn, W. Hanke, E. Arrigoni, M. Potthoff, Phys. Rev. B **70**, 245110 (2004).

29. M. Aichhorn, H.G. Evertz, W. von der Linden, M. Potthoff, Phys. Rev. B **70**, 235107 (2004).
30. M. Aichhorn, E. Ya. Sherman, H. G. Evertz, cond-mat/0409162.
31. D. Sénéchal, P.-L. Lavertu, M.-A. Marois, A.-M.S. Tremblay, cond-mat/0410162.
32. M. Aichhorn, E. Arrigoni, cond-mat/0502047.
33. M. Balzer, M. Potthoff, Physica B, in press.
34. M. Potthoff, M. Balzer, to be published.
35. M. Potthoff, cond-mat/0406671.

Sol-Gel Approaches to Photonic Crystal Systems

Frank Marlow, Denan Konjhodzic, Helmut Bretinger, and Hongliang Li

Max-Planck-Institut für Kohlenforschung
Kaiser-Wilhelm-Platz 1, 45470 Mülheim an der Ruhr Germany

Summary. A prerequisite for an efficient "molding the flow of light" by photonic crystals is the molding of materials in desired nanostructures. Very often, conventional materials and processing techniques cannot fulfill the theoretical requirements for the materials and structures. Sol-gel methods enable material processing in opal pores and the controlled introduction of sub-wavelength pores into the materials by the use of molecular or supramolecular templates. The porosity can be used for lowering the refractive index, for soft processing of the materials and for stress relaxation. Examples for this approach are ultra-low refractive index films used as support for 2D photonic crystals, inverse opals with a skeleton-like unit cell filling and ferroelectric films with high transparency. In detail, we describe mesoporous silica films with a refractive index of 1.14 for use in 2D photonic crystal waveguide systems and the structure tuning possibilities of inverse opals.

1 Introduction

Photonic crystals (PhCs) and photonic crystal devices are well structured systems using different materials. The structures are complicated and subjected to strict accuracy requirements. Many fabrication techniques cannot be applied because they are only suited for less complex structures (2D instead of 3D systems or thin instead of thick structured films), they do not allow the tuning of the materials, or they are connected with large fluctuations. Sol-gel approaches [1] to PhCs can be considered as inorganic structure formation by self-assembly on multiple nanometer length scales. This is not an advantage automatically, however, it is a worth-to-think alternative to conventional structuring or material formation processes. The potential advantages of sol-gel processes are (i) the ease of fine tuning by the sol composition, (ii) the possible use of structure-influencing additives, (iii) the possibility of structure-healing processes, and (iv) the low costs.

Sol-gel processes (Fig. 1) lead to systems which are frozen in a non-equilibrium state. These states are highly dependent on the exact reaction pathway and, therefore, on the used processing conditions such as heating and cooling rates, humidity, and diffusion paths. This dependence is a general difficulty, but also an option. The materials can be tuned by changing the processing conditions.

B. Kramer (Ed.): Adv. in Solid State Phys. **45**, 149–160 (2005)

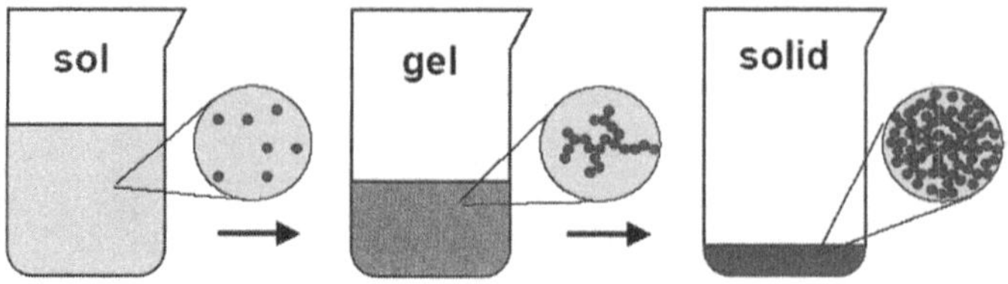

Fig. 1. Scheme for a sol-gel process. While the sol is a liquid with suspended particles, a network is formed in the gel which can be transformed into a solid

Common properties of sol-gel approaches are the large shrinkage and often the high porosity of the synthesized materials. The shrinkage leads to cracks or even to a total destruction of the morphology of the materials. The low density can have a positive effect if materials with a low refractive index are the aim as described in section 3, but it can also represent a problem if the high density material part of a photonic crystal is made by a sol-gel process. Therefore, the tuning of the porosity is an important issue during the synthesis and in post-synthesis treatments.

Sol-gel techniques have been applied in two fields of the PhC research area. Most famous are the realizations of inverse opals (e.g. [2]) which are examples for three-dimensional (3D) PhCs. In the field of two-dimensional (2D) PhC slab waveguides [3], sol-gel methods are discussed sometimes, but for the practical realization of the systems, sputtered inorganic layers or spin-coated polymers have been preferred up to now. To our knowledge, the first real attempt in this direction was published recently [4]. There, a sol-gel film was used as a support layer for the structured slab.

In the present paper we will describe sol-gel techniques to realize PhC systems. The use of these self-organization mechanisms enables a finetuning of the synthesized structures and, sometimes, structures not accessible with other techniques.

2 Inverse Opals

2.1 Different Techniques for Inverse Opals

Inverse opals [5, 6] are fabricated in two steps, opal formation and inversion. Techniques have been developed to prepare opals on a large scale, fairly accurate, and with tunable properties [7, 8, 9] (see appendix). Inversion, also called replica formation, is a nice conceptual possibility for multifunctional use of opals. The problems arise from the fact that a liquid or gaseous precursor is needed for the inversion which is less dense than the solid. Apart from this difficulty replica formation allows the molding of many materials and the incorporation of functional units into the PhC.

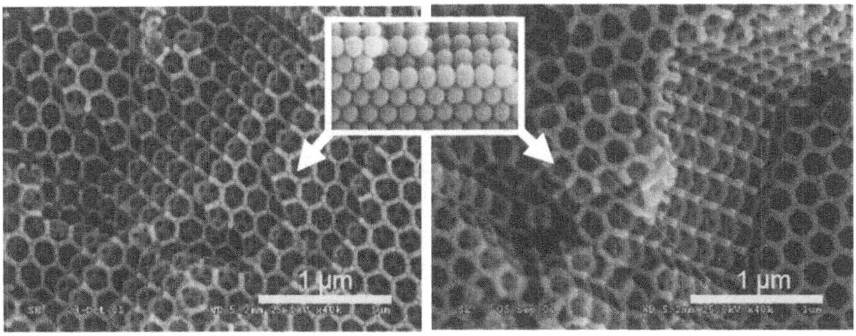

Fig. 2. SEM pictures of skeleton structures obtained by an opal inversion process [12]

Different techniques have been used to achieve inverted systems besides the sol-gel methods: chemical vapor deposition (CVD) [6], atomic layer deposition (ALD) [10], and electrochemical growth. Every technique has different drawbacks and is, therefore, suited in special cases. The kind of deposited material, its density and the fluctuations of properties have to be regarded. The fluctuations can be (i) parental opal-induced fluctuations, (ii) material density gradients, and (iii) replication-induced unit cell fluctuations. CVD and ALD allow a high material density, but fluctuations of the material thickness and gradients in vertical direction have to be expected as visible in [10]. Here, sol-gel methods may have advantages. One seeks for techniques allowing the inversion of the opal template for a variety of systems and preventing the rise of fluctuations and defects by the inversion.

2.2 Structure Types for Sol-Gel Made Inverse Opals

For the known well-shaped TiO_2 structures fabricated from a sol-forming precursor [2], diluted precursor solutions were used and titania shell systems were formed with high accuracy. Obvious indications for the accuracy are the constant shell thickness and the secondary voids visible in some works. These voids are situated between shells which have been formed around the polystyrene (PS) spheres of the opal template. They justify naming these structures as shell structures (ShS). On the other hand there are well-shaped structures known [5] without the secondary voids. They fill the whole residual volume between the opal spheres completely and can be named residual volume structures (RVS) [11].

In our works [12, 13], concentrated precursor solutions was used for a one-cycle inversion process. After calcination, the inverted product showed a structure being qualitatively different from the ShS. Under most conditions, no shells were found, but cylinders connecting the former octahedral and tetrahedral voids of the opal (see Fig. 2). A skeleton structure (SkS), which

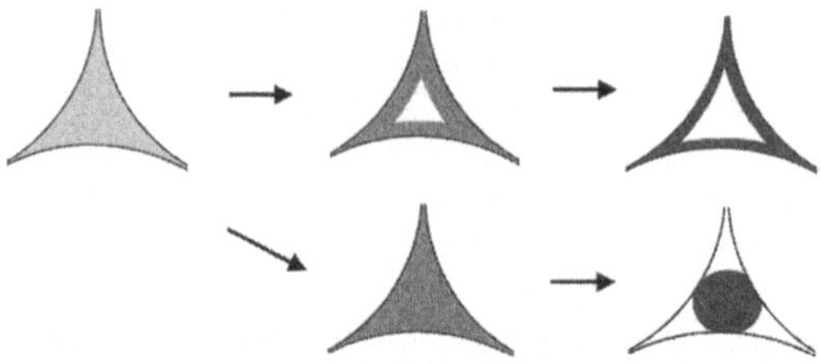

Fig. 3. Scheme analogous to Fig. 1 for the possible sol-gel processes in nanopores, e.g. an opal void. The upper path refers to the ShS and the lower one to the SkS

is not filling the whole residual opal volume, is a good description of such a product. The regularity of this structure seems to be very high, no inversion-induced increase of the fluctuations was visible.

The structural variations of the inverse opal systems can be assigned to the peculiarities of a sol-gel process in nanopores. If the gel formation is connected with much solvent evaporation, it results in a large shrinkage. Then, the gel covers the inner surface of the opal only, leading to shell systems. If the initial shrinkage is avoided by a high precursor concentration and an appropriate reaction path, the gel is formed in nearly the whole volume. Then, the drying of the gel leads to the cylinder formation driven by minimization of the surface tension (Fig. 3).

The observed structural differences and the secondary voids in ShS indicate that replica formation might be a wrong word for the investigated process. The word replica points to a more or less exact negative copy of the original. However, there are different negative copies and these copies can show even topological differences (secondary voids) and also qualitatively different physical behavior (next paragraph). Therefore, we prefer to speak of inversion or inverse systems, respectively. We are aware that also these words can be misleading without explanation. Inverted systems can, in our sense, consist of different building blocks incorporated into the opal voids [11].

2.3 Band Structures

Are the skeletons only nice peculiarities of a special synthesis or do they induce new physical properties? To answer this question, band structure calculations have been made for ideal skeleton systems (Fig. 4). Surprisingly, it turned out that the band structures of the skeletons are different from those of the ShS in some crucial aspects. There are the same bands present, but complete bandgaps can arise between different bands compared with the ShS. For the ShS and for the RVS a bandgap is known between the 8th and the 9th bands, but for the SkS the gap can be opened between the 5th and the 6th bands (5-6 gap). Lower bands are likely more resistant to structural fluctuations. Also materials with two simultaneous bandgaps can be constructed by SkSs.

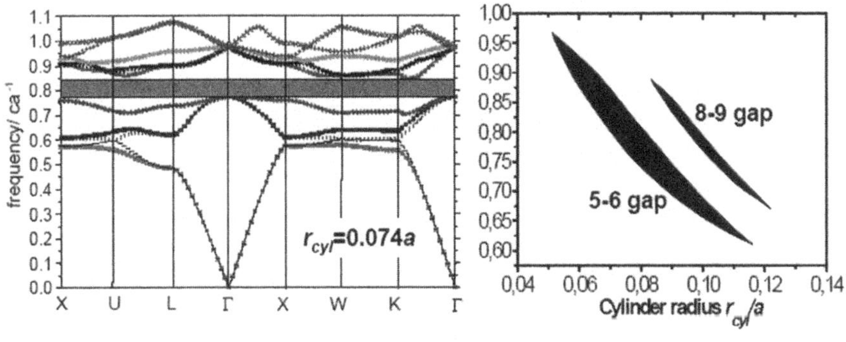

Fig. 4. Band structure and gap map of a SkS with n = 3.45. The 5-6 gap can have a width of 10% [12]

It might be surprising that relatively small structural details can influence the band structure in such a manner. However, the bandgaps occur between higher bands. This means that the eigen functions are strongly varying over a unit cell. The characteristic length of this variation determines the sensitivity of the bands to structural variations. This sensitivity can be used for band structure tuning, but is also a way how structural fluctuations can destroy bandgaps.

2.4 Structural Accuracy Requirements

The sol-gel inversion process is kinetically determined (water diffusion, hydrolysis, condensation, drying) and leads to a structure which is not the thermodynamical equilibrium. Therefore, it requires an accurate control of the processing conditions. They determine the type of inverse opal produced [13] and the exact structure parameters (e.g. the cylinder radius r_{cyl} for skeletons and the lattice constant a). The processing parameters also offer an efficient way to control these parameters. On the other hand fluctuations of the processing parameters will induce fluctuations of the structure able to destroy band structure effects. To estimate the influence of these fluctuations, we make following consideration. Let us assume a layer of skeleton structures consisting of randomly oriented domains with the same lattice constant but slightly fluctuating cylinder diameter. We are looking for the range of zero transmission of this layer which we identify with the fluctuation-reduced bandgap. This bandgap results from the overlap of the single bandgaps and can be estimated by

$$\overline{\Delta\omega} = \Delta\omega - \left| \frac{\partial\omega_+}{\partial r_{cyl}} + \frac{\partial\omega_-}{\partial r_{cyl}} \right| \Delta r_{cyl} \qquad (1)$$

where $\Delta\omega$ is the bandgap of the ideal SkS, Δr_{cyl} is the maximum deviation of the cylinder radius from the mean value, and the derivatives are the slopes

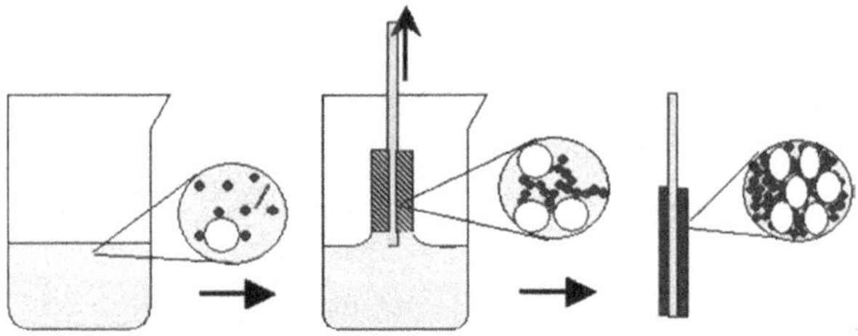

Fig. 5. Scheme describing the influence of an SDA (white) analogous to Fig. 1. The SDA can be restructured itself. It produces pores

of the upper and lower band edge with cylinder radius. Using the values obtained in the band structure calculation we find that the 10% wide 5-6 gap is reduced to zero if the fluctuations of the cylinder radius exceeds 5%. This means that for structures with $a = 300$ nm, the cylinder radius fluctuations must be below 5 nm which seems to be a fairly hard requirement. Sol-gel approaches could help to synthesize such perfect structures.

3 Low-n Films

3.1 Sol-Gel Approach to Low-n Supports

Mesoporous silica thin films are interesting for applications in separation, catalysis, and as materials with low dielectric constant or refractive index (low-k and low-n) [14]. One promising application of low-n films is their use as a substrate for 2D PhC waveguides. A high vertical refractive index contrast allows more modes to be guided. Since the air bridge type PhC waveguide, which has the highest possible vertical index contrast, is inherently fragile, a solid substrate with a refractive index close to that of air ($n = 1$) is needed. One way to reduce the refractive index of a solid substrate is to introduce porosity. For highly defined porous materials, sol-gel processes modified by surfactants can be applied. The surfactants, forming liquid crystal phases, act as structure-directing agent (SDA) as discovered in the MCM-41 synthesis [15]. Such processes can be realized by evaporation-induced self-assembly in a dip-coating process [16] (Fig. 5). An appropriate silica precursor interacts with the surfactant micelles and forms a continuous solid network simultaneously [17, 18].

We fabricated mesoporous silica films by such a method. A solution containing the triblock copolymer poly(ethylene oxide)-*block*-poly(propylene oxide)-*block*-poly(ethylene oxide) ($EO_{20}PO_{70}EO_{20}$), EtOH, water, HCl, and Rhodamine 6G was mixed with tetrabutyl orthosilicate as a silica precursor (for details see [19]). This solution formed a sol which was used in the dip-coating process. The porosity of the films was obtained by the removal of the surfactant in a heat treatment (calcination). The thickness of the resulting transparent films was possible to be tuned in the range 300-1100 nm by the drawing rate during the dip-coating. Two types of films, denoted as A and B, were found in dependence on the processing conditions, which strongly influence the self-assembly. A-type films turned out to be well suited as low-n supports and were used in ref. [4].

3.2 Film Structures

Structure-directing agents are used to control the pores of the synthesized material. In our case they form liquid crystalline phases which determine the film structure. However these structures are far away from the equilibrum because they are frozen-in by the gel formation somewhere on the reaction path. Therefore, the dependence of the sol-gel process on the processing conditions is transferred to the formation of the pore structure. Many parameters influence this process, but the most decisive one is the humidity. Although our structure-directing agent can have several equilibrium states, we restricted the processing conditions in such a way that only two film types were obtained, which turned out to be connected with different pore architectures. The A-type films were perfectly transparent. Their surface was investigated by AFM and turned out to be very smooth with RMS roughness lower than 0.5 nm [19]. The films were mechanically stable up to a thickness of 1100 nm. The refractive index of these films was $n = 1.14$ for λ=1.3 μm determined with a total reflection method [4]. This is to our knowledge the lowest measured n for a transparent material suited as waveguide support. The air filling fraction of these mesoporous films was $(61\pm2)\%$ [19]. In contrary to the A-type films, which are well-suited optical materials, B-type films appeared milky. Point-like defects of a size between 1 and 100 μm were seen with an optical microscope.

The determination of the film structures is of high practical interest because it may explain the different optical properties and may help to realize a stable synthesis. In a practical synthesis it is impossible to control all processing parameters with infinite accuracy. Most decisive parameters have to be found out and the variation limits have to be determined. To determine the structures, small angle x-ray scattering (SAXS) and TEM investigations were performed. The SAXS of as-synthesized A-type films showed a circle-like diffraction pattern (Fig. 6a). After calcination an ellipse with the half axes of 1.08° and 1.95° was found (Fig. 6b). This diffraction pattern can be attributed to a partial ordering of the films with mesostructured domains that have random orientations in the film. The structure shrinks in the direction normal to the surface upon calcination. This effect is visible in the one changed half-axis in the diffraction pattern and amounts to 55% [19]. A TEM analysis was performed on cross sections of the samples. The resulting micrograph (Fig. 6c) is typical for so-called worm-like structures [20]. This structure can be considered as a non-equilibrium state of a dense-packed channel array.

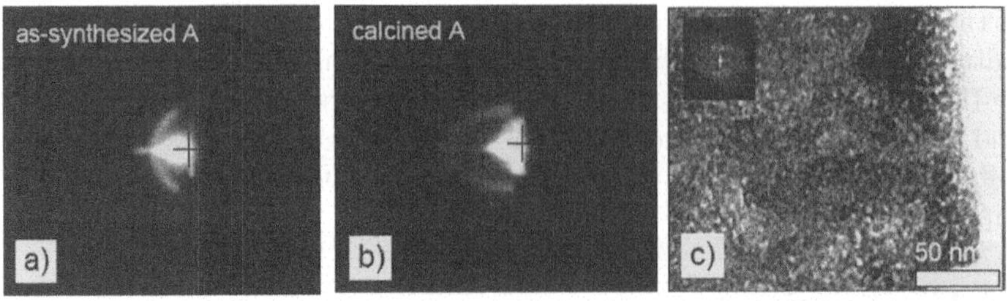

Fig. 6. Structure determination of A-type films. (a) The SAXS diffraction pattern of as-synthesized and (b) calcined films. The right sides of diffraction patterns are not accessible by the scattered waves. The sample was positioned perpendicular to the image plane in vertical direction. The primary beam position is marked by a cross. (c) TEM micrograph of the cross-section of a calcined A-type film. Scale bar: 50 nm, Inset: FFT [19]

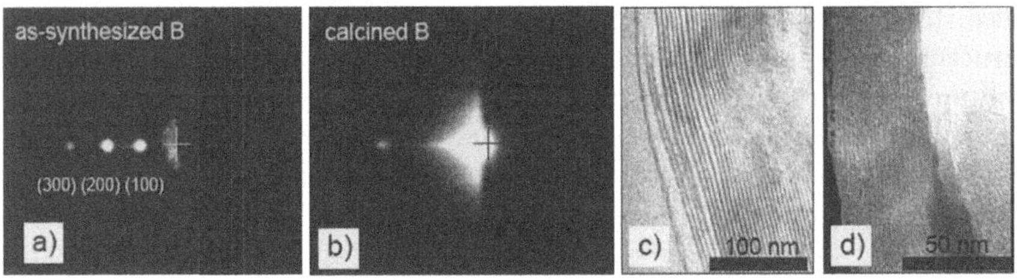

Fig. 7. Structure determination of B-type films. (a) The SAXS diffraction pattern of as-synthesized and (b) calcined films. Sharp diffraction spots out of film plane indicate layers ordered parallel to the substrate, which remain stable upon calcination. The layers are clearly resolved in the TEM micrographs of the cross-sections of (c) as-synthesized and (d) calcined B-type films

The SAXS pattern of as-synthesized B-type films exhibited very pronounced equidistant peaks diffracted perpendicular to the film plane (Fig. 7a) indicating a layer structure ordered parallel to the film surface with a d-spacing of 8 nm. In the diffraction pattern of calcined B-type films (Fig. 7b), one sharp diffraction spot was visible at $2\Theta = 3.1°$. This means that the layer structure remained stable during calcination. The layer spacing decreased to 3 nm, which means shrinkage to 38%. In the TEM images of the B-type films, the layers are also visible (Fig. 7c and 7d). A calcination-stable layer structure is a surprise, because one expects a collapse of the interlayer spacing after template removal. Such a structure can only be stable if there is a sustaining system among the layers. Although B-type films are not suitable as low-n substrates, their internal layer structure is very interesting in itself. Information on the structure of the sustaining network were found in AFM investigations [19]. The AFM phase image in the Fig. 8 shows a dark network which can be assigned to higher energy dissipation. This network might be a

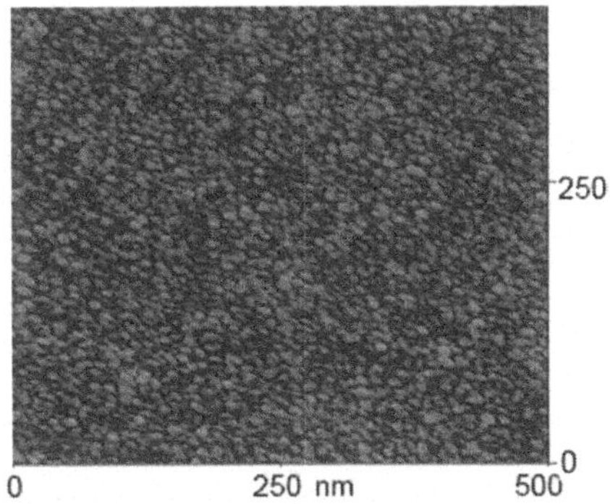

Fig. 8. AFM phase image of the surface of a calcined B-type film. The phase angle variations are in the 20° range. The distance between the dark regions is approximately 20 nm

picture of the sustainers that are supporting the separated layers. In [19] we describe a model resulting in higher dissipation on the top of the sustainers compared with the dissipation on the bridges between them.

3.3 Application of Low-n Films

The application of mesoporous films in the field of optics requires a very reproducible synthesis and film properties within a sufficiently narrow range. This requirement was met by controlling the processing conditions especially the humidity and the air circulation during dip-coating. The resulting mesoporous films turned out to be suited as ultra low refractive index substrates for 2D photonic crystal waveguides. A-type films with a thickness of 1 μm deposited onto oxidized silicon wafers were used for further processing. A slab waveguide was produced by spin coating an polymethylmethacrylate-based polymer (P(MMA-DR1)) onto this substrate (Fig. 9a). Then 2D PhC line defect resonator structures were fabricated by reactive ion etching. It consists of two finite PhCs separated by a non-structured region. The exact design of the structure was optimized by simulations as described in [4]. The transmission spectra of such a resonator are measured for different polarizations by the prism coupling method (Fig. 9b). The resonator structure on the mesoporous substrate showed a high transmission in the resonance peak (about 60% for TE and 80% for the TM polarization). These values were expected from simulations and exceed values determined on similar structures on conventional Teflon supports. However, the width of the resonance peak is larger than expected. It is currently not fully clear if this is a result of the scattering in the mesoporous film, surface roughness effects, imperfect structuring, or the device layout. Fluctuations in the shape of the PhC hole position can enhance the damping of the resonator and radiation losses at the resonator ends may also contribute in a unexpected manner. A further

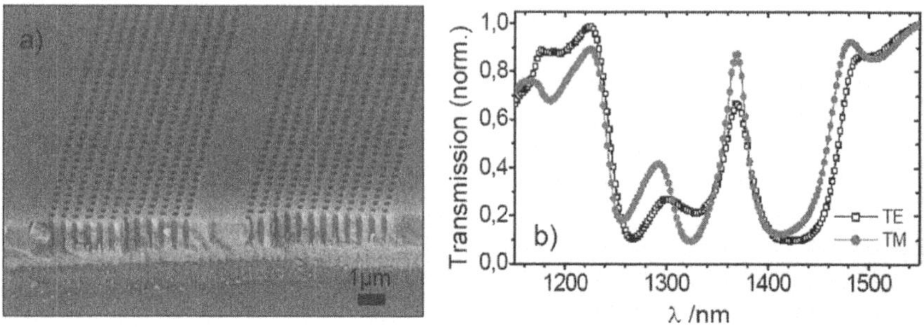

Fig. 9. (a) The SEM picture shows a resonator structure made of the polymer P(MMA/DR-1) on a A-type mesoporous film. (b) Transmission spectrum of such a resonator. The high transmission at the resonance wavelength and the low losses on the air band side indicate a strong confinement of light inside the slab waveguide core [4]

interesting property of this transmission spectrum are the seemingly absent radiation losses at the air band side of the band gap. The transmission has the same height at this band edge as at the dielectric edge. In contrast, the PhC waveguides on Teflon show strong losses at the air band edge since these states are above the light cone [4].

These optical investigations of 2D PhC on mesoporous films demonstrate that the sol-gel approach can lead to materials with extraordinary properties and with sufficiently high reliability. This approach gives possibility for the controlling of desired properties which are in this special case the refractive index, roughness, and film thickness.

4 Outlook: Ferroelectric PhC Films

A further field of strong interest is the incorporation of active materials into the PhCs. They can be switchable, nonlinear optical, or amplifying and can be incorporated as guests or as constituent of the PhC itself. The second way ensures high densities but requires solving the structuring problem for these active materials. For example, electro-optic PhC would be very interesting. However, no material with high electro-optic coefficients and a possibility for structuring in PhCs has been identified up to now. Let us have a closer look onto the problems with electro-optical 2D PhC slab waveguides. Poled polymer films are one possibility to achieve such systems; however they could suffer under stability problems and the relatively small refractive index of the polymers. Inorganic ferroelectric materials represent an interesting alternative, but they are very difficult to be structured. The structuring consists of two parts for the PhC slabs, the vertical structuring (film formation) and lateral structuring (periodic pattern formation). Already the first step is diffi-

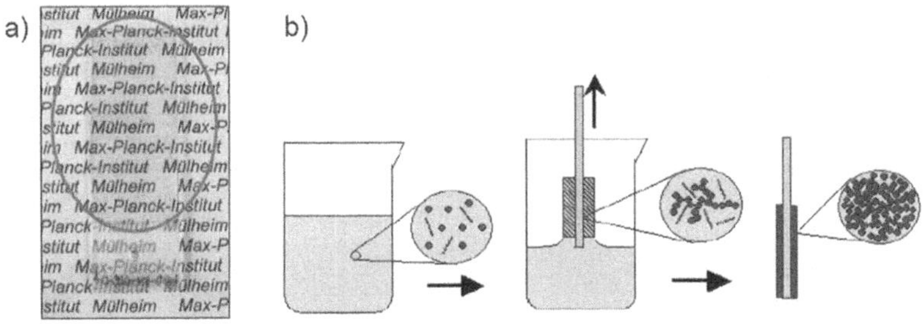

Fig. 10. (a) A perfectly clear PZT film (in circle) made with a sol-gel method. (b) Scheme describing the effect of stress-relaxing agents analogous to Fig. 1

cult for most inorganic systems. Here, the sol-gel approach can be very useful because it is offering a number of tuning possibilities.

Figure 10 shows highly transparent $Pb(Zr_xTi_{1-x})O_3$ (PZT) films made by a sol-gel technique introduced by Kozuka et al. [21]. The films are ferroelectric and may represent a step towards inorganic electro-optical PhCs. The key of the preparation method consists in the use of stress-relaxing agents modifying the sol-gel process. Also here, as for the other sol-gel made systems, we remarked a strong dependence of the film quality from the processing conditions. This enables a tuning of the optical properties of the PZT film. The lateral structuring step, which is essential for the use as PhC, will be carried out by reactive ion beam treatments. The detailed properties of the PZT films such as domain size and residual porosity will be used for fine-tuning of this process. The three examples described in this work show that sol-gel approaches can be useful to structure the materials as required for PhC. They are all not the only approaches to a specific problem, however, they can show advantages over competing techniques concerning exact shape control, fluctuation suppression, domain size control and stress relaxation. They require exact control of processing conditions which can be used for controlling the PhC fabrication.

Acknowledgements

We thank W. Dong for band structure calculations and the DFG for project support (Ma1745/7) within the program SPP 1113 "Photonic Crystals".

Appendix

The accuracy of opals is a current research topic. In our laboratory, homogenous opal films were made by newly developed method combining the advantages of Xia's [8] and Colvin's [7] methods (Fig. 11). It avoids the main drawbacks of both methods which are the need of a micro-fabricated spacer

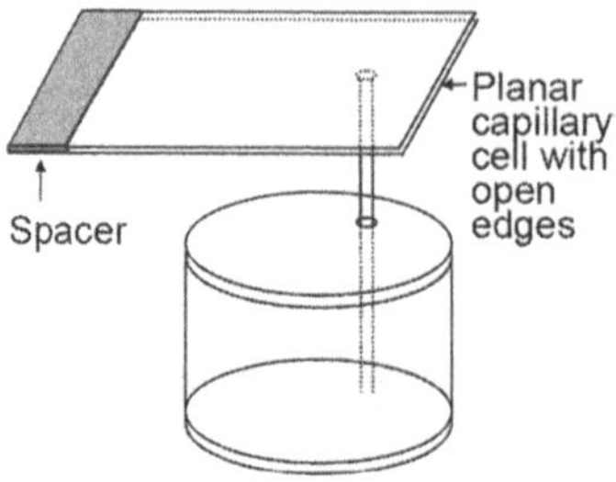

Fig. 11. Set-up of a capillary deposition method for opal films. The PS suspension is transported from a container via a tube into a planar capillary cell with open edges. Opal assembly starts at these edges and fills the whole cell in a self-regulated process. Dense spacers between 25 and 100 μm thickness were used

and the difficult thickness control, respectively. Furthermore, it allows the control of opal cracks which cannot be avoided [9].

References

1. C. J. Brinker, G. W. Scherer, *Sol-Gel Science* (Academic Press, San Diego 1989).
2. J. E. G. J. Wijnhoven, W. L. Vos, Science **281**, 802 (1998).
3. T. F. Krauss, R.M. De La Rue, S. Brand, Nature **383**, 699 (1996).
4. M. Schmidt, G. Boettger, M. Eich, W. Morgenroth, U. Huebner, H. G. Meyer, D. Konjhodzic, H. Bretinger, F. Marlow, Appl. Phys. Lett. **85**, 16 (2004).
5. A. A. Zakhidov, R. H. Baughman, Z. Iqbal, C. X. Cui, I. Khayrullin, S. O. Dan-tas, I. Marti, V. G. Ralchenko, Science **282**, 897 (1998).
6. A. Blanco et al., Nature **405**, 437 (2000).
7. P. Jiang, J. F. Bertone, K. S. Hwang, V. L. Colvin, Chem. Mat. **11**, 2132 (1999).
8. S. H. Park, D. Qin, Y. N. Xia, Adv. Mater. **10**, 1028 (1998)
9. H. Li, W. Dong, H. Bongard, F. Marlow, J. Phys. Chem. B, **109**, 9939 (2005).
10. J. S. King, C. W. Neff, C. J. Summers, W. Park, S. Blomquist, E. Forsythe, D. Morton, Appl. Phys. Lett. **83**, 2566 (2003).
11. F. Marlow, W. Dong, Chem. Phys. Chem. **4**, 549 (2003).
12. W. Dong, H. Bongard, B. Tesche, F. Marlow, Adv. Mater. **14**, 1457 (2002).
13. W. Dong, H. Bongard, F. Marlow, Chem. Mater. **15**, 568 (2003).
14. F. Schüth, W. Schmidt, Adv. Mater. **14**, 629 (2002).
15. C. T. Kresge, M. E. Leonowicz, W. J. Roth, J. C. Vartuli, J. S. Beck, Nature **359**, 710 (1992).
16. C. J. Brinker, Y. Lu, A. Sellinger, H. Fan, Adv. Mater. **11**, 579 (1999).
17. J. S. Beck et al., J. Am. Chem. Soc. **114**, 10834 (1992).
18. A. Monnier et al.s Science **261**, 1299 (1993).
19. D. Konjhodzic, H. Bretinger, U. Wilczok, A. Dreier, A. Ladenburger, M. Schmidt, M. Eich, F. Marlow, Appl. Phys. A, **81**, 425 (2005).
20. D. Grosso, F. Carnol, G. Soler-Illia, E. L. Crepaldi, H. Amenitsch, A. Brunet-Bruneau, A. Bourgeois, C. Sanchez, Adv. Funct. Mater. **14**, 309 (2004).
21. H. Kozuka, M. Kajimura, T. Hirano, K. Katayama, J. Sol-Gel Sci. Technol. **19**, 205 (2000).

Theory of Surface Optical Properties

Olivia Pulci[1], Maurizia Palummo[1],
Margherita Marsili[1], and Rodolfo Del Sole[1]

INFM, University of Rome Tor Vergata, Department of Physics,
Via della Ricerca Scientifica 1, 00133 Rome, Italy

Summary. The theory of surface optical properties is reviewed. The use of Many-Body Perturbation Theory, with Density Functional Theory calculations as zero order approximation, is now the state of the art to obtain quasi-particle excitation energies and optical spectra of surfaces. In this paper we will present the current status of theoretical and computational approach. Realistic calculations of surface optical properties within the single-particle approximation and within the Many-Body Perturbation Theory are discussed; as examples, results for InAs(110) and C(100) surfaces are given.

1 Introduction

Optical spectroscopy has become a widely used tool for studies, characterization and monitoring of surfaces [1, 2, 3]. This is due to the fact that the optical probe does not damage the sample and has an excellent energy resolution. On the other hand, and in contrast to electrons, the light has a large penetration depth, hence it is poorly sensitive to the surface region. Nevertheless, some techniques have been devised to increase its surface sensitivity. Two linear optics methods, Surface Differential Reflectance (SDR) [4] and Reflectance Anisotropy Spectroscopy (RAS) [5, 6], have been developed to this purpose. In the former case, the reflectance of the clean surface is measured first, and then that of the surface after chemisorption; the difference, the SDR, a few percent of the reflectance, surely originates at the surface. The latter technique is used on the surfaces of cubic materials: in this case the bulk contribution to reflectance, described by Fresnel formula, does not depend on the direction of light polarization, while the surface contribution, due to the lower symmetry of the surface, may depend on it. Hence, the reflectance difference between measurements carried out with two different light polarizations in the surface plane, the reflectance anisotropy (RAS), again a few percent of the total reflectance, is generated at the surface.

The interpretation of these experiments is generally not straightforward. Hence theoretical work is needed in order to fully exploit the potentiality of surface optical spectroscopy. The contribution of theorists to this collaboration can be of two types: (i) in cases where two or more structural models are hypothesized, comparison of calculated with measured optical properties

B. Kramer (Ed.): Adv. in Solid State Phys. **45**, 161–172 (2005)

may confirm one model and discard the other ones; (ii) if the structural model is well established and a reasonable agreement is found between theory and experiment, the theoretical interpretation helps in identifying the origin of the observed structures (for instance, a given peak may be due to transitions between surface states, or alternatively may involve bulk states modified by the surface).

After the atomic structure of the surface of interest has been determined by energy minimization or inferred from experiments, three main steps are involved in the theory; i) the determination of one-electron wave functions; ii) the calculation of the dielectric susceptibility, possibly including many-body effects; iii) the solution of light-propagation equations. After the pioneering work of Feibelman [7], the last problem has been solved in a quite simple way by Bagchi et al. [8] for the jellium surface at the end of the seventies. The solution has been generalized to the case of crystal surfaces by Del Sole in 1981 [9]. The surface contribution to reflectance, that is the deviation from Fresnel formulas, can be calculated from the dielectric susceptibility of the vacuum-crystal interface. This can be in turn obtained, within the one-electron approximation, from the wave fúnctions. These and their energies should be calculated ab-initio according to the Green function method, using the so-called GW approximation for the exchange-correlation self-energy. The electron-hole interaction and local-field effects altogether may be included by solving a Bethe-Salpeter equation for the interacting electron and hole (for a review, see for example [10]). Calculations carried out according to this recipe for some surfaces have yielded excellent agreement with experiments. Unfortunately such calculations are very demanding from the computational point of view. Therefore theoreticians often resort to simpler -yet less accurate- methods, allowed by the partial cancellation of self-energy, excitonic and local-field effects. The best compromise between speed and accuracy is to determine the wave functions and energies according to density functional theory [11] within the local density approximation (DFT-LDA) [12] or Generalized Gradient Approximations (DFT-GGA) [13], and to describe many-body effects through the calculation of the GW electronic structure. A cruder approximation consists in a suitable rigid upward shift of the DFT conduction bands with respect to valence bands, the so-called scissors operator. A even cruder approximation relies on the description of the wave functions and energies within semi-empirical methods, for instance using semi-empirical tight binding [14]. For its computational speed the tight binding method has been employed in 1986 in the first realistic calculations of surface optical properties [15], while it is now used only for very complex systems, where ab-initio methods can not be applied.

As we will show for the case of InAs(110), good agreement between theory and experiment can be obtained in some cases within the GW single-quasiparticle approach (section 4); but, in the cases where strong excitonic effects are present (as in C(001)2x1), one has to resort to the solution of the (cumbersome) Bethe-Salpeter equation (section 6).

2 Reflectance Anisotropy Calculations

The difference of the Reflectance for light polarized in two different directions gives, for cubic crystals, a measure of dielectric anisotropy of the surface, which has been shown to be very sensitive to the surface geometrical and electronic structure. The reflectance anisotropy is defined as,

$$\frac{\Delta R}{R} = \frac{R_y - R_x}{R} \tag{1}$$

with R being the average isotropic reflectivity and R_i (i=x,y) is the reflectivity for light polarized along the i direction. An expression for the RAS has been given in an elegant and closed form by Bagchi et al. [8] in 1979 by solving the light propagation equations at a jellium surface, and later generalized to crystal surfaces by Del Sole [9]. Realistic surface calculations are usually carried out in a slab geometry. That is, the semi-infinite crystal is replaced by a slab: if this is thick enough, each surface of the slab should be representative of the surface of the semi-infinite crystal. In practice, this is achieved by choosing the slab thickness in the range from 10 to 30 monolayers [16].

The resulting expression for the RAS in a slab geometry is

$$\Delta R/R = \frac{R_y - R_x}{R} = 4(\omega/c)Im[(\epsilon^{hs}_{yyM} - \epsilon^{hs}_{xxM})/(\epsilon^{b}_{M} - 1)] \tag{2}$$

where ϵ^b_M is the macroscopic bulk dielectric function and the macroscopic half-slab dielectric function $\epsilon^{hs}_{iiM}(\omega)$ has been introduced. All these macroscopic quantities are related to the microscopic dielectric function $\varepsilon(\omega)$ by

$$\epsilon_M(\omega) = \lim_{\mathbf{q}\to 0} \frac{1}{\varepsilon^{-1}_{\mathbf{G}=0,\mathbf{G}'=0}(q,\omega)}. \tag{3}$$

Within the single-particle (RPA) scheme, and neglecting local field effects [17], the imaginary part of the half slab macroscopic dielectric functions is related to the transition probability induced by the radiation between slab eigenstates (Fermi Golden Rule)

$$Im[\epsilon^{hs}_{iiM}(\omega)] = \frac{4\pi^2 e^2}{m^2\omega^2 A}\Sigma_{v,c,\boldsymbol{k}}| < \phi_v|p^i(\boldsymbol{k})|\phi_c > |^2\delta[E_c(\boldsymbol{k}) - E_v(\boldsymbol{k}) - \hbar\omega] \tag{4}$$

which involves the calculation of the matrix element of the i-component of the momentum operator p^i between initial (valence, v) and final (conduction, c) eigenstates $\phi_i(\boldsymbol{r})$ at the point $\boldsymbol{k}$ in the two dimensional Brillouin Zone. A is the slab area, and $E_{c,v}$ are the eigenenergies of the single particle states. The real part is computed via the Kramers-Kronig transform. From a dimensional analysis, we see that ϵ^{hs}_{iiM} has the dimension of a length. Another ingredient of (2) is the bulk dielectric function $\epsilon^b_M(\omega)$. Its imaginary part is analogous to (4), where now eigenstates and eigenvalues refer to the infinite crystal, the $\boldsymbol{k}$ vectors in the summation are three dimensional, and the area A is replaced by the volume V (as usual, $\epsilon^b_M(\omega)$ is a c-number). Again the real part is computed via the Kramers-Kronig transform. Thus, in conclusion, in order to calculate the RAS within a single particle approach, we have to determine the wavefunctions ϕ_i and eigenvalues E_i of the electrons in the slab and in the bulk.

3 DFT Single-Particle Approach

Density Functional Theory (DFT) [11] is an exact approach to the study of the ground state properties of an N-electron system. The electron charge density $\rho(\boldsymbol{r})$, rather than the electron wave function, plays here a central role.

This theory has been introduced by Hohenberg and Kohn (HK) in 1964 [11]. They showed how the ground state energy of a system of N interacting electrons in an external potential $V_{ext}(\mathbf{r})$ can be written as a functional of the ground state electronic density. In a later paper Kohn and Sham [12] reformulated the variational problem of minimizing the HK functional with respect to the charge density in terms of one-particle effective equations (Kohn-Sham equations). The most common way, nowadays, to calculate the electrons energies and wavefunction in complex systems is based on the solution of the Kohn Sham equations within the LDA or GGA approximation [12, 13], with the use of pseudopotentials. Calculations within the all electron scheme have not shown any difference in the optical spectra [18]. The Kohn-Sham single particle equations, given by

$$H^{KS}\phi_i = [p^2/2m + V_{ion}(\boldsymbol{r}) + \int d^3\boldsymbol{r}' \frac{\rho(\boldsymbol{r}')}{|\boldsymbol{r}-\boldsymbol{r}'|} + V^{KS}_{xc}(\boldsymbol{r})]\phi_i = E_i\phi_i \qquad (5)$$

is solved, thus obtaining the eigenvalues and the eigenstates. The second term on the right-hand side of (5) is the (pseudo) potential energy of the electron interaction with ion cores, the third term is the electron-electron electrostatic interaction (the Hartree term), involving the electron density $\rho(\boldsymbol{r})$, and the fourth term is the DFT exchange-correlation potential, usually given within the LDA or GGA Approximations [12, 13]. Since the electron density appears in the Hartree term, as well as in the exchange-correlation potential, the (occupied) eigenstates of the Hamiltonian must be calculated self consistently with the electron density. This is done by iterating the calculation until the input and output charge densities are equal.

4 GW: Single Quasi-Particle Approach

Density Functional Theory is strictly valid only for the ground state of a many-electron system. Excited states properties, which are of course essential in determining optical properties, should not be described within this framework, but rather in the many-body Green function formalism [19]. This implies changing $V_{xc}^{KS}(\boldsymbol{r})$ into a non-local, energy dependent and non hermitian self energy $\Sigma(\boldsymbol{r},\boldsymbol{r}';\omega)$ which makes the calculation much more cumbersome. As a matter of fact, energy bands calculated according to the density functional equations (5), involving $V_{xc}^{KS}(\boldsymbol{r})$, are qualitatively close to the experimental ones: the most serious shortcoming is that the calculated gap between filled and empty states is underestimated, typically by 50-100 %, in semiconductors and insulators [20]. As a consequence, optical spectra calculated within DFT are always redshifted with respect to experiments.

A solution of the underestimation of the electronic gaps is found with the use of the Many Body Perturbation Theory based on the Green function formalism. Within this approach, it is possible to write a so called 'quasi particle equation' [19]:

$$\left(-\frac{1}{2}\nabla^2 + V^{ext} + V^H\right)\Psi_i(\boldsymbol{r},\omega) + \int \Sigma(\boldsymbol{r},\boldsymbol{r}',\omega)\Psi_i(\boldsymbol{r}',\omega)d\boldsymbol{r}' = E_i^{QP}(\omega)\Psi_i(\boldsymbol{r},\omega), \quad (6)$$

which describes the extra hole created when an electron is extracted from the system (for example in a photoemission experiment; or, in a inverse photoemission experiment, where an electron is added to the system) and the polarization cloud that the other electrons create in order to screen it. Due to the non hermiticity of Σ, the energies $E_i^{QP}(\omega)$ are in general complex. It is worth noticing that this equation reduces to the Hartree equation when Σ=0, to the Hartree-Fock ones when $\Sigma = iGV$, and to the Kohn-Sham equations when, instead of Σ, a local, hermitian and energy independent operator is taken: $\Sigma \rightarrow V_{xc}^{KS}(r)$. This observation gives an *a-posteriori* justification of the use of the Kohn-Sham equations to calculate the excitation energies; the qualitative agreement between DFT and experimental band structure results from the fact that the Kohn-Sham equations already contain, at least approximately, exchange and correlation effects. In other words, $V_{xc}^{KS}(r)$ is often a quite good approximation for the *true* self-energy Σ.

The quasi-particle (QP) energies are calculated solving eq. (6) within the GW scheme [19], which involves the approximation $\Sigma = iGW$ with G being the single particle Green function and W the screened Coulomb interaction. This approximation comes from a first iteration of Hedin's equations [19]. Computing the excited states energies as a first order perturbation correction to the DFT bands (perturbative GW):

$$E_i^{QP} = E_i^{DFT} + < \Sigma - V_{xc}^{KS} > \quad (7)$$

gives the so called GW energies, and represents nowadays the most refined method for band structures calculations. Once the eigenvalues of the quasi-particle equation (6) are known, we can determine the absorption strength according to the golden rule formula (4) using the DFT wavefunctions and the quasi-particle energies. Hence, a single quasi-particle spectrum is produced.

5 Application to InAs(110)

InAs is a small gap semiconductor of increasing technological interest for optical infrared devices. Its cleavage face, the (110) surface, has been theoretically studied within the semiempirical tight-binding approach [21, 22, 23] and within the DFT approach [24, 25, 26]. DFT is, as already mentioned, a powerful tool in dealing with ground state properties, but suffers from the well known gap problem in the determination of the electronic bands [19, 27]. In particular, a metallic behavior is predicted for bulk InAs, since a crossing of the conduction and valence bands appears at Γ within DFT, thus giving a negative bulk gap [28]. This is an example where the importance of quasi-particle calculation is more evident: the theoretical GW gaps, shown in Tab. 1, are in very good agreement with the experimental ones and the problem of underestimation of the electronic gaps, typical of DFT calculations, is overcome. Similarly, a good agreement for the surface electronic bands is found within the GW approximation (Tab. 1) [29].

For what concerns optical properties, InAs(110) is a good example where a single-quasiparticle scheme is a sufficient approximation for RAS calculation. The Reflectance Anisotropy Spectrum [29] is shown in Fig.1, and compared with experiments. The GW results shows a remarkable agreement for all energies when compared to the experimental spectrum.

Table 1. Bulk (Γ, X, L, E_0') and surface ($\bar{X}$, $\bar{M}$, $\bar{X}'$) electronic gaps at high symmetry points.

	Γ	X	L	E_0'	$\bar{X}$	$\bar{M}$	$\bar{X}'$
DFT	-0.1	3.7	1.9	4.0	1.8	2.1	2.0
GW	0.4	4.3	2.5	4.6	2.4	2.6	2.6
exp.	$\sim$ 0.4 [30]	4.3[31]-4.7[30]	2.5-2.6 [30]	4.4-4.6 [30]	2.5 [33]	2.6 [33]	2.4 [33]

RAS shows a rich spectral structure because calculations show that contributions from all kind of electron transitions involving surface states as well as electron transitions involving modified bulk states are present. Only the small shoulder at 2.5 eV is found to be due to pure surface state to surface state transitions. In the energy range between 2.5 eV and 4.2 eV, electron transitions from occupied surface states to empty bulk states and from bulk to surface states dominate the spectra, mainly arising from transitions along the $\bar{M} - \bar{X}' - \bar{\Gamma}$ direction in the IBZ. These theoretical findings are in agreement with very recent experiments of adsorption of cesium on InAs [34] where it is shown that for increasing Cs coverage the experimental structures at 2.0, 3.1 and 4 eV disappear, thus confirming the surface-related nature of those peaks. The intense peak at about 4.5 eV is due to transitions among modified bulk states and just a few contributions from surface related transitions play

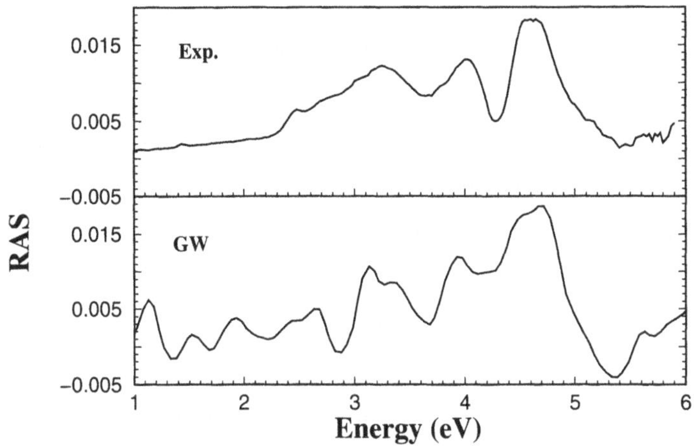

Fig. 1. Reflectance Anisotropy Spectrum of InAs(110): experiment (*top*), and calculation within the single quasiparticle approach (*bottom*). RAS is here defined as the relative difference of Reflectivity for light polarized along the [-110] and [001] directions

a role. Following Cs coverage [34], this structure has been shown to survive, although its intensity decreases somewhat. This is in agreement with the GW result that it is mainly bulk-related. The agreement between GW theory and experiment is satisfactory for all energies, thus showing that a good theoretical description of the electronic structure and consequently of the optical properties is possible within GW in this case. Excitonic effects, not included in the calculation, seem hence not to be important for this surface. But, as we will point out in the next section, this is not a general rule: excitonic effects might be important at some surfaces, and a full solution of the Bethe Salpeter Equation (BSE) is, in those cases, needed to get quantitative agreement with the experiments.

6 BSE: Excitonic Effects

Quasi-particle effects show up in band energies as shifts, with respect to one-particle levels, due to the interaction of an electron or hole with the other electrons (self-energy shifts), whose main effect is to increase the gaps between filled and empty states. But, in optical spectra, it is sometime crucial to take into account also the interaction of the electrons and holes generated in the optical transition (excitonic effects) and their screening by the other electrons. Not only the absorption lineshape can be modified with respect to independent-particle theory [35], but bound electron-hole states (excitons) can occur below the onset of interband transitions [36]. Excitonic effects can be rigorously taken into account within the Many-Body perturbative approach. If we start from the GW approximation for Σ and iterate Hedin equations again, we obtain a four-point generalized polarizability $\bar{P}$ [10] which obeys the Bethe-Salpeter equation,

$$\bar{P}(1,1';2,2') = P_0(1,1';2,2') + P_0(1,1';3,3')\Xi(3,3';4,4')\bar{P}(4,4';2,2') \quad (8)$$

Notation (1,2) stands for a pair of space and time coordinates, $(\mathbf{r}_1, t_1; \mathbf{r}_2, t_2)$. Repeated arguments are integrated over. $P_0 = -iGG$ yields the polarization of the independent quasi particles. The kernel Ξ contains two contributions,

$$\Xi(1,1',2,2') = -i\delta(1,1')\delta(2,2')\bar{v}(1,2) + i\delta(1,2)\delta(1',2')W(1,1'). \quad (9)$$

The first term $\bar{v}$ is the bare Coulomb potential without long range contribution. This term is equivalent to the inclusion of local field effects in the matrix inversion of a standard single-particle calculation [37]. When spin is not explicitly treated, $\bar{v}$ gets a factor of two for singlet excitons. In the second term, W is the screened Coulomb attraction between electron and hole. The important point is that the macroscopic dielectric function can be expressed in terms of this modified reducible polarizability $\bar{P}$ as [10, 38]:

$$\epsilon_M(\omega) = 1 - \lim_{\mathbf{q}\to 0} v(\mathbf{q})_0 \bar{P}_{\mathbf{G}=\mathbf{G}'=0}(q,\omega). \quad (10)$$

Equation (8) can be rewritten as

$$\bar{P} = [1 + iP^0\Xi]^{-1}\bar{P}^0 = [H_{exc} - I\omega]^{-1}_{(n_1,n_2),(n_3,n_4)}(f_{n_4} - f_{n_3}) \quad (11)$$

where f_i is the occupation number of the band i, and we have introduced an effective excitonic Hamiltonian defined as

$$H_{exc}^{(n_1,n_2),(n_3,n_4)} = (E_{n_2} - E_{n_1})\delta_{n_1,n_3}\delta_{n_2,n_4} - i(f_{n_2} - f_{n_1})\times$$

$$\int d\mathbf{r}_1\, d\mathbf{r}'_1\, d\mathbf{r}_2\, d\mathbf{r}'_2\; \phi_{n_1}(\mathbf{r}_1)\, \phi^*_{n_2}(\mathbf{r}'_1)\, \Xi(\mathbf{r}_1,\mathbf{r}'_1,\mathbf{r}_2,\mathbf{r}'_2)\, \phi^*_{n_3}(\mathbf{r}_2)\, \phi_{n_4}(\mathbf{r}'_2)$$

Hence, using the spectral representation for the inverse of a matrix, the interacting polarization can be obtained solving an effective eigenvalue problem

$$\sum_{(n_3,n_4)} H_{exc}^{(n_1,n_2),(n_3,n_4)} A_\lambda^{(n_3,n_4)} = E_\lambda A_\lambda^{(n_1,n_2)}. \quad (12)$$

For the calculation of absorption spectra, we can limit ourselves to transitions with positive frequency [10], i.e. (n_1, n_2) and (n_3, n_4) are pairs of one valence and one conduction band, respectively (in other words, we consider only the resonant part of Eq.11). Moreover, we build up the spectra of optical properties by considering only negligible momentum transfer, hence the same $\mathbf{k}$ for the valence and the conduction state. In this way the macroscopic dielectric function, defined by Eqs.(3, 10), reads as

$$\epsilon_M(\omega) = 1 + \lim_{\mathbf{q}\to 0} v(\mathbf{q}) \sum_\lambda \frac{\left|\sum_{v,c;\mathbf{k}} \langle v, \mathbf{k} - \mathbf{q}|e^{-i\mathbf{qr}}|c, \mathbf{k}\rangle A_\lambda^{(v,c;\mathbf{k})}\right|^2}{(E_\lambda - \omega)}. \quad (13)$$

This calculation is generally very demanding because the matrix to be calculated and diagonalized can be very large. In fact the relevant parameters which determine its size are the number of **k** points in the BZ, the number of the valence bands N_v, the number of conduction bands N_c which build the basis set of pairs of states.

Due to the complexity of this ab-initio scheme, the optical spectra of real materials including excitonic effects have started to be obtained only recently [35, 36]. The first applications appeared for clusters and bulk semiconductors, and more recently for surfaces [39, 40, 41].

All the calculations performed in insulators and semiconductors show that the inclusion of the electron-hole Coulomb interaction allows a quantitative comparison with experiments, not only below the electronic gaps, where generally bound excitons are formed, but also above the continuum edge. One of the first examples appeared in the literature is given by bulk silicon (see S. Albrecht et al. [35]) where excitonic effects have been shown to enhance the E_1 peak by almost 100% . Furthermore, the e–h interaction may also induce a redshift of the spectral peaks, which partially cancels the blue shift arising from the self–energy corrections.

7 Application to C(001)2x1

As an example of the importance of inclusion of excitonic effects in RAS calculations, we present results for the C(001)2x1 surface [42]. The theoretical RAS spectrum in Fig. 2b (dashed line) is calculated within the independent-quasiparticle (IQP) approximation neglecting excitonic effects. Two peaks are clearly apparent, at 4.1 and 4.9 eV. Both peaks are related to transitions from a filled to an empty surface state band in the gap (Fig. 3). The filled band, just above the valence band, originates from the π-bond of the two dangling bonds of a dimer, while the empty surface band, well inside the gap, originates from the π-antibond of the same orbitals. Peak A originates from transitions around J', while peak B from transitions at Γ. The theoretical RAS spectrum of Fig. 2b (dashed line), which is consistent with previous one-particle calculations [43], differs from experiments (Fig. 2a) in that the two peaks A and B are at too high in energy and the strong negative contribution above 4.5 eV is not reproduced. However, the inclusion of the electron-hole interaction in the calculation (Fig.2b, solid line) has dramatic effects: two strongly bound excitons (A_{ex} and B_{ex}) develop from the corresponding one-particle transitions (A and B). The agreement with the experiment is now improved, with the main difference being due to the larger broadening of the experimental curve (solid curve, Fig.2a). The importance of the electron-hole interaction can be seen from their effect on the RAS spectrum in the region of interband transitions where they completely dominate the spectrum giving rise to the deep negative contribution to the RAS above 4.5 eV. This is determined by the transfer of oscillator strength to lower energies for light

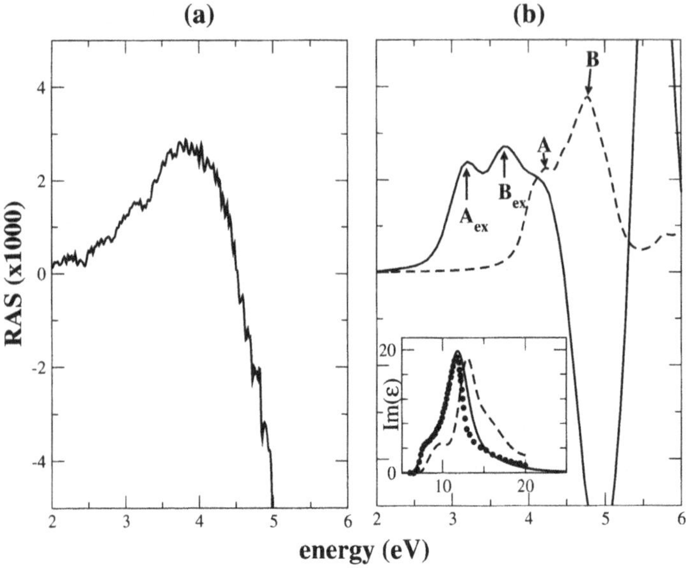

Fig. 2. (a): Experimental RAS [42]; (b) dashed line: calculated single quasi-particle spectrum (x 0.1); solid line: calculated spectrum with excitonic, self-energy (GW) and local field (LF) effects included. A Gaussian broadening of 0.15 eV has been used. RAS is defined as the relative reflectivity difference between light polarized parallel to the surface dimers and perpendicular to them. Insert: Diamond bulk dielectric function. Dotted line: experimental; dashed line: single quasiparticle (GW) calculation; solid line: with excitonic effects

polarized parallel to the dimers (due to the formation of bound excitons), with a consequent depletion at higher frequencies.

The calculated binding energy of the two excitons is 0.9 eV, the largest ever found for semiconductors. Surface-state excitons have been found so far for Si and Ge (111) (2×1) surfaces, with binding energy in both cases close to 0.25 eV [39, 40]. Even more striking is the difference with respect to a similar surface, Si(100)-(2×1), where the RAS calculated within the single quasi-particle approximation is in qualitative agreement with experiments, suggesting that excitonic effects are weak, with binding energy estimated to be smaller than 0.1 eV.

8 Conclusion

We have shown that first-principle calculations of the optical properties of surfaces are possible using a combination of DFT and Many Body Perturbation Theory. We have illustrated the self-energy equation and the GW approximation, which yields quite accurate results in many materials. Moreover, we have introduced the Bethe-Salpeter equation needed to describe the

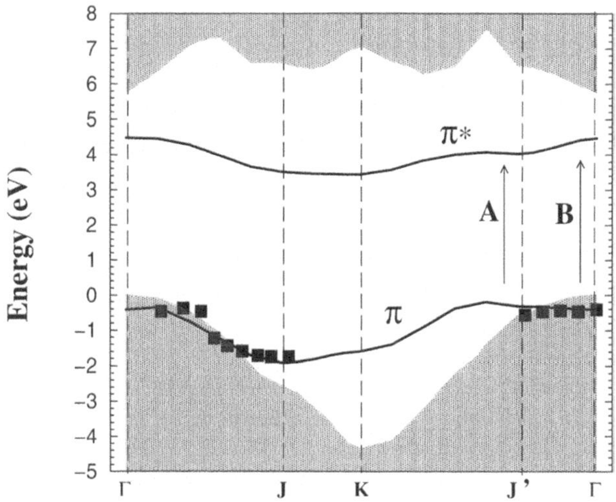

Fig. 3. GW electronic band structure of C(001)2x1. The A and B arrows indicate transitions responsible for the observed peaks in the calculated optical spectra. Experimental data from [44]

electron-hole interaction and shown how the optical spectra are strongly influenced by many body effects.

Acknowledgements

This work has been supported by the INFM PAIS project "CELEX", MIUR-COFIN 2002 and by the EU 6th Framework Programme through the Network of Excellence 'NANOQUANTA' (NMP4-CT-2004-500198). We acknowledge CINECA CPU time granted by INFM.

References

1. D. E. Aspnes, Mat.Sci. and Eng. B **30**, 109 (1995).
2. W. Richter, Appl. Phys. A **75**, 129 (2002).
3. B. G. Frederick et al., Phys. Rev. Lett. **80**, 4490 (1998).
4. S. Nannarone, et al., Solid State Commun. **33**, 593 (1980).
5. D. E. Aspnes, A. A. Studna, Phys. Rev. Lett. **54**, 1956 (1985).
6. V. L. Berkovits, et al., Solid State Commun. **56**, 449 (1985).
7. P. J. Feibelman, Phys. Rev. B **12**, 1319 (1975); ibidem 4282 (1975); ibidem **14**, 762 (1976).
8. A. Bagchi, R. G. Barrera, A. K. Rajagopal, Phys. Rev. B **20**, 4824 (1979).
9. R. Del Sole, Solid State Commun. **37**, 537 (1981).
10. G. Onida, L. Reining, A. Rubio, Rev. of Mod. Phys. **74**, 601 (2002).
11. P. Hohenberg, W. Kohn, Phys. Rev. **136**, B864 (1964).

12. W. Kohn, L. J. Sham, Phys. Rev. **140**, A1113 (1965).
13. J. Perdew et al. Phys. Rev. B **46**, 6671 (1992).
14. P. Vogl, H. P. Hjalmarson, D. J. Dow, J. Phys. Chem. Solids **44**, 365 (1983).
15. A. Selloni, P. Marsella, R. Del Sole, Phys. Rev. B **33**, 8885 (1986)
16. R. Del Sole, in *Photonic Probes of Surfaces*, edited by P. Halevi, Elsevier (1995); page 131.
17. S. L. Adler, Phys. Rev. **126**, 413 (1962); N. Wiser, Phys. Rev. **129**, 62 (1963).
18. P. Monachesi, et al., Phys. Stat. Sol. (a) **184**, 101 (2001).
19. L. Hedin, Phys. Rev. **139**, A796 (1965); L. Hedin, B.J. Lundquist in Solid State Physics. edited by H. Ehrenreich, F. Seitz and D. Turnbull (Academic press, New York, N.Y. 1969), Vol **23**, p. 1.
20. F. Aryasetiawan, O. Gunnarsson, Rep. Prog. Phys. **61**, 237-312 (1998); Aulbur W. G. Johnson and Wilkins J.W, 1999 Solid State Physics ed. H. Ehrenreich and F. Spaepen (New York Academic), and references therein.
21. C. Mailhiot, C. B. Duke, D. J. Chadi, Phys. Rev. B **31**, 2213 (1985).
22. R. P. Beres, R. E. Allen, J. D. Dow, Phys. Rev. B **26**, 5207 (1982).
23. A. I. Shkrebtii, et al., Appl. Surf. Sci. **104/105**, 176 (1996).
24. J.L. Alves, J. Hebenstreit, M. Scheffler, Phys. Rev. B **44**, 6188 (1991).
25. B. Engels, et al. Phys. Rev. B **58**, 7799 (1998).
26. J. Klijn et al, Phys. Rev. Phys. Rev. B **68**, 205327 (2003).
27. M.S. Hybertsen and S.G. Louie, Phys. Rev. B **34**, 5390 (1986); *ibid* **38**, 4033 (1988); R. Godby, M. Schlüter, L. J. Sham, Phys. Rev. B **37**, 10159 (1988).
28. X. Zhu, S. G. Louie, Phys. Rev. B 43, 14142 (1991).
29. X. Lopez-Lozano, et al. Phys. Rev. B (in press).
30. Landolt-Börnstein, *Numerical Data and Functional Relationships in Science and Technology*, New Series Vol. 17a, Ed. O. Madelung, Springer-Verlag Berlin (1982).
31. The value 4.3 eV has been evaluated using results from inverse photoemission experiment (1.9 eV [32]) and from XPS measurement (-2.4 eV [30]).
32. W. Drube, D. Straub, F. J. Himpsel, Phys. Rev. B **35**, 5563 (1987).
33. H. Carstensen, et al., Phys. Rev. B **41**, 9880 (1990).
34. K. Fleischer, et al., J. Phys.: Condens. Matter **16**, 4353 (2004).
35. S. Albrecht, et al., Phys. Rev. Lett. **80**, 4510 (1998).
36. G. Onida, et al., Phys. Rev. Lett. **75**, 818 (1995); M. Rohlfing, S. G. Louie, Phys. Rev. Lett. **80**, 3320 (1998); L. X. Benedict, E. L. Shirley, R. B. Bohn, Phys. Rev. Lett. **80**, 4514 (1998); M. Rohlfing, S. G. Louie, Phys. Rev. Lett.**82**, 1959 (1999).
37. W. Hanke, Adv. Phys. **27**, 287 (1978).
38. The two point function $\bar{P}$ can be recovered by contraction of indexes: $\bar{P}(21) = \bar{P}(2211)$.
39. M. Rohlfing. S.G. Louie, Phys. Rev. Lett. **83**, 856 (1999).
40. M. Rohlfing, et al., Phys. Rev. Lett. **85**, 5440 (2000).
41. P. H. Hahn, W. G. Schmidt, F. Bechstedt, Phys. Rev. Lett. **88** 016402 (2002).
42. M. Palummo, et al. Phys. Rev. Lett. **94**, 087404 (2005).
43. C. Kress, A. I. Shkrebtii, R. Del Sole, Surface Science **377- 379**, 398 (1997), V.I.Gavrilenko, F. Bechstedt Phys. Rev. B **56**, 3903 (1997).
44. R. Graupner et al Phys. Rev. B **57**, 12397 (1998).

Manipulations of a Qubit in a Semiconductor Quantum Dot

Artur Zrenner[1], Stefan Stufler[1], Patrick Ester[1], and Max Bichler[2]

[1] Universität Paderborn, Warburger Str. 100, 33098 Paderborn, Germany
[2] Walter Schottky Institut, Technische Universität München, Am Coulombwall, 85748 Garching, Germany

Summary. In a single self-assembled InGaAs quantum dot, the one exciton ground state transition defines a two-level system, which appears as an extremely narrow resonance of only a few μeV width. The resonant interaction of this two-level system with cw laser fields can be studied in detail by photocurrent spectroscopy, revealing the fine structure of the excitonic ground state as well as the effects of nonlinear absorption and power broadening. For the case of pulsed laser fields and in the absence of decoherence, the two-level system represents a qubit. Excitations with ps laser pulses result in qubit rotations, which appear as Rabi oscillations in photocurrent experiments. Double pulse experiments further allow us to infer the decoherence time and to perform coherent control on a two-level system.

1 Introduction

From physical point of view, all available active key components for information technology rely on incoherent phenomena in view of the associated electronic states in a solid. The use of coherent phenomena for the implementation of quantum information technology is expected to give plenty of room for advanced developments in the future [1]. Semiconductor quantum dots (QDs), often referred to as artificial atoms, are suitable entities to implement arrays of qubits for quantum information processing. One possible approach

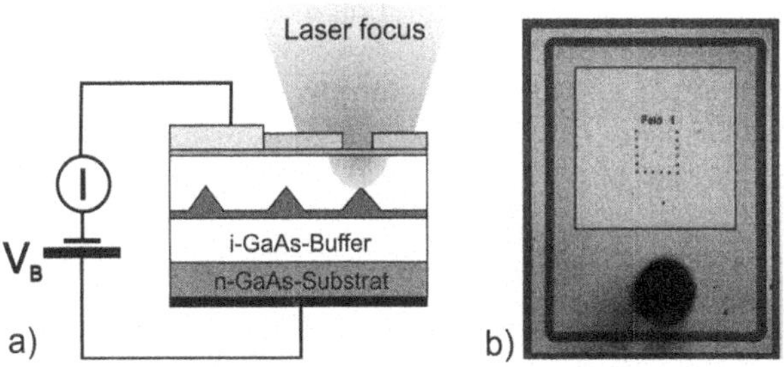

Fig. 1. Schematic view (a) and micrograph (b) of a n-i-Schottky diode with e-beam written shadow masks for optical access to single InGaAs QDs

B. Kramer (Ed.): Adv. in Solid State Phys. **45**, 173–184 (2005)

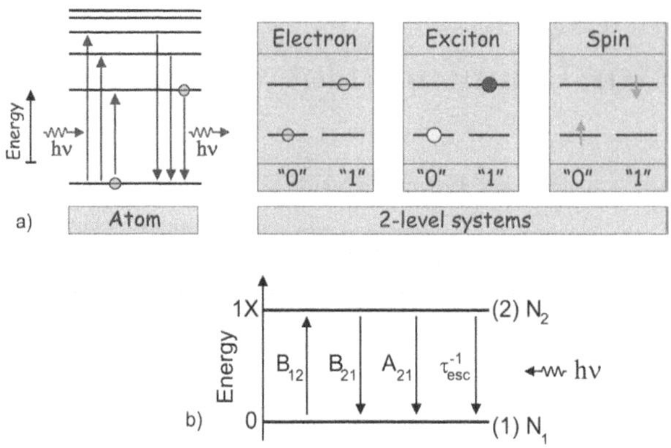

Fig. 2. (a) Schematic representation of various 2-level systems. An excitonic 2-level system is in the state "0" in the absence, in state "1" in the presence of an exciton, and otherwise in a mixed state. (b) Fundamental processes in an excitonic 2-level system in the presence of electric field: Transitions are controlled by absorption B_{12}, stimulated emission B_{21}, spontaneous emission A_{21}, and tunneling τ_{esc}^{-1}

is the use of excitonic excitations in the ground state of a QD as basis for a two-level system. Recently coherent population oscillations, so-called Rabi oscillations [2], have been demonstrated in the exciton population of single QDs [3, 4, 5, 6, 7, 8]. Low temperature dephasing times for excitons in self-assembled QDs have been shown to exceed several hundred picoseconds, basically allowing for sufficiently high numbers of coherent manipulations with picosecond pulses [9, 10].

2 Single Quantum Dot Photodiodes

The experimental results presented here have been obtained from the ground states of single self-assembled $In_{0.5}Ga_{0.5}As$ QDs. Within this contribution we further concentrate on n-i-Schottky diodes grown by molecular beam epitaxy on a (100)-oriented n^+-GaAs substrate. While based on a conventional diode structure, here a GaAs *n-i*-Schottky structure, the only optically active part is a single self-assembled $In_{0.5}Ga_{0.5}As$ QD contained in the intrinsic layer of the diode (see Figure 1). The QDs are embedded in a 360 nm thick intrinsic GaAs-layer, 40 nm above the n-doped GaAs back contact. A semitransparent Schottky contact is provided by a 5-nm-thick titanium layer. The optical selection of a single QD is done by shadow masks with apertures from 100 to 500 nm, which are prepared by electron beam lithography from a 80-nm-thick aluminum layer (see [11] for details). For resonant excitation we use a tunable Ti:Sapphire laser, which is focused on the sample by a NA = 0.75 microscope objective. All experiments were carried out at 4.2 K.

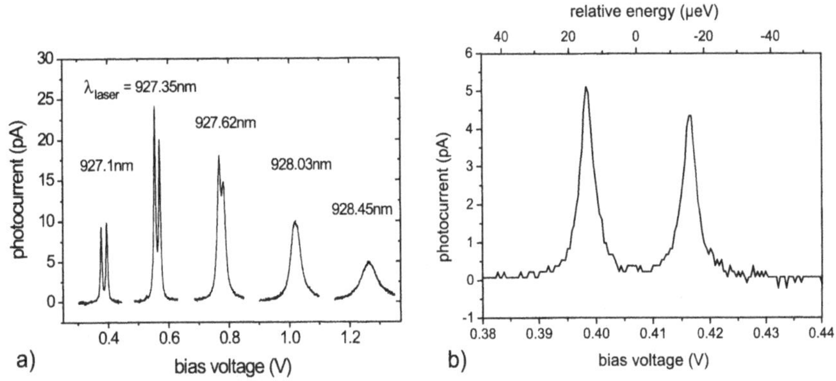

Fig. 3. (a) PC resonance for various wavelengths versus bias voltage. At low bias the fine structure splitting is fully resolved, at higher bias the linewidth is increased due to fast tunneling. (b) PC resonance at fixed wavelength. Bias voltage can be converted into an energy scale via the Stark effect, as shown on the upper axis. A linewidth below 5 μeV and an asymmetry splitting of 30 μeV are observed

A QD is an artificial atom in a semiconductor, which acts as a protective container for quantized electrons and holes. On the basis of those particles various two-level systems can be defined as indicated in Figure 2a. A single QD photodiode as used here essentially is an exciton two-level systems with electric contacts. In addition to the fundamental optical processes (B_{12}, B_{21}, A_{21}) indicated in Figure 2b, substantial tunneling escape (τ_{esc}^{-1}) appears at electric fields beyond about 35 kV/cm. The diode arrangement allows for photocurrent (PC) detection, a very sensitive and, as a matter of fact, quantitative way to determine the excitonic occupancy of the two-level system.

3 Incoherent Spectroscopic Results

In Fig. 3a we show a number of spectra for a QD one exciton (1X) resonance. The excitation wavelength was slightly increased for each spectrum, which results in a stepwise shift of the resonance towards higher bias voltage. All spectra in Fig. 3a were taken at the same excitation power of 65 nW. One immediately recognizes a notable increase in linewidth towards higher bias voltages. At low bias two sharp peaks are clearly resolved whereas at high bias only one broad peak can be observed. A detailed measurement of the Stark effect allows for accurate energy calibration versus bias voltage. Thus we are able to infer an increase in linewidth from 10 μeV at 0.4 V to 240 μeV at 1.4 V. This increase corresponds to a decrease in tunneling time with increasing electric field. The doublet line structure visible at low voltages can be further investigated by control of the polarization of the excitation beam.

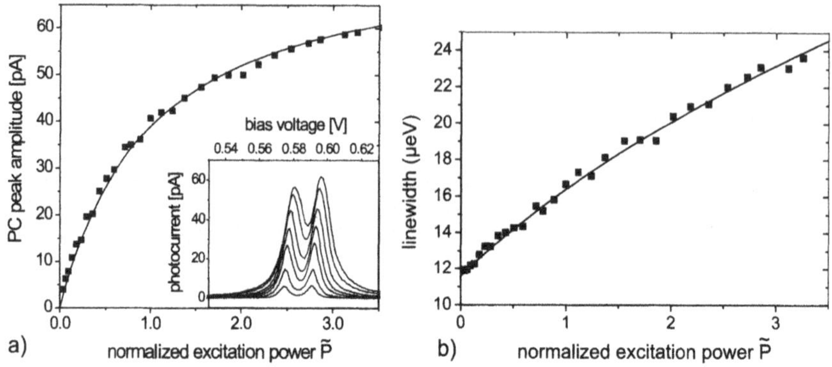

Fig. 4. (a) Analysis of the PC peak height, showing a saturation at high excitation power. Normalized power levels are displayed, according to the theoretical model described in the text. $\widetilde{P} = 1$ corresponds to an actual laser power of 90nW on the sample. Some of the original spectra used for the analysis are shown in the inset. (b) Analysis of the power broadening obtained from the same set of spectra. Since the same scaling of the x-axis is applied in both figures, the fit curve contains only one free parameter, which is the linewidth at zero excitation power

On rotating the orientation of linear polarization each peak can be clearly suppressed with respect to the other. This can be explained by a slight shape asymmetry, present in almost any self assembled QD, resulting in an energy splitting between wave functions oriented along or perpendicular to the elongation axis [12, 13]. We observe here an energy difference between both levels of about 30 μeV. Resonant laser absorption spectroscopy therefore is a very capable method for the line shape and fine structure analysis of QDs [14]. Due to improvements in sample design we were able to substantially reduce the apparent linewidth of the QD resonance (see Fig. 3b). The spacing between QDs and n^+-back contact is 40 nm in the present case as compared to 20 nm in the experiments reported in [15]. In the best case we have observed so far a purely Lorentzian linewidth as low as 3.3 μeV in PC experiments. As a consequence, we think, Coulomb interactions between QD states and fluctuating background charges have been substantially reduced, which otherwise cause considerable dephasing and therefore an increase in linewidth.

Another feature of the spectra displayed in Fig. 3a is the variation of their respective peak height. The decrease at high voltages can be explained simply by the fact that one would expect the integrated signal to be fairly constant rather than its height. The decrease at 0.4 V has to be explained otherwise. Here the tunneling time increases and gets comparable to the radiative recombination time, resulting in a quenching of PC signal as described by a simple rate equation. In general, however, the height and width of the PC peaks in controlled by the fundamental processes in a two-level system as outlined in Fig. 2b. The inset of Fig. 4a shows a series of PC spectra taken at varying laser intensities. Each spectrum has been fitted by two Lorentzian lines where data points and fit typically show a correlation of 99.95%. For further analysis the arithmetic mean of the two peak heights is used in order to reduce complexity and to minimize random fluctuations. Figure 4a shows an analysis of peak height versus excitation power. A clearly nonlinear power dependence can be determined resulting in a saturation of PC peak amplitude at high excitation according to the relation $I = I_{sat}\widetilde{P}/(\widetilde{P}+1)$ [15]. Here I denotes the PC peak amplitude, I_{sat} its saturation value and $\widetilde{P}$ corresponds to a normalized excitation power. The physics behind PC saturation value can be derived fairly easily: If the QD is already occupied by one exciton no further absorption can take place because of energy renormalization [16, 17, 18]. This also holds true if the QD is occupied only by one carrier. Under the applied conditions the first tunneling process happens fairly fast, whereas the tunneling time τ_{slow} of the slower carrier may even exceed the radiative lifetime. The observed saturation value then is given by $I_{sat} = e/2\tau_{slow}$ [15], where e is the elementary charge. Evaluation of the measured data gives $I_{sat} = 78\,\text{pA}$ and a tunneling time of $\tau_{slow} \approx 1$ ns, respectively.

This saturation behavior also has a direct effect on the linewidth of absorption peaks, independent of any other line broadening mechanisms. At exact resonance, i.e. at the center of a PC peak, the absorption naturally comes nearest to its saturation value. Therefore, with increasing excitation, the increase in absorption or in our case the increase in PC signal is weakest at the center of a peak and comparatively stronger at its sides. This results in a broadening of the absorption line known in literature as power broadening [19]. If one has a homogeneously broadened absorption peak with a Lorentzian line shape of width Γ_0 the power broadened peak again is Lorentzian but with an increased width Γ of: $\Gamma = \Gamma_0\sqrt{1+\widetilde{P}}$.

In Fig. 4b we show an analysis of linewidth versus excitation power obtained from the same set of data as in Fig. 4a. The same conversion of excitation power to $\widetilde{P}$-values is used for both diagrams. Note that therefore at a fixed value of Γ_0 the slope of the fit curve in Fig. 4b is no free parameter! From the fact that data points and fit still show very good agreement, we can conclude that no other power dependent line broadening mechanism is of any significant role here. The extrapolation to zero excitation power gives a value of $\Gamma_0 = 11.6\,\mu\text{eV}$. If one assumes the first tunneling process to be the main dephasing mechanism, this corresponds to an escape time of $\tau_{fast} = 57\,\text{ps}$.

Let us discuss the results of the analysis shown in Fig. 4a and 4b. The saturation value derived in Fig. 4a gives us a measure for the time it takes the system to go back to its initial state. In native two-level systems this would be the lifetime of the excited state T_1. In our case this is the escape time of the slower of both photoexcited carriers, as discussed above. The linewidth Γ_0 derived in Fig. 4b corresponds to the dephasing time of the system, in the context of two-level systems usually denoted as T_2. It is important to

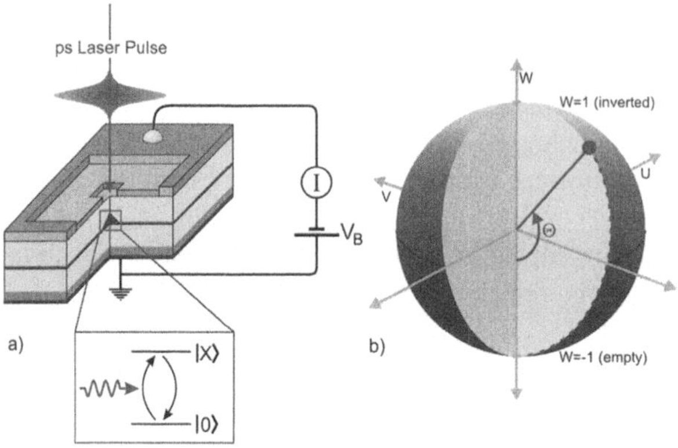

Fig. 5. (a) Coherent excitations of a single QD can be dephased by tunneling and quantitatively measured as PC. (b) Bloch sphere representation of a resonant, coherent excitation in a dot. The rotation angle Θ is proportional to the pulse area and to the oscillator strength of the ground state transition

note that the correct dephasing time can only be derived by an extrapolation to zero power. We performed the according measurements also at $V_{bias} = 0.4$ V and infer a linewidth Γ_0 as low as 4.2 μeV. This reflects a significant increase of the tunneling time towards lower bias voltages. On the other hand, a long tunneling time also means that the system needs a long time to come back to its initial state. Saturation and power broadening therefore play an important role even at a comparatively low excitation power. Thus in Fig. 3a the spectrum at $V_{bias} = 0.4$ V already is notably power broadened, resulting in a linewidth of $\Gamma = 10$μeV. This shows that particularly in systems with a long lifetime one could easily infer too short dephasing times if power broadening was neglected. The third fit parameter used in the analysis of Fig. 4a and 4b is the scaling of the x-axis. If the intensity of the light field at the spot of the QD would be well known, one could deduce the oscillator strength of the 1X transition. Due to the near field shadow mask used on our sample this is not possible here. Probably the most important information can be derived from a comparison of the saturation and power broadening analysis. Both sets of data can be fitted well with the same scaling of the x-axis. This means that the observed increase in linewidth is only a consequence of saturation and that dephasing in our system does not rise with increasing excitation power, at least in the regarded range.

4 Coherent Manipulations

In the following section we want to focus on the coherent behavior of the QD. Due to finite dephasing times we use ultrashort laser pulses for excitation. The fundamental experiment in the coherent regime is the observation of

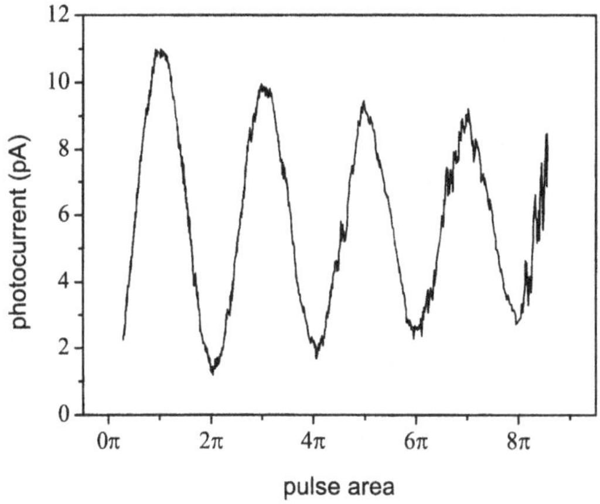

Fig. 6. Exciton Rabi oscillation at excitation with ps laser pulses (bias voltage 0.6V). The oscillation is only slightly damped towards high pulse areas

Rabi oscillations [2]. The occupancy of the upper level of a two-level system under coherent resonant excitation is given by $sin^2(\Omega t/2)$ [19], where the Rabi frequency Ω is proportional to the square root of the laser intensity and t corresponds to the pulse length. A π-pulse thereby results in a complete inversion of the two-level system. In the context of quantum computing this represents a qubit rotation analogous to the classical NOT operation. We define the pulse area, i.e. the rotation angle $\theta = \Omega t$, by adjusting the excitation amplitude rather than the pulse length (see Fig. 5). A π-pulse typically corresponds to an average laser power on the sample of about 2 μW at a pulse-length of 2.3 ps and a repetition frequency of $f_{Laser} = 80\,\text{MHz}$. If the tunnel efficiency of our device was 100%, any π-pulse would contribute to the PC with one elementary charge, resulting in a maximum value of $I = f_{Laser} \times e = 12.8\,\text{pA}$ [5]. At low bias voltage however, the tunneling time increases to values similar to the radiative lifetime, which causes a quenching of the PC. At 0.4 V for example, the maximum observed PC is only about 6 pA, as compared to 12 pA at 0.8 V. Figure 6 shows the upper level occupancy, reflected by the PC, as a function of the excitation pulse area.

At the highest excitation intensities the system undergoes here more than eight full inversions with each laser pulse. The original measurement was thereby corrected for an incoherent background, probably caused by absorption of stray light in wetting layer tail states. As the background is linear in excitation power, it is clearly distinguishable from the oscillatory coherent signal. At π-pulse excitation the incoherent part is about 6% of the total signal. We use here circular polarized light, in order to avoid biexciton generation not only by spectral separation but also by Pauli blocking [20]. The maximum rotation angle of almost 9π significantly exceeds that of any previously published data on excitonic Rabi oscillations [3, 4, 5, 6, 7, 8]. Even

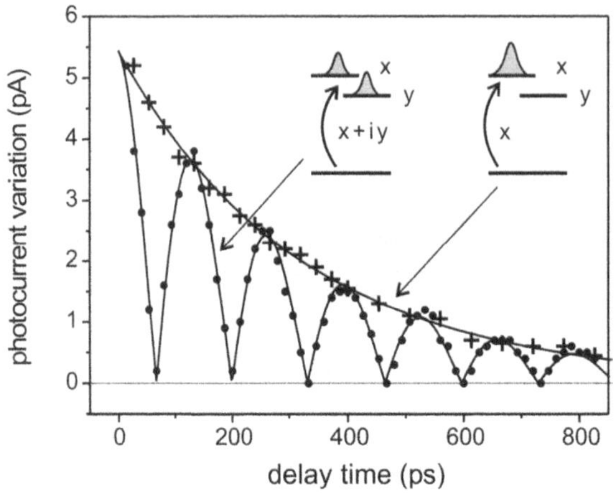

Fig. 7. Quantum interference experiment with varying delay between two $\pi/2$-pulses (bias voltage 0.4 V). Crosses correspond to measurements for which only one of the asymmetry split levels is addressed. Circles correspond to an excitation of both lines, resulting in quantum beats. Solid lines represent fit curves, from which we deduce a dephasing time of 322 ps and an asymmetry splitting of 31 μeV

more important, the observed oscillations only slightly decrease towards high pulse areas, although the excitation power at 9π is 81 times higher than at 1π. This is the first experimental proof that the generally observed strong damping at $\theta > 1\pi$ is of no principal nature, but usually is caused by sample specifics or by the measurement technique. The data displayed in Fig. 6 were measured at a bias voltage of 0.6 V, but similar results are obtained in the whole range between 0.4 V and 0.8 V. This is remarkable because the dephasing time of the system varies within this region by about a factor of 6, as discussed later in this article. The damping finally increases at voltage levels where the dephasing time is less than 30 ps, which is about 10 to 15 times the pulse-length.

While the measurement of Rabi oscillations represents the occupancy of a two-level system, we have to perform quantum interference experiments to also gain access to the phase of coherent excitations. First experiments of this kind have been done in the weak excitation regime, i.e. at pulse areas much less than 1π [9, 21]. In order to obtain relevant results on phase coherence with respect to quantum information processing, these experiments have to be extended to the strong excitation regime [4, 6, 22]. We have performed here experiments with $\pi/2$ pulses, representing a 1 qubit Hadamard transformation in context of quantum computing [23]. The first pulse thereby creates a coherent superposition of the "0" and "1" state of the QD two-level system. The second pulse then follows with a variable delay in the range of 0 to 1000 ps. The relative phase of the second pulse can be controlled via an additional fine delay with sub-fs resolution. If coherence is maintained,

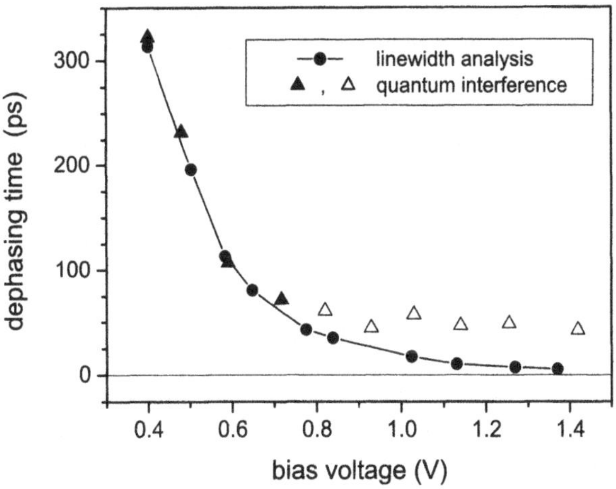

Fig. 8. Comparison of dephasing times derived from linewidth analysis and quantum interference. Both sets of data show very good agreement up to 0.7 V. At higher bias voltage, quantum beats could not be suppressed, independent of the polarization of the excitation beam

the superposition state is expected to be transferred into the pure "1" or "0" state, depending on whether the two pulses are of the same or opposite phase, respectively. When varying the phase continuously, we observe an oscillation of the PC at the same period as the optical interference at overlapping pulses. The amplitude of these oscillations versus delay time is displayed in Fig. 7. The data represented by crosses correspond to a linear polarization for which only one of the asymmetry split levels is excited. A fit to these data points reveals a purely exponential decay at delay times > 10 ps, corresponding to a dephasing time of $T_2 = 322 \pm 5$ ps.

The analysis of the first few picoseconds is complicated by the fact that an overlap of both pulses to some degree influences the measurement results. We still are able to determine some initial dephasing, though, in the best measurements, this amounts to less than 4%. Going back to Fig. 7, the circles correspond to a polarization for which both states of the asymmetry split doublet are excited. We observe here an additional beat with a period of $T = 133$ ps. As our experiment is insensitive to a phase shift of 180°, the data have to be fitted by an exponential decay modulated not by a plain cosine function but by its absolute $|\cos(\pi t/T)|$. The period T can be converted into an energy difference of $\Delta E = h/T = 31\,\mu$eV, in good agreement to the value observed in a direct cw measurement as displayed in Fig. 3b. Moreover, in expansion of previous work [24, 25, 26, 27], the modulation here amounts to the full amplitude of the reference signal.

We further are able to compare dephasing times measured by quantum interference with those derived by a linewidth analysis. At low bias and accordingly long tunneling times, however, saturation results in a broadening of the linewidth even at low excitation intensities [28]. We consequently performed a full power broadening analysis with an extrapolation to zero excitation for all measurements up to 0.6 V. At higher bias voltages the PC saturation value is high enough so that the linewidth of single low power spectra will already yield the correct results. The linewidth Γ can be converted into a dephasing time T_2 via $T_2 = 2\hbar/\Gamma$ [19]. The corresponding data are plotted versus bias voltage in Fig. 8 (full circles). The strong voltage dependence indicates that T_2 is primarily determined by the tunneling lifetime. T_2-times derived from quantum interference are shown by triangles. Both sets of data show excellent agreement up to a bias of 0.7 V. At still higher bias we observe quantum beats independent of the choice of polarization and it is therefore difficult to infer a dephasing time. The interference at the first maximum of the quantum beats (delay time 133 ps), however, is higher than expected, as indicated by open triangles. The reason for this is not fully understood up to now. Resuming the comparison of coherent versus steady state measurements, we get an agreement in several aspects: The ground state linewidth corresponds to the decay time of quantum interference. The asymmetry induced splitting of energy levels is reflected in the period of quantum beats. The polarization at which these effects are suppressed is the same in both measurements. Furthermore the power dependence of different experiments should show some kind of correlation as it is invariably determined by the transition matrix element of the QD. In saturation and power broadening measurements we indeed derive a characteristic dimensionless power level $\widetilde{P} = \Omega^2 T_1 T_2$ [19, 28]. As T_1 and T_2-times are also obtained in these measurements, we can compare the Rabi frequency Ω with a direct measurement of Rabi oscillations. In cw-measurements we typically derive a value of $\Omega \approx 0.2$ GHz at a laser power of $P = 100\,\text{nW}$. ¿From a comparison of different measurements we get the more general ratio $\Omega/\sqrt{P} = 0.19 \pm 0.4\ \text{THz}/\sqrt{\text{mW}}$. In a measurement of Rabi oscillations as shown in Fig. 6, π-pulse excitation is achieved at an *average* laser power of 2 μW. A conversion to continuous excitation[3] results in a value of $\Omega/\sqrt{P} = 0.25 \pm 0.3\ \text{THz}/\sqrt{\text{mW}}$. This is in good agreement with saturation measurements, even though the optical peak power typically differs by five orders of magnitude.

5 Summary

In summary we have performed a whole range of fundamental experiments with respect to two-level systems. All experimental results can be brought down to few basic properties of the investigated single QD, giving evidence for an almost ideal quantum system. We are able here to draw a comparison

[3] Assuming $sech^2$ pulses with an FWHM of 2.3 ps at a repetition frequency of 80 MHz.

between complementary experimental methods, so that any indirectly derived parameter can be confirmed by a direct measurement. Furthermore many results mark a major advance in experimentally proven quality, encouraging further work on this kind of quantum system. In the context of quantum computing the present work demonstrates excellent control over an exciton qubit in a semiconductor QD.

Acknowledgements

The author would like to thank T. Kuhn, V. M. Axt, and P. Machnikowski for fruitful discussions. We want to acknowledge financial support by the BMBF via 01BM466.

References

1. D. Bouwmeester, A. Ekert, A. Zeilinger, *The Physics of Quantum Information* (Springer Verlag, Berlin 2000).
2. I. I. Rabi, Phys. Rev. **51**, 652–654 (1937).
3. T. H. Stievater, Xiaoqin Li, D. G. Steel, D. Gammon, D. S. Katzer, D. Park, C. Piermarocchi, L. J. Sham, Phys. Rev. Lett. **87**, 133603 (2001).
4. H. Kamada, H. Gotoh, J. Temmyo, T. Takagahara, H. Ando, Phys. Rev. Lett. **87**, 246401 (2001).
5. A. Zrenner, E. Beham, S. Stufler, F. Findeis, M. Bichler, G. Abstreiter, Nature **418**, 612 (2002).
6. H. Htoon, T. Takagahara, D. Kulik, O. Baklenov, A. L. Holmes, Jr., C. K. Shih, Phys. Rev. Lett. **88**, 087401 (2002).
7. P. Borri, W. Langbein, S. Schneider, U. Woggon, R. L. Sellin, D. Ouyang, D. Bimberg, Phys. Rev. B **66**, 081306 (2002).
8. L. Besombes, J. J. Baumberg, J. Motohisa, Phys. Rev. Lett. **90**, 257402 (2003).
9. P. Borri, W. Langbein, S. Schneider, U. Woggon, R. L. Sellin, D. Ouyang, D. Bimberg, Phys. Rev. Lett. **87**, 157401 (2001).
10. M. Bayer, A. Forchel, Phys. Rev. B **65**, 041308 (2002).
11. F. Findeis, M. Baier, E. Beham, A. Zrenner, G. Abstreiter, Appl. Phys. Lett. **78**, 2958 (2001).
12. D. Gammon, E. S. Snow, B. V. Shanabrook, D. S. Katzer, D. Park, Phys. Rev. Lett. **76**, 3005 (1996).
13. M. Bayer, A. Kuther, A. Forchel, A. Gorbunov, V. B. Timofeev, F. Schafer, J. P. Reithmaier, T. L. Reinecke, S. N. Walck, Phys. Rev. Lett. **82**, 1748-1751 (1999).
14. Alexander Hogele, Stefan Seidl, Martin Kroner, Khaled Karrai, Richard J. Warburton, Brian D. Gerardot, Pierre M. Petroff, Phys. Rev. Lett. **93**, 217401 (2004).
15. E. Beham, A. Zrenner, F. Findeis, M. Bichler, G. Abstreiter, Appl. Phys. Lett. **79**, 2808 (2001).
16. E. Dekel, D. Gershoni, E. Ehrenfreund, D. Spektor, J. M. Garcia, P. M. Petroff, Phys. Rev. Lett. **80**, 4991 (1998).

17. L. Landin, M. S. Miller, M.-E. Pistol, C. E. Pryor, L. Samuelson, Science **280**, 262 (1998).
18. M. Bayer, O. Stern, P. Hawrylak, S. Fafard, A. Forchel, Nature **405**, 923 (2000).
19. L. Allen, J. H. Eberly, *Optical Resonance and Two Level Atoms* (Wiley, New York 1975); P. Meystre, M. Sargent III, *Elements of Quantum Optics*, 2nd ed. (Springer Verlag, Berlin 1991).
20. K. Brunner, G. Abstreiter, G. Bohm, G. Trankle, G. Weimann, Phys. Rev. Lett. **73**, 1138 (1994).
21. N. H. Bonadeo, G. Chen, D. Gammon, D. S. Katzer, D. Park, D. G. Steel, Phys. Rev. Lett. **81**, 2759 (1998).
22. P. Machnikowski, L. Jacak, Phys. Rev. B **69**, 193302 (2004).
23. P. Bianucci, A. Muller, C. K. Shih, Q. Q. Wang, Q. K. Xue, C. Piermarocchi, Phys. Rev. B **69**, 161303 (2004).
24. W. Langbein, P. Borri, U. Woggon, V. Stavarache, D. Reuter, A. D. Wieck, Phys. Rev. B **69**, 161301 (2004).
25. T. Flissikowski, A. Hundt, M. Lowisch, M. Rabe, F. Henneberger, Phys. Rev. Lett. **86**, 3172 (2001).
26. Kazuhiro Nishibayashi, Tsuyoshi Okuno, Yasuaki Masumoto, Hong-Wen Ren, Phys. Rev. B **68**, 035333 (2003).
27. A. I. Tartakovskii, J. Cahill, M. N. Makhonin, D. M. Whittaker, J-P. R. Wells, A. M. Fox, D. J. Mowbray, M. S. Skolnick, K. M. Groom, M. J. Steer, M. Hopkinson, Phys. Rev. Lett. **93**, 057401 (2004).
28. S. Stufler, P. Ester, A. Zrenner, M. Bichler, Appl. Phys. Lett. **85**, 4202-4204 (2004).

The Integrable $sl(2/1)$ Superspin Chain and the Spin Quantum Hall Effect

Holger Frahm[1], Fabian H. L. Essler[2], and Hubert Saleur[3,4]

[1] Institut für Theoretische Physik, Universität Hannover
Appelstr. 2, 30167 Hannover, Germany
[2] The Rudolf Peierls Centre for Theoretical Physics, University of Oxford
1 Keble Road, Oxford OX1 3NP, UK
[3] Service de Physique Théorique, CEA Saclay,
Gif Sur Yvette, 91191, France
[4] Dept. of Physics and Astronomy, University of Southern California,
Los Angeles, CA 90089, USA

Summary. The analysis of network models – either directly or by numerical studies of related quantum spin chains with super group symmetry – allows to gain insights into the delocalization transition of non-interacting two-dimensional electrons with disorder. Critical exponents determined this way are an important step towards the identification of the conformal field theory (CFT) describing the universal physics at the transition. To collect more information on the universal properties integrable versions of the corresponding superspin chains have been proposed. A simplified version avoiding some of the technical problems of the original problem is the $sl(2|1)$ superspin chain related to the so-called spin quantum Hall transition. Analyzing the ground state and some low-lying excitations of this integrable chain we determine the central charge of the underlying CFT and obtain evidence for an infinite degeneracy of the finite size gaps that are possibly related to a continuous spectrum of the CFT. Finally, we compare our findings for the integrable chain with known properties of the (non-integrable) network model for the spin QHE.

1 Introduction

The nature of the transitions between the quantized plateaus of the Hall conductance observed in a two-dimensional electron gas subject to strong magnetic fields has been a focus of research since the discovery of the quantum Hall (QH) effect [1]. For the integer effect, which can be described in terms of non interacting electrons, the plateau transition with divergent correlation length is a consequence of the interplay of disorder favoring localization of electrons and the Landau quantization of the kinetic energy causing delocalization at special values of the Fermi energy.

An convenient description of the physics at the transition in such systems is the sigma model approach [2, 3, 4] where the plateau transition is associated with a fixed point of the sigma model at a topological angle $\theta = \pi$. The strong coupling nature of this fixed point, however, did not allow to extract the conformally invariant effective theory at low energies by means of the

B. Kramer (Ed.): Adv. in Solid State Phys. **45**, 185–196 (2005)

perturbative methods available which then would allow for a systematic determination of critical exponents known experimentally and from numerical studies.

A different approach to this problem comes from network models [5] describing charge transport across a two-dimensional QH sample in a slowly varying random potential on a square network with random scattering at its nodes. In a suitable spatially anisotropic limit the functional integral of the disorder averaged theory can be interpreted as a coherent state path integral representation of a superspin chain [6, 7, 8] with Hilbert space of type

$$\cdots \otimes V \otimes \bar{V} \otimes V \otimes \bar{V} \otimes \cdots \tag{1}$$

(for the integer QHE V and $\bar{V}$ are an infinite-dimensional module of the super algebra $gl(2|2)$ and its dual). The Hamiltonian is given in terms of the invariant second Casimir of the algebra describing nearest neighbor interactions of "Heisenberg" type $\boldsymbol{S} \cdot \boldsymbol{S}$. The same superspin Hamiltonian can be obtained as a lattice regularized version of the sigma model at $\theta = \pi$ [9].

This mapping to a superspin chain has opened the possibility to apply well-established techniques used to study quantum spin systems in one dimension for the identification of the universality class of the quantum critical point of the plateau transition. Therefore it is natural to wonder whether the critical exponents can be obtained from an integrable deformation of the superspin Hamiltonian [10]. The building blocks for a vertex model of the required symmetry with commuting transfer matrices have been constructed but various technical problems have prevented the computation of its spectrum so far: apart from the complications arising from the representation theory of the super algebra it is in particular the infinite dimension of the modules V, $\bar{V}$ which inhibits the application of standard Bethe ansatz methods. At the same time, even if these difficulties could be resolved, the question remains whether the deformation leaves the integrable chain in the same universality class as the original delocalization problem.

To address this last question we consider a simplified version of the superspin chain defined on a Hilbert space of the type (1) with V and $\bar{V}$ being the two conjugate three-dimensional representations of $sl(2|1)$. Such a model describes the disorder transition in a system without time-reversal symmetry but unbroken $SU(2)$-spin symmetry (class C in [11]) – the so-called spin QHE. The problem has been shown to be equivalent to bond percolation on the square lattice [12] for which conformal dimensions are known [13]. An integrable deformation for this $sl(2|1)$ has been constructed [14] and the algebraic Bethe ansatz has be applied to this system [15, 16]. Here we present results for the finite size spectrum of this superspin chain obtained from the analysis of these Bethe ansatz equations. We find the central charge and some conformal exponents of the integrable chain which are consistent with the $SU(2|1)$ level-1 WZW model as the continuum limit of this model – quite different from what is known for the spin quantum Hall effect [13]. The macroscopic degeneracies of the scaling dimensions appearing in the analysis of the lattice model are discussed in view of what is known for the WZW model.

2 The Integrable Vertex Model

An integrable vertex model constructed from R-matrices satisfying Yang-Baxter equations (YBEs) related to the spin quantum Hall effect has been suggested in Ref. [14]. The R-matrices are acting on tensor products of spaces carrying the fundamental representation 3 of the super Lie algebra $sl(2/1)$ and its dual $\bar{3}$ and read

$$\begin{aligned} R_{33}(v) &= 1 - \frac{2}{v}\mathcal{P} , \qquad R_{\bar{3}\bar{3}}(v) = 1 - \frac{2}{v}\mathcal{P} , \\ R_{3\bar{3}}(v) &= 1 + \frac{2}{v}\mathcal{O} , \qquad R_{\bar{3}3}(v) = 1 + \frac{2}{v}\mathcal{O} \end{aligned} \tag{2}$$

(see also [15, 16]). Here $\mathcal{P}$ and $\mathcal{O}$ are the graded permutation and monoid operators of $sl(2/1)$, respectively. From these R-matrices two families of transfer matrices are constructed[5]

$$\begin{aligned} \tau_3(v) &\equiv \mathrm{str}_0 \left(\prod_{n=1}^{L} R_{33}^{(n)}(v) R_{3\bar{3}}^{(n)}(v-1) \right) , \\ \tau_{\bar{3}}(v) &\equiv \mathrm{str}_0 \left(\prod_{n=1}^{L} R_{\bar{3}3}^{(n)}(v-1) R_{\bar{3}\bar{3}}^{(n)}(v) \right) \end{aligned} \tag{3}$$

where $R_{\alpha\beta}^{(n)}$ acts in the tensor product of a three-dimensional "quantum space" $V^{(n_\beta)}$ and a matrix space V^α carrying representations $\alpha, \beta \in \{3, \bar{3}\}$. Products of R-matrices and the super trace are taken in the matrix space. As a consequence of the YBEs satisfied by the R-matrices the transfer matrices $\tau_\alpha(v)$ commute for arbitrary values of the variable v. The integrable vertex model is obtained by staggering of the transfer matrices (3) as shown in Fig. 1.

The corresponding Hamiltonian of a superspin chain is obtained by taking the logarithmic derivative of the two-layer transfer matrix (see also [15])

$$\mathcal{H} = -\, \partial_v \ln \left\{ \left(\frac{v}{v-2} \right)^{2L} \tau_3(v) \tau_{\bar{3}}(v) \right\} \Bigg|_{v=0} . \tag{4}$$

$\mathcal{H}$ contains a Heisenberg type coupling of the $sl(2|1)$ generators on neighboring sites proportional to $\boldsymbol{S}_j \cdot \boldsymbol{S}_{j+1}$, the invariant product of the algebra [17], just as the superspin chain for the spin quantum Hall effect [12]. For the

[5] A more general construction involving a free parameter can be related to the so-called quantum transfer matrix of the integrable supersymmetric t–J model [19, 20].

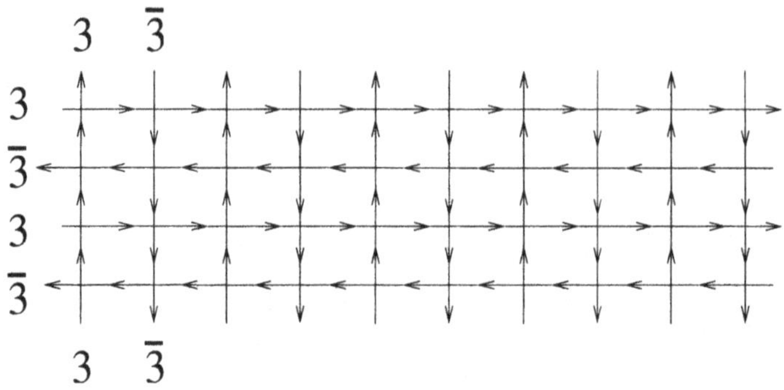

Fig. 1. The integrable vertex model: arrows pointing upwards or right (downwards or left) denote links carrying the fundamental 3 (the dual $\bar{3}$) representation of $sl(2|1)$. The vertices weights are given by the R-matrices (2)

integrability additional terms involving operators on three neighboring sites are contained in (4).

The spectrum of the transfer matrices has been obtained by means of the graded quantum inverse scattering method in Ref. [15]. For a particular choice of the grading the eigenvalues and eigenvectors with $sl(2|1)$ quantum numbers $b = (N - M)/2$ and $s^3 = L - (M + N)/2$ are parameterized by solutions to the Bethe ansatz equations

$$\begin{aligned} \left(\frac{u_j + i}{u_j - i}\right)^L &= \prod_{\beta=1}^{M} \frac{u_j - \gamma_\beta + i}{u_j - \gamma_\beta - i}, \qquad j = 1, \ldots, N \\ \left(\frac{\gamma_\alpha + i}{\gamma_\alpha - i}\right)^L &= \prod_{k=1}^{N} \frac{\gamma_\alpha - u_k + i}{\gamma_\alpha - u_k - i}, \qquad \alpha = 1, \ldots, M\,. \end{aligned} \tag{5}$$

As long as $b \neq \pm s$ the corresponding eigenstates are $sl(2|1)$ highest weight states of the typical representation $[b, s]$ with $s = s^3$ [15]. In terms of the solutions to Eqs. (5) the eigenvalues of the integrable Hamiltonian (4) are given as

$$E(\{u_k\}, \{\gamma_\beta\}) = -\sum_{k=1}^{N} \frac{2}{u_k^2 + 1} - \sum_{\beta=1}^{M} \frac{2}{\gamma_\beta^2 + 1}\,. \tag{6}$$

3 Spectrum of the Supersymmetric Spin Chain

To study the spectrum of the spin chain (4) we first have to find the relevant configurations of roots u_j, γ_α to the Bethe equations (5). From the analytical properties of the phase factors appearing on the right hand side of these equations and using results from explicit diagonalization of the transfer matrix and the Hamiltonian on small lattices with up to 10 sites (corresponding to $L = 1, \ldots, 5$) we are able to identify eigenstates of these operators and the multiplicities of the corresponding eigenvalues with solutions to Eqs. (5). Based on this information the properties of the ground state and low-lying excitations as the thermodynamic limit ($L \to \infty$ with M/L, N/L fixed) is approached can be determined by standard Bethe ansatz methods.

3.1 The Ground State

One result of this analysis which can be established rigorously is that there is only one singlet in the decomposition of the tensor product $(3 \otimes \bar{3})^{\otimes L}$ into representations of $sl(2|1)$ for any L. This does not imply however, that this state is unique, since it can always be combined with invariant states from other reducible but indecomposable representations appearing in these tensor products [17, 18] (such states are annihilated by all generators of the algebra but appear in the image of another state under their action). Hence, although the singlet appears in the spectrum only once, its exact form will depend on the Hamiltonian considered.

The eigenvalue of the transfer matrix $\tau_3(v)\tau_{\bar{3}}(v)$ corresponding to this singlet turns out to be of a particularly simple form

$$\Lambda(v) = \left(\frac{(v-2)(v+1)}{v(v-1)} \right)^{2L} . \tag{7}$$

Interestingly, there exists no non-degenerate solution of the Bethe equations (5) corresponding to this eigenvalue. By careful analysis we were able to verify that the singlet is described by the *degenerate* solution $u_k \equiv 0$, $\gamma_\alpha \equiv 0$, $k, \alpha = 1, \ldots, L$ (a similar degeneration is observed in [20]). The singlet is found to be the ground state of the Hamiltonian (4), from Eq. (6) we find that its energy, $E_0 = -4L$, is proportional to the system size without any finite size corrections. This implies that the central charge of the conformal field theory arising in the continuum limit of the lattice model vanishes, $c = 0$.

3.2 Low-Lying Excitations

A particular class of roots to the Bethe equations (5) which – based on our studies of small systems – correspond to low energy excited states, turn out to be built from complexes involving one rapidity $u = x \pm i/2$ and one rapidity $\gamma = u^* = x \mp i/2$ ("type-$\pm$ strange strings" in the classification of Ref. [19]). We proceed by studying configurations obtained by combining N_+ (N_-) strange strings of type + (−)

$$\begin{aligned} u_j^{(-)} &= x_j - \tfrac{i}{2} + \epsilon_{j,-} = \left(\gamma_j^{(-)}\right)^* , \qquad j = 1, \ldots, N_- , \\ u_k^{(+)} &= y_k + \tfrac{i}{2} + \epsilon_{k,+} = \left(\gamma_k^{(+)}\right)^* , \qquad k = 1, \ldots, N_+ \end{aligned} \tag{8}$$

with small corrections $\epsilon_\pm = \mathcal{O}(1/L)$. Applying standard Bethe ansatz methods we find that in the infinite volume limit the corresponding states – up to corrections of order $1/L$ – are degenerate with the singlet ground state for $N_+ + N_- \approx L$ and $\Delta N = N_+ - N_- = \mathcal{O}(L^0)$.

The Takhtajan-Babujian Subset of Excitations.

In the special case $\Delta N = 0$ corresponding to equal number of type-+ and type-− strings, the two sets (8) degenerate into a configuration with $x_j \equiv y_j$, $j = 1, \ldots, N$ after suitable relabeling of the indices. Using this identity the Bethe equations (5) and – up to an overall normalization – the spectrum (6) of the superspin chain in this sector becomes identical to that of the spin-1 antiferromagnetic Takhtajan-Babujian (TB) chain with L lattice sites and *antiperiodic* boundary conditions [21, 22]. In the periodic case this model is known to scale to a $SU(2)$ WZW model at level $k = 2$, the conformal dimensions in the presence of a twist have been computed by Alcaraz and Martins [22]. Thus, by taking into account the different energy scales in the TB chain and the superspin chain as well as the offset of the energies with respect to the ground state (7) of the latter we obtain an important subset of the finite size spectrum and the corresponding conformal dimensions in the superspin model studied here. Note that this approach does not allow to deduce the degeneracies of the levels in the superspin chain. From the equivalence of the Bethe ansatz equations we know the $sl(2|1)$ quantum numbers of the corresponding states in the latter. In particular, we have $b = 0$ and the $SU(2)$ spin s of the state is related to the corresponding quantum number in the TB states. The lowest scaling dimensions appearing in this subset together with these quantum numbers for odd and even length L of the TB chain are given in Table 1.

Similarly, by considering the case of antiperiodic boundary conditions for the superspin chain in the sector $N = M$, the Bethe equations are mapped onto the ones for the TB chain with *periodic* boundary conditions (note that all states in this sector have integer total $SU(2)$ spin in the $sl(2|1)$ chain, hence one needs to consider only bosonic states and no subtleties arise in the mapping the boundary conditions of the superspins to those of the $SU(2)$ spins of the TB chain). This allows to deduce the Takhtajan-Babujian part of the conformal spectrum of the superspin chain in the Neveu-Schwarz sector (see our discussion below).

Excitations with $N_+ \neq N_-$.

The conformal dimensions corresponding to the more general excitations with $\Delta N \neq 0$, i.e. different numbers $N_\pm$ of strange strings (8), cannot be obtained from the relation to the Takhtajan-Babujian chain. Again, it is useful to consider these states in the thermodynamic limit first. Computing the finite size spectrum of such excitations by standard Bethe ansatz methods (with proper regularization of the divergent kernel of the integral operator describing the

Table 1. Smallest conformal dimensions $(h, \bar{h})$ and corresponding typical $sl(2|1)$ representations $[b, s]$ in the Takhtajan-Babujian subset of the spectrum for odd/even L. Note that the $b = s = 0$ states will appear in the spectrum as invariant states within indecomposable representations rather than "true" $sl(2|1)$ singlets

$[b, s]$	$(h, \bar{h})$ for L odd
$(0, 0)$	$(0, 0)$, $(\frac{9}{8}, \frac{1}{8})$, $(\frac{1}{8}, \frac{9}{8})$, $(1, 1)$
$[0, 1]$	$(\frac{1}{8}, \frac{1}{8})$, $(1, 0)$, $(0, 1)$, $(\frac{17}{8}, \frac{1}{8})$, $(\frac{1}{8}, \frac{17}{8})$
$[0, 2]$	$(1, 1)$, $(\frac{25}{8}, \frac{1}{8})$, $(\frac{1}{8}, \frac{25}{8})$, $(4, 0)$, $(0, 4)$

$[b, s]$	$(h, \bar{h})$ for L even
$(0, 0)$	$(0, 0)$, $(\frac{1}{8}, \frac{1}{8})$, $(1, 1)$, $(\frac{9}{8}, \frac{9}{8})$, $(\frac{17}{8}, \frac{17}{8})$
$[0, 1]$	$(\frac{9}{8}, \frac{1}{8})$, $(\frac{1}{8}, \frac{9}{8})$, $(1, 0)$, $(0, 1)$, $(\frac{25}{8}, \frac{1}{8})$, $(\frac{1}{8}, \frac{25}{8})$
$[0, 2]$	$(1, 1)$, $(\frac{17}{8}, \frac{1}{8})$, $(\frac{1}{8}, \frac{17}{8})$, $(4, 0)$, $(0, 4)$

scattering between type-+ and type-− complexes) one obtains the following result for the finite size gaps of these excitations as compared to the lowest one which is expected to lie in the TB sector [19]:

$$\frac{L}{2\pi v}\Delta E = \frac{1}{2}(\Delta N_+ + \Delta N_-)^2 + 0 \times (\Delta N_+ - \Delta N_-)^2 \\ + \frac{1}{8}(D_+ + D_-)^2 + \infty \times (D_+ - D_-)^2 \,. \tag{9}$$

Here $v = \pi$ is the characteristic (Fermi) velocity of the excitations in the superspin chain, $\Delta N_\pm$ are integers counting the change in the number of strings of type $\pm$ compared to the lowest energy state of this type and $D_\pm$ are the number of solutions "backscattered" from the left to the right Fermi point for type $\pm$ strings.

While it is well known that the computation of the finite size gaps for Bethe states involving complex solutions involves certain subtleties – dimensions of other operators and additional selection rules may appear – the striking qualitative feature in Eq. (9) is expected to be present in the true spectrum of the superspin chain: replacing a finite number of type-+ strings with type-− strings in a given configuration, i.e. changing $N_\pm$ while keeping $N_+ + N_-$ constant such that only $\Delta N = \Delta N_+ - \Delta N_-$ is varied, does not lead to a different finite size gap to order $1/L$. To verify this degeneration of the finite size gaps we have performed an extensive numerical analysis of the Bethe equations (5) for large but finite systems up to $L \approx 5.000$.

The results of this finite-size scaling analysis for several low-lying $[0, 1]$ $sl(2|1)$ octet states (with $N_+ + N_- = L - 1$ for different $N_\pm$) as well as the octet and the indecomposable $[0, -1/2, 1/2, 0]$ representations (the latter with $N_+ + N_- = L$) in the Takhtajan-Babujian sector are summarized in Fig. 2.

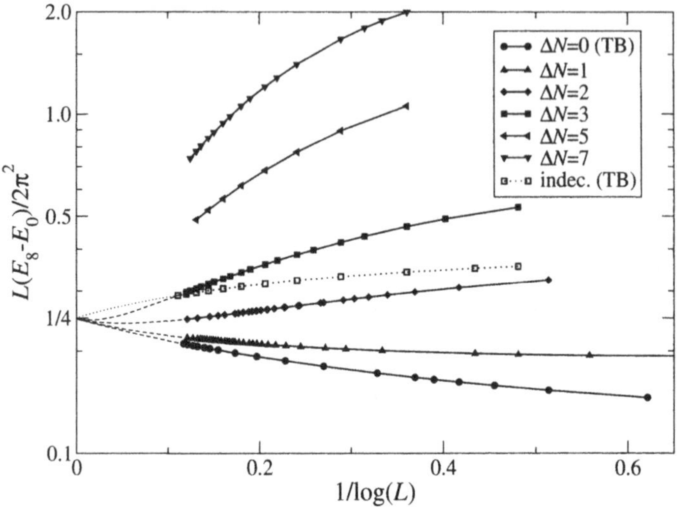

Fig. 2. Corrections to scaling of the energies (9) of some low lying eigenstates of the superspin chain: full symbols denote $sl(2|1)$ octets, open symbols are for the indecomposable representation $[0, -1/2, 1/2, 0]$. The dashed lines indicate a rational function extrapolation of the finite size data to $L \to \infty$

We plot the scaling of their energies expected from conformal invariance

$$\frac{L}{2\pi v}\left[E(N_+, N_-) - E_0\right] \xrightarrow[L\to\infty]{} h + \bar{h} = x \ , \tag{10}$$

(x is the scaling dimension) as a function of the lattice length for several different excited states. Although there are strong logarithmic corrections to the energy gaps we find within rational function extrapolation of our numerical data that the energies of the states considered all become degenerate to order $1/L$ in the thermodynamic limit – in complete agreement with the result (9) of the qualitative analysis based on the string solutions to the Bethe equations. The scaling dimension obtained from this extrapolation of (10) is

$$x = \frac{1}{4} \ . \tag{11}$$

Based on our findings we expect this scaling dimension to be macroscopically degenerate in the thermodynamic limit.

For a proper identification of the continuum limit the analysis of the finite size gaps has to be extended to the case of antiperiodic boundary conditions (Neveu-Schwarz sector). Again the relevant root configurations for the lowest excitations are built from the strings (8). The details of our numerical analysis of these states can be found in Ref. [19], here we just present the main results which give support the claim that there are many states degenerate with the lowest state in this sector (belonging to the Takhtajan-Babujian part of the

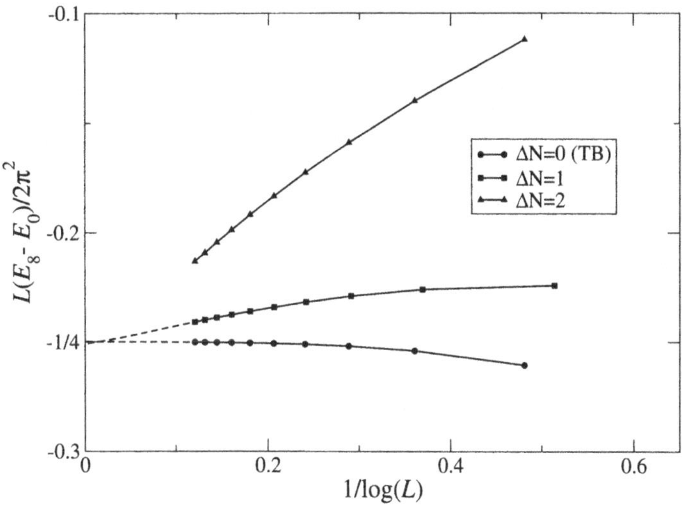

Fig. 3. Same as Fig. 2 but for antiperiodic boundary conditions. The gaps are measured with relative to the ground state energy E_0 in the sector with periodic boundary conditions

spectrum) in the thermodynamic ($L \to \infty$) limit. All of these states have $N_+ + N_- = L$, but different values of $\Delta N = N_+ - N_-$. The scaling dimensions for $\Delta N = 0, 1, 2$ are found to extrapolate to $x = -1/4$ relative to the ground state of the superspin chain with periodic boundary conditions (see Fig. 3). The resulting effective central charge of the superspin chain in this sector is thus $c_{\text{eff}} = 3$. This is twice the value of the $SU(2)$ level-2 WZW model corresponding to the continuum limit of the TB chain as a consequence of the different normalization of the $sl(2|1)$ chain.

4 Continuum Limit of the Lattice Model

Our results for the central charge, $c = 0$ in the Ramond sector and $c_{\text{eff}} = 3$ in the Neveu-Schwarz sector, and the part of the spectrum of scaling dimensions identified from the mapping to the spin-1 Takhtajan-Babujian chain are a strong indication that the integrable $sl(2|1)$ superspin chain is in the universality class of the $SU(2|1)$ WZW model at level $k = 1$. It is known that this model admits a free field representation [23, 24]: the fundamental quartet of the theory is a $[b, s] = [0, 1/2]$ representation of the algebra and can be expressed in terms of a pair of symplectic fermions [25] and a pair of bosons. Four admissible characters for this theory have been identified [26]: in the Ramond sector (periodic fermions on the cylinder) these are the characters for the field $[0, 1/2]$ with conformal weights $h = 1/8$ modulo integers and for the identity field ($h = 0 \mod$ integers). In the Neveu Schwarz sector (with antiperiodic boundary conditions for the fermions) the lowest energy state has negative conformal dimension $h = -1/8$ leading to an effective central charge $c_{\text{eff}} = 3$. The other conformal weight appearing in this sector is $h = 1/4$. Both characters in the Neveu Schwarz sector show the existence of half integer gaps, hence additional conformal weights corresponding to $h = \frac{3}{8}$ and $h = \frac{3}{4}$ modulo integers are expected. Our analysis of the superspin chain shows that they are not in the TB part of the spectrum. Their existence waits to be shown by a more detailed study of the Bethe ansatz equations (5).

On the other hand, the huge degeneracy of conformal weights established by our study of states outside the TB part of the spectrum emphasizes the need for a better understanding of the WZW theory: For one, it is easy to see that the free field representation mentioned above is incomplete: The tensor product of the $[0, 1/2]$ representation (this is the multiplet of the fundamental field of the theory) with itself [17]

$$[0, 1/2] \otimes [0, 1/2] = [0, 1] + [0, 1/2, -1/2, 0] \tag{12}$$

does not contain the true singlet $[0, 0]$. The fields in the adjoint representation $[0, 1]$ are the currents. Similarly, from the OPE associated with (12) for the free field representation one can express the fields in the indecomposable representation $[0, 1/2, -1/2, 0]$ in terms of the symplectic fermions and the bosons. The conformal dimension of the fields appearing in this representation is vanishing. However, since the partition function of the theory is normalized to 1 and the super dimension of the multiplet is also vanishing, there must exist an additional "true" identity field which is not embedded in a bigger indecomposable. Therefore we conclude that the free field representation has to be supplemented by another identity field.

In addition, many other indecomposable representations associated with fields of vanishing conformal weight can be built starting from the free fundamental fields and then be embedded in larger structures, thereby producing fields with arbitrarily large *negative dimensions* [19]. Their appearance can be related to the fact that the invariant metric of the super group is not positive definite, leading to an ill defined functional integral.

It is not clear how this presence of a spectrum which is unbounded from below in a conformal field theory defined through a free field representation of the $SU(2|1)$ current algebra can be related to the suggestion that this theory is a candidate for the continuum limit of a lattice model such as (4) for which we have established the existence of a unique ground state. Based on calculations involving modular transformations of the known characters, however, it can be argued that the infinite degeneracy of the conformal weights found in the lattice model may be the result of a hitherto unknown regularization scheme replacing the unbounded spectrum of negative dimensions by a continuous spectrum of positive ones [19].

5 Summary and Conclusions

Using both analytical and numerical methods we have analyzed the ground state and part of the low-lying excitations of an integrable superspin chain which has been introduced originally to shed some light on the critical properties at the delocalization transition in the so-called spin quantum Hall universality class. Within our finite size scaling analysis we have identified a subset of conformal dimensions based on which we conclude that the continuum limit of the model is an $SU(2|1)$ WZW model at level 1. An explanation of the infinite degeneracy of the conformal weights identified from the study of the lattice model is beyond the present understanding of the WZW theory. More work in this area is needed and is believed to be crucial for the CFT description of the plateau transition.

At the same time, it should be emphasized that there remain quite a few open questions regarding the spectrum of the superspin chain. In particular, there is no one-to-one mapping between the possible configurations of roots to the Bethe ansatz equations (5) and the eigenstates of the system. This together with the complicated representation theory of the super algebra leave the question of completeness of the Bethe states unanswered for now.

Finally, we come back to our original question concerning the universality of $3 \otimes \bar{3} \otimes 3 \otimes \bar{3} \ldots$ $sl(2/1)$ spin chains. It is easy to check that the analysis in [13] for the "Heisenberg" model (with bilinear $\boldsymbol{S} \cdot \boldsymbol{S}$ nearest neighbor coupling only) in the Ramond sector, gives a totally different spectrum. First of all, the multiplicities of conformal weights are all finite. Then, scalar operators with dimensions

$$h = \bar{h} = \frac{(3P+1)^2 - 1}{24}, \quad h = \bar{h} = \frac{4M^2 - 1}{24} \tag{13}$$

appear, as well as many other non scalar operators whose chiral weights generically read (modulo integers)

$$h = \frac{(3P/N + 2M)^2 - 1}{24}, \quad N|M \tag{14}$$

Except for the lowest ones, there is no indication that any of these appear in the spectrum of the integrable $sl(2/1)$ chain studied in this paper. In particular no excitation with gap scaling to the dimension $h = \bar{h} = \frac{5}{8}$ of the "two-hull operator" of 2D percolation is found in the integrable chain.

Hence we conclude that the integrable model describes a different fixed point. Since fine tuning of nearest neighbor and next nearest neighbor couplings is required, it is expected to be the less stable one when compared to the one describing the plateau transition of the spin QHE. To further this expectation the RG flow in the complete phase diagram of the 3-$\bar{3}$ superspin chains needs to be analyzed in more detail. Progress in this direction calls, again, for more work on the conformal field theory side. For the lattice models, different representations e.g. of the ground states with trivial energies for both the integrable and the Heisenberg model might be useful.

A result of the present study is that generic physical properties of the plateau transition can not easily be inferred by analyzing an integrable chain with the same symmetry. Although we have focused here on the spin QHE where some technical complications of the superspin chain approach are absent, this observation presumably holds for the ordinary integer QHE as well.

Acknowledgements

We are grateful to J. Germoni, G. Goetz, V. Gurarie, A. Klümper, J. Links, A.W.W. Ludwig, M. Martins, N. Read, V. Schomerus, M. Shiroishi, J. Teschner and A.M. Tsvelik for helpful discussions. This work was supported in part by the Deutsche Forschungsgemeinschaft (HF), the DOE (HS), the EPSRC (FE), the Humboldt Foundation (HS), and the TMR network EUCLID.

References

1. K. v. Klitzing, G. Dorda, M. Pepper, Phys. Rev. Lett. **45** 494 (1980).
2. A. M. M. Pruisken, Nucl. Phys. B**235**, 277 (1984).
3. D. E. Khmelńitskii, JETP Lett. **38**, 454 (1983).
4. H. A. Weidenmüller, Nucl. Phys. B**290** 87 (1987).
5. J. T. Chalker, P. D. Coddington, J. Phys. C**21**, 2665 (1987).
6. D. H. Lee, Phys. Rev. B**50**, 10788 (1994),
7. M. Zirnbauer, J. Math. Phys. **38**, 2007 (1997).
8. J. Kondev, J. B. Marston, Nucl. Phys. B**497** 639 (1997).
9. M. Zirnbauer, Ann. der Physik **3**, 513 (1994).
10. R.M. Gade, J. Phys. A**31**, 4909 (1998).
11. A. Altland, M. Zirnbauer, Phys. Rev. B**55**, 1142 (1997).
12. I. A. Gruzberg, A. W. W. Ludwig, N. Read, Phys. Rev. Lett. **82**, 4524 (1999).
13. N. Read, H. Saleur, Nucl. Phys. B**613**, 409 (2001).
14. R.M. Gade, J. Phys. A**32**, 7071 (1999).
15. J. Links, A. Foerster, J. Phys. A**32**, 147 (1999).
16. J. Abad, M. Ríos, J. Phys. A**32**, 3535 (1999).
17. M. Scheunert, W. Nahm and V. Rittenberg, J. Math. Phys. **18**, 146 (1977).
18. M. Marcu, J. Math. Phys. **21**, 1277 (1980); *ibid.* **21**, 1284 (1980).
19. F. H. L. Essler, H. Frahm, H. Saleur, Nucl. Phys. B**712**, 513 (2005).
20. G. Jüttner, A. Klümper, J. Suzuki, Nucl. Phys. B**487**, 650 (1997).
21. L. Takhtajan, Phys. Lett. A**87**, 479 (1982); H. Babujian, Nucl. Phys. B**215**, 317 (1983),
22. F.C. Alcaraz, M.J. Martins, J. Phys. A**23**, 1439 (1990); *ibid.* A**22**, 1829 (1989),
23. P. Goddard, D. Olive, G. Waterson, Comm. Math. Phys. **112**, 591 (1987).
24. A. W. W. Ludwig, A free field representation of the $Osp(2|2)$ current algebra at level $k = -2$, and Dirac fermions in a random $SU(2)$ gauge potential, cond-mat/0012189.
25. V. Gurarie, M. Flohr, C. Nayak, Nucl. Phys. B**498** 513 (1997).
26. P. Bowcock, M. Hayes, A. Taormina, Nucl. Phys. B**510**, 739 (1998).

Intrinsic Spin Hall Effect

Shuichi Murakami[1]

Department of Applied Physics, University of Tokyo,
Hongo, Bunkyo-ku, Tokyo 113-8656, Japan

Summary. A brief review is given on a spin Hall effect, where an external electric field induces a transverse spin current. It has been recognized over 30 years that such effect occurs due to impurities in the presence of spin-orbit coupling. Meanwhile, it was proposed recently that there is also an intrinsic contribution for this effect. We explain the mechanism for this intrinsic spin Hall effect. We also discuss recent experimental observations of the spin Hall effect.

1 Introduction

In the emerging field of spintronics [1], it is important to understand the nature of spins and spin current inside semiconductors. There have been many proposals for semiconductor spintronics devices, whereas their realization remains elusive. One of the largest obstacles is an efficient spin injection into semiconductors. One way is to make semiconductors ferromagnetic, such as (Ga, Mn)As [2]. The Curie temperature is, however, still lower than the room temperature, and there are still rooms for improvement towards practical use.

Spin Hall effect (SHE) can be an alternative way for efficiently injecting spin current into semiconductors. In the first proposal of the spin Hall effect by D'yakonov and Perel [3], followed by several papers [4, 5], the SHE has been considered as an extrinsic effect, due to impurities in the presence of spin-orbit (SO) coupling. Nevertheless, quantitative estimate for this extrinsic SHE is difficult, and this extrinsic effect is not easily controllable.

In 2003 two groups independently proposed an intrinsic spin Hall effect in different systems. Murakami, Nagaosa, and Zhang [6] proposed it in p-type semiconductors like p-GaAs. On the other hand, Sinova et al. [7] proposed a spin Hall effect in n-type semiconductors in two-dimensional heterostructures. This induced spin current is dissipationless, and can flow even in nonmagnetic materials. It shares some features in common with the quantum Hall effect.

Because the predicted effect is large enough to be measured, even at room temperature in principle, this intrinsic SHE attracted much attention, and many related works, have been done. Nevertheless there remain several issues, relevant also for experiments. One of the important questions is disorder effect. While there are a lot of works on disorder effect, the most striking one is by Inoue et al. [8, 9]. They considered dilutely distributed impurities

B. Kramer (Ed.): Adv. in Solid State Phys. **45**, 197–209 (2005)

with short-ranged potentials, and calculate the SHE, incorporating the vertex correction in the ladder approximation. Remarkably the resulting spin Hall conductivity is exactly zero in the clean limit. This work made many people to consider that the SHE is "fragile" to impurities; namely, only a small amount of impurities will completely kill the intrinsic SHE. However, this is not in general true. In fact, the spin Hall conductivity is in general nonzero even in the presence of disorder, as we see later.

In such circumstances, two seminal experiments on the SHE have been done. Kato et al. [10] observed spin accumulation in n-type GaAs by means of Kerr rotation. Wunderlich et al. [11] observed a circularly polarized light emitted from a light-emitting diode (LED) structure, confirming the SHE in p-type semiconductors. Separation between the intrinsic and extrinsic SHE for these experimental data is not straightforward, and is still under debate.

The paper is organized as follows. In Sect. 2 we explain basic mechanisms and features for the intrinsic SHE. Section 3 is devoted to a disorder effect on the SHE. In Section 4 we collect a number of recent interesting topics on the SHE. In Section 5 we introduce two recent experimental reports on the SHE. We conclude the paper in Sect. 6.

2 Intrinsic Spin Hall Effect

In this section we explain the two theoretical proposals for the intrinsic SHE.

2.1 Spin Hall Effect in p-Type Semiconductors

We begin with semiclassical description of the SHE, and apply it to the p-type semiconductors [6]. In this description, we introduce a "Berry phase in momentum space". The Berry phase [12, 13, 14] is a change of a phase of a quantum state caused by an adiabatic change of some parameters. As we explain later, Berry phase in momentum space gives rise to the Hall effect, as first demonstrated for the quantum Hall effect [15, 16, 17]. Here, the wavevector $\boldsymbol{k}$ is regarded as adiabatically changing due to a small external electric field. In two-dimensional systems, for example, the Hall conductivity σ_{xy} in a clean system is calculated from the Kubo formula as

$$\sigma_{xy} = -\frac{e^2}{2\pi h} \sum_n \int_{\mathrm{BZ}} d^2k \; n_F(\epsilon_n(\boldsymbol{k})) B_{nz}(\boldsymbol{k}), \tag{1}$$

where n is the band index, and the integral is over the entire Brillouin zone. $B_{nz}(\boldsymbol{k})$ is defined as a z component of $\boldsymbol{B}_n(\boldsymbol{k}) = \nabla_{\boldsymbol{k}} \times \boldsymbol{A}_n(\boldsymbol{k})$, where

$$A_{ni}(\boldsymbol{k}) = -i \left\langle n\boldsymbol{k} \left| \frac{\partial}{\partial k_i} \right| n\boldsymbol{k} \right\rangle \equiv -i \int_{\text{unit cell}} u^{\dagger}_{n\boldsymbol{k}} \frac{\partial u_{n\boldsymbol{k}}}{\partial k_i} d^2x, \tag{2}$$

and $u_{n\boldsymbol{k}}(\boldsymbol{x})$ is the periodic part of the Bloch wavefunction $\phi_{n\boldsymbol{k}}(\boldsymbol{x}) = e^{i\boldsymbol{k}\cdot\boldsymbol{x}} u_{n\boldsymbol{k}}(\boldsymbol{x})$. This $B_{nz}(\boldsymbol{k})$ represents the effect of Berry phase in momentum space. $n_F(\epsilon_n(\boldsymbol{k}))$ is the Fermi distribution function for the n-th band. This intrinsic Hall conductivity (1) was first recognized in the paper by Karplus and Luttinger [18]. This Berry phase in momentum space has been studied in the recent works on anomalous Hall effect (AHE) [19, 20, 21, 22, 23, 24], as well as those on the SHE.

By incorporating the effect of $\boldsymbol{B}(\boldsymbol{k})$, the Boltzmann-type semiclassical equation of motion (SEOM) acquires an additional term [17],

$$\dot{\boldsymbol{x}} = \frac{1}{\hbar}\frac{\partial E_n(\boldsymbol{k})}{\partial \boldsymbol{k}} + \dot{\boldsymbol{k}} \times \boldsymbol{B}_n(\boldsymbol{k}), \quad \hbar\dot{\boldsymbol{k}} = -e(\boldsymbol{E} + \dot{\boldsymbol{x}} \times \boldsymbol{B}(\boldsymbol{x})). \tag{3}$$

The term $\dot{\boldsymbol{k}} \times \boldsymbol{B}_n(\boldsymbol{x})$ represents the effect of Berry phase, and it is called an anomalous velocity. Under the external electric field, the anomalous velocity becomes perpendicular to the field, and gives rise to the Hall effect. This Hall current is distinct from the usual Ohmic current, which comes from the shift of the Fermi surface from its equilibrium. This Hall effect comes from all the occupied states, not only from the states on the Fermi level. By summing up the anomalous velocity over the filled states, one can reproduce the Kubo formula result (1). Given the Hamiltonian, the vector field $\boldsymbol{B}(\boldsymbol{k})$ is calculable, and we can get the intrinsic Hall conductivity, as in the ab initio calculation of the AHE in [23, 24]. Due to remarkable similarity of the two equations in Eq. (3), $\boldsymbol{B}_n(\boldsymbol{k})$ can be regarded as a "magnetic field" in $\boldsymbol{k}$-space. $\boldsymbol{B}_n(\boldsymbol{k})$ can have monopoles, and such monopoles can give nontrivial topological structure for magnetic superconductors [25].

This anomalous velocity leads to the SHE in semiconductors with diamond structure (e.g. Si) or zincblende structure (e.g. GaAs). The valence bands consist of two doubly degenerate bands: heavy-hole (HH) and light-hole (LH) bands. They are degenerate at $\boldsymbol{k} = 0$ as shown in Fig. 1. Near $\boldsymbol{k} = 0$, the valence bands are described by the Luttinger Hamiltonian [26]

$$H = \frac{\hbar^2}{2m}\left[\left(\gamma_1 + \frac{5}{2}\gamma_2\right)k^2 - 2\gamma_2(\boldsymbol{k}\cdot\boldsymbol{S})^2\right], \tag{4}$$

where $\boldsymbol{S}$ is the spin-3/2 matrices representing the total angular momentum. For simplicity, we employed the spherical approximation for the Luttinger Hamiltonian, while a calculation without it is also possible [27]. In this Hamiltonian, a helicity $\lambda = \frac{\boldsymbol{k}\cdot\boldsymbol{S}}{k}$ is a good quantum number, and can be used as a label for eigenstates. The HH and LH bands have $\lambda_H = \pm\frac{3}{2}$ and $\lambda_L = \pm\frac{1}{2}$, respectively. The SEOM reads as

$$\dot{\boldsymbol{x}} = \frac{1}{\hbar}\frac{\partial E_\lambda(\boldsymbol{k})}{\partial \boldsymbol{k}} + \dot{\boldsymbol{k}} \times \boldsymbol{B}_\lambda(\boldsymbol{k}), \quad \hbar\dot{\boldsymbol{k}} = e\boldsymbol{E}. \tag{5}$$

Because we are considering holes, the sign of the charge has been changed. By straightforward calculation, we get $\boldsymbol{B}_\lambda(\boldsymbol{k}) = \lambda(2\lambda^2 - \frac{7}{2})\boldsymbol{k}/k^3$. Hence, the anomalous velocity due to Berry phase is along the direction $\boldsymbol{E} \times \boldsymbol{k}$. By integration in terms of the time t, we get a trajectory of the holes as shown in Fig. 2. This shows the motion projected on a plane perpendicular to $\boldsymbol{E}$.

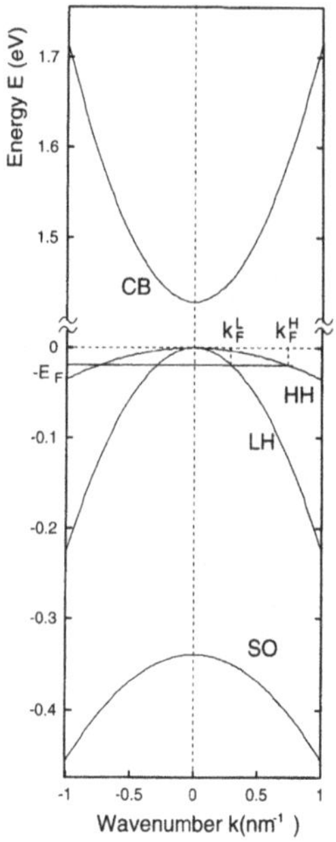

Fig. 1. Schematic band structure for GaAs. CB, HH, LH, SO represent the conduction, heavy-hole, light-hole and split-off bands, respectively.

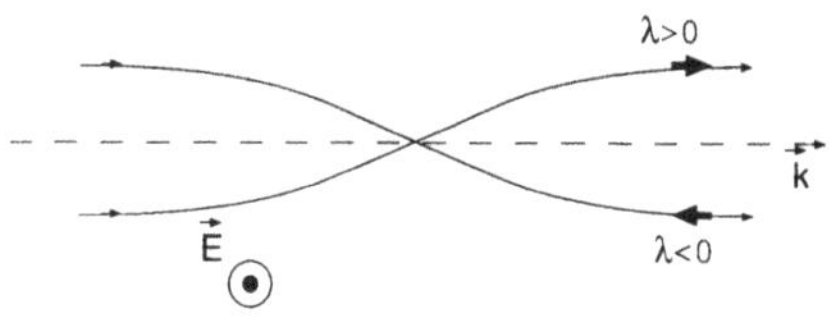

Fig. 2. Trajectory of holes from the semiclassical equation of motion with Berry-phase terms. This is a projection on the plane perpendicular to the electric field $\boldsymbol{E}$. The transverse shift of the trajectory is to the opposite direction, depending on the sign of the helicity $\lambda = \hat{k} \cdot \boldsymbol{S}$. The bold arrows represent the direction of spin $\boldsymbol{S}$

We note that a semiclassical trajectory can be calculated directly from the Heisenberg equation of motion; the resulting trajectory agree well with the one from (5), but with small rapid oscillations [28]. This justifies validity of the SEOM (5) for adiabatic transport, which was questioned in [29].

Due to the anomalous velocity, the motion of the holes is deflected from an otherwise straight motion along $\boldsymbol{k}$ (dashed line). The shift of the motion is opposite for the signs of the helicity λ, referring to whether the spin $\boldsymbol{S}$ and the wavevector $\boldsymbol{k}$ are parallel or antiparallel. This shift amounts to the SHE. By summing up this shift over the occupied states, we can calculate a spin current in response to the electric field. If the electric field is along l-axis, the spin current with S^i spin flowing toward the j-direction is [6]

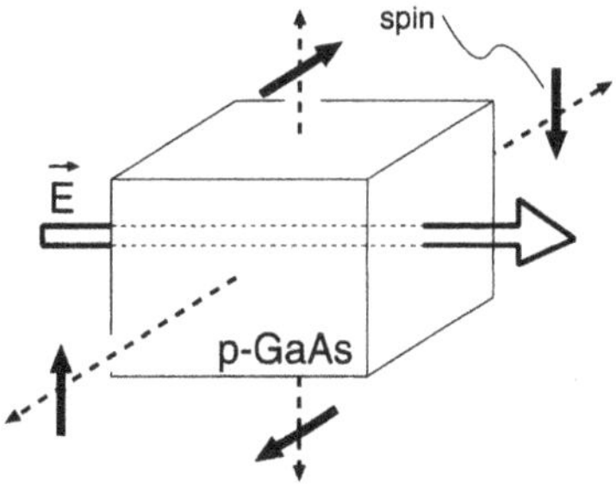

Fig. 3. Schematic of the spin current induced by an electric field

$$j_j^i = \frac{e}{12\pi^2}(3k_F^{\mathrm{H}} - k_F^{\mathrm{L}})\epsilon_{ijl}E_l, \tag{6}$$

where k_F^{H} and k_F^{L} are the Fermi wavenumber for the HH and LH bands, respectively. This is schematically shown in Fig. 3. Nominal values obtained spin Hall conductivity for p-GaAs are of the similar order of magnitude as the condutivity at room temperature [6]. In GaAs, the nominal energy difference of the two bands is larger than the room temperature, and the effect can in principle survive even at room temperature.

2.2 Spin Hall Effect in n-Type Semiconductors in Heterostructures

In n-type semiconductors with diamond or zincblende structure, the SO coupling is small. On the other hand, however, if they are incorporated into two-dimensional heterostructure, the inversion symmetry is broken, and the SO coupling becomes relevant. The Hamiltonian is approximated as

$$H = \frac{k^2}{2m} + \lambda(\boldsymbol{\sigma} \times \boldsymbol{k})_z, \tag{7}$$

where σ_i is the Pauli matrix. The second term is called the Rashba term [30, 31], representing the SO coupling. The coupling constant λ can be experimentally determined, and can be controlled by the gate voltage [32].

Sinova et al. applied the Kubo formula to this Rashba Hamiltonian [7]. For this procedure, they defined the spin current J_y^z to be a symmetrized product of the spin S^z and the velocity $v_y = \frac{\partial H}{\partial k_y}$. By assuming no disorder, the resulting spin Hall conductivity is $e/(8\pi)$, which is independent of the Rashba coupling λ. The Rashba term in (7) can be regarded as a $\boldsymbol{k}$-dependent effective Zeeman field $\boldsymbol{B}_{\mathrm{eff}} = \lambda(\hat{z} \times \boldsymbol{k})$. In an equilibrium the spins are pointing either parallel or antiparallel to $\boldsymbol{B}_{\mathrm{eff}}$ for the lower and upper bands, respectively. An external electric field $\boldsymbol{E} \| \hat{x}$ changes the wavevectors $\boldsymbol{k}$ of Bloch wavefunctions, and $\boldsymbol{B}_{\mathrm{eff}}$ also changes accordingly. The spins will then precess around $\boldsymbol{B}_{\mathrm{eff}}$, and tilt to the $\pm z$-direction, depending on the sign of k_y. This appears as the SHE, and the spin Hall conductivity is calculated to be $\frac{e}{8\pi}$, in agreement with the Kubo formula. We can incorporate also the Dresselhaus term representing the bulk inversion-symmetry breaking of the zincblende structure. The Hamiltonian becomes

$$H = \frac{k^2}{2m} + \lambda(k_x\sigma_y - k_y\sigma_x) + \beta(k_x\sigma_x - k_y\sigma_y). \tag{8}$$

The spin Hall conductivity σ_s is given as follows; $\sigma_s = e/(8\pi)$ for $\lambda^2 > \beta^2$, and $\sigma_s = -e/(8\pi)$ for $\lambda^2 < \beta^2$ [33, 34].

When a perpendicular magnetic field is applied, the spin Hall conductance will have a resonant behavior as a function of the magnetic field [35, 36]. The Zeeman splitting induces degeneracies between different Landau levels, and if this degeneracy occurs at the Fermi level, the spin Hall conductance is divergent. An in-plane magnetic field also affects the SHE, as studied in [37]

Close relationship between the SHE and the Pauli spin susceptibility [38, 39, 40] or the dielectric function [41] has been argued for these models. An effect of electron-electron interaction has been investigated by use of this relationship [40]. In the Luttinger model, on the other hand, these three have different frequency dependence [42]. Thus it is not clear whether this relationship remains for disordered case or for generic SO-coupled models.

2.3 General Properties of the Intrinsic Spin Hall Effect

We consider the intrinsic SHE to be robust against spin relaxation. Momentum relaxation induces rapid spin relaxation via the SO coupling. Namely, if the Fermi surface deviates from its equilibrium position, the momentum and spin distributions will rapidly relax. Nevertheless, because the spin Hall current comes from the anomalous velocity, it will survive even when the momentum and spin relaxation is in equilibrium. On the other hand, near sample boundaries, the spin Hall current will induce spin accumulation. Hence, the spin distribution is deviated largely from equilibrium, and the spin relaxation becomes effective. The amount of accumulated spins is roughly estimated as a product of spin current and spin relaxation time τ_s [6]. The spin accumulation affects also the spin current itself near the boundaries [43, 44].

Because spin current is even under time-reversal, it can be induced even when the time-reversal symmetry is preserved. It also implies that the spin Hall current is dissipationless. In doped semiconductors, however, the longitudinal conductivity is finite and the system undergoes Joule heating. Nevertheless, there exist spin-Hall insulators, which are band insulators with nonzero SHE; in such systems the longitudinal conductivity is zero, and the SHE accompanies no dissipation.

3 Disorder Effect and Extrinsic Spin Hall Effect

One can include an effect of the self-energy broadening $1/\tau$ by disorder [45, 46]. The intrinsic SHE is reduced as expected. In the clean limit the spin Hall conductivity reproduces its intrinsic value. Inoue et al. [8, 9] found an important result. They assumed dilutely distributed impurities with a δ-function potential. They calculated the self-energy within the self-consistent Born approximation, and the vertex correction within the ladder approximation. In a clean limit, they obtained a vertex-correction contribution $-\frac{e}{8\pi}$ to the spin Hall conductivity, exactly cancelling the intrinsic value $\frac{e}{8\pi}$. This result was rediscovered and generalized by several people [47, 48, 49, 50, 51, 52, 53, 54, 55, 56, 57]. In particular, by the Keldysh formalism, it is found that the spin Hall current appears *only near the electrodes* whereas in the bulk of the sample the spin Hall current vanishes irrespective of the lifetime τ [47]. The Keldysh formalism is used for more generalized cases in the Rashba model and related models [54, 55, 56, 57].

One may wonder whether the SHE vanishes in other systems, and there remain some controversies in this respect. It is well-established that the SHE vanishes in the Rashba model with δ-function impurities in the clean limit. Meanwhile, even for the Rashba model, it is still under debate whether it vanishes for finite τ [47, 48, 49, 50, 51, 54, 57, 58, 59] or for finite-ranged impurities [9, 50, 51, 54]. Here we note that the Rashba model is exceptional, in that the SHE vanishes rather accidentally, namely because the spin current operator J_y^z is proportional to $\dot{S}_y = i[H, S_y]$ [40, 48, 49, 52, 55]. In fact one can check that the SHE does not vanish in general models; for example, when the Rashba model is generalized to include a higher-order term in k [60], the spin Hall conductivity no longer vanishes. In addition, there are some models where the vertex correction does not cancel the intrinsic value [56], or even vanishes by symmetry [60, 61].

In retrospect, extrinsic SHE has been considered since more than thirty years ago [3, 4, 5], as mentioned in the Introduction. The relationship with the disorder effect on the intrinsic SHE is, however, unclear at present. We note that there have been a similar debate in the disorder effect on the AHE over decades. To summarize, the studies on the disorder effect are so far restricted to special models; general and exhaustive understanding for the disorder effect on the SHE is still lacking.

4 Discussions

In this section we discuss several topics on the SHE, which are still currently under intensive research.

4.1 Definition of the Spin Current

In the presence of the SO coupling, the total spin is not conserved. Hence there is no unique way to define a spin current. Naively we expect that the "spin current" $\boldsymbol{J}^s$ should satisfy the equation of continuity $\frac{\partial S_i}{\partial t} + \nabla \cdot \boldsymbol{J}_i^s = 0$; this relationship requires the conservation of total spin, namely,

$$0=\frac{\partial}{\partial t}\int S_i d^d x = -i\left[\int S_i d^d x, H\right]. \tag{9}$$

In the cases relevant for the SHE, the SO coupling violates this conservation of total spin. In other words, due to the nonconservation of spin, Noether's theorem is not applicable for a definition of spin current.

One can adopt the symmetrized product $\frac{1}{2}(v_i S_j + S_j v_i)$ between the velocity $\boldsymbol{v}$ and the spin $\boldsymbol{S}$ as a definition of the spin current as in [7]. The result calculated by the Kubo formula with this definition is in general different from that by semiclassical theory described above [62]. This difference comes from noncommutatibity between the spin $\boldsymbol{S}$ and the velocity $\boldsymbol{v}$. In other words, this comes from the non-uniqueness of the definition of spin. One can modify the semiclassical theory to give the same result as the Kubo formula, by adding three contributions: spin dipole, torque moment, and change of wavepacket spins due to electric field [63].

An alternative way is to separate the spin $\boldsymbol{S}$ into conserved (intraband) part $\boldsymbol{S}_{(\mathrm{c})}$ and nonconserved (interband) part $\boldsymbol{S}_{(\mathrm{n})}$[62]. As $[\boldsymbol{S}_{(\mathrm{c})}, H]=0$, spin current can be uniquely defined for $\boldsymbol{S}_{(\mathrm{c})}$. The resulting spin current is different from (6), and this difference is considered as a quantum correction to (6). The reason why we only take $\boldsymbol{S}_{(\mathrm{c})}$ is because relaxation by impurities will rapidly smear out the non-conserved part $\boldsymbol{S}_{(\mathrm{n})}$. Another attempt for defining conserved spin current is done by introducing torque dipole moment [64]; this definition ensures the Onsager relation. Because there is no unique definition for the spin current, we have to choose one definition which matches the considered experimental setup to measure the spin current.

4.2 Landauer-Büttiker Formalism

The four-terminal Landauer-Büttiker formalism can be used to study the SHE. In [65, 66], the authors used a tight-binding Hamiltonian with SO coupling on a square lattice, and used the four-terminal Landauer-Büttiker formalism to study mesoscopic SHE for system size up to $\sim 100\times 100$. In the bottom of the band the tight-binding model reduces to the Rashba model. They first studied the SHE without any disorder, by changing the system size and the SO coupling. The resulting spin Hall conductivity is not equal to the universal value of $\frac{e}{8\pi}$, and is critically dependent on the strength of the SO coupling. They also studied the dependence on on-site disorder. Even in the presence of disorder, the SHE remains nonzero, and depends on the disorder strength. The Luttinger model was studied in a similar fashion [67]. We remark that because of the nonuniqueness of the definition of spin current, the comparison between the results by Landauer-Büttiker formalism and those by Kubo formula is not straightforward.

Non-equilibrium spin accumulation has been studied in a two-terminal geometry [68]. The Keldysh non-equilium Green's function is combined with the Landauer formalism to study numerically the spin accumulation. The spins accumulate at the both edges, with their direction along z-axis opposite for the two edges. Spin accumulation at the edges of ballistic systems is also studied [69]. This accumulation at the edges is qualitatively similar to the experimental result [10].

In [70], an H-shaped structure is proposed for a measurement of the SHE via dc-transport porperties. With use of the Landauer-Büttiker formalism, the dc voltage response is calculated with realistic parameters. In [71], the Landauer-Büttiker formalism is applied to a mesoscopic ring with Rashba SO coupling; by tuning the Rashba coupling, the spin Hall current oscillates due to the Aharanov-Casher phase around the ring.

4.3 Criterion for Nonzero SHE, Spin Hall Insulator

The SHE is induced by the SO coupling, which is inherent in every material. However, if the bands within the same multimplet are all filled or all empty, the bands do not contribute to the SHE. Thus, the criterion for nonzero SHE is that the bands within the same multiplet have different fillings. For example, in GaAs, the valence bands ($J = 3/2$) consist of the LH and the HH; therefore, p-doping brings about the difference of fillings between the LH and the HH, giving nonzero SHE. On the other hand, the conduction band ($J = 1/2$) is doubly degenerate (if we ignore the bulk inversion symmetry breaking from the zincblende structure), and n-doping does not give rise to nonzero SHE. If we incorporate it into heterostructure, the degeneracy of the conduction band is lifted, and n-doping induces nonzero SHE.

According to this criterion, some band insulators have nonzero SHE, even though the charge conductivity is zero [72]. Two classes of materials have been proposed for such "spin Hall insulators" in [72]. One is zero-gap semiconductors such as HgTe and α-Sn. By introducing uniaxial anisotropy the gap becomes finite. The other is narrow-gap semiconductors such as PbTe. In these semiconductors the gaps come from the SO coupling, and the SHE is nonzero even though they are band insulators. In [73], on the other hand, a graphene sheet is proposed to be a quantum spin Hall insulator.

One can consider a Hall effect for the orbital angular momentum (OAM) instead of spin [74]. In the Rashba model, the resulting intrinsic Hall conductivity for the OAM is $-e/(8\pi)$, exactly cancelling the intrinsic SHE. Thus the intrinsic Hall effect for the *total angular momentum* vanishes. It follows from the conservation of the total angular momentum $s_z + L_z$ [74], i.e. from a continuous rotation symmetry around the z-axis. Nevertheless, for general systems it is not true, and the cancellation does not take place in general.

Several first-principle calculations have been done [75, 76]. In [76], the intrinsic SHE is calculated for Si, GaAs, W and Au for various values of the Fermi level. It is found that even without doping, Si and GaAs show a small but finite SHE. It is due to a small hybridization. In this sense, these undoped semiconductors are spin Hall insulators. In [75], on the other hand, the intrinsic SHE is calculated for n-type Ge, GaAs and AlAs as a function of the hole concentration and strain. They also calculated the Hall effect for the OAM, and showed that they do not cancel with the spin Hall effect.

5 Experiments

Kato et al. [10] observed the SHE in n-type semiconductors by measuring spin accumulation at the edges of the sample by Kerr rotation. The spin accumulation is uniformly distributed along the both edges. They evaluated from experimental data the amount of spin accumulation and spin lifetime as a function of an external magnetic field. The measured spin Hall resistivity is 2 MΩµm. They concluded the observed SHE to be extrinsic for the following reasons; (i) spin splitting is negligibly small in the sample, and (ii) the effect has no dependence on crystal orientation.

Nevertheless, Bernevig and Zhang argued that the observed SHE can be intrinsic, coming from the Dresselhaus term representing bulk inversion-symmetry breaking [61]. They showed that even if the spin splitting due to the Dresselhaus term is negligibly small, the SHE can be as large as the experimental data. It can also account for the absence of dependence on crystal orientation. Thus, the source of the observed SHE is still to be resolved.

The observed distribution of spin accumulation is clearly different from the Keldysh formalism calculation on the Rashba model [47]. As discussed in the previous section, the Rashba model may not be general enough to be useful for comparison with experiments.

On the other hand, Wunderlich et al. [11] observed the SHE in a 2D p-type system, using a p-n junction light-emitting diode. They applied an electric field across the hole channel, and observed a circular polarization of the emitted light, whose sign is opposite for the two edges of the channel. The circular polarization is $\sim 1\%$ at maximum. They argued that it is near to the clean limit, and the obtained SHE is mostly intrinsic. More refined argument, by showing vanishing vertex correction, also supports this conclusion [77].

As we have seen in Sect. 3, disorder effect on the SHE is still under intensive studies. It will take some time to determine whether the above experimental results for the SHE is mostly intrinsic or extrinsic (or both). To pursue this issue experimentally, it will be ideal to change systematically the disorder, and measure the SHE in the same line as in [78].

6 Summary

In this review we summarize recent results for the SHE. In these two years there has been much progress in this field, both in theories and in experiments. Nevertheless, as many results have accumulated, we come up with new questions to be solved. In the present stage there are a lot of ways to approach the problem theoretically and experimentally, and the results from different methods have not yet satistactorily converged into a unified picture of the SHE. In particular, disorder effect is the key issue to enable comparison between theories and experiments in a systematic fashion.

The author would like to thank N. Nagaosa, S. Onoda, N. Sugimoto, and S.-C. Zhang for collaborations. He is grateful to G. E. W. Bauer, D. Culcer, Z. Fang, J. Inoue, A. H. MacDonald, B. Nikolić, Q. Niu, K. Nomura, M. Onoda, S.-Q. Shen and J. Sinova for fruitful discussions. This work has been supported by a Grant-in-Aid (No. 16740167) for Scientific Reserach from the Ministry of Education, Culture, Sports, Science and Technology of Japan.

References

1. S. A. Wolf et al., Science **294**, 1488 (2001).
2. H. Ohno, Science **281**, 951 (1998) and references therein.
3. M. I. D'yakonov, V. I. Perel', ZhETF Pis. Red. **13**, 657 (1971) [JETP Lett. **13**, 467 (1971)].
4. J. E. Hirsch, Phys. Rev. Lett. **83**, 1834 (1999).
5. S. Zhang, Phys. Rev. Lett. **85**, 393 (2000).
6. S. Murakami, N. Nagaosa, S. C. Zhang, Science **301**, 1348 (2003).
7. J. Sinova, D. Culcer, Q. Niu, N. A. Sinitsyn, T. Jungwirth, A. H. MacDonald, Phys. Rev. Lett. **92**, 126603 (2004).
8. J. Inoue, G. E. W. Bauer, L. W. Molenkamp, Phys. Rev. B **67**, 033104 (2003).
9. J. Inoue, G. E. W. Bauer, L. W. Molenkamp, cond-mat/0402442.
10. Y. K. Kato, R. C. Myers, A. C. Gossard, D. D. Awschalom, Science, **306**, 1910 (2004).
11. J. Wunderlich, B. Kaestner, J. Sinova, T. Jungwirth, Phys. Rev. Lett. **94** 047204 (2005).
12. M. V. Berry, Proc. R. Soc. London, Ser. A **392**, 45 (1984).
13. A. Shapere, F. Wilczek, *Geometric Phases in Physics* (World Scientific Publishers, Singapore 1989).
14. A. Bohm, A. Mostafazadeh, H. Koizumi, Q. Niu and J. Zwanziger, *The Geometric Phase in Quantum Systems* (Springer Verlag, Berlin 2003).
15. D. J. Thouless, M. Kohmoto, M. P. Nightingale, and M. den Nijs, Phys. Rev. Lett. **49**, 405 (1982).
16. M. Kohmoto, Ann. Phys. (N.Y.) **160**, 343 (1985).
17. G. Sundaram, Q. Niu, Phys. Rev. B **59**, 14915 (1999).
18. R. Karplus, J. M. Luttinger, Phys. Rev. **95**, 1154 (1954).
19. J. Ye et al., Phys. Rev. Lett. **83**, 3737 (1999).
20. K. Ohgushi, S. Murakami, and N. Nagaosa, Phys. Rev. B **62**, R6065 (2000).
21. Y. Taguchi et al., Science **291**, 2573 (2001).
22. T. Jungwirth, Q. Niu, A. H. MacDonald, Phys. Rev. Lett. **88**, 207208 (2002).
23. Z. Fang et al., Science **302**, 92 (2003).
24. Y. Yao et al., Phys. Rev. Lett. **92**, 037204 (2004).
25. S. Murakami and N. Nagaosa, Phys. Rev. Lett. **90**, 057002 (2003).
26. J. M. Luttinger, Phys. Rev. **102**, 1030 (1956).
27. B. A. Bernevig, J. P. Hu, E. Mukamel, and S. C. Zhang, Phys. Rev. **B70**, 113301 (2004).

28. Z. F. Jiang, R. D. Li, S.-C. Zhang, and W. M. Liu, Phys. Rev. **B72**, 045201 (2005).
29. X. Wang and X.-G. Zhang, J. Magn. Magn. Mater. **288**, 297 (2005).
30. E. I. Rashba, Sov. Phys. Solid State **2**, 1109 (1960).
31. Y. A. Bychkov, E. I. Rashba, J. Phys. C **17**, 6039 (1984).
32. J. Nitta, T. Akazaki, H. Takayanagi, T. Enoki, Phys. Rev. Lett **78**, 1335 (1997).
33. S.-Q. Shen, Phys. Rev. B**70**, 081311 (2004).
34. N. A. Sinitsyn, E. M. Hankiewicz, W. Teizer, and J. Sinova, Phys. Rev. B**70**, 081312 (2004).
35. S.-Q. Shen et al., Phys. Rev. Lett. 92, 256603 (2004).
36. S.-Q. Shen et al., Phys. Rev. **B71**, 155316 (2005).
37. M.-C. Chang, Phys. Rev. B**71**, 085315 (2005).
38. O. V. Dimitrova, cond-mat/0407612.
39. S. I. Erlingsson, J. Schliemann, D. Loss, Phys. Rev. **B71**, 035319 (2005).
40. A. Shekhter, M. Khodas, A. M. Finkel'stein, Phys. Rev. **B71**, 165329 (2005).
41. E. I. Rashba, Phys. Rev. B **70**, 161201 (2004).
42. B. A. Bernevig, Phys. Rev. B**71**, 073201 (2005).
43. X. Ma et al., Phys. Rev. B**70**, 195343 (2004).
44. L. Hu, J. Gao, S. Q. Shen, Phys. Rev. B**70**, 235323 (2004).
45. J. Schliemann, D. Loss, Phys. Rev. B **69**, 165315 (2004).
46. J. Schliemann, D. Loss, Phys. Rev. B **71**, 085308 (2005).
47. E. G. Mishchenko, A. V. Shytov, and B. I. Halperin, Phys. Rev. Lett. **93**, 226602 (2004).
48. O. V. Dimitrova, cond-mat/0405339.
49. O. Chalaev, D. Loss, Phys. Rev. **B71**, 245318 (2005).
50. A. Khaetskii, cond-mat/0408136.
51. R. Raimondi, P. Schwab, Phys. Rev. B**71**, 033311 (2005).
52. E. I. Rashba, Phys. Rev. B **70**, 201309 (2004).
53. A. G. Mal'shukov, K. A. Chao, Phys. Rev. B**71**, 121308 (2005).
54. S. Y. Liu, X. L. Lei, cond-mat/0411629.
55. S. Y. Liu, X. L. Lei, cond-mat/0502392.
56. S. Y. Liu, X. L. Lei, cond-mat/0503352.
57. N. Sugimoto, S. Onoda, S. Murakami and N. Nagaosa, cond-mat/0503475.
58. K. Nomura et al., Phys. Rev. B**71**, 041304 (2005).
59. D. N. Sheng, L. Sheng, Z. Y. Weng, F. D. M. Haldane, cond-mat/0504218.
60. S. Murakami, Phys. Rev. B**69**, 241202 (2004).
61. B. A. Bernevig, S.-C. Zhang, cond-mat/0412550.
62. S. Murakami, N. Nagaosa, S. C. Zhang, Phys. Rev. B**69**, 235206 (2004).
63. D. Culcer et al., Phys. Rev. Lett. **93**, 046602 (2004).
64. P. Zhang et al., cond-mat/0503505.
65. B. K. Nikolic, L. P. Zârbo, S. Souma, cond-mat/0408693.
66. L. Sheng, D. N. Sheng, C. S. Ting, Phys. Rev. Lett. **94**, 016602 (2005).
67. M. W. Wu and J. Zhou, cond-mat/0503626.
68. B. K. Nikolić, S. Souma, L. P. Zârbo, J. Sinova, Phys. Rev. Lett. **95**, 046601 (2005).
69. G. Usaj and C. A. Balseiro, cond-mat/0405065.
70. E. M. Hankiewicz, L. W. Molenkamp, T. Jungwirth, J. Sinova, Phys. Rev. B**70**, 241301 (2004).
71. S. Souma, B. K. Nikolić, Phys. Rev. Lett. **94**, 106602 (2005).

72. S. Murakami, N. Nagaosa, S.-C. Zhang, Phys. Rev. Lett. **93**, 156804 (2004).
73. C. L. Kane, E. J. Mele, cond-mat/0411737.
74. S. Zhang and Z. Yang, Phys. Rev. Lett. **94** 066602 (2005).
75. G. Y. Guo, Y. Yao, Q. Niu, Phys. Rev. Lett. **94**, 226601 (2005).
76. Y. Yao, Z. Fang, cond-mat/0502351.
77. B. A. Bernevig, S.-C. Zhang, Phys. Rev. Lett. **95**, 016801 (2005).
78. W.-L. Lee et al., Science **303**, 1647 (2004).

Part III

Superconductivity

Point-Contact Spectroscopy on Conventional and Unconventional Superconductors

Gernot Goll

Physikalisches Institut, Universität Karlsruhe, 76128 Karlsruhe, Germany

Summary. In the past decades it has become evident that a growing number of metals exhibit exotic types of superconductivity as manifested by their thermodynamic and transport properties. Multi-band superconductivity has attracted renewed interest triggered by experimental results on MgB_2 and Nb_3Sn. Unconventional superconductivity is observed in many U- and Ce-based heavy-fermion superconductors as well as in the oxide superconductors. Among other methods, point-contact spectroscopy is an important tool to study the symmetry and the nodal structure of the energy gap Δ. Andreev reflection of charge carriers at the normal metal/superconductor interface leads to minima at $V \approx \pm\Delta/e$ in the differential resistance dV/dI as a function of applied bias V and thus allows determination of the gap size. In addition, Andreev reflection causes an excess current through metallic (low-barrier) point contacts and the temperature dependence of the excess current can be analyzed with respect to the order-parameter symmetry. Here, recent developments in this field will be reviewed with a particular focus on investigations of the heavy-fermion superconductor $CeCoIn_5$, a possible *d*-wave superconductor, and the oxide superconductor Sr_2RuO_4, a candidate for *p*-wave superconductivity.

1 Introduction

In the history of superconductivity it was always a challenge for both experimentalists and theorists to reveal the nature of superconductivity. After the discovery of the first superconductor by Kamerlingh Onnes in 1911 [1] it took almost half a century until a pairing theory was developed which was able to account for many properties of superconductors known at that time. In 1957, Bardeen, Cooper and Schrieffer (BCS) [2] pointed out that even a weak attraction between fermions will lead to simultaneous formation and condensation of so-called Cooper pairs. In classical superconductors, first considered by the BCS theory, these pairs have a total spin $S = 0$ and total orbital momentum $L = 0$ (singlet pairing), and are described, in analogy to quantum mechanics, by an isotropic pair wave function, denoted as *s*-wave pairing. The weak attractive interaction arises from the virtual exchange of phonons and the screened Coulomb repulsion between electrons. The bosonic condensate forms a macroscopic quantum state which breaks the gauge invariance by setting a macroscopic phase relationship. As a consequence, BCS

B. Kramer (Ed.): Adv. in Solid State Phys. **45**, 213–225 (2005)

theory predicts a temperature-dependent excitation gap $\Delta(T)$ in the density of states at the Fermi level. The gap opens below the superconducting transition temperature T_c and in weak-coupling superconductors it reaches an isotropic maximum width of $2\Delta_0/k_B T_c = 3.53$. S-wave pairing mediated by electron-phonon interaction is the key mechanism of conventional superconductivity as found in most element and alloy superconductors. Superconductivity is denoted unconventional, if below the transition temperature T_c additional symmetries are broken besides the gauge symmetry, or if the pairing mechanism is mediated by non-phononic interactions. The hallmark of the significance and universality of the concepts of BCS pairing theory is the fact that a generalized BCS theory [3, 4] can account for unconventional superconductivity.

Point-contact spectroscopy has been used to study the energy gap of a large variety of conventional superconductors [5] and unconventional superconductors [6]. Here I will focus on investigations of the multi-band superconductors MgB_2 and Nb_3Sn and the possible unconventional superconductors $CeCoIn_5$ and Sr_2RuO_4.

2 Experimental

Point-contact spectroscopy is a method which probes the excitations of a solid with the use of ballistic electrons. These electrons are accelerated within the potential drop V along a micro contact between two electrodes gaining the energy eV. The contact is called point contact if its lateral dimensions are smaller than the characteristic electronic length scales of the electrodes materials [5]. An additional mechanism of transport through the contact occurs when the temperature is lowered below the superconducting transition temperature T_c of one electrode of a metal-metal point contact: Andreev reflection at the normal metal-superconductor interface. As a conventional superconductor is characterized by a gap Δ in its single-particle spectrum, single electrons cannot be injected into the superconductor for energy $|eV| < \Delta$, but two electrons forming a Cooper pair can be injected. This scattering process in which an electron is injected and a hole is retro-reflected is called *Andreev reflection*. Andreev reflection leads to a characteristic increase in the differential conductance dI/dV of a point contact at bias voltages of the order of the gap energy $\pm\Delta/e$ [7, 8], i. e. a minimum in the differential resistance dV/dI vs V. A small fraction of normally reflected electrons due to an interface barrier increases the zero-bias resistance and leads to the characteristic double-minimum feature in dV/dI. Using the above model the differential resistance dV/dI as a function of applied voltage V can be calculated, and the gap value Δ can be derived from the position of the minima in the spectrum, $V = \pm\Delta/e$.

Minima in the differential resistance are expected for both conventional superconductors [7, 8] and unconventional superconductors [9]. However, for the latter, information on the gap symmetry and width cannot be obtained straightforward as in the case of ordinary superconductors by following reasons: (1) Because of the self-consistency problem (see below) Δ cannot simply be replaced by $\Delta(\mathbf{k})$. (2) The gap value Δ/e cannot directly be read from the position of the minima in dV/dI. The position of the minima depends on the choice of the order-parameter symmetry, the microscopic details of the contact and the barrier strength [10]. In general, the Andreev equations which describe point-contact experiments theoretically contain a **k**-dependent and spatially dependent pair potential $\Delta(\mathbf{k}, \mathbf{r})$ which has to be determined self-consistently, because a boundary of an unconventional superconductor will be pair-breaking [9].

For the experiments reviewed below normal metal (N)-superconductor (S) heterocontacts between a sharpened Pt needle as a counter electrode and the investigated superconductor were realized inside the mixing chamber of a ^{3}He/^{4}He dilution refrigerator for measurements below 1 K, or inside the 1 K pot of a ^{4}He bath cryostat for measurements above $T \approx 1.5$ K. In addition, Nb_3Sn has been investigated by the break-junction technique where a bar-shaped sample with four leads attached has been broken in a three-point bending mechanism at low T under He atmosphere in order to avoid surface degradation. The differential resistance dV/dI vs. V was recorded by a standard lock-in technique. The differential conductance, dI/dV, is obtained by numerical inversion of the measured dV/dI data. Mechanical feed throughs allow to establish new contacts *in situ* without warming up the sample. The counter electrode was mounted on a spring in order to stabilize the contact and to minimize applied pressure to the contact.

3 Conventional Superconductors

Multi-band superconductivity has attracted renewed interest as triggered by the discovery of superconductivity at 39 K in MgB_2 [11] which immediately raised the question of its nature and the origin of its remarkable high transition temperature. MgB_2 is a classical non-oxide intermetallic compound with a surprisingly high T_c. A particularly interesting feature is the existence of two sets of electronic bands crossing the Fermi level which give rise to distinct superconducting gaps [12]. The presence of two gaps or groups of gaps was assessed by various techniques probing the surface as well as the bulk of the sample (see [13] for a review). The two-gap feature is particularly evident in MgB_2 because the density of states at the Fermi level is almost equally shared between the two sets of bands and because the gap widths are sufficiently different to be easily resolved.

Two-band superconductivity in MgB_2 lead to the question whether multi-band superconductivity is an exception or whether it does occur in other superconductors as well. Therefore, it has triggered the search for other materials which might exhibit multi-band superconductivity. The $A15$ compound

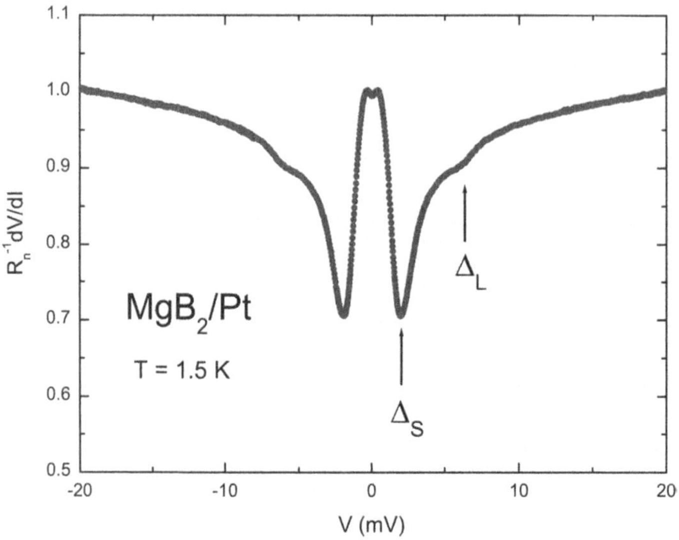

Fig. 1. Point-contact spectrum dV/dI vs V of a MgB_2-Pt contact at $T = 1.5\,K$. Two structures in dV/dI for each polarity of the bias can be assigned to the occurrence of two energy gaps Δ_S and Δ_L

Nb_3Sn is a well known, technically relevant type-II superconductor used e.g. in high field magnet coils. Nb_3Sn is a strong coupling s-wave superconductor with $T_c \approx 18\,K$ and an upper critical field $H_{c2} \approx 25\,T$. Recent measurements of the specific heat in the superconducting state [14] show a deviation from the predictions of the BCS theory [2]. The expected exponential decay of the specific heat for low temperatures $T < T_c$ has not been observed. Instead, a plateau below $T_c/4$ occurs. This extra specific heat amounts $\approx \gamma/13$ where γ is the Sommerfeld constant in the normal state and is reminiscent of the excess specific heat found in MgB_2. The data can be fitted by a phenomenological two-gap model and have been interpreted as the signature of the opening of a second gap on a band or set of bands that represent 7.5 % of the total density of states [14].

3.1 MgB_2

An example of a point-contact spectrum of a MgB_2-Pt contact is displayed in Fig. 1. Gap-related structures in dV/dI vs V vary in width and position from contact to contact depending on the microscopic direction of current flow and the Fermi velocity of the probed Fermi-surface sheet. However, the distribution of energy gaps shows a distinct accumulation around 1.7 and 7 meV which is associated with the occurrence of a small and a large energy gap in MgB_2 [15].

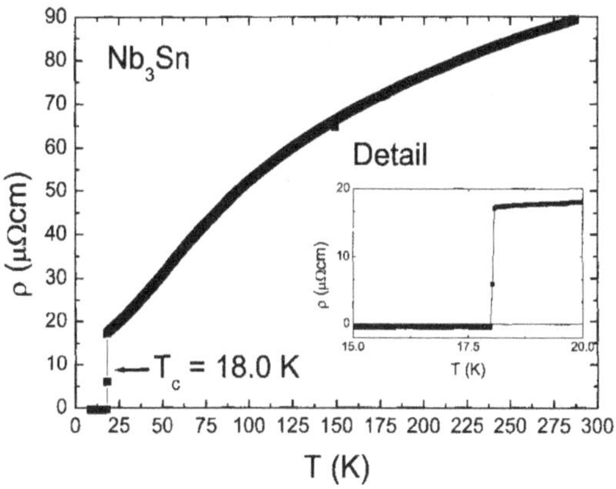

Fig. 2. Temperature dependence of the resistivity of the Nb_3Sn sample used for the break-junction experiment

3.2 Nb_3Sn

Polycrystalline samples have been prepared at the Forschungszentrum Karlsruhe by hot isostatic pressing (HIP), for details see [14]. Figure 2 displays the resistivity ρ vs. temperature T for such a sample. The sharp step at T_c and the high transition temperature $T_c = 18.0\,\mathrm{K}$ in $\rho(T)$ (Fig. 2), as well confirmed by measurements of $\chi_{ac}(T)$ and $C(T)$ [14] prove the high quality of the sample. We applied the break-junction technique in order to avoid impurities or oxide layers in the contact region.

For superconductor-superconductor contacts with identical material on both sides one characteristic peak in the differential conductance dI/dV vs. voltage V spectra at $V = 2\Delta/e$ is expected for each polarity. For break-junctions this is naturally fulfilled. Whereas, in most of our measurements we find two of these characteristic peaks for each polarity [16]. For all measurements at $T < 2\,\mathrm{K}$ we read the voltage values and plot them into a histogram. The cluster points can be identified as two superconducting gaps with

$$\Delta_l \approx 3.3\,\mathrm{meV} \text{ and } \Delta_s \approx 0.6\,\mathrm{meV}\,.$$

These values are in accordance with the results of the specific-heat measurements.

One typical spectrum of dI/dV vs. V at $T = 2\,\mathrm{K}$ is shown in Fig. 3. The filled dots represent the data. The triangles represent a simple fit to these data. We consider the currents caused by the two superconducting gaps as independent from each other and neglect interband scattering, so we can write the total current as a weighted sum of each current through the contact with a weighting parameter α, $0 < \alpha < 1$,

$$I_{tot}(V) = \alpha I_{\Delta_l}(V) + (1-\alpha) I_{\Delta_s}(V)\,.$$

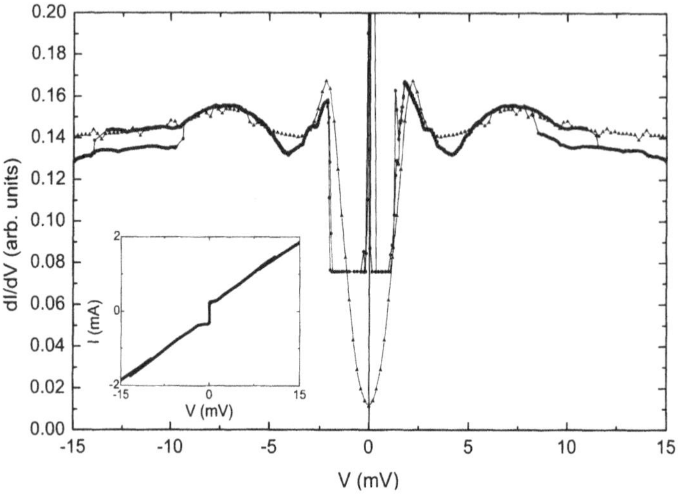

Fig. 3. dI/dV vs. V spectra of a Nb_3Sn break junction at $T = 2.0\,\mathrm{K}$. The filled dots show the data and the triangles represent a fit according to the theory discussed in the text. The flat region between $V \approx \pm 2\,\mathrm{mV}$ is due to saturation of the Lock In amplifier caused by the large changes in dI/dV when the Josephson current (see inset) at $V = 0$ breaks down

The current through a single-band superconductor-superconductor contact is given by [17]

$$I_\Delta(V) \propto \int_{-\infty}^{+\infty} \frac{N_s(E) \cdot N_s(E+eV)}{N_n^2(0)} \left(f(E) - f(E+eV)\right) dE\,.$$

In order to describe the data in Fig. 3 we used

$$\Delta_l = 3.4\,\mathrm{meV} \text{ and } \Delta_s = 1.0\,\mathrm{meV}$$

and a weighting parameter $\alpha = 0.22$. The fit accounts also for inelastic scattering in the contact region parameterized by $\Gamma_{l,s} = 0.32 \cdot \Delta_{l,s}$.

In addition to the gap structure at finite voltage the dI/dV vs. V curve shows a huge maximum at $V = 0$ caused by Josephson current. The corresponding I-V-curve is displayed in the inset of Fig. 3. According to theory, the product of the contact resistance R_N and the critical Josephson current I_c is proportional to the superconducting energy gap, $I_c R_N = \pi\Delta/2\mathrm{e}$. An analysis of the data yields values of $I_c R_N$ between $\pi\Delta_l/2\mathrm{e}$ and $\pi\Delta_s/2\mathrm{e}$ as upper and lower limits, respectively which gives further evidence for multi-band superconductivity in Nb_3Sn.

4 Unconventional Superconductors

The heavy-fermion superconductors are the class of superconductors, where first hints at unconventional superconductivity in solids appeared. Heavy-fermion compounds are mostly metals containing rare-earth or actinides ions like cerium and uranium, where the f electrons of the incomplete inner f shell can couple to the conduction electrons. The electronic properties of these compounds can simply be described in a Fermi-liquid picture of "heavy fermions", i. e. quasiparticle of immensely enhanced effective masses of 100-1000 free-electron masses. Superconductivity in these systems arises from the pair formation of these heavy quasiparticles. The ternary rare-earth compound $CeCoIn_5$ is a heavy-fermion superconductor with $T_c = 2.3\,K$ [18], the highest transition temperature among the heavy-fermion superconductors, and a linear coefficient of the specific heat $\gamma(T_c) = 350\,mJ/molK^2$ [19]. Superconductivity in $CeCoIn_5$ is built out of a normal state displaying non-Fermi-liquid behavior as evidenced by a resistivity nearly linear in T [18], a logarithmically diverging specific-heat coefficient [18, 20], and a power-law behavior in ac-susceptibility [18, 20] and nuclear spin-lattice relaxation rate [21]. On the other hand, under pressure the ground state evolves into a conventional Fermi liquid. Both points to the existence of a pressure-tuned quantum critical point close to ambient pressure [22]. In the superconducting state, power-law behavior of the specific heat close to T^2 [19, 23] and of the thermal conductivity close to T^3 [23] give evidence that the superconducting order parameter has line nodes. From the observation of a clear fourfold symmetry of the thermal conductivity in the superconducting state an order-parameter with the angular position of the nodes along the $(\pm\pi, \pm\pi)$ direction has been proposed [24] which belongs most likely to $d_{x^2-y^2}$ symmetry. From NMR measurements [21] which showed a vanishing spin susceptibility for $T \ll T_c$, a spin-singlet-pairing state is inferred as well.

Metal-oxide superconductors are a second class of superconductors with possible unconventional behavior. Up to now, the high-temperature superconductor $YBa_2Cu_3O_7$ is the only superconductor where d-wave superconductivity has unambiguously been assigned by phase-sensitive probes [25]. Sr_2RuO_4 [26] has the same layered perovskite structure, but behaves otherwise very different: Sr_2RuO_4 is metallic, displays Fermi-liquid behavior and becomes superconducting at the rather low transition temperature $T_c \approx 1.5\,K$ [27]. Shortly after the discovery of superconductivity below $T_c \sim 1.5$ K it has been suggested that it might form odd-parity (spin-triplet) Cooper pairs in contrast to even-parity pairing in cuprate systems [28, 29]. The basis for this claim was partially the analogy to ^3He and the presence of ferromagnetism in related compounds. The enhanced specific heat, magnetic susceptibility, and electronic mass indicate the presence of significant correlations [30]. The mechanism leading to the superconducting state remains unresolved. In the normal state orbital dependent spin correlations have been reported by NMR experiments [31], but neutron scattering did not detect ferromagnetic correla-

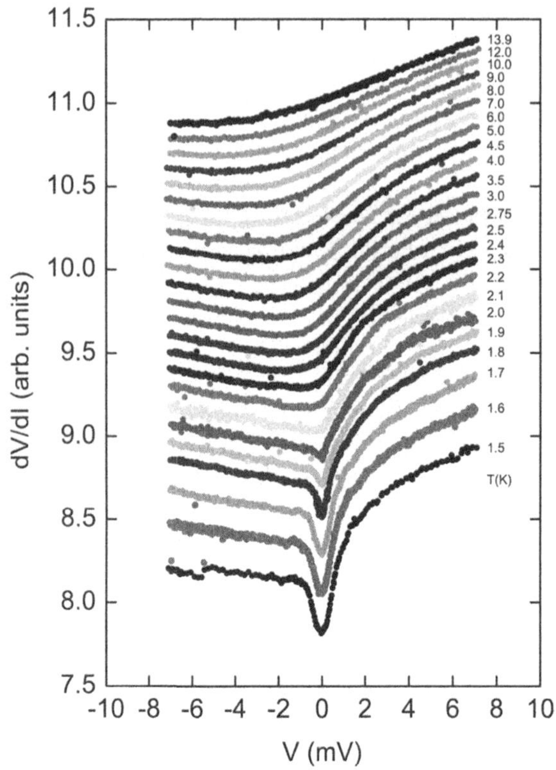

Fig. 4. Point-contact spectra for a Pt-$CeCoIn_5$ contact with $R_0 = 4\,\Omega$ in the temperature range $1.5 \leq T \leq 13.9\,K$, i.e., in the normal and superconducting states

tions so far, which would help to explain a spin-triplet superconducting state in a simple picture. In fact, incommensurable antiferromagnetic fluctuations above [32] and below [33] T_c are reported by neutron scattering. The exact symmetry of the superconducting order parameter [34, 35, 36] and notably the pairing mechanism [32, 37] are still controversial as well.

4.1 $CeCoIn_5$

Point-contact spectra, i. e. the differential resistance dV/dI vs. V have been recorded by standard lock-in technique from $T = 1.5\,K$ to temperatures well above T_c. An example is shown in Fig. 4. In the normal state a pronounced asymmetry of the differential resistance dV/dI as a function of applied bias V is observed which becomes more pronounced when T is reduced. For a contact in the ballistic regime the asymmetry can be attributed to the emergence of the coherent heavy-fermion liquid. Below T_c characteristic minima in the dV/dI vs V spectra are expected due to Andreev reflection of quasiparticles

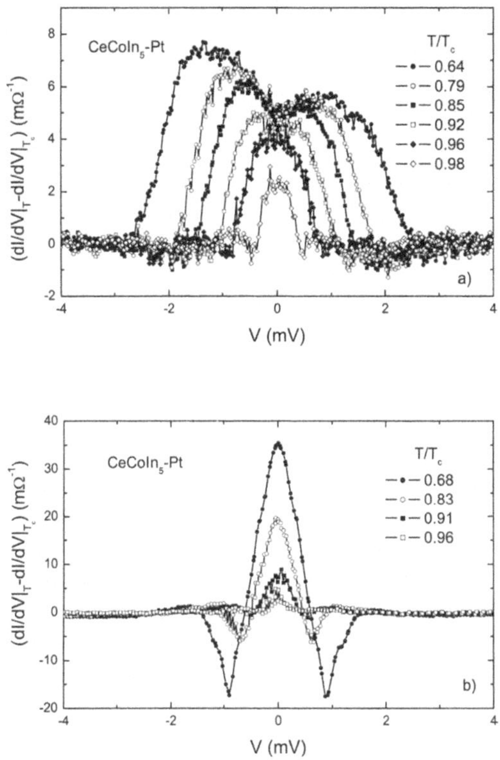

Fig. 5. Temperature dependence of the differential conductance dI/dV vs V of $CeCoIn_5$-Pt point contacts showing either an enhanced conductance for bias $|V| < \Delta/e$ (a) or a zero-bias anomaly (b)

at a normal metal/superconductor interface. For further analysis, the spectra have been inverted in order to obtain dI/dV and the asymmetric background has been subtracted by choosing the curve just above T_c as a reference for the normal-state background. We measured dI/dV curves which show either an enhanced conductance for bias $|V| < \Delta/e$ or a single maximum of dI/dV for $V = 0$, i.e. a zero-bias anomaly [39]. Figure 5 displays the T dependence of representative spectra of each type.

The observation of a zero-bias conductance anomaly (ZBA) is expected only if the order parameter exhibits a sign change as a function of $\boldsymbol{k}$. Andreev reflection of the quasiparticles at the surface leads to a surface bound state which is detected in the dI/dV vs V spectra. It was shown [38] that for the $d_{x^2-y^2}$ symmetry, the occurrence of a ZBA depends on the direction of current flow with respect to the crystallographic axis. For current injection along the [100] direction, no zero-bias anomaly is detected because no surface bound state is formed. In contrast, along [110], the order parameter exhibits a sign change and, therefore, a surface bound state at zero energy is formed which is seen in the spectra as a enhanced conductance at zero bias. However, the directional dependence of the $CeCoIn_5$-Pt contacts has not yet been investigated in detail, and up to now we have been restricted to $T/T_c \approx 0.6$. Therefore, a definite assignment of the results with a specific order-parameter symmetry is pending.

4.2 Sr_2RuO_4

Point-contact measurements on Sr_2RuO_4 were performed in different configurations with respect to the predominant current injection relative to the crystallographic axis of Sr_2RuO_4, the applied magnetic field, and the surface treatment [40]. In the following I will focus on the measurements with current injection parallel to the *ab*-planes which are in the ballistic limit. In the spectra with current perpendicular to the *ab*-planes evidence for non-spectroscopic effects can be found [41] which can be explained in terms of a heating model [42]. In the superconducting state at zero field two distinctly different types of structures in dV/dI vs V spectra are observed: either a double-minimum structure or a zero-bias anomaly, i. e. a single-minimum structure centered at $V = 0$. The occurrence of a zero-bias anomaly is believed to have its origin in an Andreev-bound surface state hinting at an unconventional order parameter of the superconducting phase which changes sign as function of the **k**-vector. The shape of the spectra were satisfactorily reproduced by an analysis of a *p*-wave pairing state with order parameter $\mathbf{d}(\mathbf{k}) = \hat{\mathbf{z}}(k_x \pm ik_y)$. Within this model, amongst other things, information about the transmittivity of the contact can be extracted [43]. Contacts showing the zero-bias anomaly are described by a low transmission probability, the ones with double-minimum structures are described by a high transmission probability. In contrast to the *d*-wave pairing state discussed above where a ZBA occurs only for certain directions, for the chiral *p*-wave state the ZBA occurs as a function of transmittivity.

The most remarkable consequence of Andreev reflection is the occurrence of an extra current I_{exc} in addition to the normal-state current $I = V/R_{\text{N}}$, where R_{N} is the normal-state resistance of the contact. It has been shown that the excess current in *s*-wave superconductors is proportional to the superconductive energy gap [7, 8] and consequently contains further information on the superconducting state. For a clean metallic N-S contact the excess current is given by

$$I_{\text{exc}} = \frac{4\Delta}{3eR_{\text{N}}},$$

but it decreases rapidly with increasing barrier strength Z. The proportionality between Δ and I_{exc} requires that both quantities exhibit the same functional dependence on temperature and magnetic field, respectively. For unconventional superconductors, however, this relation is altered and the excess current is not necessarily proportional to the order parameter [44]. In general, one obtains a scaling relation near T_c

$$I_{\rm exc} = \text{constant} \times \Delta^{1/\nu} ,$$

where ν is defined by the order-parameter symmetry.

The excess current was determined by numerical integration of dI/dV vs. V of high transmission contacts, which exhibit a double-minimum structure in the differential resistance, i.e., a double-maximum in the differential conductance dI/dV vs. V. In this metallic regime, in contrast to the tunneling limit, excess current via the mechanism of Andreev reflection occurs. Before integration a normal-conducting background has been subtracted. The p-wave analysis to the point-contact spectra applied in [43] yields the BCS temperature dependence for the superconducting gap in Sr_2RuO_4, i.e., $\nu = 0.5$. The resulting temperature dependence of the excess current determined in the framework of the p-wave analysis for a mean free path of 10 coherence lengths ($\xi_0 = v_f/2\pi T_c$) obviously is insufficient to describe the experimental data (see [44], Fig. 2). However, the more general conclusion is that unlike in the s-wave case in unconventional superconductors the excess current is not necessarily proportional to the order parameter. This is a result of the fact that impurities and disorder strongly affect the surface properties of unconventional superconductors. To reconcile the measured data an additional pair-breaking channel has to be considered. Excellent agreement is obtained if coupling to a low-frequency bosonic fluctuation mode is assumed [44].

In conclusion, point-contact spectroscopy is an important tool to derive information on size and directional dependence of the energy gap of both, conventional and unconventional superconductors. This information is necessary for an assignment of the order-parameter symmetry, and tightly connected with the question of the mechanism which leads to an unconventional superconducting ground state in a surprisingly large number of compounds.

Acknowledgements

Finally, I thank all people who contributed to this work in many respects during the last years, first of all at the Physikalisches Institut M. Marz, T. Brugger, S. Kontermann, and F. Laube. I am grateful to H. v. Löhneysen for continuous support, encouragement, and many stimulating discussions. I would also like to acknowledge many fruitful discussions with M. B. Maple, M. Eschrig, M. Fogelström, R. Werner and the collaboration with Y. Maeno, Z. Q. Mao, F. Lichtenberg, V. Zapf, E. Bauer, T. Wolf, A. Junod and W. Goldacker. Financial support by the Deutsche Forschungsgemeinschaft through Sonderforschungsbereich 195 "Lokalisierung von Elektronen in makroskopischen und mikroskopischen Systemen", and Graduiertenkolleg "Anwendungen der Supraleitung" is acknowledged.

References

1. H. Kamerlingh Onnes, Comm. Phys. Lab. Univ. Leiden **120b, 122b, 124c**, (1911).
2. J. Bardeen, L. N. Cooper, J. R. Schrieffer, Phys. Rev. **108**, 1175 (1957).
3. P. W. Anderson, P. Morel, Phys. Rev. **123**, 1911 (1961).
4. R. Balian, N. R. Werthamer, Phys. Rev. **131**, 1553 (1963).
5. Y. G. Naidyuk and I. K. Yanson, *Point-contact Spectroscopy* (Springer-Verlag, Berlin 2004).
6. For a recent review, see G. Goll, *Unconventional superconductivity* (Springer 2005), to be published.
7. G. E. Blonder, M. Tinkham, T. M. Klapwijk, Phys. Rev. B **25**, 4515 (1982).
8. A. V. Zaitsev, JETP **59**, 1015 (1984).
9. Chr. Bruder, Phys. Rev. B **41**, 4017 (1990).
10. G. Goll, Chr. Bruder, H. v. Löhneysen, Phys. Rev. B **52**, 6801 (1995).
11. J. Nagamatsu *et al.*, Nature **410**, 63 (2001).
12. A. Y. Liu, I. I. Mazin, J. Kortus, Phys. Rev. Lett. **87**, 087005 (2001).
13. F. Bouquet *et al.*, Physica C **385**, 192 (2003).
14. V. Guritanu *et al.*, Phys. Rev. B **70**, 184526 (2004).
15. F. Laube *et al.*, Europhys. Lett. **56** 296 (2001).
16. G. Goll *et al.*, to be published in AIP LT24 Conference Proceedings.
17. M. Tinkham, *Introduction to superconductivity* (Dover Publ. 2004).
18. C. Petrovic *et al.*, J. Phys.: Condens. Matter **13**, L337 (2001).
19. V. S. Zapf, *et al.*, Phys. Rev. **B**65, 014506 (2001).
20. J. S. Kim *et al.*, Phys. Rev. B **64**, 134524 (2001).
21. Y. Kohori *et al.*, Phys. Rev. B **64**, 134526 (2001).
22. V. A. Sidorov *et al.*, Phys. Rev. Lett. **89**, 157004 (2002).
23. R. Movshovich *et al.*, Phys. Rev. Lett. **86**, 5152 (2001).
24. K. Izawa *et al.*, Phys. Rev. Lett. **87**, 057002 (2001).
25. D. J. van Harlingen, Rev. Mod. Phys. **67**, 515 (1995); C. C. Tsui and J. R. Kirtley: Rev. Mod. Phys. **72**, 969 (2000).
26. Y. Maeno *et al.*, Nature **372**, 532 (1994).
27. Y. Maeno, Physica C **282-287**, 206 (1997).
28. T. M. Rice, M. Sigrist, J. Phys.: Condens. Matter **7**, L643 (1995).
29. G. Basakaran, Physica B **223 & 224**, 490 (1996).
30. C. Bergemann *et al.*, Phys. Rev. Lett. **84**, 2662 (2000); A. Damascelli *et al.*, Phys. Rev. Lett. **85**, 5194 (2000); S. Nishizaki*et al.*, J. Phys. Soc. Jpn. **67**, 560 (1998); J. Low Temp. Phys. **117**, 1581 (1999); J. Phys. Soc. Jpn. **69**, 572 (2000); C. Bergemann *et al.*: Adv. Phys. **52**, 639 (2003).
31. T. Imai *et al.*, Phys. Rev. Lett. **81**, 3006 (1998).
32. Y. Sidis *et al.*, Phys. Rev. Lett. **83**, 3320 (1999).
33. F. Servant *et al.*, Phys. Rev. B **65**, 184511 (2002).
34. C. Honerkamp and M. Sigrist, J. Low. Temp. Phys. **111**, 895 (1998).
35. M. Sigrist *et al.*, Physica C **317-318**, 134 (1999).
36. M. Graf and A.V. Balatsky, Phys. Rev. B **62**, 9697 (2000); M. Eschrig *et al.*, Phys. Rev. B **63**, 220509(R) (2001); H. Kusunose and M. Sigrist, Europhys. Lett. **60**, 281 (2002); R. Werner: cond-mat/0208301 (2002).
37. I. I. Mazin and D. J. Singh, Phys. Rev. Lett. **82**, 4324 (1999); I. Eremin *et al.*, Europhys. Lett. **58**, 871 (2002); A. A. Ovchinnikov and M. Y. Ovchinnikova, cond-mat/0205529, Sov. Phys. JETP, to be published (2002).

38. M. Fogelström *et al.*, Phys. Rev. Lett. **79**, 281 (1997).
39. G. Goll *et al.*, Acta Phys. Pol. **34**, 575 (2003).
40. F. Laube *et al*, to be published.
41. F. Laube *et al.*, J. Low Temp. Phys. **117**, 1575 (1999).
42. I. O. Kulik, Sov. J. Low Temp. Phys. **18**, 302 (1992).
43. F. Laube *et al.*, Phys. Rev. Lett. **84**, 1595 (2000).
44. F. Laube *et al.*, Phys. Rev. B **69**, 014516 (2004).

Ordering Phenomena in Cuprates

Rudi Hackl[1], Leonardo Tassini[1], Francesca Venturini[1,2], Christine Hartinger[1], Andreas Erb[1], Naoki Kikugawa[3,4], and Toshitsu Fujita[3,5]

[1] Walther Meissner Institute, Bavarian Academy of Sciences, 85748 Garching, Germany
[2] Bruker Biospin AG, 8117 Fällanden, Switzerland
[3] ADSM, Hiroshima University, Higashi-Hiroshima 739-8526, Japan
[4] Department of Physics, Kyoto University, Kyoto 606-8502, Japan
[5] Institute of Spatial Science for Regional and Global Culture, Waseda University, Tokyo 169-8555, Japan

Summary. Results of light-scattering experiments in $Bi_2Sr_2CaCu_2O_{8+\delta}$ (Bi-2212), $La_{2-x}Sr_xCuO_4$ (LSCO), and $YBa_2Cu_3O_{6+x}$ (Y-123) single crystals are presented. The emphasis is placed on the low doping range where the evolution of the metallic state from the antiferromagnetic one is studied. Along the diagonals of the CuO_2 planes the response is metallic even at small doping levels well below $p = 0.05$ where superconductivity disappears. In contrast, along the principal axes the carriers exhibit insulator-like characteristics up to approximately $p = 0.21$. In the underdoped range charge-ordering fluctuations are observed manifesting themselves as quasielastic peaks at low temperatures. The symmetries at which the new response is found are compatible with the neutron results on the spin system. Possible connections between the localization phenomena, the pseudogap and the charge and spin-ordering instabilities as well as the implications for superconductivity are discussed.

The copper-oxygen systems have a complex phase diagram. Except for the well established antiferromagnetic Néel state and the superconducting phase there is still a lot of room for controversies. A great part of the debate focuses on the normal state at low and optimal doping close to the maximal transition temperature $T_c^{\max}$. Motivated mainly by the linear resistivity, the metallic state above the superconducting dome was interpreted in terms of fluctuations leading to marginal behavior of the electronic properties [1].

At lower doping, $p < 0.16$, the metallic regime with linear resistivity $\rho(T)$ still prevails at high temperature while there are deviations below a doping dependent cross over line characterized by $T^*(p)$ [2]. For $T < T^*$, not only $\rho(T)$ exhibits qualitatively new features [3] but also the spin-lattice relaxation time T_1 [4], the thermopower, the optical conductivity, and the Raman spectra [5] deviate from what one expects for conventional metals [2, 7]. T^* is not very well defined and depends on the experimental method, in particular on the characteristic time of a probe. This motivated the conclusion that fluctuations dominate the entire phase diagram implying that the different values for T^* reflect the low-frequency cutoff of the experiment [6]. In the overdoped range, $p > 0.16$, the electronic properties are expected to become

B. Kramer (Ed.): Adv. in Solid State Phys. **45**, 227–238 (2005)

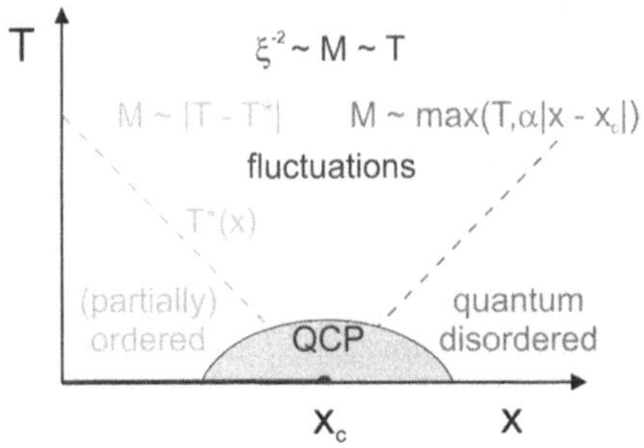

Fig. 1. Sketch of the phase diagram close to a quantum-critical point. ξ is the correlation length of the order. M is the mass in the fluctuation propagator approaching zero for long-range order. T and x are the temperature and the control parameter such as doping, respectively. Below x_c and for $T > T^*$ the mass is proportional to the temperature, above x_c M is proportional to temperature or to the control parameter

more conventional at least at elevated temperatures. At lower temperatures, a second crossover line is predicted which separates the Fermi-liquid-like state from a quantum-disordered one.

Figure 1 shows the typical phase diagram for the scenario described above. At $T = 0$ the ordered and the quantum disordered states are separated by the quantum critical point (QCP) at a critical value x_c of the control parameter x. The scenario is quite general, and x is not necessarily the doping level but can be also the pressure or the magnetic field [8, 9]. Normally one cannot approach the QCP arbitrarily closely, but rather a new phase is established which competes with the order at $x < x_c$. In the case of the cuprates the new phase is superconductivity. For the explanation of the high transition temperatures it is very important to identify the type of order at $x < x_c$ and $T < T^*$ and to clarify the interrelation of the two phenomena. To this end the fluctuation regime at sufficiently high temperatures has to be studied.

The purpose of this paper is to explore the phase diagram by studying electronic transport properties with inelastic scattering of light. We focus on the normal state but include also a discussion of earlier results on the superconducting state to address a possible relation between fluctuations and pairing. To this end spectra are being studied in the temperature range between 10 and 330 K as a function of polarization $\mathbf{e}^{\mathrm{I}}$, $\mathbf{e}^{\mathrm{S}}$ corresponding to different excitation symmetries $\mu = B_{1g}$, B_{2g} etc. and doping p.

The electronic transport at finite energies is usually measured by infrared (IR) spectroscopy. Similarly as with conventional dc transport one obtains an average over the entire Fermi surface. Putative anisotropies of the electronic properties are not easily uncovered. In the cuprates one captures therefore only part of the story since the pronounced $\mathbf{k}$ dependence of the electronic properties observed, for instance, by angle-resolved photoemission experi-

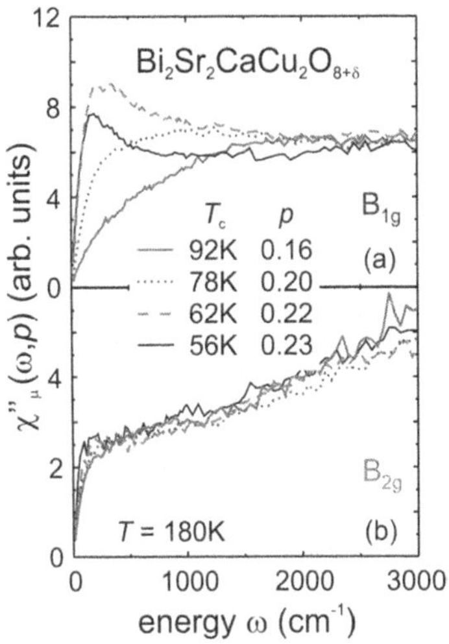

Fig. 2. Raman spectra $\chi''_\mu(\omega, p)$ as a function of doping on an extended energy scale. For clarity the contributions from lattice vibrations have been subtracted out

ments (ARPES) cannot be resolved by IR. In contrast, Raman scattering can partially overcome this problem and returns transport properties with some **k** resolution since the polarizations of the incoming and the outgoing photons can be adjusted independently [10, 11]. The type of response is density-like but is closely related to the conductivity via the continuity equation, and one can show that $\Omega \mathrm{Re}\sigma \propto \mathrm{Im}\chi$ [12] with the conductivity σ and the Raman response χ. This implies that the slope of the Raman spectra in the dc limit is proportional to the conductivity. Usually we plot the inverse slope which, consequently, can be considered a "Raman resistivity".

For $Bi_2Sr_2CaCu_2O_{8+\delta}$ (Bi-2212) the most striking effect of decreasing carrier concentration is the suppression of the B_{1g} spectral weight over an energy range of at least 1000 cm^{-1} (Fig. 2). As a consequence, the initial slope decreases towards lower carrier concentration. Since the B_{2g} spectra are essentially doping independent the suppression in B_{1g} symmetry indicates a big difference between the nodal and the anti-nodal regions of the Fermi surface. This observation is interpreted in terms of an unconventional metal-insulator transition where the transport is suppressed only for quasiparticles moving along the principal axes while diagonal directions remain metallic [13, 14].

The temperature dependence supports this view. In Fig. 3 the inverse slopes of the spectra in the dc limit, $\Gamma_0^\mu(T)$, corresponding to a relaxation rate $1/\tau$ or a "Raman resistivity" are plotted as a function of temperature. At $p = 0.23$, $\Gamma_0^{B_{1g}}(T)$ is essentially equal to $\Gamma_0^{B_{2g}}(T)$ and exhibits the same temperature dependence. A substantial difference between B_{1g} and B_{2g} de-

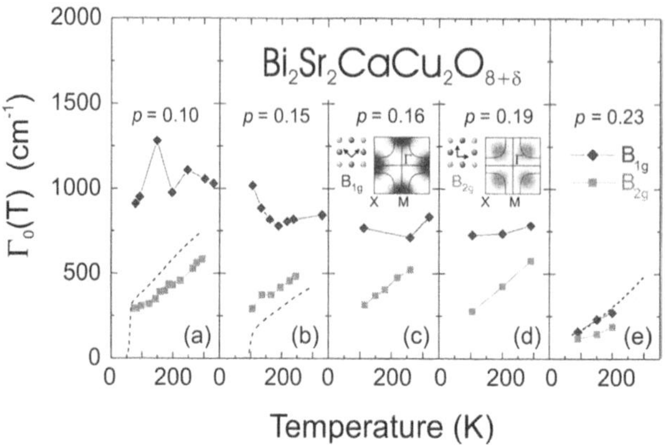

Fig. 3. Static electron relaxation rates $\Gamma_0^\mu(T)$ of Bi-2212 at various doping levels as measured by Raman scattering. The relationship between symmetry and sensitivity is indicated. The dashed lines represent conventional dc resistivities being directly related to $\Gamma_0^\mu(T)$ in a Drude picture [11, 15]

velops at lower doping, and $\Gamma_0^{B_{1g}}(T)$ becomes constant or even insulating (Fig. 3).

The situation is fundamentally different in $La_{2-x}Sr_xCuO_4$ (LSCO) (Fig. 4). At $p = x = 0.10$, $\Gamma_0^{B_{1g}}(T)$ decreases stronger with temperature than $\Gamma_0^{B_{2g}}(T)$ and saturates at a very low energy in the limit $T \to 0$ (Fig. 4 (b)). However, $\Gamma_0^{B_{2g}}(T)$ still follows closely the resistivity. Hence, something new occurs in B_{1g} symmetry as can be seen immediately in the spectra (Fig. 5 (c)). Instead of the flat and essentially temperature independent response one would anticipate from Bi-2212 or $YBa_2Cu_3O_{6+x}$ (Y-123) at low doping (Fig. 6) the intensity and, consequently, the dc slope of the spectra increase dramatically. Apparently, a new scattering channel opens up in LSCO at low doping. There is actually already an indication in the relaxation rates at $p = 0.15$ (Fig. 4 (c)) while the expected isotropic behavior is found at $p = 0.26$ (Fig. 4 (d)).

The comparison of the three compounds studied suggests that there is a superposition of two anomalies in B_{1g} symmetry: (1) below $p \simeq 0.22$ antinodal quasiparticles become localized in all compounds at sufficiently high temperature leading to a reduction of the dc slope in the spectra; (2) with decreasing temperature a new well-defined peak develops in LSCO at low energy leading to a strong increase of the dc slope. Although nothing comparable can be resolved in Y-123 and Bi-2212, the additional response must not necessarily be absent. It rather can mean that the new features are broader here and shifted to higher energies hence becoming very similar in shape to the response from the 2D planes.

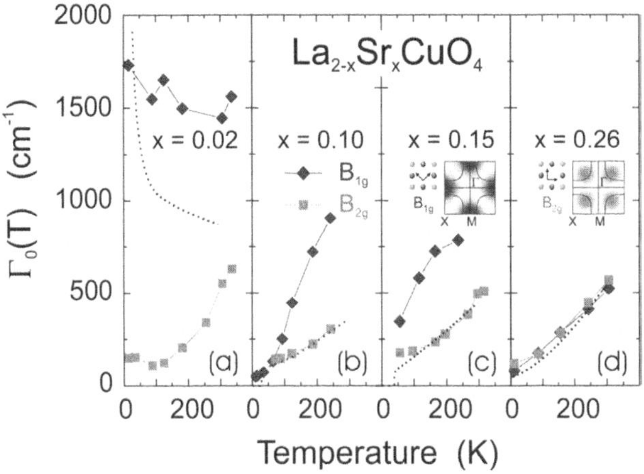

Fig. 4. Static electron relaxation rates $\Gamma_0^{\mu}(T)$ of LSCO. In LSCO p is equal to x. The dashed lines represent conventional dc resistivities

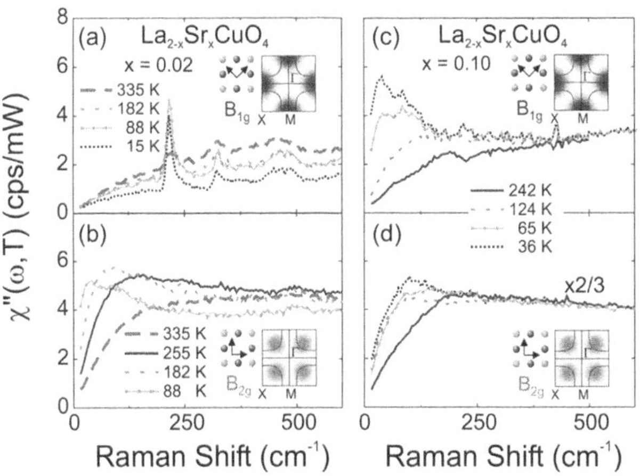

Fig. 5. Raman response $\chi''_{\mu}(\omega, T)$ of $La_{1.98}Sr_{0.02}CuO_4$ (a,b) and $La_{1.90}Sr_{0.10}CuO_4$ (c,d). Areas around the M point and the center of the quadrant are projected out in B_{1g} and B_{2g} symmetry, respectively, on a quadratic two-dimensional lattice. The selection rules for 1D structures are explained below in Fig. 7

At $p = x = 0.02$, the new B_{1g} peak at low energy does not show up any more (Fig. 5 (a)). Instead, the dc slope decreases as expected for an insulator. More or less the same behavior should be found at B_{2g} symmetry which, according to all we know from the other compounds, resembles the resistivity [5, 11] (see also Figs. 3 and 4 (b)–(d)). However, upon cooling the relaxation rate decreases (Fig. 4 (a)) and the slope increases as strongly as in B_{1g} at $p = 0.10$ (compare Fig. 5 (b) and (c)). As a complication, no peak can be resolved in the raw data at $p = 0.02$. Nevertheless, the relaxation rates

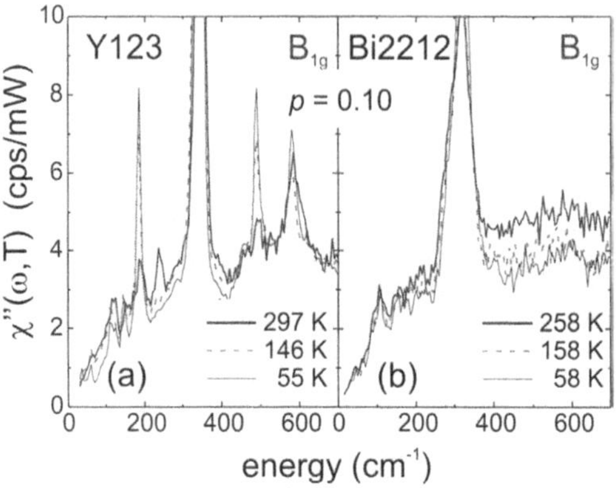

Fig. 6. Raman response of $YBa_2Cu_3O_{6.5}$ (a) and $Bi_2Sr_2(Ca_{0.62}Y_{0.38})Cu_2O_{8+\delta}$ (b) at B_{1g} symmetry. In either case the doping level is close to $p = 0.10$. Due to Y substitution the B_{1g} phonon in Bi-2212 is strong and inhomogeneously broadened

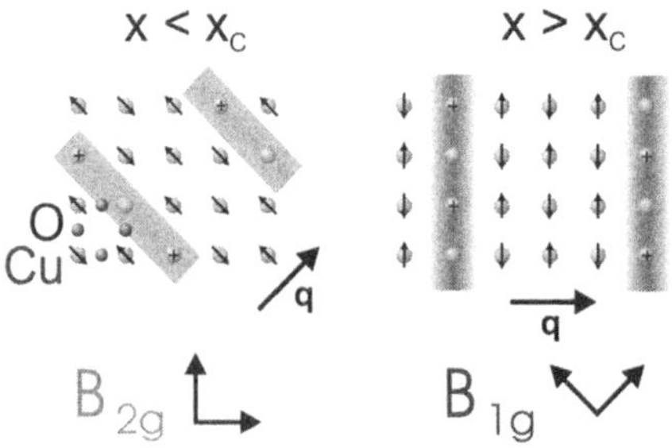

Fig. 7. Sketch of spin-charge-ordered states in the copper-oxygen plane (adopted from ref. [20]). There are antiferromagnetic insulating areas and charged spin-free "stripes". The modulation is characterized by the vector **q**. If the pattern fluctuates the correlation length ξ_s is finite and can be as small as a few lattice constants. The response of 1D objects perpendicular to the modulation direction **q** can only be observed by Raman scattering if the polarization vectors of both the incoming and the outgoing photons have a finite projection on either **q** or the stripe direction for transverse or longitudinal excitations, respectively. This implies that stripes parallel to the principal axes can be observed only in B_{1g} and diagonal ones only in B_{2g} symmetry

demonstrate clearly the difference between the B_{2g} spectra at $x = 0.02$ and $x = 0.10$ (compare Fig. 4 (a) and (b)).

Unexpected support comes from the results in Y-123 at $p = 0.02$ where the low-temperature peak is well separated from the continuum (Fig. 8 (b)). The 2D response shows a loss of the spectral weight by approximately 40 % in the range between 200 and 600 cm^{-1} due to the reduction of the density of states at low T. A similar loss of spectral weight is found in LSCO, and,

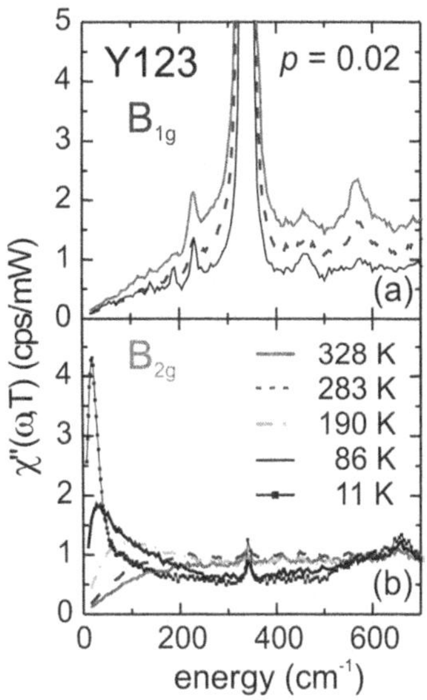

Fig. 8. Raman response $\chi''_\mu(\omega, T)$ of $(Y_{1.97}Ca_{0.03})Ba_2Cu_3O_{6.05}$ in B_{1g} (a) and B_{2g} (b) symmetry. The doping level is close to $p = 0.02$

for various reasons such as inhomogeneity and pinning, the peak is weaker and broader. Therefore, it cannot be observed as easily. The B_{1g} response (Fig. 8 (a)) is identical to the one in LSCO if the differences in the phonon spectra are taken into account.

The most direct way to properly interpret the new structures is via the selection rules. As explained in Fig. 7 we have to deal with order parallel and at 45° to the Cu–O bonds in B_{1g} and B_{2g} symmetry, respectively. Therefore, the observation of the low-energy Raman peaks in LSCO at B_{2g} and B_{1g} symmetry for $x = 0.02$ and $x = 0.10$ (Fig. 5 (b) and (c)), respectively, is clear evidence for the formation of 1D structures at low temperature which flip by 45° as a function of doping. The orientation is in full agreement with neutron results on the spin superstructure [16]. However, in the experiment here the *charge* response is measured, hence the dynamics of the charge "rivers" between the antiferromagnetically ordered spin domains is observed.

Recently, narrow modes have been observed in the energy range 10 – 20 cm^{-1} in LSCO at $x = 0$ and $x = 0.01$ [17]. For the doping, temperature, and symmetry dependence they can clearly be identified to be of magnetic origin. The features we find here exist well outside the Néel state where the magnetic modes rapidly disappear [17] and have the opposite temperature dependence. Even if the evidence is indirect the interpretation of the spectra observed here in terms of charge fluctuations is supported. The explicit calculation of the fluctuation diagrams predicts the observed spectral response

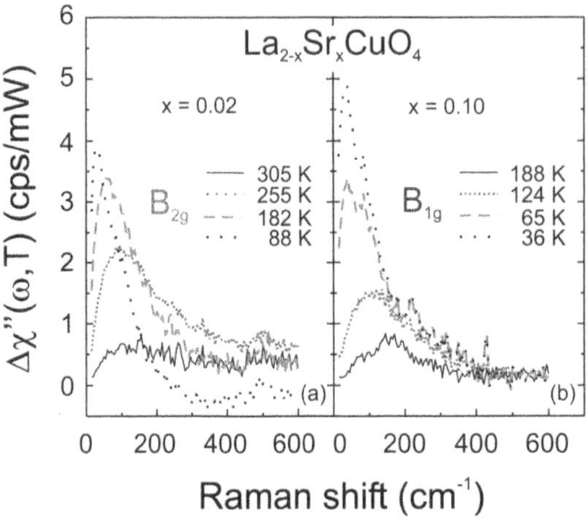

Fig. 9. Response of fluctuating charge order. A Drude-like peak [15] with a characteristic energy $\Omega_c(x,T)$ is revealed after subtraction of the 2D response of the CuO_2 planes. At $x = 0.02$ (a) and 0.10 (b) the additional response is observed in B_{2g} and B_{1g} symmetry, respectively. The styles of the lines (colors online) do *not* correspond to similar temperatures but rather highlight the *scaling* of the response with temperature: similar spectra are obtained if the temperatures differ by approximately a factor of 2

[18, 19]. Therefore, we found solid evidence for the formation of charged stripes and could observe the dynamics.

In Y-123 at $p = 0.02$ and in LSCO at $p = 0.10$ the determination of the peak frequency Ω_c is straightforward. In LSCO at $p = 0.02$, however, a peak appears only after the subtraction of the continuum from the 2D CuO_2 planes (Fig. 9). The results for Ω_c are summarized in Fig. 10. At $p = 0.02$ the results in LSCO and Y-123 are practically identical. At $p = 0.10$ the new response starts to develop only at half the temperature as at $p = 0.02$. As demonstrated in the inset of Fig. 10, temperature is the only scale for Ω_c. This temperature dependence of Ω_c is reminiscent of that of the mass term in the fluctuation propagator for doping levels below the quantum critical point [18, 19].

At the moment it is only qualitatively clear why the low-energy response cannot be observed in Y-123 and Bi-2212 at $p \simeq 0.10$ (Fig. 6). We assume that different correlation lengths of the stripes ξ_s are responsible: if ξ_s is comparable or larger than the electronic mean free path ℓ the response becomes visible. This condition seems to be qualitatively compatible with the STM results [21, 22, 23, 24]. On the other hand, since $\Omega_c \propto (\xi_s)^{-2}$ [18], Ω_c would increase with decreasing ξ_s. If T_c indeed depends on the fluctuations the trend would be correct. In turn, if the stripes become static superconductivity disappears completely [25]. LSCO could be somewhere in the middle where the lattice helps to stabilize the stripes without pinning them completely. In any

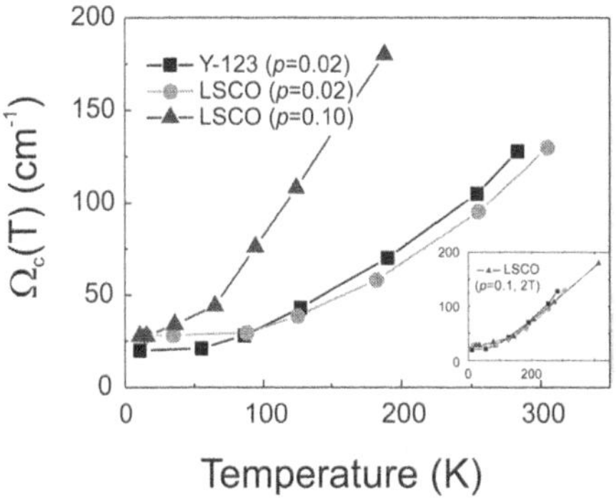

Fig. 10. Characteristic (*peak*) energies Ω_c of the low-energy response as a function of temperature. In the inset, Ω_c of LSCO at $p = 0.10$ is replotted as a function of $2T$ to demonstrate the scaling with T

case, the relationship between spin, charge, and lattice degrees of freedom and superconductivity seems to be subtle and most likely needs to be treated on equal footing.

It is also worthwhile to explore the possible connection between the metal-insulator transition and the formation of stripes. To this end we recall the respective doping ranges: (1) The anisotropy of the Raman relaxation rates starts to develop below $p \simeq 0.22$. At this carrier concentration the mean field transition to the fluctuation regime is expected. [6] (2) At a renormalized doping level around $p \simeq 0.16$ stripes start to form [6] but are not necessarily observable in a Raman experiment as outlined above. Once ξ_s exceeds the electronic mean free path ℓ new peaks appear in the light scattering response. The condition $\xi_s \geq \ell$ depends on the lattice and on doping.

So far mainly low doping levels have been studied. For further clarification of a possible importance of quantum criticality we will briefly look at the overdoped range of the phase diagram. As already discussed above the Raman relaxation rates $\Gamma_0^\mu(T)$ do not depend on symmetry any longer for $p \geq 0.22$ (Figs. 3 (e) and 4 (d)). Naively one might expect that the spectra look alike, too. However, this is not the case even in samples which do not become superconducting any longer.

In LSCO at $p = x = 0.26$, there is a substantial anisotropy in the spectral shape at room temperature (Fig. 11). While the B_{1g} spectra exhibit a pronounced Fermi-liquid peak which strengthens upon cooling the B_{2g} spectra are and remain relatively flat. The real surprise occurs at the lowest temperature where the Fermi-liquid peak disappears completely in B_{1g} symmetry (Fig. 11 (b)). Although the reduction of intensity is mild at B_{2g} symmetry it can be well resolved (Fig. 11 (a)). At $T = 10$ K the response is finally isotropic but not at all Fermi-liquid like. The trends are similar in Bi-2212 at $p = 0.23$ although the the collapse of the Fermi-liquid peak cannot be

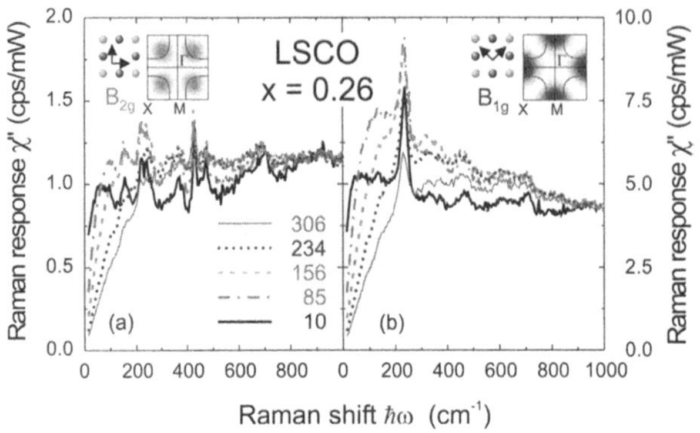

Fig. 11. Raman spectra of LSCO at $p = 0.26$. The symmetries and temperatures are indicated

observed most likely because of the transition into the superconducting state at $T_c \simeq 60$ K.

Given these results we cannot clearly conclude that we have entered the quantum disordered state (see Fig. 1). However, there is evidence that even at $p = 0.26$ the Fermi liquid does not survive down to low temperatures as one might expect on the basis of the Raman and the conventional resistivities (Fig. 5 (d)).

For $p < 0.26$ superconductivity apparently prevents the QCP to be accessed. It is therefore interesting to find out whether or not the vicinity of the quantum critical point has an influence on the superconducting properties. As a possible scenario there could be a transition from a pure $d_{x^2-y^2}$ to a $d_{x^2-y^2} + id_{xy}$ state [26]. Since the gap parameter can be accessed in a Raman experiment a careful inspection of the data is worthwhile. As an example, we show the doping dependence of the B_{2g} spectra deep in the superconducting state [5] and of the peak frequencies of the pair-breaking features at both B_{1g} and B_{2g} symmetry (Fig. 12). Apart from an artifact at the highest doping level due to improper orientation of the sample the normalized spectra collapse perfectly on top of each other indicating scaling between Δ_0 and T_c at this symmetry (Fig. 12 (a)). There is no scaling of the peak frequencies at B_{1g} symmetry, $\omega_{\text{peak}}^{\text{B}_{1g}}$, with T_c; in agreement with earlier results [27, 28, 29] one rather finds $\omega_{\text{peak}}^{\text{B}_{1g}} \propto (1-p)$.

Neither in the spectra nor in the peak frequencies any type of anomaly is found for $0.18 < p < 0.22$. This at the first glance disappointing result is not supportive of an interaction between superconductivity and quantum criticality induced by the charge fluctuations. However, the intensity of the B_{1g} peaks depends strongly on doping [28, 29] and disappears below $p \approx 0.14$ [30]. In agreement with scanning tunneling microscopy [31] one possible interpretation is therefore that the coherence around the $\pi, 0$ points is being

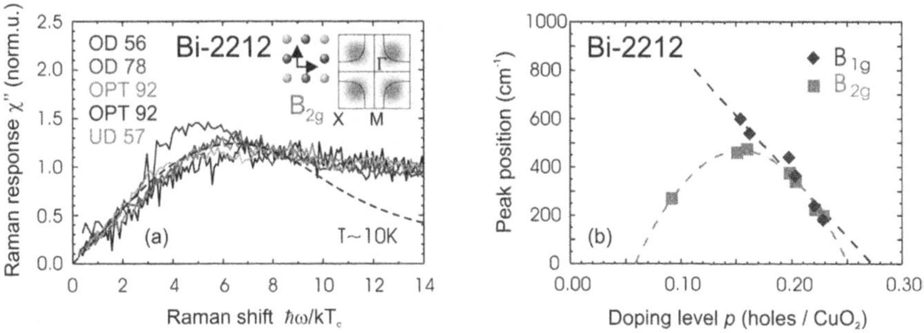

Fig. 12. (a) B_{2g} Raman spectra of superconducting Bi-2212 at $0.10 < p < 0.23$. The energies are normalized to the respective transition temperatures T_c. The intensities are normalized to 1 at the highest energy. The dashed line represents the theoretical prediction for a $d_{x^2-y^2}$ gap [10] with $\Delta_0/k_BT_c = 4.5$. (b) Peak frequencies of the pair breaking peaks at B_{1g} (*diamonds*) and B_{2g} (*squares*) symmetry. The best fit to the B_{2g} data follows closely the variation of T_c with doping

destroyed progressively below $p \approx 0.22$. This information taken together with the normal-state (Raman) transport strongly favors a singular interaction being at work. Charge fluctuations or a Fermi surface instability [32] are certainly candidates but also spin fluctuations have the proper **k** dependence. The fluctuation spectrum at low p would favor a charge instability. Whether or not the strong coupling supports pairing cannot be answered at present.

In summary, the electronic Raman spectra provide substantial new information about the transport properties in the cuprates. We clearly identify at least one ordering phenomenon which competes with superconductivity. The selection rules and a comparison to studies of charge [33] and magnetic order [17] provide evidence that we are dealing with a charge-ordering instability which strongly renormalizes the electronic properties of the cuprates. There is no obvious and direct influence on the superconducting order parameter but the correlation length ξ_s of the ordering phenomenon seems to influence T_c.

Acknowledgements

We gratefully acknowledge enlightening discussions with C. Di Castro, T. P. Devereaux, and M. Grilli. The work has been supported by the DFG via the Research Unit FOR538 under grant no. Ha 2071/2-1, Ha 2071/2-2, and Ha 2071/3-1.

References

1. C. M. Varma, P. B. Littlewood, S. Schmitt-Rink, E. Abrahams, A. E. Ruckenstein, Phys. Rev. Lett. **63**, 1996 (1989).

2. T. Timusk, B. W. Statt, Rep. Prog. Phys. **62**, 61 (1999).
3. Y. Ando, S. Komiya, K. Segawa, S. Ono, and Y. Kurita Phys. Rev. Lett. **93**, 267001 (2004).
4. S.J.L. Billinge, M. Gutmann and E.S. Božin, Int. J. Mod. Phys. B **17**, 3640 (2003) and M. Gutmann *et al.*, cond-mat/0009141.
5. M. Opel *et al.*, Phys. Rev. B **61**, 9752 (2000).
6. S. Andergassen, S. Caprara, C. Di Castro, M. Grilli, Phys. Rev. Lett. **87**, 056401 (2001).
7. J. L. Tallon and J. W. Loram, Physica C **349**, 53 (2001).
8. S. Sachdev, Science **288**, 475 (2000).
9. M. Vojta, Physik Journal **1** (3), 55 (2002).
10. T. P. Devereaux *et al.*, Phys. Rev. Lett. **72**, 396 (1994).
11. T. P. Devereaux, Phys. Rev. B **68**, 094503 (2003).
12. B. S. Shastry and B.I. Shraiman, Phys. Rev. Lett. **65**, 1068 (1990).
13. F. Venturini *et al.*, Phys. Rev. Lett. **89**, 107003 (2002).
14. F. Venturini, Adv. in Solid State Phys. **43**, 253 (2003).
15. A. Zawadowski and M. Cardona, Phys. Rev. B **42**, 10732 (1990).
16. S. Wakimoto *et al.*, Phys. Rev. B **60**, 769 (1999); M. Fujita *et al.*, Phys. Rev. B **65**, 064505 (2002).
17. A. Gozar *et al.*, Phys. Rev. Lett. **93**, 027001 (2004).
18. C. Di Castro, M. Grilli, S. Caprara, and D. Suppa, accepted for publication in J. Phys. Chem. Solids, cond-mat/0408058 (2004).
19. S. Caprara, C. Di Castro, M. Grilli, and D. Suppa, accepted for publication in Phys. Rev. Lett., cond-mat/0501671 (2005).
20. J. M. Tranquada, B. J. Sternlieb, J. D. Axe, Y. Nakamura, S. Uchida, Nature **375**, 561 (1995).
21. K.M. Lang *et al.*, Nature **415**, 412 (2002).
22. C. Howald *et al.*, Phys. Rev. B **67**, 014533 (2003).
23. J. E. Hoffman *et al.*, Science **297**, 1148 (2004).
24. M. Vershinin *et al.*, Science **303**, 1995 (2004).
25. H.-H. Klauss *et al.*, Phys. Rev. Lett. **85**, 4590 (2000).
26. G. Sangiovanni, M. Capone, and S. Caprara, Int. J. Mod. Phys. B **17**, 614 (2003).
27. C. Kendziora, R. J. Kelley, and M. Onellion, Phys. Rev. Lett. **77**, 727 (1996).
28. S. Sugai and T. Hosokawa, Phys. Rev. Lett. **85**, 1112 (2000).
29. K.C. Hewit and J.C. Irwin, Phys.Rev. B **66**, 054516 (2002).
30. The controversy about the B_{1g} intensity below T_c in underdoped cuprates is still not settled. Without claiming this is beyond any doubt we have good reasons to believe that it disappears in homogeneous samples. However, this is not the appropriate place to clean up this issue.
31. K. McElroy *et al.*, Phys.Rev. Lett. **94**, 197005 (2005).
32. W. Metzner, D. Rohe and S. Andergassen, Phys. Rev. Lett. **91**, 066402 (2001).
33. A. Gozar *et al.*, Phys. Rev. Lett. **91**, 087401 (2003).

Two-Gap Superconductivity in Magnesium Diboride

Thomas Dahm

Institut für Theoretische Physik, Universität Tübingen,
Auf der Morgenstelle 14, 72076 Tübingen, Germany

Summary. It is by now generally accepted that the recently discovered superconductor MgB_2 is a superconductor with two energy gaps. Here we discuss how this happens and address some of its unusual implications. The anisotropy ratio of the upper critical field in MgB_2 single crystals shows an unusual strong temperature dependence. Measurements of the microwave conductivity on MgB_2 thin films have shown an anomalous coherence peak appearing at significantly lower temperatures than in conventional superconductors. We argue that both of these observations are natural consequences of the existence of the two gaps, if the special Fermi surface structure of MgB_2 is taken into account. We also examine consequences for the expected intrinsic nonlinear microwave response in MgB_2. Calculations including impurity scattering indicate that under certain circumstances impurity scattering might improve the microwave properties of this material.

1 Introduction

The discovery of superconductivity in the binary compound magnesium diboride (MgB_2) with a surprisingly high transition temperature of about 40 K in 2001 [1] created a flurry of research activities shortly afterwards. Although this transition temperature is much smaller that that of the high-T_c cuprate superconductors (currently reaching up to 142 K at ambient pressure) there are several advantages of MgB_2 which make it an interesting candidate for superconducting applications: it can be much easier produced and handled than the high-T_c cuprates, it is comparatively cheap, it can be grown on cheap substrates, and it is the highest T_c conventional s-wave superconductor known so far. There is now common agreement that superconductivity in MgB_2 is due to conventional strong electron-phonon interaction. More excitingly, however, it appears to be well established that there exist two superconducting gaps of distinct size, making MgB_2 the clearest example of a two-gap superconductor to date. This two-gap nature makes MgB_2 an interesting object for study of fundamental questions, but it also has important consequences for applications of this material, which still have to be explored. Here, we want to review some of the more unusual implications of the two-gap nature of MgB_2 and discuss its relevance for some applications, in particular microwave applications.

B. Kramer (Ed.): Adv. in Solid State Phys. **45**, 239–249 (2005)

2 Two-Gap Superconductivity

In BCS theory the zero temperature value of the superconducting gap Δ_0 is related to the superconducting transition temperature T_c via the BCS ratio $\Delta_0 = 1.76 k_B T_c$. For a T_c =40 K superconductor one would thus expect a gap of the order of $\Delta_0 \approx 6$ meV. In strong electron-phonon coupling superconductors the gap can be larger than this value, but usually not smaller [2]. Gap values found in early experiments on MgB_2 largely scattered between 2 and 8 meV with some clustering around 2.5 meV and 7 meV (for an early review see Ref.[3]). Tunneling experiments often observed two-gap structures [4], while infrared absorption [5] and penetration depth measurements [6] could only detect a small gap of the order of 2.5 meV. Specific heat measurements yielded a shoulder-like feature at low temperatures which could be associated with such a small gap [7]. Such observations raised the question whether there are two gaps or a strongly anisotropic gap in MgB_2. Other, related inconsistencies were observed in the vortex state: scanning-tunneling microscopy of the vortex core in MgB_2 found a core size of about 50 nm, while the coherence length as extracted from the upper critical field value was about 10 nm [8], indicating that the core size may be governed by a smaller gap than the upper critical field. Another problem concerns the anisotropy of the upper and the lower critical fields. According to anisotropic Ginzburg-Landau theory the ratio of the upper critical field for field direction along the c-axis of the crystal divided by the upper critical field for field direction perpendicular to the c-axis should be inversely proportional to the corresponding ratio of the lower critical fields. Experimentally, it has been established, however, that this relation is violated in MgB_2[9, 10, 11]. Again, the presence of two different gaps could explain the violation of this relation [12].

Important insight from the theory side into the question of the gap in MgB_2 was provided by band structure calculations [13] and solutions of Migdal-Eliashberg equations based on the band structure. In MgB_2 there are four bands crossing the Fermi level, which leads to four disconnected parts of the Fermi surface (see Fig. 3 in Ref. [13]). These four bands can be grouped into two classes: two of them are π bands mainly composed of the Boron p_z orbitals, the other two are σ bands mainly composed of the Boron p_x and p_y orbitals. The σ bands possess mainly two dimensional character, i.e. their electrons mainly move within the Boron planes, while the π bands possess three dimensional character [13]. The presence of all four bands has been confirmed by de-Haas-van-Alphen experiments [14].

Calculations of the phonon modes and their coupling strength to the electrons at the Fermi level in this system [15, 16, 17] have shown that the E_{2g} phonon mode, corresponding to a Boron-Boron in-plane bond-stretching vibration of the Boron sub lattice, couples very strongly to the electrons on the σ bands. This leads to a larger pairing energy and thus to a larger gap on the σ bands than on the π bands, as was shown by detailed calculations of the gap distribution on the Fermi surface sheets using Migdal-Eliashberg theory

[18, 19]. These calculations were consistent with a large gap value around 7 meV on the σ bands and a small gap value around 2 meV on the π bands.

An important question concerns the stability of these two different gaps. In conventional superconductors impurity scattering is known to equalize the gaps on the Fermi surface at the same time reducing T_c. This is one of the reasons why two-gap superconductivity or anisotropic s-wave superconductivity is rarely observed. This does not seem to be the case in MgB_2, though. Samples with highly different residual resistivities were seen to possess almost the same critical temperature and the small gap could be observed even in seemingly dirty samples. A theoretical explanation for this unexpected behavior was given in Ref. [20]: the electronic wave functions on the σ bands and π bands possess different parity with respect to reflection at the Boron plane. This makes the impurity scattering matrix element between these two types of bands much smaller than the impurity scattering matrix element within each of these bands. This means that impurity scattering equalizes the gaps within each of the two types of bands, but not very much between them, making the two gaps in MgB_2 exceptionally stable against impurity scattering.

Taken all together, these experimental and theoretical results convincingly show that MgB_2 is a two-gap superconductor and add up to a consistent picture of why and how this arises.

3 Upper Critical Field Anisotropy

In MgB_2 at low temperatures the upper critical field B_{c2}^{c} for field applied along the crystal c-axis is much smaller than the upper critical field B_{c2}^{ab} for field applied perpendicular to the c-axis. Usually, in conventional superconductors the anisotropy ratio $\Gamma_{c2} = B_{c2}^{ab}/B_{c2}^{c}$ does not change very much as a function of temperature. In MgB_2, however, this quantity was observed to be strongly temperature dependent, decreasing by more than a factor of 2 when the temperature was raised up to T_c [21, 22] (see the solid circles in Fig 1.). Such a strong temperature dependence is difficult to understand within a single gap model. It is possible to obtain such a strong temperature dependence within an anisotropic s-wave model [23]. However, the gap anisotropy would have to be larger than a factor of 10, which is inconsistent with the gap values found experimentally. It is also clear, though, that the mere presence of two gaps in MgB_2 is not sufficient to understand the strong temperature dependence of the anisotropy ratio either. It is necessary to include the anisotropies of the Fermi velocities on the σ and π bands also.

From the band structure calculations mentioned above detailed information about the Fermi surface structure and Fermi velocities of the σ and π bands is available. Therefore, the natural question arises whether this specific Fermi surface structure together with the two-gap scenario is capable to explain the strong temperature dependence of the anisotropy ratio. In an

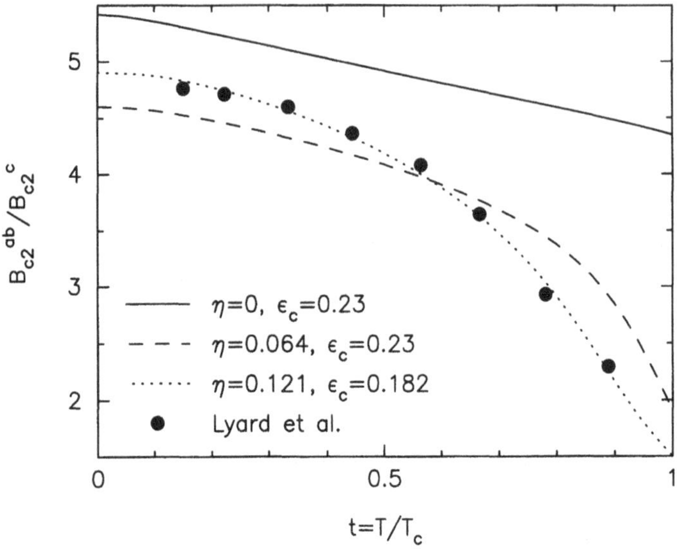

Fig. 1. Temperature dependence of the anisotropy ratio $\Gamma_{c2} = B_{c2}^{ab}/B_{c2}^{c}$ for the band structure based two-gap model of Ref. [24]. Results are shown for different interband pairing strengths η and c-axis dispersions ϵ_c. Solid circles are experimental results from Lyard et al. [22] (adapted from Ref. [24], ©2003 The American Physical Society)

effort to answer this question we constructed a simple but realistic model of the Fermi surface [24]. The σ band Fermi surface can be modeled as a distorted cylinder, where the distortion is directly related to the c-axis dispersion of the electrons on the σ band. The π band Fermi surface can be approximated as a half torus [25]. The Fermi velocities and their anisotropies were taken from band structure calculations. Based on this model it is possible to calculate the temperature dependence of the upper critical field using a two-gap generalization of Eilenberger theory. This was done in Ref. [24]. The result for the temperature dependence of the anisotropy ratio is shown in Fig. 1. Solid circles are experimental results on MgB_2 single crystals from Ref. [22]. The dashed line shows the result obtained using parameters from band structure calculations. For comparison, the solid line shows the result, if only the σ band with the large gap is included in the calculation (single gap model). The dotted line illustrates that a perfect fit to the experimental data can be obtained, if two parameters η (interband pairing strength) and ϵ_c (c-axis dispersion) are allowed to vary from their band structure values. More details about the calculation and the parameters η and ϵ_c can be found in Refs. [24, 25].

These results show that the two-gap nature of MgB_2 in conjunction with its specific Fermi surface structure are sufficient and also necessary ingredients to understand the strong temperature dependence of the anisotropy

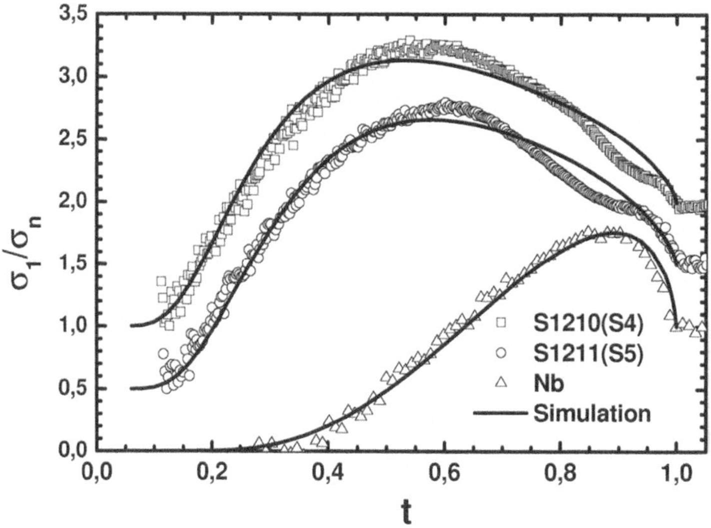

Fig. 2. Temperature dependence of the microwave conductivity in MgB_2 thin films (circles and squares). For comparison, the microwave conductivity of a Niobium thin film in the dirty limit is also shown (triangles). The curves are offset by 0.5 for clarity. (Adapted from Ref. [27], ©2003 The American Physical Society)

ratio. An analysis of these calculations allows the following interpretation of this effect: at low temperatures and high magnetic fields the small gap on the π band is suppressed and the upper critical field is thus dominated by the large gap on the σ band. Since the cylindrical σ band is strongly anisotropic this results in a strong anisotropy of the upper critical field at low temperatures. At higher temperatures the influence of the π band with the small gap gradually increases. Since the Fermi velocity on the π band is fairly isotropic, the anisotropy of the upper critical field is gradually reduced. This means that the strong temperature dependence of the anisotropy ratio not just needs the presence of two gaps, but also the fact that these gaps exist on Fermi surfaces with significantly different anisotropies. As the calculations show, the known Fermi velocity anisotropies and gap values result in a temperature dependence of the upper critical field anisotropy consistent with the experimental observations.

4 Microwave Conductivity Peak

One of the peculiar consequences of BCS theory is the existence of coherence effects. The first experimental confirmation was the observation of the Hebel-Slichter peak in the temperature dependence of the NMR relaxation rate and it was regarded as one of the key triumphs of BCS theory to explain this

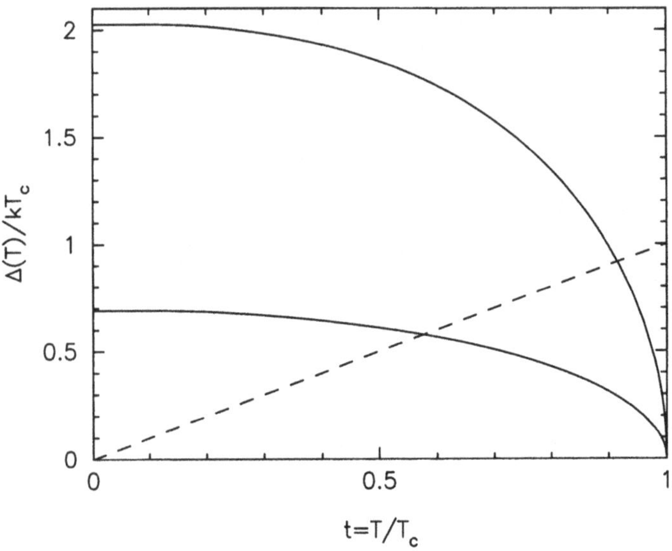

Fig. 3. Temperature dependence of the two gaps in MgB_2 found from a solution of the two-by-two gap equation. The dashed line shows $\Delta = k_B T$. Its crossing point with the small gap appears at a much lower temperature than the one with the large gap. (Adapted from Ref. [27], ©2003 The American Physical Society)

effect [26]. Such a coherence peak can also be observed in the temperature dependence of the microwave conductivity, if the impurity scattering rates in the sample are large enough. An example for this can be seen in Fig. 2 for the microwave conductivity of a Niobium thin film measured at 18 GHz (triangles). Decreasing the temperature below T_c the conductivity initially increases, going through a peak value at about $0.9T_c$, and then decreasing exponentially to zero at low temperatures.

In Ref. [27] the temperature dependence of the microwave conductivity of MgB_2 thin films was measured for the first time. It was shown that also in MgB_2 a coherence peak can be observed. However, this peak appears at much lower reduced temperatures than in conventional superconductors. Fig. 2 shows the temperature dependence of the microwave conductivity at 18 GHz for two MgB_2 samples (circles and squares). A conductivity peak appears around $0.5T_c$.

This anomalous position of the coherence peak can be understood as a consequence of the presence of two gaps, too. In order to see this, it is useful to have some rough criterion for the position of the peak. At low temperatures the exponential suppression of the conductivity sets in, when the temperature dependent gap $\Delta(T)$ becomes larger than $k_B T$. Thus, a rough criterion for the position of the peak is given by the relation $\Delta(T) = k_B T$. The temperature dependence of the two gaps in MgB_2 is shown in Fig. 3. The crossing point with the dashed line roughly denotes the expected position of the coherence peak. From this figure it is clear that in the σ band with the large gap one should expect a "conventional" coherence peak appearing around $0.9T_c$, while in the π band one can expect a coherence peak around $0.5T_c$, because the exponential suppression sets in later in this band. The total conductivity in such a two band system is given as the sum of the partial conductivities of the two bands (to the extend that interband impurity scattering can be neglected [25]). Thus, if we assume that the total conductivity in these MgB_2 films is dominated by the π band, we can understand the experimental observation qualitatively. This assumption means that the impurity scattering rate in the σ band has to be larger than the one in the π band, which is consistent with recent point contact tunneling experiments [28].

The solid lines in Fig. 2 show calculations of the temperature dependence of the conductivity using the dirty limit Mattis-Bardeen formula for a two-gap superconductor [25, 27]. These calculations confirm that the theoretically expected temperature dependence is in good quantitative agreement with the experimental results.

5 Nonlinear Microwave Response

Superconducting thin films can be used to construct passive microwave devices like resonators and filters. Such devices based on high-T_c cuprates are currently being used in base stations for wireless communication. The advantage of using superconductors instead of metals, for example, are their low losses in the microwave regime. However, superconductors also possess a comparatively large nonlinear microwave response, which leads to undesirable harmonic generation and intermodulation. This is one of the limiting factors in these kind of applications. Due to their d-wave superconducting state the nonlinear response in the high-T_c cuprates is particularly strong, even increasing at low temperatures [29, 30, 31] due to the nonlinear Meissner effect [32]. For this reason there is interest in assessing the suitability of MgB_2 for such applications. Due to its s-wave superconducting state one can expect a much lower nonlinear response at low enough temperatures. However, the question arises what the impact of the two-gap nature on the nonlinear response is. This question has been studied in Ref. [33].

The intrinsic nonlinear microwave response is due to the nonlinear change of the superfluid density n_s with the current density j. In lowest order in j^2 we can write

$$n_s(T,j) = n_{s0}(T)\left[1 - b(T)\frac{j^2}{j_c^2}\right] \tag{1}$$

Here, $b(T)$ is a nonlinear coefficient determining the strength of the nonlinear effects and can be calculated from the gap distribution. j_c is the critical pair-breaking current density and was found to be roughly equal in MgB_2 and

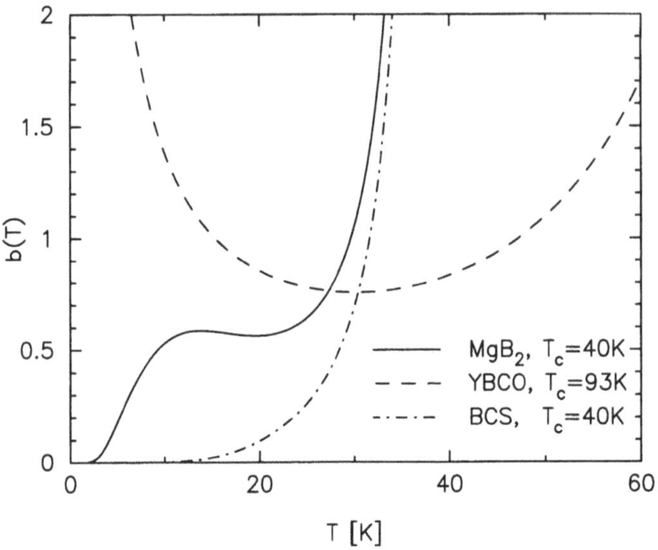

Fig. 4. Temperature dependence of the intrinsic nonlinear coefficient $b(T)$ for MgB_2 (solid line), for a d-wave superconductor with $T_c = 93$ K (YBCO, dashed line), and a hypothetical BCS single gap superconductor with $T_c = 40$ K (dashed-dotted line). (Adapted from Ref. [33], ©2004 American Institute of Physics)

$YBa_2Cu_3O_7$ (YBCO) [33, 34]. In Fig. 4 the calculated temperature dependence of b is shown for three different cases in the clean limit. The dashed line shows its temperature dependence for a d-wave superconductor with a critical transition temperature $T_c = 93$ K, corresponding to YBCO. The upturn at low temperatures is due to the nonlinear Meissner effect in d-wave superconductors [29]. The dashed-dotted line shows the temperature dependence of a hypothetical single gap s-wave superconductor with $T_c = 40$ K. In this case the nonlinear response is suppressed exponentially at low temperatures due to the opening of the gap. The solid line shows the temperature dependence expected for MgB_2 due to its two-gap nature. One can see that the presence of the small gap dominates the nonlinear response in the temperature range below about 25 K leading to a plateau-like behavior down to about 10 K. It is only below 10 K when an exponential suppression due to the small gap starts to set in. For microwave applications this means that the intrinsic nonlinear response of MgB_2 is not really much smaller than the one of YBCO unless one is willing to go to temperatures below about 20 K.

The comparison shown in Fig. 4 was made in the clean limit, i.e. in the absence of impurity scattering. Usually, the presence of impurities increases the nonlinear response. However, the effect of impurities in MgB_2 is more complex than in single gap superconductors because there are two intraband scattering rates, one for each band. This complexity can in fact be used to reduce the nonlinear response, at least in a certain temperature range. This

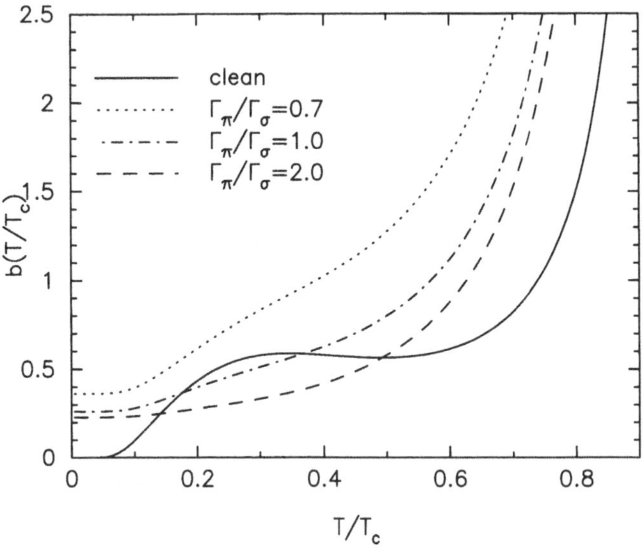

Fig. 5. Temperature dependence of $b(T)$ in the clean (solid line) and dirty limit. In the dirty limit $b(T)$ depends on the relative scattering rates in the two bands of MgB_2. Results are shown for Γ_π/Γ_σ =0.7 (dotted), 1 (dashed-dotted), and 2 (dashed). (Adapted from Ref. [33], ©2004 American Institute of Physics)

is illustrated in Fig. 5. The dotted, dashed-dotted, and dashed curves show calculations of $b(T)$ in the dirty limit for different ratios of the two intraband scattering rates [33]. The solid line shows the clean limit result. One can see that in the temperature range from 0.2 up to 0.4 T_c the nonlinear response in the dirty limit can become smaller than the one in the clean limit, if the scattering rate in the π band becomes twice as large as in the σ band. The physical reason for this is that a larger scattering rate in the π band reduces its superfluid density relative to that of the σ band and thus short-circuits the response of the π band. Then, the total response is dominated by the large gap.

This example shows that impurities could be used to improve microwave properties of MgB_2, if the scattering rate in the π band can be made larger than the one in the σ band. This could be achieved by substitutional doping at the Mg site, for example with Aluminum [20]. This example also shows that a better understanding of impurity scattering in MgB_2 could greatly help improving the material properties of this compound for applications.

6 Conclusions

Both from theoretical as well as from experimental side it appears now well established that MgB_2 is a two-gap superconductor. The two significantly different gaps arise due to the special band structure of this compound consisting of two dimensional σ bands and three dimensional π bands possessing different parity. The E_{2g} phonon mode couples more strongly to the electrons on the σ bands leading to a larger gap on these Fermi surface sheets.

The influence of the two gaps can be observed in a large number of different experimental results. Here, we have discussed some of the more peculiar consequences for the upper critical field anisotropy and linear as well as nonlinear microwave properties. The fact that the two gaps reside on Fermi surface parts with strongly differing anisotropies results in a strong temperature dependence of the upper critical field anisotropy ratio, being large at low temperatures and decreasing to more isotropic values at higher temperatures. The temperature dependence of the microwave conductivity displays an anomalous coherence peak, which appears at much lower reduced temperatures than in conventional superconductors. We have shown that this effect results from the fact that exponential suppression due to the small gap sets in at lower temperatures than usual.

We have calculated the nonlinear microwave response coefficient as a function of temperature. In the clean limit the presence of the small gap leads to a comparatively large nonlinear response in the temperature range from 10 to 25 K, not much smaller as compared with high-T_c cuprate superconductors at the same temperatures. However, impurity scattering can be used to reduce nonlinearities in MgB_2, if the π band can be made more dirty than the σ band.

Acknowledgements

The author would like to thank O. V. Dolgov, O. Eibl, S. Graser, R. B. Hammond, C. Iniotakis, B. B. Jin, N. Klein, S.-I. Lee, K. Maki, A. I. Posazhennikova, M. Salkola, D. J. Scalapino, N. Schopohl, and B. A. Willemsen for valuable discussions about this and related topics.

References

1. J. Nagamatsu, N. Nakagawa, T. Muranaka, Y. Zenitani, J. Akimitsu, Nature **410**, 63 (2001).
2. J. P. Carbotte, Rev. Mod. Phys. **62**, 1027 (1990).
3. C. Buzea T. Yamashita, Superconductor Science and Techn. **14**, R115 (2001) Fig. 35.
4. R. S. Gonnelli, D. Daghero, G. A. Ummarino, V. A. Stepanov, J. Jun, S. M. Kazakov, J. Karpinski, Phys. Rev. Lett. **89**, 247004 (2002).
5. R. A. Kaindl, M. A. Carnahan, J. Orenstein, D. S. Chemla, H. M. Christen, H.-Y. Zhai, M. Paranthaman, D. H. Lowndes. Phys. Rev. Lett. **88**, 027003 (2002).
6. B. B. Jin, N. Klein, W. N. Kang, H.-J. Kim, E.-M. Choi, S.-I. Lee, T. Dahm, and K. Maki: Phys. Rev. B **66**, 104521 (2002).

7. F. Bouquet, R. A. Fisher, N. E. Phillips, D. G. Hinks, J. D. Jorgensen, Phys. Rev. Lett. **87**, 047001 (2001).
8. M. R. Eskildsen, M. Kugler, S. Tanaka, J. Jun, S. M. Kazakov, J. Karpinski, Ø. Fischer, Phys. Rev. Lett. **89**, 187003 (2002).
9. L. Lyard, P. Szabo, T. Klein, J. Marcus, C. Marcenat, K. H. Kim, B. W. Kang, H. S. Lee, S. I. Lee, Phys. Rev. Lett. **92**, 057001 (2004).
10. R. Cubitt, M. R. Eskildsen, C. D. Dewhurst, J. Jun, S. M. Kazakov, J. Karpinski, Phys. Rev. Lett. **91**, 047002 (2003).
11. M. Xu, H. Kitazawa, Y. Takano, J. Ye, K. Nishida, H. Abe, A. Matsushita, N. Tsujii, G. Kido, Appl. Phys. Lett. **79**, 2779 (2001).
12. P. Miranovic, K. Machida, V. G. Kogan, J. Phys. Soc. Japan **72**, 221 (2003).
13. J. Kortus, I. I. Mazin, K. D. Belashchenko, V. P. Antropov, L. L. Boyer, Phys. Rev. Lett. **86**, 4656 (2001).
14. A. Carrington, P. J. Meeson, J. R. Cooper, L. Balicas, N. E. Hussey, E. A. Yelland, S. Lee, A. Yamamoto, S. Tajima, S. M. Kazakov, J. Karpinski, Phys. Rev. Lett. **91**, 037003 (2003).
15. Y. Kong, O. V. Dolgov, O. Jepsen, O. K. Andersen, Phys. Rev. B **64**, 020501(R) (2001).
16. K.-P. Bohnen, R. Heid, B. Renker, Phys. Rev. Lett. **86**, 5771 (2001).
17. A. Y. Liu, I. I. Mazin, J. Kortus, Phys. Rev. Lett. **87**, 087005 (2001).
18. A. A. Golubov, J. Kortus, O. V. Dolgov, O. Jepsen, Y. Kong, O. K. Andersen, B. J. Gibson, K. Ahn, R. K. Kremer, J. Phys.: Condens. Matter **14**, 1353 (2002).
19. H. J. Choi, D. Roundy, H. Sun, M. L. Cohen, S. G. Louie, Nature (London) **418**, 758 (2002).
20. I. I. Mazin, O. K. Andersen, O. Jepsen, O. V. Dolgov, J. Kortus, A. A. Golubov, A. B. Kuz'menko, D. van der Marel, Phys. Rev. Lett. **89**, 107002 (2002).
21. M. Angst, R. Puzniak, A. Wisniewski, J. Jun, S. M. Kazakov, J. Karpinski, J. Roos, H. Keller, Phys. Rev. Lett. **88**, 167004 (2002).
22. L. Lyard, P. Samuely, P. Szabo, T. Klein, C. Marcenat, L. Paulius, K. H. P. Kim, C. U. Jung, H.-S. Lee, B. Kang, S. Choi, S.-I. Lee, J. Marcus, S. Blanchard, A. G. M. Jansen, U. Welp, G. Karapetrov, W. K. Kwok, Phys. Rev. B **66**, 180502(R) (2002).
23. A. I. Posazhennikova, T. Dahm, K. Maki, Europhys. Lett. **60**, 134 (2002).
24. T. Dahm N. Schopohl, Phys. Rev. Lett. **91**, 017001 (2003).
25. T.Dahm, Superconductivity of Magnesium Diboride: Theoretical Aspects, in *Frontiers in Superconducting Materials*, A. V. Narlikar (Ed.), pp. 983-1009 (Springer 2005).
26. L. C. Hebel C. P. Slichter, Phys. Rev. **107**, 901 (1957); **113**, 1504 (1959).
27. B. B. Jin, T. Dahm, A. I. Gubin, E.-M. Choi, H. J. Kim, S.-I. Lee, W. N. Kang, N. Klein, Phys. Rev. Lett. **91**, 127006 (2003).
28. Y. Bugoslavsky, Y. Miyoshi, G. K. Perkins, A. D. Caplin, L. F. Cohen, A. V. Pogrebnyakov, X. X. Xi, cond-mat/0502153.
29. T. Dahm D. J. Scalapino, Appl. Phys. Lett. **69**, 4248 (1996).
30. G. Benz, S. Wunsch, T. A. Scherer, M. Neuhaus, W. Jutzi, Physica C **356**, 122 (2001).
31. D. E. Oates, S.-H. Park, G. Koren, Phys. Rev. Lett. **93**, 197001 (2004).
32. S. K. Yip J. A. Sauls, Phys. Rev. Lett. **69**, 2264 (1992).
33. T. Dahm D. J. Scalapino, Appl. Phys. Lett. **85**, 4436 (2004).
34. J. C. Booth, K. T. Leong, S. Y. Lee, J. H. Lee, B. Oh, H. N. Lee, S. H. Moon, Supercond. Sci. Technol. **16**, 1518 (2003).

Part IV

Semiconductors

Electron Spin Relaxation in Semiconductors

Daniel Hägele, Stefanie Döhrmann, Jörg Rudolph, and Michael Oestreich

Universität Hannover, Institut für Festkörperphysik, Abteilung Nanostrukturen, Appelstr. 2, 30167 Hannover

Summary. We review recent progress in the understanding of electron spin relaxation mechanisms in zinc-blende semiconductors. Increased spin lifetimes are obtained in structures with special orientation of crystal axes making them suitable for room temperature spintronic devices. In such structures the electron spin lifetime is found to critically depend on spin orientation, which directly affects the design rules for spin devices. We present new calculations concerning the recently introduced intersubband spin relaxation (ISR) mechanism. The mechanism becomes effective at elevated temperatures as scattering between higher electronic subbands starts to occur.

1 Introduction

In 1990, Datta and Das suggested a transistor based on spin effects in a gated two dimensional channel of electrons whose spin precession depends on gate voltage [1]. Over the last 15 years, an increasing number of groups started to investigate spin effects in semiconductors with respect to potential applications. In almost all of those devices - suggested or partially implemented - it is necessary to create and maintain a non-equilibrium spin distribution of the electrons for a sufficiently long time. A spintronic vertical cavity surface emitting laser (VCSEL) [2] requires for example spin lifetimes of the order of 1 ns, i.e. approximately the lifetime of an electron before it decays via optical recombination with a hole. Electron spin lifetimes are in bulk GaAs at room temperature typically around 100 ps, i.e. too short for potential applications [3]. Also two-dimensional heterostructures grown on standard (100) oriented GaAs substrates do not exhibit significantly longer lifetimes. The Dyakonov-Perel (DP) spin relaxation limits the spin lifetimes in these structures as will be discussed below [4]. In 1999, Ohno *et al.* demonstrated experimentally that in symmetric quantum wells grown in the [110] direction the DP mechanism is suppressed for spins in z-direction [5] as predicted in Ref. [6]. Room temperature lifetimes of up to 20 ns were demonstrated in such structures making them attractive for potential applications [7]. In this paper, we review recent progress in the understanding of spin relaxation in (110) grown quantum wells [8]. We find a strong dependence of spin lifetime on the spin orientation. A new spin relaxation mechanism will be discussed that limits spin-lifetime even for spins oriented in the [110] direction.

B. Kramer (Ed.): Adv. in Solid State Phys. **45**, 253–261, (2005)

2 Anisotropic Dyakonov Perel Spin Relaxation

At elevated temperatures, the most important source of electron spin relaxation in GaAs structures is spin-orbit coupling. In a gas of spin polarized electrons, each electron experiences an effective magnetic field caused by the spin orbit coupling that forces the electron to precess. The effective magnetic field is different for individual spins since electrons are in different momentum states. This destroys the coherence between electron spins leading to spin relaxation. This relaxation mechanism was introduced by Dyakonov and Perel in 1972 [4]. The relaxation rate towards spin equilibrium depends on the structure dependent spin orbit coupling, the electronic momentum scattering times, the nature of the scattering events (e.g. isotropic or forward scattering), and - as we will discuss in more detail - the initial electron spin orientation.

The starting point of our discussion is the Dresselhaus-Hamiltonian which describes spin-orbit coupling in the conduction band of a non-inversion symmetric bulk semiconductor (like GaAs, InP, CdTe, ZnSe,...)

$$H_{\mathrm{D}} = \Gamma \sum_i \sigma_i k_i (k_{i+1}^2 - k_{i+2}^2) = \frac{1}{2}\hbar \sum_i \Omega_i(\boldsymbol{k})\sigma_i, \tag{1}$$

where $i = x, y, z$ are the principal crystal axes with $i + 3 \rightarrow i$, Γ is the bulk spin-orbit coefficient ($\Gamma = 19.55\,\mathrm{eV\AA}^3$ in GaAs [9]), and σ_i are the Pauli spin matrices. The quantity $\Omega_i(\boldsymbol{k})$ is easily interpreted as a $\boldsymbol{k}$ dependent magnetic field acting on the electron spin.

The relaxation rate of spins in z-direction by spin-orbit coupling can heuristically be derived as follows. Let us assume that the electron scatters N times between different momentum states in a time interval Δt. Since $\boldsymbol{\Omega}$ is an odd function in $\boldsymbol{k}$ the average precession frequency $\langle\boldsymbol{\Omega}\rangle$ is zero. The mean deviation from zero, however, is given by $\delta_x^2 = \langle\Omega_x^2\rangle/N$ and $\delta_y^2 = \langle\Omega_y^2\rangle/N$. Only the perpendicular components of the precession vector, i.e. Ω_x and Ω_y, are relevant for relaxation of z spins. Consequently the spin polarization in z direction after the time Δt is approximately given by

$$s_z = \cos\left(\sqrt{\delta_x^2 + \delta_y^2}\,\Delta t\right) \tag{2}$$

$$\approx 1 - \frac{1}{2}\left(\langle\Omega_x^2\rangle + \langle\Omega_y^2\rangle\right)\frac{\Delta t^2}{N(\Delta t)} \tag{3}$$

Identifying $\Delta t/N(\Delta t)$ with the electron scattering time $\tau = \gamma_p^{-1}$ we find a decay of s_z that for early times depends linearly on Δt with a spin relaxation rate

$$\gamma_z = \frac{1}{2}\left(\langle\Omega_x^2\rangle + \langle\Omega_y^2\rangle\right)\gamma_p^{-1}. \tag{4}$$

This is the most simplistic formula for Dyakonov-Perel spin relaxation. Here, only the average electron scattering time enters disregarding its temperature dependence, its dependence on the initial state, and the nature of the scattering event. Eqn. (4) is further limited to the case of fast scattering where $\Omega_i \ll \tau^{-1}$. The spin relaxation rate decreases for faster momentum scattering rates γ_p, an effect known as motional narrowing. Increasing temperatures T lead, nevertheless, to faster spin relaxation because occupation of higher electronic k-states dramatically increases $\langle \Omega_i^2 \rangle \propto T^3$.

In cubic bulk material the spin relaxation rate is a scalar quantity as it does not depend on the relative orientation of crystal axis and spin orientation. In lower dimensional systems like quantum wells and quantum wires the relaxation rates can be anisotropic and have to be described by a tensor γ_{ij}. The dynamics of the decaying spin components s_i is given by $\dot{s}_i = \sum_j \gamma_{ij} s_j$. An illustrative example for a system with large relaxation anisotropy are (110) grown quantum wells where the electrons in the lowest subband only experience effective magnetic fields in the z-direction $\boldsymbol{\Omega} = (0, 0, \Omega_z(\boldsymbol{k}))$, where $\Omega_z(k_x, k_y) = -\Gamma k_y[\langle k_z^2 \rangle - (k_y^2 - 2k_x^2)]/\hbar$, and $z \parallel [110]$, $x \parallel [001]$, $y \parallel [1\bar{1}0]$. Spins in z-direction are conserved as they are not subject to spin precession. Only spins in x and y direction relax via the DP mechanism. The relaxation tensor obtains the form

$$\gamma_{(110)} \propto \begin{pmatrix} \langle \Omega_z^2 \rangle & 0 & 0 \\ 0 & \langle \Omega_z^2 \rangle & 0 \\ 0 & 0 & 0 \end{pmatrix} . \tag{5}$$

For completeness we quote the generalized form of the relaxation tensor that was given by Titkov as

$$\gamma_{ij} = \tau_p \left[\delta_{ij} \left(\langle \boldsymbol{\Omega}^2 \rangle - \langle \Omega_i \Omega_j \rangle \right) - (1 - \delta_{ij}) \langle \Omega_i \Omega_j \rangle \right] \tag{6}$$

$$= \tau_p \left(\delta_{ij} \langle \boldsymbol{\Omega}^2 \rangle - \langle \Omega_i \Omega_j \rangle \right) . \tag{7}$$

Here, we corrected a wrong sign between the Ω-terms in Eqn. (6) which occurred in the frequently quoted reference [10].

3 Experiments

In the following, we demonstrate that a huge spin relaxation anisotropy can in fact be observed in (110) oriented quantum wells. The γ_z and γ_x component of the spin relaxation tensor are measured via time and polarization resolved photoluminescence (PL) spectroscopy. The sample under investigation is a symmetrically modulation doped $GaAs/Al_{0.4}Ga_{0.6}As$ multiple quantum well with ten wells of 20 nm thickness and an electron sheet density of $n = 1.1 \times 10^{11}$ cm^{-2} per well. Optical excitation of the sample with circularly polarized laser pulses (100 fs duration) creates additional spin polarized electrons and holes with a density of about 2×10^{10} cm^{-2}. After excitation, the holes lose their spin orientation within the temporal resolution of our setup (20 ps). The spontaneous emission of the optical recombination of spin polarized electrons with the heavy holes exhibits a degree of circular polarization

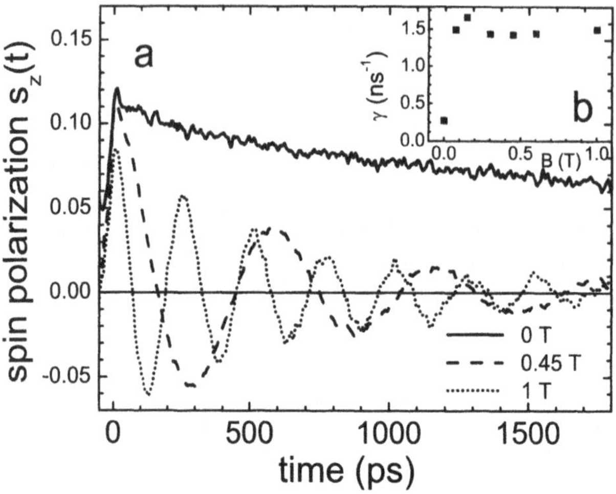

Fig. 1. (a) Spin oscillations $s_z(t)$ for different magnetic fields. (b) Spin relaxation rate γ.

$p(t)$ that at the heavy hole resonance is proportional to the degree of spin polarization $s_z(t)$ with a ratio close to unity [11]. A fit of $p(t)$ to $p_0 e^{-\gamma_z t}$ yields directly the γ_z component of the relaxation tensor [compare Fig. 1]. Unfortunately, the in-plane component γ_x cannot be directly measured by excitation and detection of spins in x direction because according to the in-plane optical selection rules the heavy hole transition does not depend on spin [12]. However, γ_x can be inferred from $s_z(t)$ after application of a weak magnetic field in y direction which makes the spin rotate in the xz-plane. Fitting $s_z(t)$ to $e^{-(\gamma_x+\gamma_z)t/2}\cos(\omega_L t)$ yields the second component of the spin relaxation tensor. Fig. 1(b) shows that the measured values for $\gamma = (\gamma_x + \gamma_z)/2$ do not depend on magnetic field in the range from 0.075 to 1 T. We can therefore exclude an influence of the magnetic field on spin relaxation rates.

Fig. 2 displays the measured relaxation rates γ_z and γ_x in dependence of sample temperature. Spin relaxation is clearly anisotropic for all temperatures reaching a maximum of $\gamma_x/\gamma_z \approx 10$ above 160 K. To the best of our knowledge this is the largest spin relaxation anisotropy reported in semiconductors so far. The weaker anisotropy towards low temperatures can be explained by spin relaxation of electrons via spin scattering with photocreated holes. This so-called Bir-Aronov-Pikus (BAP) mechanism is based on shortrange exchange interaction between electrons and holes, which is most efficient in 1s excitons [13]. The weak but significant anisotropy of $\gamma_x/\gamma_z \approx 2$ is probably due to a stronger exchange interaction in the confinement direction of the quantum well than in the lateral direction [14]. At elevated temperatures BAP loses its influence due to ionization of excitons resulting in slower relaxation rates γ_x and γ_z. Above 40 K, relaxation in x direction

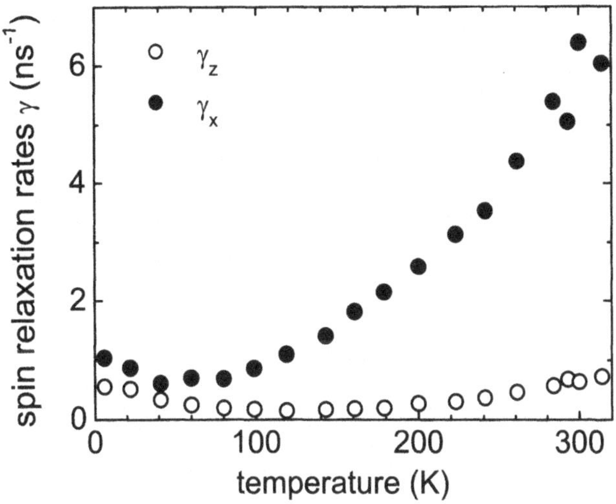

Fig. 2. Spin relaxation rates in an n-doped 20 nm (110) GaAs quantum well for spins in growth direction (γ_z) and spins oriented in plane (γ_x).

starts to increase due to the DP mechanism which gains influence as higher electron k-states get populated.

4 Intersubband Spin Relaxation

The spin relaxation rate γ_z exhibits a minimum at 120 K of $1/7\ \mathrm{ns}^{-1}$. Beyond 120 K the relaxation rate increases to about $1/1.4\ \mathrm{ns}^{-1}$ at 314 K. This increase is very surprising as the BAP mechanism should further lose influence at higher temperatures. The DP relaxation mechanism does also not apply to spins in z-direction because the effective magnetic field is parallel to the growth direction, i.e. $\boldsymbol{\Omega} \parallel \boldsymbol{e}_z$ (compare Eqn. (5) where $\gamma_z = 0$). Also in higher electronic subbands, which start to become thermally occupied at elevated temperatures, the effective magnetic field is strictly oriented in z-direction. This is easily verified by transforming (1) into the (110) coordinate system

$$H_{\mathrm{Dressel}} = \frac{1}{2}\hbar\left[\sigma_x\Omega_x + \sigma_y\Omega_y + \sigma_z\Omega_z\right], \tag{8}$$

where $(\Omega_x, \Omega_y, \Omega_z) = \Gamma/\hbar(4k_xk_yk_z,\ -k_y^2k_z - 2k_x^2k_z + k_z^3,\ k_y^3 - 2k_yk_x^2 - k_yk_z^2)$. The effective Hamiltonian for an electron in the νth subband $H_{2\mathrm{D},\nu} = \langle H_{\mathrm{Dressel}}\rangle = \langle\psi_{z,\nu}|H_{\mathrm{Dressel}}|\psi_{z,\nu}\rangle$ exhibits only components in z-direction because $\langle k_z\rangle = \langle k_z^3\rangle = 0$ for all subbands. Consequently, DP spin relaxation never applies to spins along the [110] growth direction (provided there are no Rashba contributions). Recently, we resolved the puzzle of increasing relaxation rates by a novel mechanism that is based on the interplay of spin orbit coupling and intersubband scattering. In the absence of spin orbit coupling, a simultaneous spin flip during a scattering event, like $|\boldsymbol{k},1,\uparrow\rangle \longrightarrow |\boldsymbol{k}',2,\downarrow\rangle$, is forbidden, because neither phonons nor electrons act on the electron spin. Here, $\boldsymbol{k}$ and $\boldsymbol{k}'$ describe the in-plane electron momentum, the second index is the subband number $\nu = 1,2,...$, and the third index describes the electron spin state in z-direction. If, however, spin and orbit are coupled, the corresponding transition is modified according to first order perturbation theory into $|\boldsymbol{k},1,\uparrow\rangle + \epsilon_1|\boldsymbol{k},2,\downarrow\rangle + ... \longrightarrow |\boldsymbol{k}',2,\downarrow\rangle + \epsilon_2|\boldsymbol{k}',1,\uparrow\rangle + ...$, where initial and final states now contain small contributions of states from other subbands with opposite spins. This allows for spin flip transitions initiated by phonons or by other scattering events.

The transition rate for spin flips is given by Fermi's golden rule

$$p = \frac{2\pi}{\hbar}|\langle\psi_f|A|\psi_i\rangle|^2\delta(E_i - E_f), \tag{9}$$

where A describes some spin conserving interaction, e.g. electron-electron scattering or optical phonon scattering. For the simplest case of only two subbands with $|\psi_i\rangle = |\boldsymbol{k}_i,\nu_i,\uparrow\rangle + \varepsilon_i|\boldsymbol{k}_i,\mu_i,\downarrow\rangle$ and $|\psi_f\rangle = |\boldsymbol{k}_f,\nu_f,\uparrow\rangle + \varepsilon_f|\boldsymbol{k}_f,\mu_f,\downarrow\rangle$, we can distinguish the two cases of intraband scattering and interband scattering.

In the *interband* case ($\nu_i = 1$, $\mu_i = 2$, $\nu_f = 2$, $\mu_f = 1$), we find

$$p = \frac{2\pi}{\hbar}|\varepsilon_f^*\langle\boldsymbol{k}_f,1,\uparrow|A|\boldsymbol{k}_i,1,\uparrow\rangle + \varepsilon_i\langle\boldsymbol{k}_f,2,\downarrow|A|\boldsymbol{k}_i,2,\downarrow\rangle|^2\delta(E_i - E_f), \tag{10}$$

[in the notation we suppressed a further index that labels the quantum state of the scattering partner (phonons, other carriers)] i.e. only interactions A that couple electrons within the same subband can contribute to intersubband spin flip scattering (ISR). Scattering of electrons with LO-phonons that act on the z-wavefunction will not contribute to this kind of spin-scattering as they change parity. Scattering with other electrons on the other hand is parity conserving and will contribute to intersubband spin flip scattering.

For *intraband* scattering ($\nu_i = 1$, $\mu_i = 2$, $\nu_f = 1$, $\mu_f = 2$), we find

$$p = \frac{2\pi}{\hbar}|\varepsilon_f^*\langle\boldsymbol{k}_f,2,\uparrow|A|\boldsymbol{k}_i,1,\uparrow\rangle + \varepsilon_i\langle\boldsymbol{k}_f,1,\downarrow|A|\boldsymbol{k}_i,2,\downarrow\rangle|^2\delta(E_i - E_f). \tag{11}$$

In this case, A is required to change the parity of the wavefunction. For low temperatures, where Coulomb scattering and deformation potential scattering with acoustic phonons dominates, this means that spin flips via intraband scattering do not occur as both mechanisms are parity conserving. At elevated and high temperatures, LO phonon scattering will allow for intraband spin flips. But since the LO phonon energy in GaAs is high (about 36 meV), also a large change in k of the electron is required. LO phonons with large k, however, couple only inefficiently to electrons (scattering probability $\propto k^{-2}$).

We therefore conclude that ISR will dominate over the intraband spin flip mechanism. We note that intraband spin flip scattering is very similar to the Elliott-Yafet (EY) mechanism [15]. EY however does not treat real interband transitions and is therefore different from ISR.

In order to estimate ISR spin flip rates from Eqn. (10), we recall that the expression $\langle \boldsymbol{k}_f, 1, \uparrow |A| \boldsymbol{k}_i, 1, \uparrow \rangle$ also occurs in the expression for spin conserving intraband momentum scattering ($p = \frac{2\pi}{\hbar} |\langle \boldsymbol{k}_f, 1, \uparrow |A| \boldsymbol{k}_i, 1, \uparrow \rangle|^2 \delta(E_i - E_f)$). At room temperature, this expression leads to scattering rates of typically $\gamma_p = 1/0.1\ \mathrm{ps}^{-1}$. In the case of ISR, we estimate a spin flip rate

$$\gamma_s^{(\mathrm{ISR})} = \langle |\varepsilon_f|^2 + |\varepsilon_i|^2 \rangle \gamma_p e^{-E_{\mathrm{gap}}/k_{\mathrm{B}}T}, \tag{12}$$

which regards the additional coefficients ε_i and ε_f in Eqn. (10) by a thermal average of their squared modulus. The Boltzmann-factor was added to take into account the additional amount of energy that is necessary in the intersubband scattering case to lift an electron from the first into the second subband. We are aware that this estimate for ISR is of course not strictly correct and should better be based on a microscopic theory. Only recently, Weng and Wu published a microscopy theory where they included occupation of higher subbands [16]. They, however, neglected explicitly the intersubband coherence, i.e. just that physical quantity which forms the basis for ISR. Other recent theories treat spin relaxation only in the lowest subband [17, 19].

We proceed to estimate the effectiveness of the ISR mechanism by calculating the first order contributions ε for an ideal quantum well with width d, infinite barriers, and effective electron mass m^*. The first order coefficient is given by

$$\varepsilon_{\boldsymbol{k}\nu\mu} = \frac{\langle \boldsymbol{k}, \nu, \uparrow |H_{110}| \boldsymbol{k}, \mu, \downarrow \rangle}{E_{\mathrm{gap},\nu,\mu}} \tag{13}$$

$$= \frac{1}{2} \Gamma \left\{ 4k_x k_y \langle \nu | k_z | \mu \rangle \langle \uparrow |\sigma_x| \downarrow \rangle + \left[(-k_y^2 - 2k_x^2) \langle \nu | k_z | \mu \rangle + \langle \nu | k_z^3 | \mu \rangle \right] \langle \uparrow |\sigma_y| \downarrow \rangle \right\} E_{\mathrm{gap},\nu,\mu}^{-1}. \tag{14}$$

Using the z-wavefunction $\langle z | \nu \rangle = \sqrt{\frac{2}{d}} \sin(z\pi\nu/d)$ we find the off-diagonal matrix element of k_z^n

$$\langle \nu | k_z^n | \mu \rangle = \left(-\frac{i\mu\pi}{d} \right)^n \frac{1}{\pi} \left(\frac{(-1)^{\nu+\mu} - 1}{\nu + \mu} - \frac{(-1)^{\mu-\nu} - 1}{\mu - \nu} \right) \Re(i^{n+1}). \tag{15}$$

For $d = 20$ nm, $E_{\mathrm{gap}} = 31$ meV, and $\Gamma = 19.55\ \mathrm{eV}\mathring{A}^3$ we find

$$\varepsilon_{\boldsymbol{k},1,2} \approx 4.1 \times 10^{-3} - 8.4\mathring{A}^2 k_x^2 + i16.8\mathring{A}^2 k_x k_y - 4.2\mathring{A}^2 k_y^2. \tag{16}$$

At room temperature all terms are of the same order as $\langle k_x^2 \rangle = \langle k_y^2 \rangle \approx 2.25 \times 10^{-4} \mathring{A}^{-2}$.

Figure 3 shows the experimental values for γ_z and γ_x along with theoretical fits using the ISR formula [Eq. (12)] and the DP formula [Eq. (4)], respectively. The fitting parameters to obtain agreement with experiment at 300 K were the scattering rate $\gamma_p = 1/0.04\ \mathrm{ps}^{-1}$ and the spin splitting constant $\Gamma = 26.1\ \mathrm{eV}\mathring{A}^3$. Although Γ is different from the theoretical kp-value [9], it coincides with an experimental value by Dresselhaus [18]. We obtain excellent agreement in the temperature range from 220 to 300 K. Better

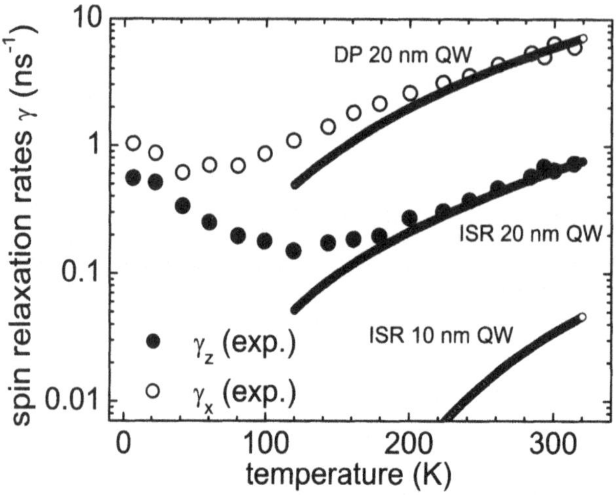

Fig. 3. Comparison of measured spin relaxation rates in an n-doped 20 nm (110) GaAs quantum well with calculated Dyakonov-Perel (DP) spin relaxation rates and Intersubband-Spin-Relaxation (ISR) rates.

agreement towards 120 K could be obtained by a temperature dependent scattering rate, that is 70 % lower at 120 K compared to the 300 K value. Such a decrease is in reasonable agreement with mobility measurements in other n-doped quantum wells [19]. We also plotted calculated ISR relaxation rates for a 10 nm QW to demonstrate the strong dependence of ISR on well width. The larger subband spacing of about 120 meV leads to a weak population of the second subband. Relaxation rates remain below 1/30 ns^{-1} up to room temperature.

5 Conclusion

In conclusion, we like to emphasize that the design of spintronic semiconductor devices needs to carefully take into account that spin relaxation tensors can be highly anisotropic. Furthermore, the advantage of a suppressed DP spin relaxation mechanism (e.g. in (110) grown quantum wells) is partly lost by the novel intersubband spin relaxation mechanism as soon as scattering between electronic subbands becomes important at high temperatures. A unified microscopic theory of spin relaxation that puts the BAP, DP, ISR and EY mechanisms on common ground remains a major challenge to theorists.

We like to acknowledge most valuable discussions with R. Winkler, E. L. Ivchenko, and S. Ganichev. We thank D. Schuh and M. Bichler for high quality samples. We thank BMBF and DFG (German Science Foundation) for financial support.

References

1. S. Datta and B. Das, Appl. Phys. Lett. **56**, 665 (1990).
2. J. Rudolph *et al.*, Appl. Phys. Lett. **82**, 4516 (2003).
3. J. M. Kikkawa and D. D. Awschalom, Phys. Rev. Lett. **80**, 4313 (1998).
4. M. Dyakonov and V. Perel, Sov. Phys. Solid State **13**, 3023 (1972).
5. Y. Ohno *et al.*, Phys. Rev. Lett. **83**, 4196 (1999).
6. M. I. Dyakonov and V. Y. Kachorovskii, Sov. Phys. Semicond. **20**, 110 (1986).
7. T. Adachi, Y. Ohno, F. Matsukura, and H. Ohno, Physica E **10**, 36 (2001).
8. S. Döhrmann *et al.*, Phys. Rev. Lett. **93**, 147405 (2004).
9. R. Winkler, *Spin-Orbit Coupling Effects in Two-Dimensional Electron and Hole Systems* (Springer, Berlin, 2003).
10. F. Meier and B. P. Zakharchenya, *Optical orientation* (Elsevier Science Publ., Amsterdam, 1984).
11. S. Pfalz *et al.*, Phys. Rev. B **71**, 165305 (2005).
12. M. Oestreich *et al.*, Solid State Comm. **108**, 753 (1998).
13. G. L. Bir, A. G. Aronov, and G. E. Pikus, Sov. Phys. JETP **42**, 705 (1976).
14. E. Blackwood *et al.*, Phys. Rev. B **50**, 14246 (1994).
15. R. J. Elliott, Phys. Rev. **96**, 266 (1954).
16. M. Q. Weng and M. W. Wu, Phys. Rev. B **70**, 195318 (2004).
17. M. M. Glazov and E. L. Ivchenko, Journal of Superconductivity **16**, 735 (2003).
18. P. D. Dresselhaus *et al.*, Phys. Rev. Lett. **68**, 106 (1992).
19. J. Kainz, U. Rössler, and R. Winkler, Phys. Rev. B **70**, 195322 (2004).

Donor Levels in ZnO

Holger von Wenckstern, Swen Weinhold, Gisela Biehne, Rainer Pickenhain, Heidemarie Schmidt, Holger Hochmuth, and Marius Grundmann

Universität Leipzig, Linnéstraße 5, 04103 Leipzig Germany

Summary. We have investigated and compared the electrical properties of ZnO grown by four different methods. From temperature dependent resistivity and Hall measurements the parameters of the dominant donors are obtained. Deep level transient and thermal admittance spectroscopy have been applied to study deep states in ZnO. The energetic position of deep donors has been identified for ZnO thin films for the first time. We have found the levels E1 ($E_d = 110$ meV) and E3 ($E_d = 290$ meV). Semi-insulating ZnO thin films grown by PLD have been demonstrated by three methods: avoiding the incorporation of shallow donors, compensation with acceptors, and co-doping. Further we report a Schottky diode structure enabling the use of ohmic back contacts for ZnO grown on insulating substrates with cut-off frequencies > 10 MHz. Schottky contacts on ZnO thin films are characterized by current voltage measurements for temperatures ranging from 210 K to 293 K and the results are evaluated considering the lateral fluctuations of the barrier height. The Schottky barrier height of Pd/ZnO has been determined to be 1.1 eV with a standard deviation of the lateral fluctuations of 130 meV.

1 Introduction

ZnO is a II-VI semiconductor with a band gap of 3.3 eV at room temperature. It has a high exciton binding energy of 61 meV which can be enhanced by confining the electron-hole pairs in quantum well structures. This opens the possibility of optoelectronic devices based on the radiative recombination of excitons working efficiently at room temperature. The controlled and reproducible doping of ZnO is a key issue nowadays. ZnO is naturally n-type due to the low formation energies of donor-like intrinsic defects and the easily achievable incorporation of extrinsic donors like hydrogen or the group III elements Al, Ga, and In. This and strong self-compensation effects lead to enormous difficulties in producing p-type ZnO. It is therefore important to study the density and thermal activation energies of shallow and of deep donors formed during growth of ZnO. Experimental techniques yielding this information with high precision and used in this work are temperature dependent Hall (TDH) measurements (information about dominating donors) and depletion zone spectroscopies like deep level transient spectroscopy (DLTS) (information about electron traps lying clearly below Fermi level) and thermal admittance spectroscopy (TAS) (information about shallow and deep

B. Kramer (Ed.): Adv. in Solid State Phys. **45**, 263–274, (2005)

levels). Due to the lack of high quality p-type ZnO, Schottky diodes are in need to conduct depletion zone spectroscopies.

In this paper we report the fabrication of Schottky contacts on n-type ZnO single crystals and n-type ZnO thin films. We further investigated ZnO single crystals and ZnO thin films by TDH, DLTS, and TAS measurements.

2 Samples

We have also investigated commercially available single crystals grown by three different methods. The samples are referred to from now on as Sample A, Sample B, and Sample C. Sample A was grown by seeded chemical vapor deposition (*Eagle Picher Inc.*), Sample B by pressure melt growth (*Cermet Inc.*), and Sample C by hydrothermal growth (*Crystec*). The electrical properties of these samples are summarized in Table 1. The ZnO thin films are grown by pulsed laser deposition (PLD) on $(11\bar{2}0)$ oriented sapphire substrates. For that we used the 248 nm line of a KrF excimer laser to ablate polycrystalline ZnO targets prepared by standard ball milling, pressing, and sintering. Using a laser repetition rate of 10 Hz, a 1 μm thick thin film is grown in about 45 minutes. Details about the growth and related properties of the ZnO thin films are published in e. g. [1] and [2].

3 Transport Properties of ZnO

The transport properties of PLD thin films depend on the oxygen partial pressure $p(O_2)$ applied during growth. Figure 1 depicts the Hall mobility μ_H and the free electron concentration n in dependence on $p(O_2)$.

The free electron concentration decreases with increasing $p(O_2)$. This can be attributed to an increasingly stoichiometric growth leading to a lower density of oxygen vacancies which form single (V_O^-) or double (V_O^{2-}) charged donor levels in ZnO. The Hall mobility exhibits a maximum at $p(O_2) \sim 0.03$ mbar. A reduced growth temperature results in an increase of n and a decrease of μ_H as shown for $p(O_2) = 0.02$ mbar in Fig. 1. The reduction of μ_H can be explained by considering the morphology of the thin films. Lower growth temperatures lead to a higher density of extended charged defects. The reduction of μ_H is then caused by the increasing number of scattering events at these extended defects that the charge carriers undergo for current transport. The increase of n is attributed to an increased density of donor-like intrinsic defects which form with higher probability at lower growth temperatures.

The dominant donor in nominally undoped PLD thin films is Al which diffuses during growth from the used sapphire substrates into the thin films as confirmed by secondary neutral mass spectroscopy (SNMS) as shown in Fig. 2. Also the spectral position of (D^0,X) photoluminescence found in our

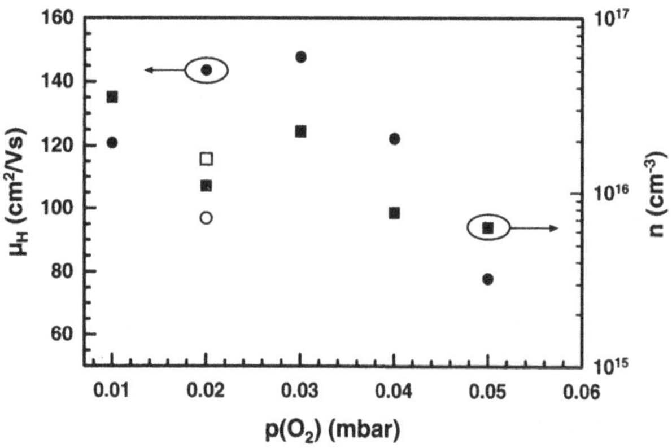

Fig. 1. Dependence of Hall mobility (circles) and free electron concentration (squares) on the oxygen pressure during growth, samples depicted with solid symbols are grown at about 730°C, open symbols correspond to a growth temperature of 650°C

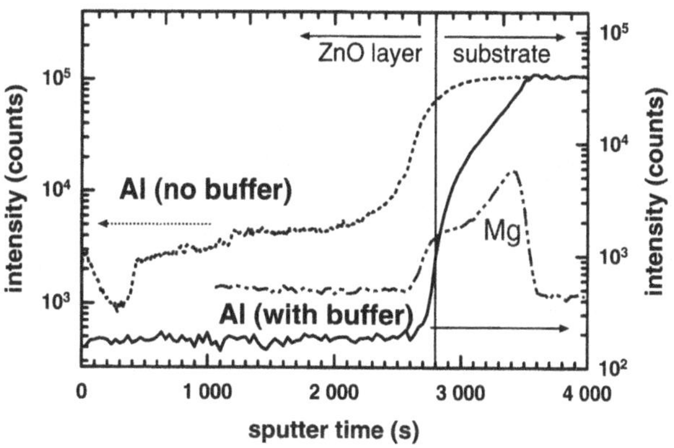

Fig. 2. Aluminum diffusion profiles for samples grown with (solid line) and without (dashed line) MgO buffer layer. ZnO-Al level

samples hints on Al [3]. This diffusion can be strongly reduced by using a MgO buffer layer. Figure 2 compares the diffusion profiles of Al in ZnO thin films grown with or without MgO buffer layer.

For samples grown without MgO buffer layer, the Al contamination is detectable up to the surface (sputter time = 0 s). A strong reduction of the Al diffusion is achieved by the insertion of the MgO buffer layer. The Al concentration falls below the detection limit of the SNMS setup within the buffer layer and remains below in the entire thin film. Subsequently the free electron concentration at room temperature is as low as 10^{14} cm^{-3}. With two other methods (compensation by acceptors or co-doping) we could also obtain such low carrier concentration in thin films. The doping of ZnO

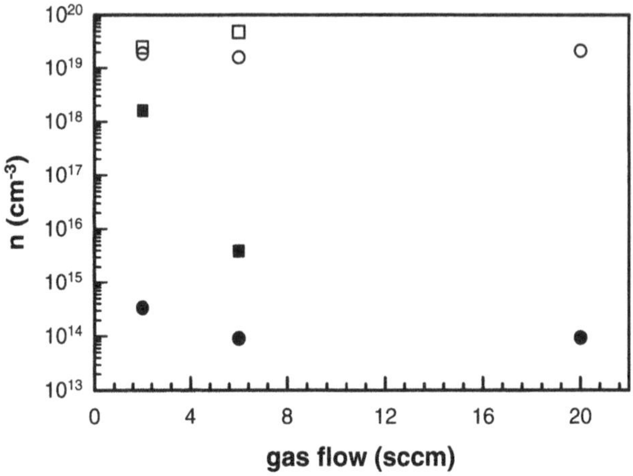

Fig. 3. Free carrier concentration of co-doped ZnO thin films grown with ZnO targets containing 10 ppm Ga (squares) or 0.5 ppm Ga (circles) in N_2O (solid symbols) or O_2 ambient (open symbols), respectively

with P yields after thermal activation of the dopant insulating ZnO layers. Co-doping by Ga (from Ga_2O_3 in the ZnO target) and N (from N_2O in the ambient atmosphere) and growth at low temperatures to achieve high nitrogen incorporation also leads to high resistive ZnO layers. Figure 3 depicts the dependence of n of co-doped samples grown at about 370°C in dependence on the flow of N_2O used as background gas. The gas was introduced into the growth chamber by a radio frequency plasma source to decompose the N_2O in its constituents. Samples grown under the same conditions but with O_2 as background gas are shown in Fig. 3 as reference, too. The free carrier concentration of the samples grown under oxygen atmosphere is 10^{19}cm^{-3} and nearly independent of the O_2 gas flow. For samples grown in N_2O n is much lower. For these co-doped thin films n shows a strong dependence on the N_2O flow. If a target containing 10 ppm Ga is used, n decreases by more than two orders of magnitude if the N_2O flow is increased from 2 sccm to 6 sccm. If a Ga content of only 0.5 ppm is used, n decreases for low gas flows and saturates at about 10^{14} cm^{-3} for higher gas flows. This behavior can be well understood within the co-doping theory [4], [5]. Firstly the much lower values of n of the co-doped samples confirms the effective incorporation of N_O in presence of a group III element, here Ga. Secondly a higher Ga concentration in the target leads to larger electron concentrations, especially for low gas flows, since not all Ga_{Zn} is bound to N_O. Thirdly n decreases with higher N_2O flow since the number of Ga_{Zn} bound to at least 2 N atoms increases because more N atoms are disposed during growth.

In summary we have demonstrated three possibilities of growing semi-insulating ZnO layers having low electron concentrations of 10^{14} cm^{-3}.

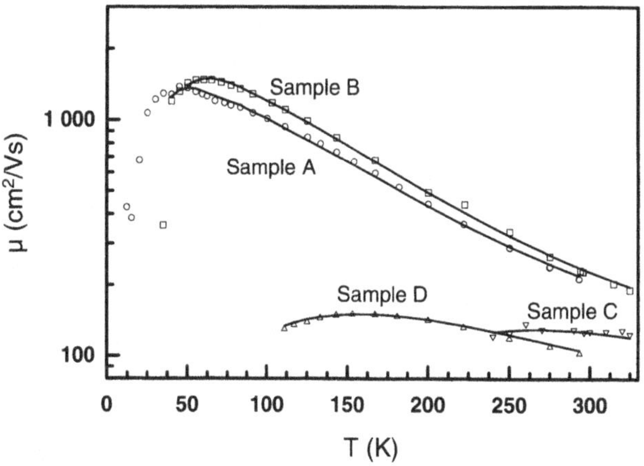

Fig. 4. Hall mobility of Samples A - D (symbols) and corresponding fits (solid lines)

3.1 Temperature Dependent Measurements of Single Crystals and Thin Films

We have investigated Samples A – C and a typical, nominally undoped PLD thin film (referred from now on as Sample D) with TDH measurements in a temperature range between 10 and 325 K [6]. For the analysis of μ_{H} we have used Matthiessen's rule and accounted ionized impurity scattering, deformation potential scattering, piezo-electric potential scattering, and polar optical scattering. The experimental data and the corresponding fits are depicted in Fig. 4. Samples A and B show the highest Hall mobilities of the investigated samples which is due to the comparatively low density or compensating acceptors (see Table 1). Sample C exhibits the lowest μ_{H} of the investigated single crystals. This sample is compensated for $T < 240$ K suggesting that is has the highest concentrations of acceptors which is confirmed by the fit of μ_{H}. The temperature dependence of μ_{H} of Sample D could not be fitted considering only the above scattering mechanisms. The existence of extended charged defects produces the dominant scattering channel in the PLD thin films. Therefore we have added scattering at small angle grain boundaries in Matthiessen's rule for Sample D and could reproduce the experimental data. We introduce two additional fitting parameters, the grain size L_G and the barrier potential height V_B between the grains. The values obtained from the fit are $L_G \sim 40$ nm and $V_B \sim 13$ meV. The density of compensating acceptors obtained from the fit of the Hall mobilities are 3×10^{15}, 1.8×10^{15}, 3×10^{17}, and $1.5 \times 10^{15}\,\mathrm{cm}^{-3}$ for Samples A, B, C, and D, respectively.

The temperature dependence of n was modelled by solving the charge balance equation taking one dominant donor and compensating acceptors into account. The experimental data and the corresponding fits are shown in

Fig. 5. The slope of $n(T)$ is different for each of the samples. The lowest slope is observed for Sample A corresponding to a thermal activation energy of 37 meV. The chemical origin of this defect is most likely H_i. Photoluminescence measurements reveal a transition at 3.363 eV which is commonly attributed to the recombination of excitons bound to H [3]. The density determined from the fit is $N_d = 1.3 \times 10^{17}\,cm^{-3}$. The dominating donor of Sample B has a density of $4.4 \times 10^{16}\,cm^{-3}$ and a thermal activation energy of $E_D = 46\,meV$. Hutson [7] has found a defect with a thermal activation energy of 46 meV in ZnO single crystals doped with Zn. Therefore we conclude that the ratio of Zn and O in Sample B is slightly non-stoichiometric resulting in the formation of Zn_i representing the dominating donor in this sample. The dominant donor in Sample D has $N_d = 6 \times 10^{16}\,cm^{-3}$ and a thermal activation energy of about 65 meV which corresponds to that of an effective mass donors (using $\epsilon = 8.12$ and $m^*/m_e = 0.32$) in ZnO. The chemical origin of this donor is Al which diffuses into the thin films from the used sapphire substrate during growth.

The dominant donor of Sample C is much deeper than those of the other investigated samples. Its density, as obtained from the fit, is $N_d = 1.8 \times 10^{17}\,cm^{-3}$. The thermal activation energy of this donor is 300 meV. Later on (in section 4.3) this level will be identified in other samples as E3 level. Photoluminescence measurements reveal also the existence of effective mass donors (Al and Ga) but they do not contribute free electrons due to compensation. With these findings the determined density of the dominant donor in Sample D must be critically reconsidered. The existence of shallow donors which are completely ionized by donating an electron to the energetically favorable acceptors states alters the charge balance equation to $n = N_d^+ + N_{ds} - N_a$, where N_d^+ is the density of ionized dominant donors and N_{ds} is the density of all shallow donors. The density of donors obtained from the fit of $n(T)$ is therefore the sum of N_d and N_{ds}. It is not possible to separately determine N_d or N_{ds} from the data presented here. We can only state that N_{ds} has to be smaller than the density of compensating acceptors and that the lower limit of the concentration of the deep dominant donor is about $10^{16}\,cm^{-3}$.

4 ZnO Schottky Diodes

We have produced circular Schottky contacts (SC) on samples similar to Sample A, B, C, and D referred to from now on as Schottky diodes of Type A, B, C, or D, respectively. The contacts were realized by thermal evaporation of Pd (Sample A, B and D) or Au (Sample C) on the sample surfaces. Prior to that, the samples were cleaned for 5 min in an ultrasonic bath of acetone. Ohmic contacts were realized by an eutectic mixture of In and Ge (Sample A, B and C) or thermal evaporation of Al (Sample D) and showed linear current voltage (I-V) characteristics for all samples. The ohmic contacts were realized

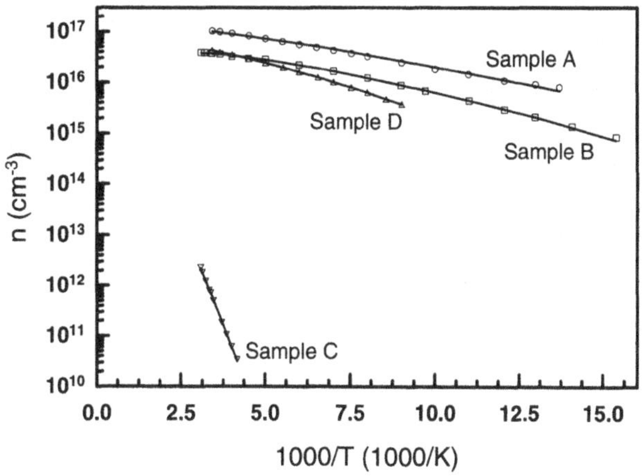

Fig. 5. Free electron concentration of Samples A - D (symbols) and corresponding fits (solid lines)

at the back surfaces of the single crystals. Due to the insulating sapphire substrate the ohmic contacts for the PLD thin films were made in the form of a ring around the SC.

4.1 Current–Voltage Measurements

A Schottky diode of Type D (PLD film) was investigated by temperature dependent I–V measurements. The characteristic curves are depicted in Fig. 6. The current density at room temperature at -3 V is -2×10^{-5} A/cm^2 at -5 V it is -4×10^{-5} A/cm^2. Considering thermionic emission as the dominating current transport process across the Schottky barrier, we determined the barrier height $\Phi_{\mathrm{B,eff}}$ and the ideality factor η to be 0.76 eV and 1.46, respectively. We determined the saturation current density j_d from which $\Phi_{\mathrm{B,eff}}$ is obtained and η for all graphs depicted in Fig. 6.

The Richardson plot, i.e. $\ln(j_d/T^2)$, from which $\Phi_{\mathrm{B,eff}}$ at 0 K and the Richardson constant can be determined however is curved. Therefore we analyzed the temperature dependence of $\Phi_{\mathrm{B,eff}}$ by taking lateral fluctuations of $\Phi_{\mathrm{B,eff}}$ into account. We follow Werner *et al.* [8] and modelled these fluctuations by a Gaussian distribution with standard deviation σ around a mean barrier height $\Phi_{\mathrm{B,m}}$. The relation between the mean barrier height and the effective barrier height is $\Phi_{\mathrm{B,eff}} = \Phi_{\mathrm{B,m}}(T = 0\,K) - e\sigma^2/(2kT)$ By fitting the experimental data depicted in Fig. 7(a), we were able to determine the mean barrier height at 0 K to be 1.10 eV and the standard deviation $\sigma = 0.13$ eV. Capacitance voltage measurements on a similar diode gave a value of $\Phi_{\mathrm{B,m}} = 1.08$ eV for three different measuring frequencies. This value is in good agreement with the value determined using the model of $\Phi_{\mathrm{B,m}}$ obtained from the temperature dependence of the I–V characteristics. We

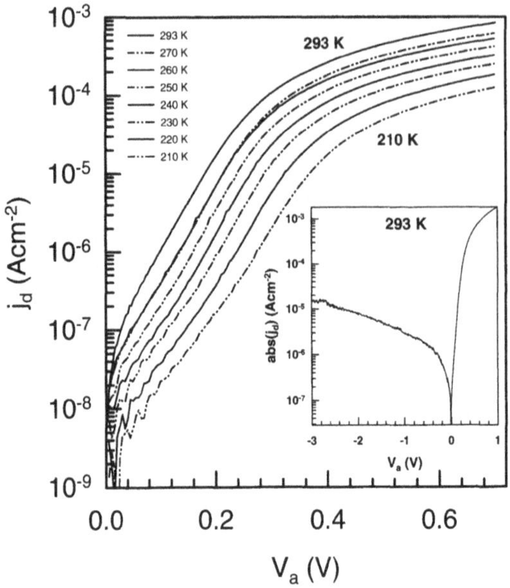

Fig. 6. Current voltage measurements for different temperatures of diode of Type D. Inset shows 293 K characteristic in different scale

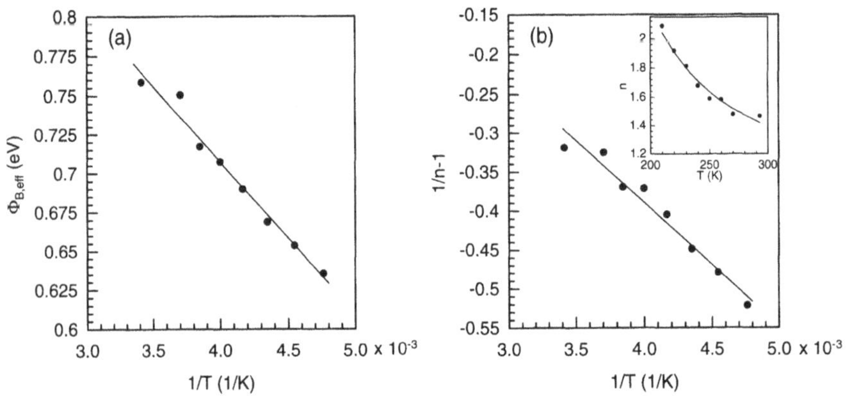

Fig. 7. (a) Effective barrier height versus reciprocal temperature and corresponding fit with $\Phi_{\mathrm{B,m}}(T = 0\,K) = 1.10$ eV and $\sigma = 0.13$ eV and (b) η versus reciprocal temperature and corresponding fit with $\rho_2 =$ -0.25 and $\rho_3 = -0.028$ V

note that the model of [8] also allows to model the temperature dependence $\eta = (1 - \rho_2 + e\rho_3/(2kT))^{-1}$. The coefficients ρ_2 and ρ_3 are a measure of the voltage dependence of $\Phi_{\mathrm{B,m}}$ or σ^2, respectively [8], and are determined from the fit shown in Fig. 7(b) to be -0.25 or -0.028 V, respectively.

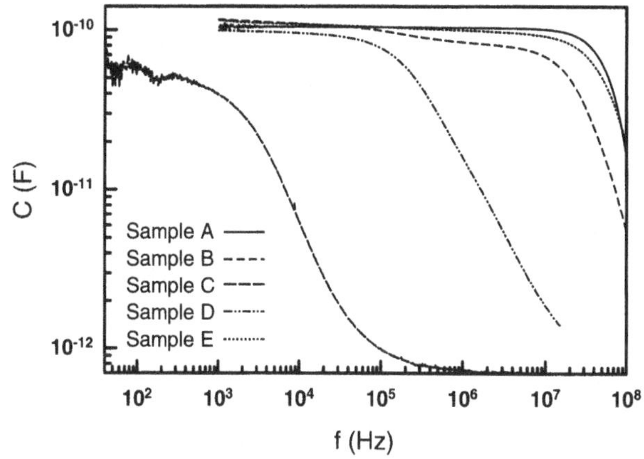

Fig. 8. Admittance of diodes of Type A, B, C, D, and E

4.2 Admittance Spectroscopy

The Schottky diodes were investigated by admittance spectroscopy at room temperature which is depicted in Fig. 8. The vanishing capacitance at higher frequencies is due to the series resistance of the diodes; the lower the series resistance, the higher the 3 dB cut-off frequency f_c. Capacitance voltage (C–V) measurements must always be conducted at frequencies lower than f_c. Since most of the capacitance bridges work with frequencies of 1 MHz it is desirable to decrease the series of the diode such that $f_c > 1$ MHz. Therefore we have developed a diode structure that allows a front-to-back contact geometry for ZnO thin films grown on insulating substrates. A degenerately doped ZnO:Al layer is grown firstly on the substrate acting as ohmic back contact. Secondly the nominally undoped ZnO layer in grown on top. The ZnO:Al layer is then contacted from the side with Au by sputtering. A diode realized in this manner is referred to as Type E. The behavior of C vs. f is shown in Fig. 8. The cut-off frequency is much higher than that of Type D diode and exceeds 1 MHz by more than one order of magnitude. Diodes of Type A, B and E having $f_c > 1$ MHz are used for the investigations by TAS and DLTS described in following section.

4.3 Deep Level Transient Spectroscopy and Thermal Admittance Spectroscopy

Deep level transient spectroscopy measurements and thermal admittance spectroscopy were carried out between 4 K and 300 K. For the DLTS we used rate windows between 64 and 4632 Hz. The oscillator frequencies used for the TAS measurements are in the range 10^3 - 10^7 Hz. For the diode of Type A TAS and DLTS measurements revealed the existence of a level labelled E1 lying about 110 meV below the conduction band minimum. Further

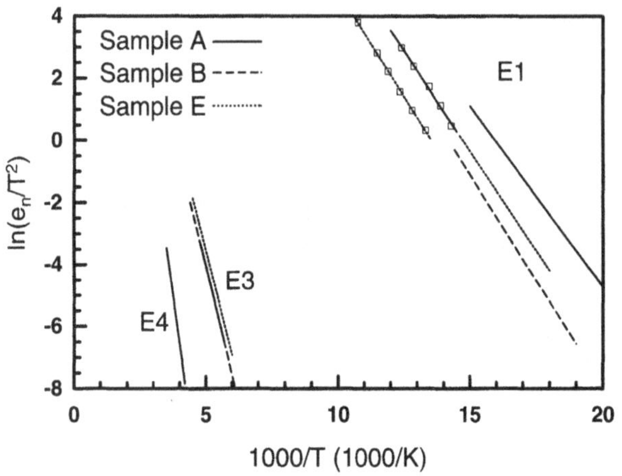

Fig. 9. Arrhenius plot of defects determined with TAS (squares) or DLTS (all others) of diodes of Type A, B and E. Data points are only shown for two measurements for clarity

the existence of the levels labelled E3 and E4 having activation energies of 300 and 540 meV [9] was confirmed by DLTS measurements. The diode of Type B exhibits the defects E1 and E3 as confirmed by DLTS. The levels E1 and E3 are present in the diode of Type E (PLD film). Again the level E1 could be detected by DLTS and TAS. The emission rates determined from DLTS and TAS are shown in an Arrhenius plot in Fig. 9 for all investigated diodes. Please note that the activation plots of the level E3 are almost similar for diode A and B. The parameters determined from the evaluation of the DLTS and TAS measurements are summarized in Table 1. The origin and microscopic nature of these defects is so far not known. The existence of the level labelled E3 in all types of samples (A, B, C, E) grown with four different methods feeds the speculation that it is an intrinsic defect possibly the oxygen vacancy.

5 Summary

We have demonstrated three possible ways of producing semi-insulating ZnO thin films ($n < 10^{14}\,\mathrm{cm}^{-3}$, $\rho > 10^{4}\,\Omega\mathrm{cm}$) with pulsed laser deposition on a-plane sapphire. The first uses a MgO buffer layer to reduce the Al diffusion from the substrate to the thin film during growth. The second uses doping with compensating acceptors and the third relies on the co-doping principle. Thin films doped with Ga and N were grown at about 370°C. We have determined the dominant donor levels in ZnO grown by four different methods. We suggest hydrogen, zinc on interstitial site and effective mass donors (Al) as possible origins of the shallow levels. High quality ZnO Schottky diodes

Table 1. Parameters of donor levels for Samples A – E. *this number corresponds to $N_{\rm d}$ + $N_{\rm ds}$ (see text)

Sample Type	Exp.	Defect	$N_{\rm d}$ (10^{16} cm^{-3})	$E_{\rm c} - E_{\rm d}$ (meV)	$N_{\rm a}$ (10^{15} cm^{-3})	σ (10^{-14} cm^2)
	TDH	H_i	13 ± 2	37 ± 2	3 ± 0.7	-
	TAS	E1	0.15 ± 0.02	110 ± 20	-	21 ± 5
A	DLTS	E1	0.14 ± 0.02	100 ± 20	-	12 ± 5
	DLTS	E3	0.022 ± 0.004	300 ± 30	-	0.062 ± 0.007
	DLTS	E4	0.018 ± 0.004	-	540 ± 40	10 ± 7
	TDH	Zn_i	4.4 ± 0.2	46 ± 3	1.8 ± 0.3	-
B	DLTS	E1	0.14 ± 0.02	120 ± 10	-	13.3 ± 3
	DLTS	E3	0.9 ± 0.04	300 ± 12	-	0.02 ± 0.003
C	TDH	V_O?	19 ± 0.6 *	299 ± 5	180 ± 4	-
D	TDH	Al_{Zn}	6 ± 1	65 ± 2	1.5 ± 0.6	-
E	TAS	E1	0.13 ± 0.05	120 ± 20	-	15 ± 5
E	DLTS	E1	0.14 ± 0.02	110 ± 20	-	9.3 ± 2
E	DLTS	E3	0.62 ± 0.07	290 ± 30	-	0.062 ± 0.007

were realized on samples grown by three different methods. A new diode high-frequency structure for ZnO thin films grown on insulating substrates was demonstrated. A degenerately doped ZnO:Al layer was used as ohmic back contact allowing the determination of the depletion width capacitance at oscillator frequencies larger than 10^7 Hz. The consideration of lateral fluctuations of the barrier height of ZnO Schottky diodes allows to explain the results of temperature dependent current measurements. The mean barrier height of a Pd Schottky contact on a ZnO thin film is 1.1 eV. The existence of the donor-like level E3 lying about 300 meV below the conduction band minimum was confirmed for all investigated samples by TDH or DLTS, respectively.

References

1. M. Lorenz, E. M. Kaidashev, H. v. Wenckstern, V. Riede, C. Bundesmann, D. Spemann, G. Benndorf, H. Hochmuth, A. Rahm, H.-C. Semmelhack und M. Grundmann: Solid-State Electronics **47**, 2205 (2003).
2. E. M. Kaidashev, M. Lorenz, H. v. Wenckstern, A. Rahm, H.-C. Semmelhack, K.-H. Han,G. Benndorf, C. Bundesmann, H. Hochmuth und M. Grundmann: Appl. Phys. Lett. **82**, 3901 (2003).
3. B. K. Meyer, H. Alves, D. M. Hofmann, W. Kriegseis, D. Forster, F. Bertram, J. Christen, A. Hoffmann, M. Straßburg, M. Dworzak, U. Haboeck, and A. V. Rodina: phys. Stat. solidi **B 241**, 231 (2004).
4. T. Yamamoto and H. Katayama-Yoshida: Jpn. J. Appl. Phys. **38**, L166 (1999).
5. L. G. Wang and A. Zunger: Phys. Rev. Lett. **90**, 256401 (2003).
6. H. v. Wenckstern, M. Grundmann *et al.*, unpublished.
7. A. R. Hutson: Phys. Rev. **108**, 222 (1957).
8. J. H. Werner and H. H. Güttler: J. Appl. Phys. **69**, 1522 (1990).
9. F. D. Auret, S. A. Goodman, M. J. Legodi, W. E. Meyer, and D. C. Look: Appl. Phys. Lett. **80**, 1340 (2002).

Excitonic Properties of ZnO

C. Klingshirn[1], H. Priller[1], M. Decker[1], J. Brückner[1], H. Kalt[1], R. Hauschild[1], J. Zeller[1], A. Waag[2], A. Bakin[2], H. Wehmann[2] K. Thonke[3], R. Sauer[3], R. Kling[3], F. Reuss[3], and Ch. Kirchner[3]

[1] Institut für Angewandte Physik and Center for Functional Nanostructures der Universität Karlsruhe, Karlsruhe, Germany
[2] Institut für Halbleitertechnik der Universität Braunschweig, Braunschweig, Germany
[3] Abteilung Halbleiterphysik der Universität Ulm, Ulm, Germany

Summary. In the first chapters we give some short comments on the history of ZnO research and on growth, doping, transport, and deep centers, topics which will be covered in more detail by other contributions to this symposium "ZnO-rediscovered". Then we concentrate in a first main chapter on the band-structure, excitons, and polaritons in the regime of linear optics. The second main chapter deals with many particle effects and lasing. It is followed by a short conclusion and outlook.

1 Introduction and Historic Remarks

The research on ZnO sees presently a vital renaissance. The data collection INSPEC gives about 13500 ZnO related entries from 1969 to 2004, among which 375 fall into the three year interval from 1969 to 1971 and 3300 from 2002 to 2004.

The wide gap II-VI semiconductor ZnO has gone through various periods of active research, which are reviewed e.g. in [1, 2, 3, 4]. The emphasis of research during the last active period from roughly the seventies to the mid eighties was on bulk materials, including growth, transport, doping, band structure, and excitons, bulk and surface polaritons, high excitation or many particle effects and lasing. For data collections of bulk material and quantum structures see [5]. Examples of the optical properties entered also in [6, 7, 8].

ZnO is used presently by many hundreds of tons per year as an additive to concrete or to rubber, in smaller quantities as an additive to human or pig food, as a UV blocker in sun creams or as one component of mixed oxide varistors for voltage limiting. It was used in the electrofax copy process.

The present renaissance is based on the possibility to grow epitaxial layers, quantum wells, nano rods and related objects or quantum dots and on the hope to obtain

- a material for blue/UV lasers and optoelectronics in addition to (or instead of) the GaN based structures
- a radiation hard material for electronic devices in a corresponding environment

B. Kramer (Ed.): Adv. in Solid State Phys. **45**, 275–287 (2005)

- a diluted or ferromagnetic material, when doped with Co, Mn, Fe, V etc. for spintronics
- a transparent, highly conducting oxide (TCO), when doped with Ga, Al, In etc. as a cheaper alternative to ITO.

For several of the above mentioned applications a stable, high, and reproducible p-doping is obligatory, which is however still a major problem.

The emphasis of the present active period of ZnO research is essentially on the same topics as before, but including nanostructures. For first reviews of this new ZnO research period, see e.g. [9]. We shall present or cite in the following deliberately both old and new results.

2 Growth, Doping, Transport, and Deep Centers

Bulk samples in the form of hexagonal needles or platelets can be grown by gas transport and/or oxidation on zinc. Examples are found e.g. in [10]. Hydrothermal growth under high pressure in LiOH + KOH solutions [11] may result in samples of several ten cm^3. For growth from melt or flux, see [12].

Textured or epitaxial layers and quantum wells (either ZnO or $Cd_{1-y}Zn_yO$ wells between $Zn_{1-y}Mg_yO$ or ZnO barriers, respectively) have been grown by a variety of techniques. For early examples of oxidation on Zn or for vapor phase transport, see [13, 14]. For modern techniques we mention only a few examples out of the tremendous number of publications and refer the reader otherwise to [9] and the other contributions to this symposium. For pulsed laser deposition (PLD) see [15], for molecular beam epitaxy (MBE) [16] or for metal organic chemical vapor deposition (MOCVD, MOVPE) [17].

A hot topic is presently the growth of nano rods, -combs, -brushes, -nails, -tubes, -rings, -belts, -wool, -walls, -tetrapods or -flowers by various techniques, see [9, 18] and other contributions to this symposium. It became already possible to produce $Zn_{1-y}Mg_yO/ZnO$ superlattices in the nanorods [19]. Quantum dots have been prepared e.g. by sol-gel techniques or by spray combustion [20].

While p-doping is still a major problem, as mentioned above and detailed in the following contributions, n-type doping with Ga, In or Al is possible beyond $n = 10^{20}\,cm^{-3}$. Such highly doped samples look bluish, because the tail of the free carrier absorption extends into the red and the plasmon frequency approaches values close to 0.5 eV [8, 21]. The diffusion constants of various dopants, including Mn and Co, have been summarized in [22]. The value of the Hall mobility of electrons at room temperature is long known and amounts about $200\,cm^2/Vs$ [23]. It is limited by the intrinsic process of LO-phonon scattering, resulting in a relaxation time τ of 2×10^{-14}s. See also the contribution by M. Grundmann to this symposium.

As most wide gap semiconductors and insulators, ZnO shows a large variety of absorption and emission bands, situated energetically deep in the gap. Two different luminescence bands in the green are known, one is related to Cu, the other to oxygen vacancies [8, 24]. Emission bands in the yellow range are due to deep Li or Na acceptors [25]. Recently a band in the orange spectral range has been observed, which expands after the steplike onset of band-to-band excitation gradually over distances of several mm in times of several hundreds of ms [26]. More details on deep centers including the para- or ferromagnetic properties induced by dopants like Mn, Co, V, Fe are reviewed in the contribution by A. Hoffmann to this symposium or in [9].

3 Band Structure, Excitons, and Polaritons

ZnO has rather strong ionic binding. Consequently, the conduction band (CB) arises essentially from the Zn^{++} 4s orbitals (symmetry Γ_7) and the upper valence bands (VB) from the O^{--} 2p states with an admixture of Zn^{++} 3d levels. The VB splits due to the hexagonal crystal field Δ_{cr} and the spin orbit coupling Δ_{so} into three subbands always labeled A, B, C from higher to lower energies. The gaps are at low temperature $E_g^A = 3.437\,\mathrm{eV}$, $E_g^B = 3.442\,\mathrm{eV}$, $E_g^C = 3.481\,\mathrm{eV}$ [4, 5, 6, 7]. In contrast to other wurtzite II-VI compounds $\Delta_{cr} \gg \Delta_{so}$ due to the small nuclear charge of oxygen. This fact in turn results in selection rules such that the transitions from the A and B VB to the CB have considerable oscillator strength only for $\boldsymbol{E} \perp \boldsymbol{c}$ (Γ_5) and from the C VB for $\boldsymbol{E} \parallel \boldsymbol{c}$ (Γ_1). The other, in principle also dipole allowed transitions involve a spin flip with a drastic reduction of their oscillator strength. Group theory allows a k-linear term for Γ_7 bands for $\boldsymbol{k} \perp \boldsymbol{c}$, but not for Γ_9 ones.

Exciton series exist for all three combinations of VB and CB. The exciton binding energies E_x^b are very similar with $E_x^b = (60 \pm 1)\,\mathrm{meV}$ [5, 6, 7, 8]. The lower values in [27] for B and C excitons arise from the rather unusual procedure to count from the reflection minima, i.e. from the longitudinal eigenenergies.

The value of Δ_{SO} in ZnO is not only small, but the interaction with the close lying Zn^{++} 3d states shifts one Γ_7 VB above the Γ_9 resulting in a VB-ordering A Γ_7, B Γ_9, C Γ_7 in contrast to the usual ordering A Γ_9, B Γ_7, C Γ_7 . This concept has been introduced for ZnO in [28] and later also for the cubic CuCl [29]. Theoretical justifications have been given in [30]. The unusual ordering has been questioned in [31]. Since then many experiments have confirmed the inverted VB structure e.g. in absorption [32], luminescence, $\boldsymbol{k}$-space spectroscopy, under $\boldsymbol{B}$ or strain fields [33, 34, 35]. Recently the inverted assignment has been questioned again e.g. by [27, 36, 37]. Some arguments against this approach have been given already in [38].

In Fig. 1 we show reflection spectra for the $n_B = 1$ A and B Γ_5 and the C Γ_1 excitons in their respective polarizations together with fit curves. Such spectra, including also states with higher main quantum numbers $n_B > 1$ of the envelope function have been reported by many authors [4, 5, 6, 7, 27, 28, 34, 39]. The longitudinal transverse splittings are $\Delta_{LT}^A = 2\,\mathrm{meV}$,

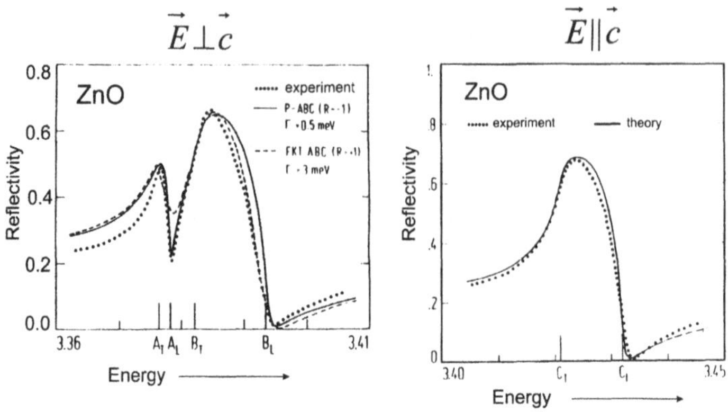

Fig. 1. Reflection spectra of the $n_B = 1$ A and B (Γ_5) and the C (Γ_1) exciton together with fit curves. From K. Hümmer [4]

$\Delta_{LT}^B = 10\,\text{meV}$, $\Delta_{LT}^C = 12\,\text{meV}$ fulfilling the claim [28] that the sum of the oscillator strength of A and B excitons should equal that of the C exciton. In [37], the different values of Δ_{LT}^A and Δ_{LT}^B have been used as an argument to doubt the band structure. However, in the case of close lying resonances the value of Δ_{LT} of the lower one is reduced, that of the upper one increased due to their interaction as has been detailed for exciton resonances already in [39] and in a more didactical manner in [8]. A small reflection feature seen for $\boldsymbol{E} \parallel \boldsymbol{c}$ in [27] at the position of B Γ_5^L is attributed to a B Γ_1 state. Actually this feature is a mixed mode polariton [8, 40] resulting from the slight mismatch to $\boldsymbol{k} \perp \boldsymbol{c}$ ($< 15°$). The $n_B = 1$ A or B $\Gamma_1 \oplus \Gamma_2$ or Γ_6 exciton states are spin triplets, independent of the band structure. The triplet states are always situated energetically slightly below the corresponding transverse eigenstates. For the investigation of the exciton resonances in reflection and absorption in a $\boldsymbol{B}$-field see e.g. [4, 5, 6, 7, 8, 33, 34, 35], and for the Landau levels and the resulting effective masses [41]. The free excitons in ZnO have been observed also in absorption. For the "forbidden" polarizations, see [32, 33]. A rather early example of the allowed transitions at RT is shown in [13], reproduced also in [2, 8]. The peak at 3.3 eV is probably the first observation of an exciton resonance in a semiconductor [8]. The Urbach-tail of the fundamental absorption edge has been investigated in [26]. The resulting variation of the gap with temperature is shown in Fig. 2.

The luminescence spectra of high quality ZnO samples are dominated at low temperature and excitation density by bound exciton complexes (BEC), especially neutral donor bound (D^0X). See e.g. [4, 8, 24, 38]. We give in Fig. 3 an example including the $\boldsymbol{B}$ dependence. More information can be found in the contributions to this symposium.

The zero-phonon luminescence of the free excitons is at low temperatures very weak but increases relative to the BEC when the latter become thermally ionized from their defect centers in the temperature range around (80 ± 20) K.

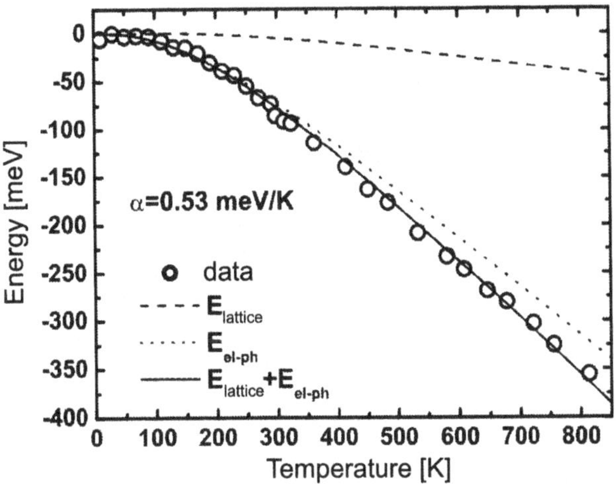

Fig. 2. The decrease of band gap with increasing temperature From [26].

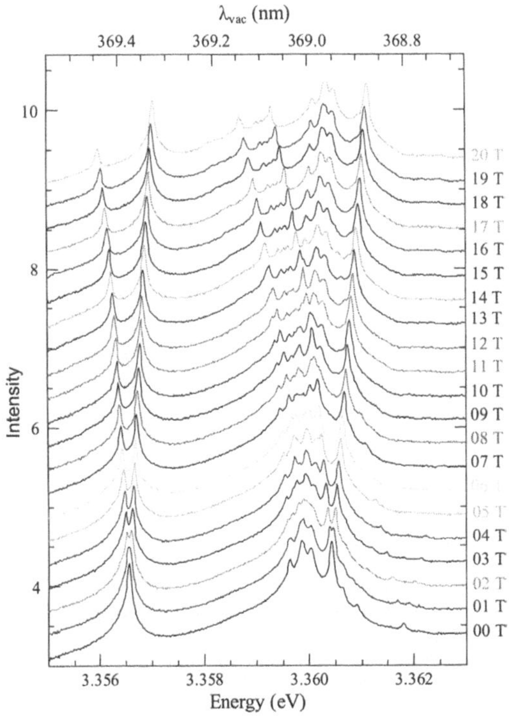

Fig. 3. Bound exciton luminescence for magnetic fields up to 20T. From G. Prinz [18] and K. Thonke [38]

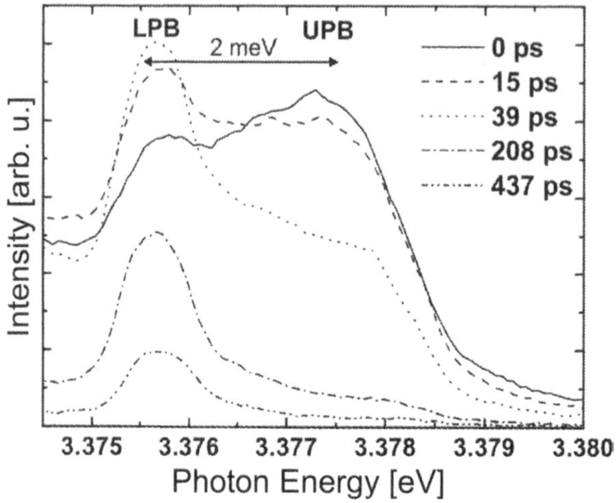

Fig. 4. Fine structure and luminescence dynamics in the exciton polariton. From [26]

The zero phonon line of the free A exciton luminescence has been investigated thoroughly in [34] and later e.g. in [26, 27, 36]. See [42]. The influence of the $\boldsymbol{k}$-linear term is nicely observed for the A Γ_5 excitons in $\boldsymbol{E} \perp \boldsymbol{c}$, for $\boldsymbol{k} \perp \boldsymbol{c}$ but not for $\boldsymbol{k} \parallel \boldsymbol{c}$ as expected and the luminescence of the A Γ_1 is seen in $\boldsymbol{E} \parallel \boldsymbol{c}$ [34]. This emission is also reported in [27], but is used only to claim "...that the crystal is of good quality.", but ignoring that this observation involves the presence of Γ_1 symmetry.

In Fig. 4 we show time resolved unpolarized emission spectra of the $n_B = 1$ A exciton polaritons. The features due to emission from the lower A Γ_5 polariton branch (LPB) around 3.3755 eV superimposed by the A Γ_1 exciton, from the intermediate A Γ_5 polariton branch caused by the $\boldsymbol{k}$-linear term around 3.377 eV and the upper A Γ_5 polariton branch (UPB) around 3.378 eV are clearly visible. The luminescence dynamics shows the relaxation from the UPB and the feeding into the LPB and the A Γ_1 state. Further examples for the luminescence dynamics of excitons in ZnO can be found in [26] and references given therein.

The temperature dependence of the near edge luminescence spectra have been observed by many groups e.g. [24, 26, 38] and a consistent picture arose. At low temperature one observes depending on the sample quality more or less resolved BEC luminescence and its LO-phonon replica. Around (80 ± 20) K BEC are thermally ionized from their centers and the free exciton luminescence, including the LO-phonon replica takes over. At RT these bands merge to an unstructured band of 100 meV FWHM, a value exceeding considerably that of many other semiconductors including GaAs and quantum structures based on this material. Furthermore it involves a homogeneous broadening γ_h around 20 meV at RT which makes room temperature lasing on Bose-condensed exciton polaritons in a μ-cavity predicted in [43] highly improbable, since the damping exceeds the values of Δ_{LT} and the resulting T_2 time falls below the round trip time in a μ-cavity in contrast to low temperature measurements. The luminescence of a ZnO/$Zn_{1-y}Mg_yO$ quantum well shows for low excitation emission from the ZnO buffer layer at 3.36 eV from the $Zn_{1-y}Mg_yO$ barrier from 3.44 to 3.48 eV and two peaks from the QW at 3.40 and 3.415 eV. The higher of these two peaks is attributed to (localized) excitons the lower one to a defect state. See [26, 44]. Surface exciton polaritons have been detected by attenuated total reflection (ATR) and by nonlinear generation [45]

4 High Excitation Phenomena

In this section we treat first so-called many particle- or high excitation phenomena in the intermediate density regime where excitons are still good quasi particles, then the transition to an electron-hole plasma at the highest density. The transition between the two regimes is not sharp but for didactic reasons this separation is useful. Finally we consider some aspects of lasing in ZnO. For some reviews of these topics in general and more specifically for ZnO, see [6, 7, 8, 46].

Typical processes in the first regime are inelastic scattering processes and biexciton formation and decay. While elastic scattering between excitons results essentially in an excitation induced increase of homogeneous broadening, the inelastic processes give frequently rise to new emission bands. In the inelastic exciton-exciton (X-X) or more precisely polariton-polariton scattering, two exciton like polaritons interact via their dipoles. One of them is scattered under energy and momentum conservation onto the photon-like part of the dispersion curve while the other reaches a state with higher quantum number $n_B = 2, 3, \ldots \infty$, resulting in new emission bands labeled P_2, P_3, ...,P_∞ [6, 8, 46].

Other inelastic scattering processes are known between free excitons and free carriers (X-e), LO-phonons (X-nLO) or between bound exciton complexes and free carriers or acoustic phonons [6, 8, 46]. In Fig. 5 we show examples for the luminescence from nanorods and from an epilayer.

The emission is dominated at low excitation by BEC recombination. With increasing ns excitation intensity I_{exc}, this line broadens on its low energy side due to the appearance of the so-called M-band, which may be due to biexciton decay, but also due to the process involving BEC mentioned above [6]. With further increasing excitation the epilayer shows at 3.32 eV the appearance of the P_∞ band. Surprisingly this band is absent in thin ($\approx 50nm$) but not in thick ($\approx 200nm$) nanorods [18, 26]. These rods are bulk like for excitons because both their length ($> 1\mu m$) and their diameters are large

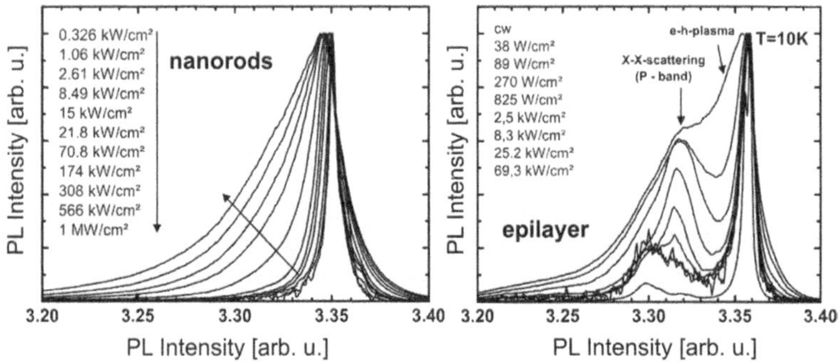

Fig. 5. The luminescence of ZnO nanorods (a) and of an epilayer (b) for increasing excitation. From [26]

compared to the excitonic Bohr radius $a_B \approx 1.8nm$. But for photon-like polaritons the thin rods act as one dimensional wave-guides. This reduction of the phase-space in one of the outgoing channels in the X-X process may quench this process. Further work is necessary to verify or falsify this hypothesis. With increasing temperature, the X-X process goes over to the X-e process, if a sufficient number of excitons are or remain thermally ionized to favour this dipole-monopole interaction [6, 8, 46]. For the biexciton, values

Table 1. Biexciton energies and binding energies with respect to two A excitons and with respect to two A/B excitons. From [6, 7, 48]. The last column gives corresponding data for A/B excitons from [50], see [51]

Biexciton energy and symmetry	Binding energy with respect to two A excit.	Binding energy with respect to A / B excit. [48]	[50]
$E_{AA\,\Gamma_1} = 6.7355\,\mathrm{eV}$	$14.7 \pm 1\,\mathrm{meV}$	$14.7 \pm 1\,\mathrm{meV}$	15.6 meV
$E_{AB\,\Gamma_{5,6}} = 6.7407\,\mathrm{eV}$	9.5 "	14.7 "	16.6 meV
$E_{BB\,\Gamma_1} = 6.7469\,\mathrm{eV}$	3.3 "	13.7 "	4.7 meV

of its binding energy relative to two free A Γ_1 or A Γ_5^T excitons between 10 and 20 meV have been deduced from luminescence or ns four wave mixing (FWM) experiments. The solution came, when two groups observed in samples from different sources independently three biexciton levels in a modified two photon absorption experiment namely luminescence assisted two photon spectroscopy (LATS)[47]. In a joint publication [48] the binding energy of the ground-state biexciton has been found to be 15 meV and the two higher states are due to biexcitons, in which one or two holes from the A VB are replaced by B-holes. See also [49].

In table 1 we give a listing of the respective biexciton energies and of the binding relative to two A excitons or to the involved A and/or B excitons from [48, 50]. The binding energies of the AA Γ_1 and AB $\Gamma_{5,6}$ biexcitons agree within experimental error. Since the data of both excitons are very similar, the deviation for the BB Γ_1 exciton claimed in [50] is difficult to understand. As mentioned in [48], FWM data may result in erroneous results. Similar arguments may hold for the claimed biexciton binding energy of a CC Γ_1 biexciton of 1.4 meV [50].

In ZnO/$Zn_{1-y}Mg_yO$ quantum wells a new emission band appears with increasing excitation at 3.385 eV [26]. If it is due to a biexciton decay, a binding energy around 20 meV results, which would make some sense bearing in mind table 1 and the enhancement of exciton and biexciton binding energies in quantized systems [5, 8]. Further examples of ZnO QW luminescence under high excitation are found e.g. in [52].

Some selected results on two photon absorption (TPA), TPA- spectroscopy or second and third harmonic generation are found in [35, 53], the dynamics of the luminescence under high excitation has been investigated e.g. in [26, 54]. For the investigation of effects in the regime of extreme nonlinear optics, see [55]. The dephasing time T_2 and/or the homogeneous broadening have been investigated at low temperatures by an analysis of the reflection spectra [4, 33, 34, 39, 45] yielding values of the order of or slightly below 1 meV and also in beautiful time resolved FWM experiments [50] giving similar values.

If the generation rate is increased to sufficiently high values, excitons cease to exist as individual quasi-particles and a new collective phase is formed, the electron-hole plasma (EHP). For details of this transition, see e.g. [6, 8]. In ZnO the EHP has been observed in gain spectroscopy, e.g. in [26, 56]. Simultaneously with the appearance of EHP gain, the excitonic reflection features started to disappear. Densities exceeding $10^{18}cm^{-3}$ or $10^{19}cm^{-3}$ have been deduced in [26, 56].

The investigation of stimulated emission and lasing of ZnO started in the mid sixties. For a few early examples, see e.g. [57] and references therein and for tunable laser emission [58]. The understanding of the processes developed in parallel (see [6, 46]). It turned out, that all of the inelastic scattering and recombination processes mentioned above, including the two-electron transitions, may result in stimulated emission. Most of them can be mapped on a four level laser system, with correspondingly low thresholds for inversion. The X-LO emission process is inverted if one exciton is created and no LO-phonon present i.e. at $\hbar\omega_{LO} \gg k_B T$. Similarly the X-X processes, e.g. the P_2 band is inverted if two excitons are present and none of them in the state $n_B = 2$ [6, 46]. Nowadays such processes come also with names like thresholdless lasing or lasing without inversion.

The band - to - band recombination in an EHP requires $\mu_{eh}(n_p, T_p) > E'_g(n_p, T_p)$ for population inversion [6, 8, 56] and is therefore more like a three level system. However, inelastic scattering processes between carriers or the emission of phonons or plasmons may lead to thresholds below that given above. Generally it has to be stated that the transition from excitons to the EHP is continuous in direct gap semiconductors [59]. Recent theories including this continuous transition are found e.g. in [60] and references given therein.

In the present phase of ZnO activity, stimulated emission and gain are generally attributed to the processes known already in bulk but now for epitaxial layers [26], in quantum wells obtaining partly gain up to 550 K [52, 61] and most recently in nanorods and related structures [62].

A new topic is lasing in random media i.e. here in ZnO powders, relying on weak localization of light due to enhanced backscattering [63].

5 Conclusion and Outlook

Evidently ZnO sees a very vital renaissance of basic and applied research. As is usual in such situations, beautiful new effects are found, but also long known phenomena are being reinvented. We tried to give a good mixture of both aspects. The question how long this renaissance will last, depends to a large extend on the question if a stable, high, and reproducible p-doping can be realized [64]. Since prophecy is a notoriously difficult task, expecially if it concerns the future, the authors do not want to give an answer here.

Acknowledgements

The authors are grateful to the Deutsche Forschungsgemeinschaft and the Landeskompetenznetz Baden-Württemberg for financial support.

References

1. M. E. Brown (Ed.), *ZnO - Rediscovered* (The New Jersey Zinc Company, New York 1957)
2. G. Heiland et al., Solid State Physics **8**, 191 (1959).
3. W. Hirschwald et al., Current Topics in Materials Science **7**, 143 (1981).
4. R. Helbig, *Freie und gebundene Excitonen in ZnO*, Habilitation Thesis, (Erlangen 1976); K. Hümmer, *Excitonische Polaritonen in einachsigen Kristallen*, Habilitation Thesis (Erlangen 1978).
5. Landolt-Börnstein: *New Series Group III, Vols. 41 B* and *34 C2*.
6. C. Klingshirn, H. Haug, Physics Reports **70**, 315 (1981).
7. B. Hönerlage et al., Physics Reports **124**, 161 (1985).
8. C. Klingshirn, *Semiconductor Optics*, 2nd ed. (Springer, Heidelberg, Berlin, 2005).
9. Ü. Özgür et al., J. Appl. Phys. Rev. (2005) (in press); C. Klingshirn et al.: Physik Journal (2005) (in press).

10. E. Scharowski, Z. Physik **135**, 318 (1953); E. M. Dodson, J. A. Savage, J. Mat. Sci. **3**, 19 (1968); R. Helbig: J. Crystal Growth **15**, 25 (1972).
11. R. A. Laudise, A. A. Ballmann, J. Phys. Chem. **64**, 688 (1960); E. Ohshima et al., J. Crystal Growth **260**, 166 (2004).
12. J. W. Nielsen, E. F. Dearborn, J. Phys. Chem. **64**, 1762 (1960); D. C. Reynolds et al., J. Appl. Phys. **95**, 4802 (2004).
13. E. Mollwo, Reichsber. Physik **1**, 1 (1943),
14. H. Schneck, R. Helbig, Thin Solid Films **27**, 101 (1975).
15. M. Lorenz et al., Annalen der Physik **13**, 59 (2004).
16. T. Yao et al., J. Vac. Sci. Technol. B **18**, 2313 (2000); J. Crystal Growth **209**, 816 (2000).
17. W. I. Park et al., Appl. Phys. Lett. **79**, 2022 (2001); Th. Gruber et al., Phys. Stat. Sol. a **192**, 166 (2002) and Appl. Phys. Lett **84**, 5359 (2004); D. Gerthsen et al., Appl. Phys. Lett. **81**, 3972 (2002); T. Riemann et al., Phys. Stat. Sol. b (to be published).
18. Z. W. Pan et al., Science **291**, 1947 (2001); H. Yan et al., Adv. Mater. **15**, 402 (2003); D. Banerjee et al., Appl. Phys. Lett. **83**, 2061 (2003); X. Y. Kong et al., Science **303**, 1348 (2004); R. Kling et al., Nanotechno. **15**, 1043 (2004); G. Prinz et al., to be published, R. Hauschild et al., to be published.
19. W. I. Park et al., Adv. Mater. **15**, 526 (2003).
20. L. Spanhel et al., J. Am. Chem. Soc. **113**, 2826 (1991) and J. Sol-Gel Sci. and Technol. **26**, 499 (2003); L. Mädler et al., J. Appl. Phys. **92**, 6537 (2002).
21. M. Göppert et al., J. Luminesc. **72-74**, 430 (1997); C. Klingshirn and E. Mollwo, Z. Physik **254**, 437 (1972).
22. F. W. Kleinlein, R. Helbig, Z. Physik **266**, 201 (1974).
23. A. R. Hutson, Phys. Rev. **108**, 222 (1957); H. Rupprecht, J. Phys. Chem. Sol. **6**, 144 (1958); M. A. Seitz, D. H. Whitmore, J. Phys. Chem. Sol. **29**, 1033 (1968); J. A. Savage, E. M. Dodson, J. Mat. Sci. **4**, 809 (1968); P. Wagner, R. Helbig, J. Phys. Chem. Sol. **35**, 327 (1974)
24. Chr. Solbrig, Z. Physik **211**, 429 (1968); R. Dingle, Phys. Rev. Lett. **23**, 579 (1969); D. C. Reynolds et al., Phys. Rev. **140**, A1726 (1965) and Phys. Rev. **185**, 1099 (1969); E. Tomzig, R. Helbig, J. Luminesc. **14**, 403 (1976); R. Kuhnert, R. Helbig, J. Luminesc. **26**, 203 (1981); K. A. Vanheusden et al., Appl. Phys. Lett. **68**, 403 (1996).
25. D. Zwingel, F. Gärtner, Sol. State Commun. **14**, 45 (1974).
26. H. Priller, Ph.D Thesis, Karlsruhe (2005), partly published in H. Priller et al., Phys. Stat. Sol. b **241**, 587 (2004) and Appl. Phys. Lett. **86**, 111909 (2005); Proc. ICPS, Flagstaff 2004 (in press); Proc. EXCON, Krakau 2004 J. Luminesc. **112**, 173 (2005).
27. S. F. Chichibu et al., J. Appl. Phys. **93**, 756 (2003).
28. D. G. Thomas, J. Phys. Chem. Sol. **15**, 86 (1960); J. J. Hopfield, J. Phys. Chem. Sol. **15**, 97 (1960),
29. M. Cardona, Phys. Rev. **129**, 69 (1963) and J. Phys. Chem. Sol. **24**, 1543 (1963).
30. K. Shindo et al., J. Phys. Soc. Japan **20**, 2054 (1965); B. Segall, Phys. Rev. **163**, 769 (1967); U. Rössler, Phys. Rev. **184**, 733 (1969).
31. D. C. Reynolds, C. W. Litton, T. C. Collins, Phys. Rev. **140**, A1726 (1965).
32. J. J. Hopfield, D. G. Thomas, Phys. Rev. Lett. **15**, 22 (1965).
33. G. Blattner et al., Phys. Rev. B **25**, 7413 (1982).

34. K. Hümmer, R. Helbig, M. Baumgärtner, Phys. Stat. Sol. b **86**, 527 (1978); R. Kuhnert, R. Helbig, K. Hümmer, Phys. Stat. Sol. b **107**, 83 (1981).
35. M. Fiebig et al., Phys. Stat. Sol. b **177**, 187 (1993); J. Wrzesinski, D. Fröhlich, Phys. Rev. B **56**, 13087 (1997); Sol. State Commun. **105**, 301 (1998).
36. D. C. Reynolds et al., Phys. Rev. B **60**, 2340 (1999).
37. B. Gil, Phys. Rev. B **64**, 201310 R (2001).
38. W. R. L. Lambrecht et al., Phys. Rev. B **65**, 075207 (2002); B. K. Meyer et al., Phys. Stat. Sol. b **241**, 231 (2004); A. V. Rodina et al., Phys. Rev. B **69**, 125206 (2004); R. Sauer, K. Thonke in *Optics of Semiconductors and Their Nanostructures*, H. Kalt and M. Hetterich (Eds.), Springer Series in Solid State Sciences **146**, 73 (2004),
39. J. Lagois, K. Hümmer, Phys. Stat. Sol. b **72**, 393 (1975); J. Lagois: Phys. Rev. B **16**, 1699 (1977).
40. R. I. Weiher, W. C. Tait, Phys. Rev. **185**, 1114 (1969) and Phys. Rev. **B5**, 623 (1972); K. Hümmer, P. Gebhardt, Phys. Stat. Sol. b **85**, 271 (1978).
41. K. Hümmer, Phys. Stat. Sol. b **56**, 249 (1973).
42. The references [34] show , that the claim in [36] to observe "...free exciton emission... for the first time" is clearly unjustified.
43. A. Kakovin, M. Zamfirescu et al., Phys. Stat. Sol. a **192**, 212 (2002) and Phys. Rev. B **65**, 161205 R (2002).
44. T. Makino et al., Appl. Phys. Lett. **77**, 975 (2000); H. D. Sun et al., Appl. Phys. Lett.**78**, 2464 (2001); S. Giemsch et al., to be published.
45. J. Lagois, B. Fischer, Phys. Rev. Lett. **36**, 680 (1976); J. Lagois, Phys. Rev. B **23**, 5511 (1981); I. Hirabayashi et al., J. Phys. Soc. Japan **51**, 2934 (1982); M. Fukui et al., J. Phys. Soc. Japan **53**, 1185 (1984); F. DeMartini et al., Phys. Rev. Lett. **38**, 1223 (1977); M. Fukui et al., Phys. Rev. B **22**, 1010 (1980),
46. C. Klingshirn, Phys. Stat. Sol. b **71**, 547 (1975); J. M. Hvam, Sol. State Commun. **12**, 95 (1973); Phys. Stat. Sol. b **63**, 511 (1974); S. W. Koch et al., Phys. Stat. Sol. b **89**, 431 (1978).
47. H. Schrey, V. Lyssenko, C. Klingshirn, Sol. State. Commun. **32**, 897 (1979).
48. J. M. Hvam et al., Phys. Stat. Sol. b **118**, 179 (1983).
49. H. J. Ko, Y. F. Chen, T. Yao, Appl. Phys. Lett. **77**, 537 (2000).
50. K. Hazu et al., Phys. Rev. B **68**, 332051 (2003); J. Appl. Phys. **95**, 5498 (2004); J. Appl. Phys. **96**, 1270 (2004).
51. The absolute values of the exciton and biexciton energies differ e.g. in [4-7,32-34,48,50] and other papers by one or two meV. This is no serious discrepancy. If spectrometers are not frequently calibrated with low pressure spectral lamps, they may easily show deviations of the order of 1 meV.
52. H. D. Sun et al., Appl. Phys. Lett. **77**, 4250 (2000).
53. E. Mollwo, G. Pensl, Z. Physik **228**, 193 (1969); R. Dinges et al., Phys. Rev. Lett. **25**, 922 (1970); C. Klingshirn, Z. Physik **248**, 433 (1971); W. Kaule, Sol. State Commun. **9**, 17 (1971); G. Pensl, Sol. State Commun. **11**, 1277 (1972); G. Koren, Phys. Rev. B **11**, 802 (1975); G. Wang et al., Appl. Optics **40**, 5436 (2001); C. Y. Liu et al., Optics Com. **237**, 65 (2004); Appl. Phys. B **79**, 83 (2004); U. Neumann et al., Appl. Phys. Lett. **84**, 170 (2004).
54. J. Collet, T. Amand, Phys. Rev. B **33**, 4129 (1986); J. Gutowski, A. Hoffmann, Adv. Mat. Opt. and Electr. **3**, 15 (1994).
55. O. D. Mücke, T. Tritschler, M. Wegener, Opt. Lett. **27**, 2127 (2002).
56. K. Bohnert, G. Schmieder, C. Klingshirn, Phys. Stat. Sol. b **98**, 175 (1980).

57. F. H. Nicoll, Appl. Phys. Lett. **9**, 13 (1966); J. R. Packard et al., J. Appl. Phys. **38**, 5255 (1967); S. Irwai, S. Namba, Appl. Phys. Lett. **16**, 354 (1970); J. M. Hvam: Phys. Rev. B **4**, 4459 (1971); W. D. Johnston jr., J. Appl. Phys. **42**, 2731 (1971); C. Klingshirn, Sol. State Commun. **13**, 297 (1973),
58. W. Wünstel, C. Klingshirn, Optics Commun. **32**, 269 (1980).
59. K. Bohnert et al., Z. Physik B **42**, 1 (1981).
60. M. F. Pereira, K Henneberger, Phys. Stat. Sol. b **202**, 751 (1997); Phys. Stat. Sol. b **206**, 477 (1998); T. J. Inagaki, M. Aihara, Phys. Rev. B **65**, 205204 (2002) and references therein.
61. G. Tobin et al., Physica B **340-342**, 245 (2003); D. M. Bagnall et al., Appl. Phys. Lett. **73**, 1038 (1998); P. Yu, J. Crystal Growth **184/185**, 601 (1998) and references therein.
62. M. H. Huang et al., Science **292**, 1897 (2001); J. C. Johnson et al., J. Phys. Chem. B **105**, 11387 (2001) and Nano Letters **4**, 197 (2004); Y. G. Wang et al., Chem. Phys. Lett. **377**, 329 (2003); Th. Nobis et al., Phys. Rev. Lett. **93**, 103903 (2004).
63. H. Cao, Waves Random Media **13**, R 1 (2003); G. Hackenbroich, Physik Journal **3** (7), 25 (2004).
64. D. C. Look et al., Phys. Stat. Sol. a **201**, 2203 (2004) and Semicond. Science and Technol. **20**, S55 (2005); A. Dadgar et al., E-MRS Meeting, Strasbourg (2005), Superl. and Microstr., in press.

Incorporation of Acceptors in ZnO

B. K. Meyer, J. Sann, and A. Zeuner

Physics Institute, Justus Liebig University Giessen, Heinrich-Buff-Ring 16, 35392 Giessen, Germany

Abstract. Diffusion experiments of N, Li, and Na into single-crystal substrates of ZnO were performed. The diffused layers were studied using mass spectroscopy and low temperature photoluminescence. Based on earlier investigations Li and Na produce deep acceptor states which participate in shallow donor to deep acceptor recombinations in the visible spectral region. We will demonstrate that in the case of group-I elements also shallow acceptors are introduced having binding energies around 300 meV. The corresponding donor–acceptor pair recombinations have weak electron–phonon coupling analogous to the "edge" emissions in other II-VI compounds. With 160 meV nitrogen so far has the smallest ionisation energy. We compare our results with ZnO epitaxial films where the respective elements are incorporated during growth.

1 Introduction

Many attempts have been undertaken to dope ZnO p-type, and still now the subject is very controversial and contradictory. *Lander* [1] reported in 1960 on the donor and acceptor action of Li, and already gave hints as to the possible defect structures. For the donor centre he proposed a molecular ion like defect LiO^-, and for the acceptor Li substitutional on Zn site. In 1963 *Schneider* et al. [2] reported on the electron paramagnetic resonance (EPR) of ZnO:Li which was extended in 1968 by a very detailed work of *Schirmer* [3]. The acceptor seen in EPR was indeed Li_{Zn}, but the analysis of the hyperfine structure showed a substantial relaxation of the Li-O bonds resulting in a deep centre with a binding energy of 800 meV. Na_{Zn} behaved in a very similar way, its binding energy is around 600 meV [4]. Both defects produce broad luminescence bands in the visible spectral region and participate in shallow donor to deep acceptor recombinations. Up to now there are no reports as to shallow acceptor levels induced by the group-I elements. In this report we will shortly summarize the status of the bound exciton recombinations in ZnO, we will in detail discuss the properties of Li and Na which give rise to deep and shallow acceptor levels in ZnO and address the incorporation of nitrogen by diffusion and in-situ doping.

B. Kramer (Ed.): Adv. in Solid State Phys. **45**, 289–299 (2005)

2 Experimental Details

We used state of the art ZnO bulk crystals from CrysTec. Salts containig group-I elements (mainly nitrates and carbonates) were brought into aequeous solutions, the surface of the crystals was covered with the solutions and carefully dried with a fan before being put into the diffusion oven. Diffusions were carried out in nitrogen atmosphere in a temperature range between 300 °C and 800 °C for 10 min. The samples were removed from the oven at the diffusion temperatures and quenched to room temperature. Li_3N was evaporated under vacuum conditions and deposited as a thin film (3 nm to 30 nm) on the ZnO crystals which were subsequently diffused without removing them from the vacuum chamber. For the nitrogen diffusion we worked with ammonium nitrate and lithium nitrate (diffusion conditions as above). The low temperature and temperature dependent luminescence studies were performed in an Oxford cryostat, the 325 nm line of a HeCd laser was used as excitation source. Secondary ion mass spectroscopy (Cameca/Riber) was employed to check the diffusion and incorporation of the group-I elements. For the in-situ doping the group-I elements were used in their metallic state, for nitrogen doping ammonia was used. Growth temperatures were between 550 °C and 650 °C and GaN/Sapphire templates were employed. Metallic zinc and NO_2 were the respective precursors, the depositions were performed at pressures between 4 mbar and 20 mbar. The typical sample thickness is around 2 μm.

3 Experimental Results

3.1 As-grown ZnO Bulk Crystals

The photoluminescence spectra of two bulk ZnO crystals before diffusion are shown in Fig. 1.

If the bulk crystals contain hydrogen i.e. grown by the hydrothermal method, then I_4 is the prominent bound exciton line, followed by $I_{6/6a}$ and I_8. Al causing the $I_{6/6a}$ recombination could be identified by implantation experiments, I_8 by in-situ doping with Ga, and I_9 by In diffusion studies (see [5]). The determination of the donor binding energies is based on the observation of two-electron-satellite (TES) lines which showed additional fine structure. During the recombination of an exciton bound to a neutral donor the donor final state can be the 1 s state (normal $D^\circ X$ line) or the 2 s, 2 p state (TES-line). In a simple hydrogen like effective mass approach (EMA) the energy separation between $n = 1$ and $n = 2$ states would be equal to 3/4 of the donor binding energy R^*. It was thus possible to correlate the energetic distance of the TES transitions from their respective bound exciton lines with the donor binding energies.

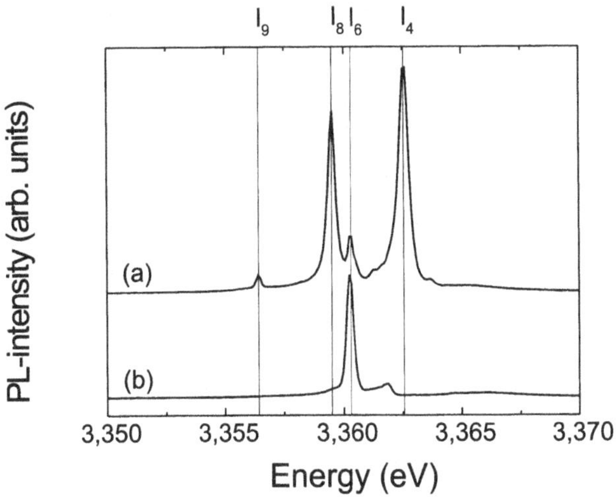

Fig. 1. Photoluminescence spectra of a bulk ZnO sample from different vendors in the as-grown state (T = 4.2 K, HeCd laser excitation)

The donor binding energies E_D show a linear relation to the bound exciton localization energies $E_{loc} = (E_{FX} - E^0_{DX})$ known as Haynes rule $E_{loc} = \alpha E_D$ [6]. More general E_{loc} is given by

$$E_{loc} = A + BE_D \tag{1}$$

and A and B are to be determined from the experiment. The linear dependence of the exciton localization energies on the donor binding energies can be fitted with $A = -3.8\,\text{meV}$ and $B = 0.365\,\text{meV}$.

3.2 ZnO Doped with Group-I Elements

Figure 2 shows a low-temperature photoluminescence measurement of a sample where Li was introduced from the gas phase from the decomposition of Li_3N at 450 °C. Here we focus on the energy range below 3.2 eV (the excitonic features are described in detail in [5]). Upon diffusion no new excitonic lines are induced by the group-I elements. The luminescence starts at around 3.1 eV and shows pronounced coupling to longitudinal optical (LO) phonons separated by 72 meV. The shape is typical for a donor–acceptor pair (DAP) recombination. The coupling strength given by the Huang–Rhys factor S is estimated from the Poisson distribution $I = I_0 S^n/n!$ where n runs from $0, 1, 2 \ldots$ and I_0 is the intensity of the first, the zero-phonon line (ZPL). For S one obtains a value of 0.35 ± 0.05 which can be compared with the value of 0.55 ± 0.05 obtained for the nitrogen acceptor with a binding energy of 165 meV (since the DAP pair band overlaps with the green emission band, we used for analysis the PL spectrum obtained from the CVD grown ZnO:Li film, see below).

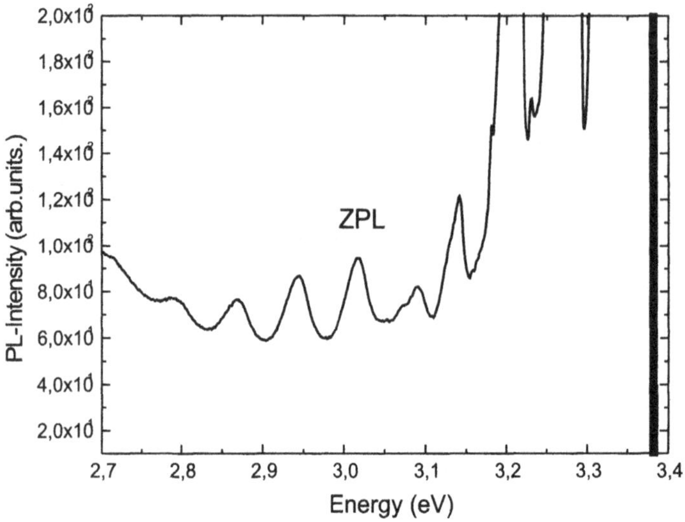

Fig. 2. Photoluminescence spectrum of a bulk ZnO sample after Li diffusion with low Li content (T = 4.2 K, HeCd laser excitation). The position of the zero phonon line (ZPL) of the donor acceptor pair band is indicated

If we assume an energy value of 50 meV for the donor binding, and neglect the coulombic interaction one obtains in ZnO:Li an acceptor binding energy around 300 meV. In order to gain further insight into the diffusion properties of Li we performed diffusions at different temperatures. As a measure we take the intensity at 4.2 K of the DAP band at 3.1 eV. Its intensity is strongest for the temperature interval between 450 °C and 650 °C (see Fig. 3). Indeed a rapid thermal annealing at 820 °C for 5 s leads to complete disappearance of the DAP band (see Fig. 4). In order to clarify whether Li is responsible or Li and/or N from the source Li_3N for the DAP band we performed diffusion from Li containing solutions ($LiNO_3$, Li_2CO_3). Within experimental error we obtained the same results. This is further confirmed by the in-situ doping experiments of Li in ZnO epitaxial films grown by chemical vapour deposition (see Fig. 5).

For the doping of ZnO with Na we used solutions of the respective salts ($NaCO_3$, NaOH). Diffusions were performed at temperatures between 500 °C and 600 °C for 20 min at N_2 pressures of 2 bar. A donor acceptor–pair band is induced at 3.045 eV (see Fig. 6). It has practically the same phonon coupling. The line positions of the ZPL in ZnO:Li and in ZnO:Na nearly coincide. Since the line position depends on the coulombic interaction term, and the background n-type carrier concentration is not known, the binding energies may actually be very close to each other. Secondary ion mass spectroscopy was used to check the incorporation of the group-I elements. For the Li doped samples the Na content was at least two orders of magnitude lower and

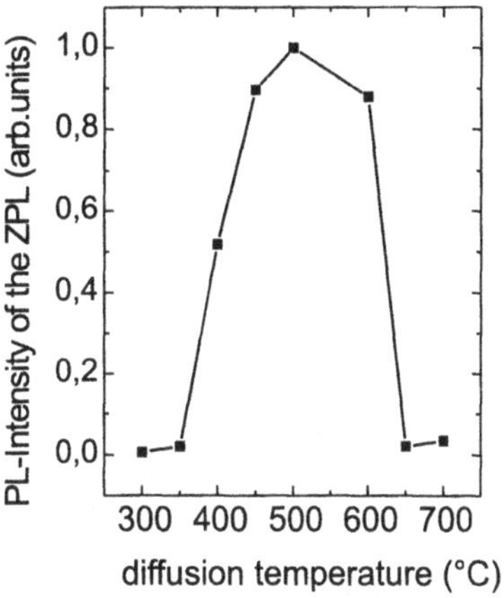

Fig. 3. Intensity of the zero phonon line at 3.05 eV in Li diffused ZnO samples as a function of the diffusion temperature

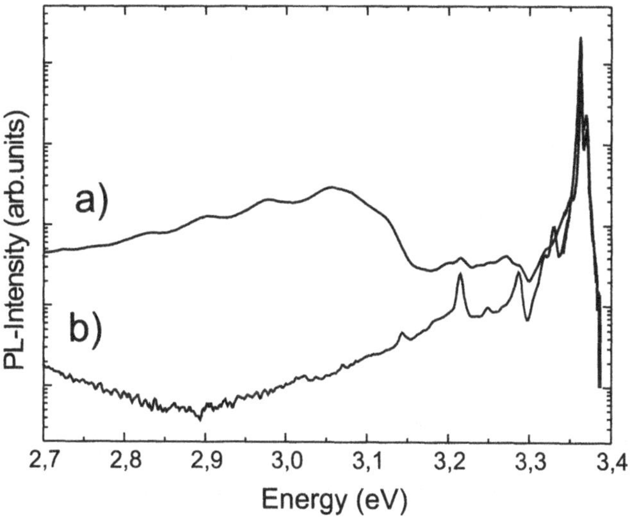

Fig. 4. Photoluminescence spectra of a ZnO bulk crystal (**a**) after Li diffusion and (**b**) subsequent rapid thermal annealing at 850 °C in N_2 atmosphere at 2.5 bar (T = 4.2 K, HeCd laser excitation)

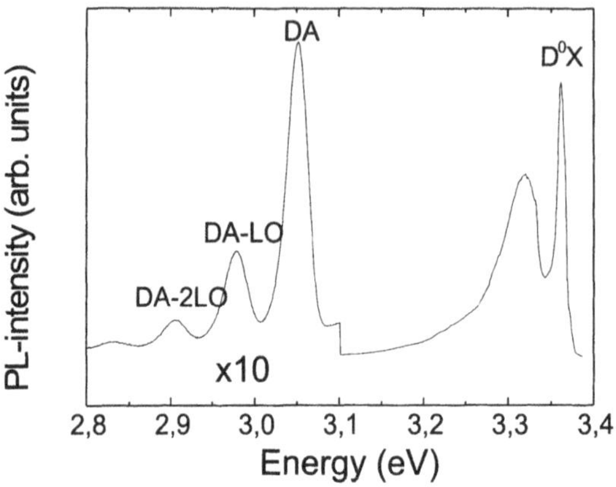

Fig. 5. Photoluminescence spectra of CVD grown ZnO on GaN templates with in-situ doping of Li (T = 4.2 K, HeCd laser excitation). The position of the zero phonon line (ZPL) of the donor–acceptor pair band plus phonon replica are indicated

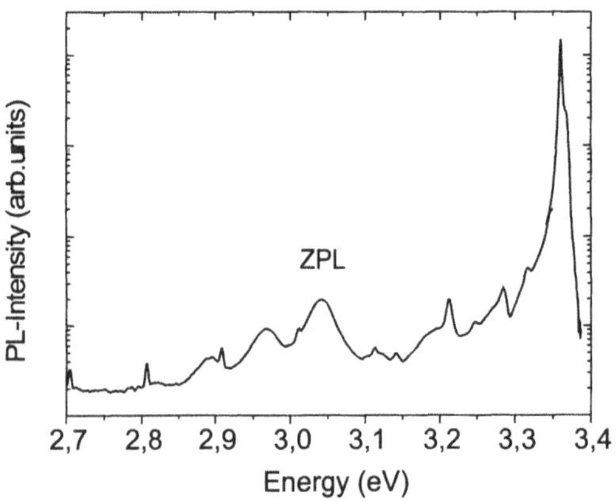

Fig. 6. Photoluminescence spectrum of a bulk ZnO sample after Na diffusion. The position of the zero phonon line (ZPL) of the donor–acceptor pair band is indicated (T = 4.2 K, HeCd laser excitation)

vice versa. But even with the source materials with the highest purity (5 N) contaminations with group-I elements not used for doping cannot be avoided.

If we take the ratio of the bound exciton recombination to the strongest line of the respective DAP transition there are obvious differences between the two group-I elements. For the case of ZnO:Li the DAP band can have comparable intensity (see Fig. 7), whereas for ZnO:Na it is two orders of magnitude lower in intensity (see Fig. 6 for a comparison). One would conclude that this reflects the limited solubility of Na in ZnO compared to Li. Our findings are in line with the results of in-situ doping of ZnO thin films

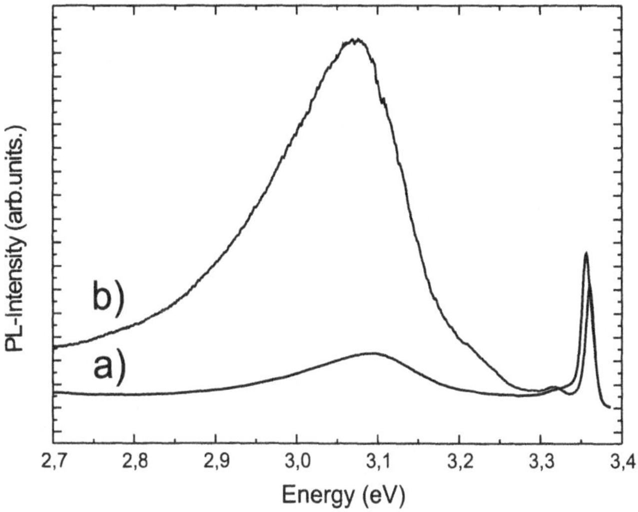

Fig. 7. Photoluminescence spectrum of bulk ZnO samples after Li diffusion with 7 nm (**a**) and 160 nm (**b**). Li_3N evaporated on the sample and diffused at 450 °C under N_2 atmosphere at 2.5 bar for 30 min (T = 4.2 K, HeCd laser excitation)

grown by chemical vapour deposition on GaN templates. As doping sources we used the respective metals, the ratio Zn to Na was chosen to provide doping concentrations above $10^{20}\,cm^{-3}$. There was no significant enhancement of the Na DAP band compared to the diffused samples, for which the doping concentrations were varied over three orders of magnitude. A recent theoretical work [7] calculated the solubility limits for Li_{Zn} and Na_{Zn} acceptors in their negative charge states. The numbers are about $10^{19}\,cm^{-3}$ for Li and about $10^{17}\,cm^{-3}$ for Na in line with our conclusions. We were able to make the 3.1 eV band the dominating transition (see Fig. 7) in ZnO:Li, and under HeCd laser excitation (325 nm) at low temperatures the samples emit a bright bluish white color.

3.3 Nitrogen Doping of ZnO

Further possible candidates acting as acceptors could come from the group-V elements, N, P and As. Theoretical reports on the behaviour of N in ZnO indicated a complex behaviour by the interaction with intrinsic defects, the passivation by hydrogen and nitrogen molecules [8, 9, 10]. Experiments used ion implantation also along the co-doping approach [11], in-situ doping [12, 13] of epitaxial films and diffusion [14]. There are only a few reports connecting a donor–acceptor pair recombination with a shallow nitrogen acceptor level [13, 15, 16].

Following the investigation of *Rommeluère* et al. [14] we studied the incorporation of N into ZnO from a thermal decomposition of ammonium and

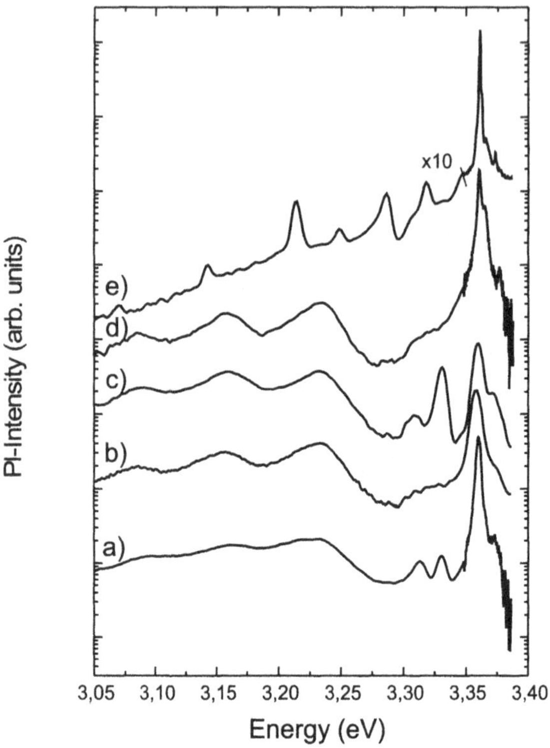

Fig. 8. Photoluminescence spectra of bulk ZnO after ammonium nitrate (**a**) and lithium nitrate (**b**) diffusion, and a ZnO epitaxial film doped by ammonia during growth (**c**). The photoluminescence spectra of the N, Zn co-implanted and activated (900 °C) ZnO sample (**d**) and in the as-grown state (**e**) are shown for comparison (T = 4.2 K, HeCd laser excitation)

lithium nitrate salts. NH_4NO_3 starts to decompose at temperatures above 370 °C. We studied the reaction for temperatures between 500 °C and 700 °C reaching pressures between 2 bar and 4 bar. The diffusion of N from the gas phase into ZnO results in the very same donor–acceptor pair recombination as found for implanted and in-situ doped samples [15,16]. Similar results were obtained for $LiNO_3$ (see Fig. 8). It decomposes into Li_2O, NO_2 and H_2O at temperatures above 500 °C. Best results were obtained for a diffusion temperature of 750 °C in the pressure range between 4 bar and 6 bar. The band edge luminescence and the DAP recombination have comparable intensities.

4 Discussion

Historically the group-I elements Li and Na were the first ones to be tested in order to achieve p-type conduction. Already in 1960 *Lander* [1] reported on the donor and acceptor properties of Li and gave first indications about the possible defect structure. He proposed the interplay between interstitial Li donors and substitutional Li_{Zn} acceptors. In the EPR investigations reported by *Schirmer* in 1968 [3] the g-anisotropy and the hyperfine interactions of the Li_{Zn} acceptor were analysed. The interpretation of the hyperfine data showed that the hole was primarily located on one neighbouring oxygen atom along the *c*-axis and induces a substantial distortion. It is the reason why Li_{Zn} is a deep acceptor with a binding energy of 800 meV. Na_{Zn} behaves similar, the binding energy is 600 meV. Both defects are responsible for broad strong phonon coupled luminescence bands in the visible spectral range with maxima at 2.1 eV (Li) and 2.17 eV (Na). The acceptors show up in shallow donor to deep acceptor recombinations as demonstrated by optically detected magnetic resonance experiments [17, 18].

Lander [1] observed a decrease in conductivity which was dependent on the oxygen pressure during doping and concluded that for the realisation of p-type ZnO:Li very high (and not achievable) oxygen pressures were needed. The main reason for the conversion from n-type to high resistive ZnO:Li was presented in the work of *Zwingel* [19]. He showed that the conductivity decreased in inverse proportion to the square of the Li concentration. Only 1 % of the total Li concentration was effective in the compensation of the donors already present in the crystal. He assumed equal concentrations for the interstitial Li donors and Li_{Zn} acceptors and postulated the existence of another Li acceptor, needed to model the reaction kinetics. It should arise from the pairing of two Li centres.

For the epitaxial films Li and Na doping in all cases enhanced significantly the luminescence in the yellow spectral range (compared to undoped samples where the Cu emission with well resolved phonon structure is observed). This can be used as an indication that the deep Li acceptor is formed and that there is a branching ratio between the shallow and deep Li. Note that our findings do not allow to conclude whether one or two Li atoms are part of the shallow acceptor.

The DAP bands at around 3 eV showed weak electron–phonon coupling which indicates very small lattice relaxation. If we assume that also for the shallow Li acceptor Li is incorporated on a Zn site, we have the same core for two significant different recombinations and defect properties – a shallow centre without significant relaxation and a deep one with significant relaxation. Similar arguments hold for ZnO:Na. First principles calculations predict shallow levels for the group-I elements Li: 0.09 eV and Na: 0.17 eV [8] with almost no relaxation around the Li_{Zn} acceptor, and a small outward relaxation for the surrounding oxygen atoms in the case of Na_{Zn}.

The observation of the same donor–acceptor pair band in samples where nitrogen was introduced by three different techniques, implantation, in–situ doping and diffusion from the gas phase, provides strong evidence that nitrogen on oxygen site is the acceptor. Its small binding energy may give hope to reach p-type conductivity in near future as presented in a recent publication [20].

5 Conclusions

The doping and diffusion experiments of ZnO with Li, Na and N provide new insight into the acceptor properties of the respective elements. Li and Na induce shallow and deep acceptor levels. By analogy to Cu doping and the interaction of the copper acceptor with hydrogen [21] one might speculate that the shallow levels of Li and Na arise from a complex where hydrogen is an essential part. Especially for deep acceptor Li°_{Zn} the EPR investigations in [3] provided clear evidence that the hole is bound to a nearest neighbor oxygen atom since Li prefers to be in the $1\,s^2$ configuration. The localization of the hole in the framework of small polaron coupling causes the large distortion and the substantial change in bond length – an increase of appr. 30 %. This distortion may be reduced if hydrogen forms an OH-group with one of the oxygen lattice atoms and hybridizes with the Li on Zn site. More experiments together with theoretical calculations are needed to substantiate this working idea. Despite all shortcomings, nitrogen, as shown in [20], is still the most promising towards the achievement of p-type conductivity in ZnO.

Acknowledgements

Part of this work was supported by BMBF (contract no 01 BM 151) and DFG (ME 898/17-2).

References

[1] J.J. Lander, J. Phys. Chem. Solids **15**, 324 (1960)
[2] J. Schneider and O. Schirmer, Z. Naturforsch. A **18**, 20 (1963)
[3] O.F. Schirmer, J. Phys. Chem. Solids, **29**, 1407 (1968)
[4] D. Zwingel and F. Gärtner, Solid State Communications, **14**, 45 (1974)
[5] B.K. Meyer, H. Alves, D.M. Hofmann, W. Kriegseis, D. Forster, F. Bertram, J. Christen, A. Hoffmann, M. Straßburg, M. Dworzak, U. Haboeck, and A.V. Rodina, phys. stat. sol. (b) **241**, 2, 231-260 (2004)
[6] J.R. Haynes, Phys. Rev. Lett. **4**, 361 (1960)
[7] E.-C. Lee, and K.J. Chang, Phys. Rev. B **70**, 115210 (2004)
[8] C.H. Park, S.B. Zhang, and S.-H. Wei, Phys. Rev. B **66**, 073202 (2002)
[9] S.B. Zhang, S.-H. Wie, and A. Zunger, Phys. Rev. B **63**, 075205 (2001)
[10] E.-C. Lee, K. Y.-G. Jin, and K.J. Chang, Phys. Rev. B **64**, 085120 (2001)
[11] M. Joseph, H. Tabata, and T. Kawai, Jpn. J. Appl. Phys. **38**, L1205 (1999)
[12] K. Minegishi, Y. Koiwai, Y. Kikuchi, K. Yano, M. Kasuga, and A. Shimizu, Jpn. J. Appl. Phys. **36**, L1453 (1997)

[13] D.C. Look, D.C. Reynolds, C.W. Litton, R.L. Jones, D.B. Eason, and G. Cantwell, Appl. Phys. Lett. **81**, 1830 (2002)

[14] J.F. Rommeluére, L. Svob, F. Jomard, J. Mimila-Arroyo, A. Lusson, V. Sallet, and Y. Marfaing, Appl. Phys. Lett. **83**, 2, 287 (2003)

[15] F. Reuss, C. Kirchner, Th. Gruber, R. Kling, S. Maschek, W. Limmer, A.Waag, and P. Ziemann, J. Appl. Phys. **95**, 7 (2004)

[16] A. Zeuner, H. Alves, D.M. Hofmann, B.K. Meyer, A. Hoffmann, U. Haboeck, M. Strassburg, and M. Dworzak, phys. stat. sol. (b) **234**, R 7 (2002)

[17] R.T. Cox, D. Block, A. Hervé, R. Picard, and C. Santier, Solid State Communications **25**, 77 (1978)

[18] F.H. Leiter, Optisch detektierte magnetische Resonanz an intrinsischen Defekten und Akzeptoren in ZnO und GaN, Thesis, Giessen, 2003

[19] D. Zwingel, J. of Luminescence **5**, 385 (1972)

[20] A. Tsukazaki, A. Ohtomo, T. Onuma, M. Ohtani, T. Makino, M. Sumiya, K. Ohtani, S.F. Chichibu, S. Fuke, Y. Segawa, H. Ohno, H. Koinuma, and M. Kawasaki, Nature Materials, Vol. **4**, 42 (2005)

[21] D. Zwingel, phys. stat.sol. (b) **67**, 507 (1975)

Optical Analysis of Monolayers at Surfaces and Interfaces

Veit Wagner

International University Bremen, 28759 Bremen, Germany
v.wagner@iu-bremen.de

Summary. Surface as well as interface properties are crucial for tailoring advanced materials, especially if the structure size in one or more dimensions approaches the nanoscale regime. Thus, analysis and modification of the last atomic layer is of crucial importance. Optical probes for analysis offer the advantage of high spectral resolution and the possibility to analyze under liquid, vapor and vacuum environments. With emphasis on Raman spectroscopy various examples of analysis of structural and binding properties of such monolayers are presented. Especially materials used in electronic applications are addressed. Examples are determination of the surface reconstructions of II-VI semiconductors and overgrowth analysis by Raman spectroscopy and the modification of bonding properties of organic molecules, i.e. 3,4,9,10-perylene-tetracarboxylic-dianhydride (PTCDA), at metallic surfaces.

Surface conditions and chemical reactions at surfaces are known to be of crucial importance for heterostructure growth and catalysis processes in a wide field of applications. Especially in present semiconductor technology with structure sizes in the nanometer scale, corresponding to few atomic layers, interface control is a major issue. Advanced growth techniques like molecular beam epitaxy (MBE) and metal organic chemical vapor deposition (MOCVD) allow growth control at the single atomic layer level. However, each material system has its characteristic surface/interface chemistry and requests generally its own optimized growth sequence. This optimization requires the knowledge of the possible chemical surface reactions.

1 Introduction

Surface sensitive methods involve typically scattered particles or scanning probe techniques. Latter, while offering extreme lateral spatial resolution even down to the atomic level, typically shield the sample in a standard growth set-up. Furthermore, spectroscopic information about the chemical properties are not easy to obtain by a scanning probe approach. Electron based scattering techniques offer the advantage of high surface sensitivity, while requiring vacuum conditions. A chemical analysis may be deduced from chemical shifts in photoemission experiments (XPS/UPS) which reveals the charging of individual atomic species. Vibrational spectroscopy methods, like

B. Kramer (Ed.): Adv. in Solid State Phys. **45**, 301–312 (2005)

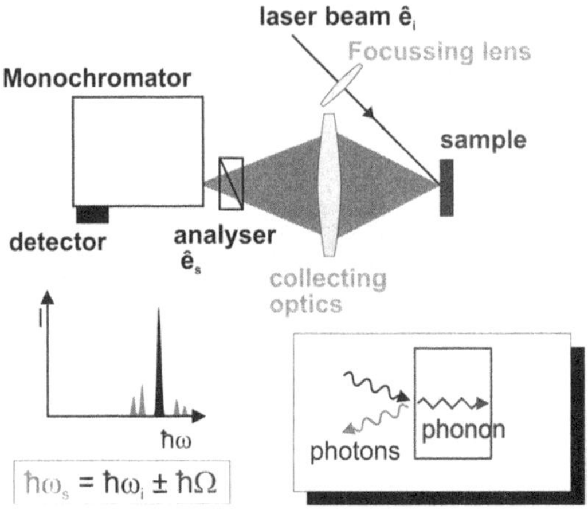

Fig. 1. Sketch of the principal set-up for Raman measurements (UHV vessel around the sample not shown) Raman spectrum and quasi particle exchange picture for the Raman process

high-resolution electron energy-loss spectroscopy (HREELS) or Helium atom scattering (HAS), in principle allow a direct measurement of the chemical bond strength via its vibration frequency. An intrinsic advantage of these vibrational spectroscopics applied in surface analysis is the ease of theoretical description. Quite accurate descriptions can be obtained already by very simple models using empirical parameters on the basis of (charged) point masses interconnected by harmonic springs. In the framework of density functional theory the vibrational eigenmodes and eigenfrequencies of microscopic atomic ensembles can be obtained even without empirical parameters. Since only electronic ground state properties are needed for first principles vibrational mode calculations [1], these calculations need much less computer resources than simulations for other surface techniques, which rely on electronic excitations [2], e.g. reflectance anisotropy spectroscopy (RAS) [3] or second harmonic generation (SHG). In contrast to these electron/helium based tools, optical methods offer the advantage of high spectral resolution (e.g. 3 cm^{-1} for Γ-point surface phonons) and the possibility to analyze under liquid, vapor and vacuum environments. With emphasis on Raman spectroscopy various examples of analysis of structural and binding properties of such monolayers are presented here. Optical vibrational spectroscopy yields more specific selection rules by varying the polarisation of the two photons involved in the light scattering process. Thus, additional information of the symmetry of the vibrational mode under investigation is obtained. Furthermore, Raman scattering is neither restricted to dipole-allowed vibrations like HREELS in its standard application, nor to low lying phonon bands like HAS.

As already reported in literature, Raman spectroscopy can offer sufficient sensitivity to analyze single monolayers as for Sb/InP(110) [4]. Even vibrations at a clean InP(110)-(1x1) surface are detectable. A review about those III-V and silicon surfaces was given recently by Esser and Richter [5] and an introduction to the basics of Raman spectroscopy and its application to analyze II-VI nanostructures by the author [6]. In this paper surface analysis of MBE grown II-VI semiconductors is reported. In a second step this analysis method is extended to the epitaxial growth of organic molecules on single crystal metal surfaces.

The experimental set-up used for back scattering Raman measurements is sketched in Fig. 1 together with a schematic of a Raman spectrum showing the Stokes- and Antistokes lines as well as the elastic peak from the laser line. Due to their higher intensity only Stoke-side scattering is considered in this paper [6]. For clarity the UHV vessel around the sample is not shown.

2 Inorganic Surfaces and Interfaces

One of the largest demands for material control and new designs originates from needs for electronic applications like laser diodes or transistors. These demands are mainly fulfilled by IV-IV, III-V and II-VI semiconductor materials today. The majority of these materials crystallize in the zinc blende lattice. In this section we concentrate on wide-bandgap II-VI semiconductors, which are interesting materials for transparent contacts, light emitting devices and more recently by doping with isoelectric magnetic ions for spintronics applications [7]. Due to the possibility of high quality layer growth and the availability of extreme physical properties among the II-VI semiconductors this material class is specially suited for interface tailoring. E.g. epitaxial growth of materials with extremely different intrinsic bond polarities. Among II-VI compounds, Be chalcogenides offer especially interesting properties. Due to the combination of smallest group-II element Be (atomic mass is only 9) with very large group-VI elements, Se or Te, the charge transfer that results from cation to anion is significantly reduced with respect to the more common II-VI compounds such as ZnSe. This implies a lower bond polarity, in particular for BeTe, which allows for band-offset tailoring in heterostructures and increased material stability, e.g. for optoelectronics. For high quality heterostructures the close lattice matching to GaAs and ZnSe is important, which makes it compatible to commercially available GaAs(100) substrates. BeTe/ZnSe interfaces show type II band alignment with an extremely high conduction band offset of 2.3 eV, which was exploited in resonant tunneling and allowed for interface induced polarized luminescence [8] and electromodulation spectroscopy [9]. Furthermore, the large difference in mass between the cation and the anion results in quite unique lattice dynamics, which results in an extraordinarily large gap between the optical and acoustical phonon branches. This gap provides a wide frequency window, which enables localized vibrations within or close to the BeTe host material.

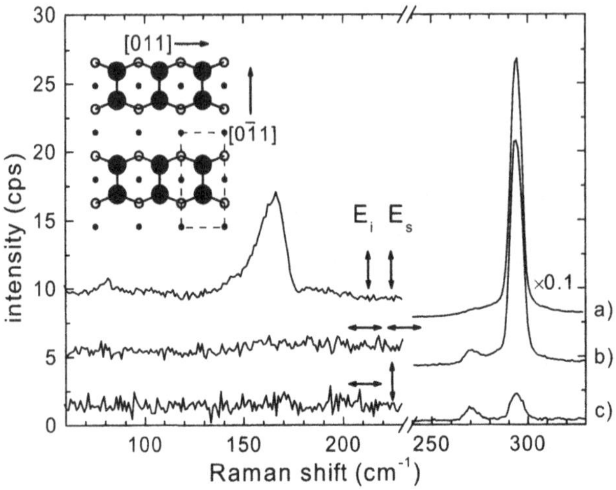

Fig. 2. Surface phonons of a (2×1) reconstructed BeTe(100) surface. Measurements for various polarization configurations at T=80K for laser excitation at 476.5 nm

2.1 BeTe(100) Surfaces

Prior to the investigation of BeTe containing heterostructures the initial clean BeTe surface will be characterized in detail. Due to the close lattice matching of BeTe ($a_{BeTe} = 5.626$ Å) and GaAs ($a_{GaAs} = 5.653$ Å) high quality BeTe-layers can be grown on GaAs(100) substrates. In the following BeTe(100) surfaces are prepared by MBE in growing a BeTe layer of 25 nm thickness on GaAs(100) at 350°C substrate temperature. Electron diffraction analysis of those surfaces reveals (2×1) surface reconstruction. Depending on the growth conditions (3×1) or (4×1) reconstructions can be prepared as well [10]. Our aim is a closer investigation of these surfaces by Raman spectroscopy and supporting first-principles calculations. Figure 2 shows the Raman spectra of freshly grown BeTe layer under Te-rich conditions, which exhibits the (2×1) surface reconstruction. After growth the sample was transferred from the MBE via a moveable UHV-chamber to the Raman measurement laboratory. Besides the GaAs substrate related phonons above 250 cm^{-1}, i.e. TO-GaAs at 271 cm^{-1} and LO-GaAs at 294 cm^{-1}, a structure appears at 165 cm^{-1}. Note that BeTe is transparent for the exciting laser line. The selection rules of this peak are not compatible with the cubic bulk symmetry of the zinc blende lattice. With the polarization of the exciting laser light being parallel to the polarization of the scattered light the same intensity is expected for polarization along [011] and [0$\bar{1}$1] direction for bulk symmetry. But this is not the case here. The origin of the structure must be therefore located at the GaAs/BeTe interface or the BeTe surface, where the full cubic bulk symmetry is broken. Oxidation experiments proof, that the 165 cm^{-1} peak must be attributed to the BeTe(100) surface [11], as well as the Selenium flux

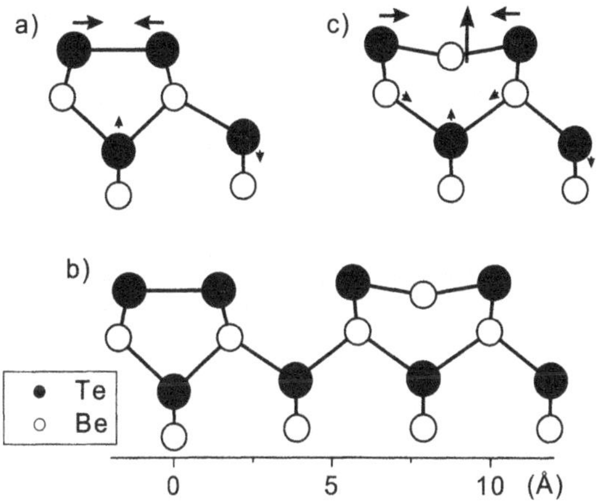

Fig. 3. Sketch of calculated surface phonon modes with view along the [011] direction. First pictures show (2×1) reconstructed BeTe(100) surface for (a) Te-rich and (c) Be-rich conditions. (b) depicts a model for a (4×1) reconstruction. Arrows indicate atomic motions corresponding to the surface vibration

exposure experiment later in section 2.3. Therefore we have to attribute the 165 cm^{-1} peak to the BeTe surface. The inset of Fig. 2 depicts the predicted atomic arrangement of a (2×1)-reconstructed BeTe(100) surface forming Te-dimers, which is typically found for II-VI semiconductors and was confirmed by density functional theory (DFT) calculations in the local density approximation (LDA). Large Raman scattering intensities are observed for the given atomic arrangement only, if the light polarization is aligned parallel to the Te-dimer axis. Indeed, intuitively it is expected that polarization along the bond axis should be larger than perpendicular to the bond axis. These findings motivated a theoretical investigation of the atomic arrangements and frozen phonon calculations of vibrational modes for the most likely surface reconstructions of BeTe(100) by first principles calculations [11]. The resulting three lowest energy surface reconstructions are shown in Fig. 3. The pictures show the atomic arrangements as well as the vibrational surface mode pattern. Note that the surface composition changes from Te-rich a) over case b) to Be-rich for case c) in figure 3. The vibrational mode for the Te-rich case a) with (2×1) reconstruction calculates to 166.8 cm^{-1} and has A_1 symmetry, which is in excellent agreement to the experimental value observed at 165 cm^{-1} at T=80 K for samples prepared under Te rich conditions and the experimental Raman selection rules.

Thus, we observe here vibrations of a clean reconstructed (100) surface by Raman spectroscopy. These findings were further supported by measuring Be-rich surfaces which show two peaks at different frequencies (157 cm^{-1} and 188 cm^{-1}) [11, 12] and by determining the surface atomic positions by x-ray diffraction in agreement with our theoretical calculations [13].

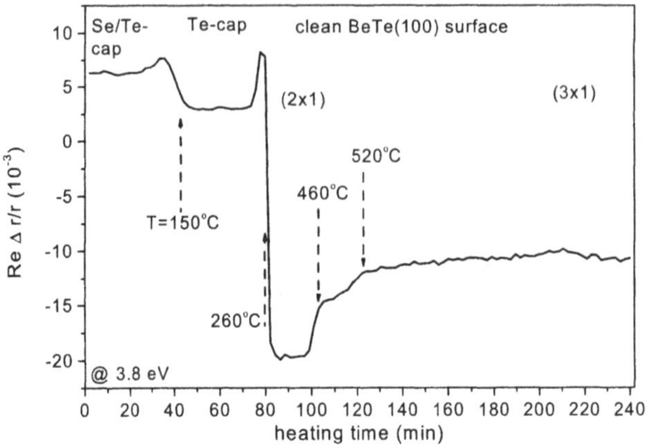

Fig. 4. RAS monitoring of removing the protective Se/Te-capping layer of a BeTe(100) surface by heating

2.2 Surface Protection by Capping Layers

Since only very few labs are equipped with Be chalcogenides growth facilities, techniques which enable more widespread availability of Be chalcogenide surfaces, either for overgrowth or for surface analytical studies, e.g., by synchrotron radiation, are highly desired. An obvious option is sample transport in an UHV-transfer vessel to desired locations. However, long term storage problems of the surfaces and the requirement for appropriate docking stations ask for alternative procedures. A second option is to deposit an employment layer immediately after MBE growth, followed by sample transfer to a vacuum system a distance away, and subsequent removal of the cap layer by thermal annealing. This has been reported for several III-V systems and also for ZnSe (see references in [14]). For tellurium based compounds an employment layer of elemental Te is the first choice to avoid anion exchange reactions at the surface. Because elemental Te growth results in microcrystallites (with voids) an additional Se layer is used on top to achieve a protective closed top layer. The two-layer cap approach is known e.g. from phosphorous caps, which are protected by a 2nd layer of As against reaction with ambient air [15]. After transfer of the capped samples under ambient atmosphere, the decapping procedure is monitored and controlled by Raman spectroscopy and/or reflectance anisotropy spectroscopy (RAS) [15]. A typical RAS transient is shown in Fig. 4. The complete desorption of the upper Se-cap at 150°C and the lower Te-cap at 260°C are clearly observed in the transient. Furthermore, a specific BeTe(100) surface reconstruction, (2×1) or (3×1), can be prepared by further thermal annealing. E.g. heating above 520°C can be used to prepare a Be-rich (3×1) reconstructed BeTe(100) surface. The different preparations were checked by RAS and Raman spectroscopy as well as by standard low energy electron diffraction (LEED) as control experiment. Further details can be found in [14].

This example demonstrates, that optical methods as RAS and Raman spectroscopy (spectra not shown) are able to monitor and control the deposition and removal of suitable protective capping layers, which enable surface preparation without direct access to growth facilities. Furthermore, transport and storage of semiconductor surfaces through/at air ambient is possible by this approach.

2.3 Overgrowth Analysis: Converting Surfaces to Interfaces

During growth of heterostructures the switching sequence between different compounds is of crucial importance for the quality of their interfaces. During the switching process the surface chemistry will be altered by the arrival of the constitutions of the next compound. This is especially true if both, anions and cations, are exchanged, i.e. binary compound CD is grown on top of a binary layer AB. In this case either interface bonds of the kind C–B or the kind A–D are formed. Direct observation of those interface bonds allows to check, if changes in the deposition conditions are able to change the interface chemistry [16, 17]. E.g. for CdSe/BeTe superlattices it was found, that by choosing proper interface growth sequences, i.e. growing a fractional monolayer of either CdTe or BeSe before starting the CdSe growth, the interface for CdSe overgrowth can be tuned. In contrast, the other interface for BeTe overgrowth can hardly be changed by this approach [18, 19]. Optimized fractional interface layers can be realized by observing insitu the RHEED intensity during growth as a guiding signal for the shutter sequence [20]. The relevance of the material combination BeTe/CdSe considered in the following is given by the combination of very different direct band gap values (4.2 eV/1.74 eV) and the type-II band alignment, which results in a valence band offset of 0.66 eV and a conduction band offset as high as 3 eV. The relative lattice mismatch of CdSe to BeTe amounts to $\Delta a/a > 7\%$, which leads to a pronounced compressive strain in the CdSe layer. Due to the large mismatch the critical thickness for 2D growth is estimated to be 3 monolayers (ML) only. The interface bonds might contribute to the strain field as well due to the large lattice mismatch for their corresponding binary materials with respect to the GaAs substrate: BeSe $= -10\%$, CdTe $= +15\%$. In the following we will observe the growth of a CdSe layer on top of a BeTe(100) surface from zero thickness until full relaxation has occurred. Raman spectra for the clean Te-rich (2x1)-reconstructed BeTe(100) surface as well as covered by CdSe layers of thickness 1 ML, 2 ML, 3.5 ML, and 5 ML are shown in Fig. 5 in thick lines. For comparison Raman spectra of a clean GaAs-wafer (no Te-dimer signal at 165 cm^{-1}) and a BeTe(100) surface exposed to 30 seconds Se-flux are shown as well in thinner lines. Obviously the short Se-flux destroys the Te-dimers at the BeTe surface immediately.

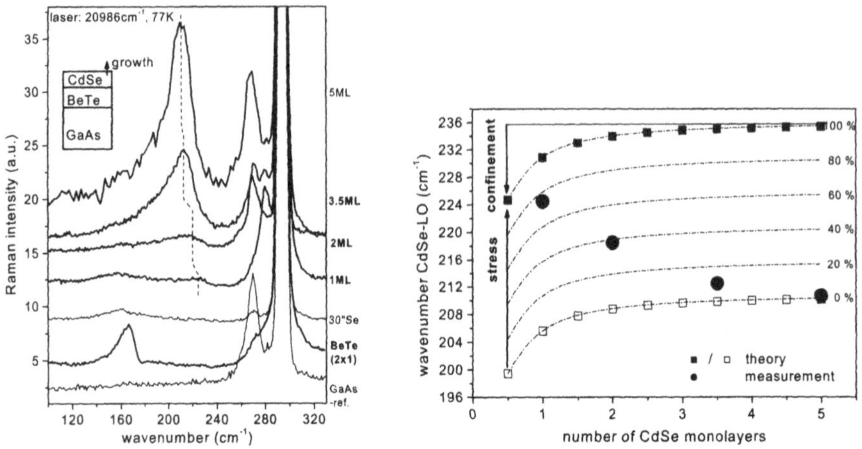

Fig. 5. Left side: Raman spectra for CdSe overgrowth of a BeTe(100) surface (Laser 476.5 nm, T=80 K, light polarizations along [011]). Right side: CdSe LO-phonon position versus layer thickness. Squares show theoretical calculations for the phonon confinement effect for no strain (full squares) and fully strained (open squares). Full circles show the experimental data

The strain of the growing CdSe layer is reflected in the frequency shift oft CdSe-LO phonon in the Raman spectra. For strain analysis two contributions have to be taken into account. First there is the strain in the CdSe layer causing an up shift and the second contribution is the phonon confinement in the CdSe layer causing a down shift of the LO phonon frequency. The up shift due to the strained layer calculates to

$$\Delta\omega = \omega_0 \cdot (p \cdot \varepsilon_{zz} + q \cdot (\varepsilon_{xx} + \varepsilon_{yy}))/2 \tag{1}$$

with the in-plane strain $\varepsilon_{xx} = \varepsilon_{yy}$ given by the lattice mismatch and the out-of-plane strain by the elastic deformation $\varepsilon_{zz} = -2\varepsilon_{xx}C_{11}/C_{12}$. The strain constants for CdSe are $C_{11} = 74.6$ GPa and $C_{12} = 46.1$ GPa and the phonon deformation potentials are $p = -0.97\,\omega_0^2$ and $q = -2.21\,\omega_0^2$ [21]. The second contribution is the phonon confinement. As boundary conditions the last Te-layer at the interface is assumed at fixed positions and the CdSe surface as an open end, i.e.

$$a_{Cd-Te} + d_{CdSe} = \lambda/4 \tag{2}$$

with a_{Cd-Te} the inter-layer distance of the Cd-Te interface layer, d_{CdSe} the CdSe layer thickness and λ the wavelength of the CdSe LO-phonon mode consistent with the boundary conditions. This wavelength λ is used to extract the expected frequency from the LO-phonon dispersion curve along [100] direction and thus the confinement induced down shift [21]. Following this calculations the expected theoretical phonon frequency with CdSe layer thickness is shown in the right side of Fig. 5. The experimental data (full circles) extrapolate to a fully strained stated for initial CdSe layer growth.

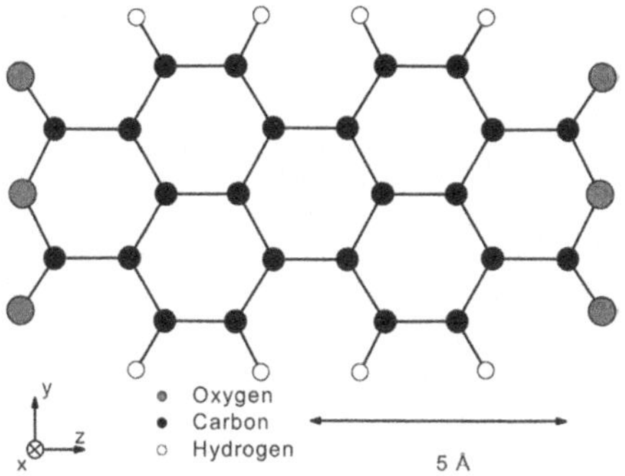

Fig. 6. Molecule structure of 3,4,9,10-perylene-tetracarboxylic-dianhydride (PTCDA)

With increasing layer thickness, their position gradually develops from the fully strained to the relaxed case. Nominally, the strain amounts to 38% for 2 ML and 10% for 3.5 ML, while the 5 ML-layer is essentially relaxed.

In summary, starting with (2×1)-BeTe(100) surface Raman spectra revealed the Te-dimers at this surface. Upon CdSe growth Raman spectra revealed at the interface the formation of Cd–Te bonds, followed by Cd–Se ones. Local vibration modes from the latter bonds developed into the LO phonon of CdSe, positioned at 211 cm^{-1}. The evaluation of the thickness dependence of the phonon position revealed the gradual strain relaxation in the CdSe layer over the first 5 ML [21].

3 Organic Surfaces and Interfaces

A very rapidly developing field for fundamental physics as well as for applications are organic semiconductors. They are cheap to produce, mountable onto flexible substrates and offer new physical effects, e.g. due to very strong phonon-electron-coupling in the organic molecules. In solid state qualitative new bonding properties occur, i.e. crystalline structures created by weak van der Waals bonding, which can form epitaxial systems [22]. In applications, like thin film transistors [23], interfaces of organic layers with metals are of crucial importance. While band alignments at those interfaces can be determined by photoemission experiments [24, 25], the arrangement and bonding details of the first molecular layers with the metal surface remains an open question. For crystalline and epitaxial systems distances for the first monolayer can be accessed by near edge x-ray absorption fine structure (NEXAFS) [26]. However, like for inorganic materials vibrational spectroscopies allow access to bonding details of the first layers deposited on metal surfaces. As a model system for an organic material 3,4,9,10-perylene-tetracarboxylic acid-dianhydride (PTCDA) is used due to the large data basis available in

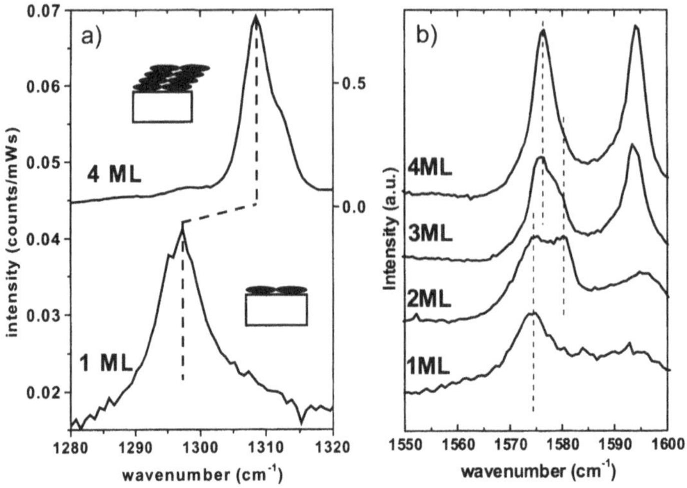

Fig. 7. Raman spectra of PTCDA/Ag(111) deposited at 180K after annealing at 450K. The spectra were recorded at T=20K by excitation at 514.5 nm

literature [22, 26, 27, 28, 29, 30] and the formation of epitaxial layers on various substrates. A sketch of the PTCDA is shown in Fig. 6. The aromatic rings of the central carbon skeleton of PTCDA form an extended conjugate π-electron system, which makes PTCDA a semiconductor with a band gap of about 2.5 eV. Ag(111) is used as substrate in the following, since PTCDA is known to form epitaxial layers on Ag(111) [30]. PTCDA layers with various thicknesses were deposited under UHV conditions (1e-7 Pa) at 180 K and Raman spectra were recorded in-situ at 20 K after various annealing steps. PTCDA gives rise to many Raman active modes with frequencies up to 3300 cm^{-1}. The carbon skeleton modes are located in the frequency range 1300$\cdots$1700 cm^{-1}. In the following we concentrate on two intense A_g-modes at 1310 cm^{-1} and 1575 cm^{-1}. Figure 7 shows Raman spectra of 1 ML to 4 ML PTCDA deposited at 180 K followed by an annealing step at 450 K. Due to the interaction with the metal substrate the peaks of the first monolayer at 1310 cm^{-1} and 1575 cm^{-1} show a down shift of 12 cm^{-1} and 2 cm^{-1} with respect to the thicker layers, respectively. First principle calculations of the corresponding mode pattern in PTCDA reveal, that the down shift are correlated to the vibration of the central carbon ring [31, 32] indicating a molecular region strongly affected by adsorption on the metal. Recent NEXAFS investigations and calculations confirm this interpretation [26], while the much larger shifts observed in HREELS remain an open question [29, 32]. Beside the modification of the first layer Fig. 7b show also a clear modification of the second monolayer (6 cm^{-1} up shift) with respect to the first layer. While thermal programmed desorption (TPD) experiments confirm this not bulklike bonding of the second monolayer, so far only high resolution techniques as Raman spectroscopy can be used for the second monolayer analysis [31, 32].

In summary it was found, that the first as well as the second monolayer have modified bonding properties due to the influence of the silver substrate. Upon annealing the crystalline order of the molecules is increased. Furthermore it was found, that the third and higher monolayers transform to the crystalline α-modification upon annealing (not shown), while the first and second monolayer remains in their specific substrate bonded state [32, 31].

Acknowledgements

The support by J. Geurts, E. Umbach, L. Molenkamp and G. Landwehr at the University Würzburg for the fruitful discussions and the excellent experimental environment is gratefully acknowledged. Many thanks to J.W. Wagner, T. Muck, and M. Schneider for doing most of the experiments presented here, S. Gundel for the support in the DFT/LDA calculations and L. Hansen for MBE sample growth. Furthermore I want to thank P. Vogt, N. Esser, and W. Richter for the support during the Raman and RAS measurements in Berlin and S.V. Ivanov at the Ioffe Instiute for his valuable input. This work was supported financially by the Deutsche Forschungsgemeinschaft (SFB 410, SP 1121) and the Volkswagenstiftung, as well as by the INTAS Grant No. 97-31907 and RFBR, which is gratefully acknowledged.

References

1. J. Fritsch, U. Schröder, Phys. Rep. **309**, 209 (1999).
2. M. Rohlfing, J. Pollmann, Phys. Rev. B **63**, 125201 (2001).
3. W. G. Schmidt, F. Bechstedt, J. Bernholc, Phys. Rev. B **63**, 45322 (2001).
4. M. Hünermann, J. Geurts, W. Richter, Phys. Rev. Lett. **66**, 640 (1991).
5. N. Esser, W. Richter, Raman Scattering from Surface Phonons, in *Light Scattering in Solids VIII, Topics in Applied Physics Vol. 76*, M. Cardona, G. Güntherodt (Eds.), p. 96 (Springer Verlag, Berlin 2000).
6. V. Wagner, J. Geurts, W. Kiefer, Raman Spectroscopy on II-VI Semiconductor Interfaces, Quantum Dots, and QuantumWires, Encyclopedia of Nanoscience and Nanotechnology, H. Nalwa (ed.), 73 pages (American Scientific Publishers, ISBN: 1-58883-001-2, 2004).
7. R. Fiederling, M. Keim, G. Reuscher, W. Ossau, G. Schmidt, A. Waag, L. W. Molenkamp, Nature **402**, 787 (1999).
8. S. V. Zaitsev, A. A. Maksimov, V. D. Kulakovskii, I. I. Tartakovskii, D. R. Yakovlev, W. Ossau, L. Hansen, G. Landwehr, A. Waag, J. Appl. Phys. **91**, 652 (2002).
9. V. Wagner, M. Becker, M. Weber, M. Korn, M. Keim, A. Waag, J. Geurts, Appl. Surf. Sci. **166**, 30 (2000).
10. E. Tournie, V. Bousquet, and J. P. Faurie, Appl. Phys. Lett. **72**, 2859 (1998).
11. V. Wagner, J. Wagner, S. Gundel, L. Hansen, J. Geurts, Phys. Rev. Lett. **89**, 166103 (2002).
12. T. Muck, J.W. Wagner, L. Hansen, V. Wagner, J. Geurts, J. Appl. Phys. **95**, 5403 (2004).

13. C. Kumpf, A. Müller, W. Weigand, E. Umbach, J. Wagner, V. Wagner, S. Gundel, L. Hansen, J. Geurts, O. Bunk, J. H. Zeysing, F. Wu, R. L. Johnson, Phys. Rev. B **68**, 035339 (2003).
14. J. W.Wagner, V. Wagner, L.Hansen, G.Schmidt, J.Geurts, P. Vogt, N. Esser, W. Richter, J. Appl. Phys. **93**, 1511 (2003).
15. C. Goletti, N. Esser, U. Resch-Esser, V. Wagner, J. Foeller, M. Pristovsek, W. Richter, J. Appl. Phys. **81**, 3611 (1997).
16. J. Spitzer, A. Höppner, M. Kuball, M. Cardona, B. Jenichen, H. Neuroth, B. Brar, H. Kroemer, J. Appl. Phys. **77**, 811 (1995).
17. A. Dinger, M. Göppert, R. Becker, M. Grün, S. Petillon, C. Klingshirn, J. Liang, V. Wagner, J. Geurts, Phys. Rev. B **6** 245310 (2001).
18. T. Muck, J. W. Wagner, L. Hansen, V. Wagner, J. Geurts, S.V. Ivanov, Phys. Rev. B **69**, 245314 (2004).
19. V. Wagner, J. Geurts, Phys. Stat. Sol. (a) **184**, 29 (2001).
20. V. Wagner, J. Wagner, T. Muck, G. Reuscher, A. Waag, J. Geurts, N. Sadchikov, S.V. Sorokin, S.V. Ivanov, P.S. Kop'ev, Appl. Surf. Sci. **175-176**, 169 (2001).
21. V. Wagner, J. Wagner, L.Hansen, S.Gundel, G.Schmidt, J.Geurts, Appl. Surf. Science **212-213**, 901 (2003).
22. S. R. Forrest, Chem. Rev. **97**, 1793 (1997).
23. T. Muck, V. Wagner, U. Bass, M. Leufgen, J. Geurts, L.W. Molenkamp, Synthetic Metals, **146**, 317 (2004).
24. I.G. Hill, D. Milliron, J. Schwartz, A. Kahn, Appl. Surf. Sci. **166**, 354 (2000).
25. L. Chkoda, C. Heske, M. Sokolowski, E. Umbach, Appl. Phys. Lett. **77**, 1093 (2000),
26. A. Hauschild, K. Karki, B. C.C. Cowie, M. Rohlfing, F.S. Tautz, M. Sokolowski, Phys. Rev. Lett. **94**, 036106 (2005).
27. D. A. Tenne, S. Park, T. U. Kampen, A. Das, R. Scholz, D. R. T. Zahn, Phys. Rev. B **61** 14564 (2000).
28. G. Salvan, D. A. Tenne, T. U. Kampen, R. Scholz, G. Jungnickel, T. Frauenheim, D. R. T. Zahn, Appl. Surf. Sci. **179**, 113 (2001).
29. F. S. Tautz, M. Eremtchenko, J. A. Schäfer, M. Sokolowski, V. Shklover, E. Umbach, Phys. Rev. B **65**, 125405 (2002).
30. K. Glöckler, C. Seidel, A. Soukopp, M. Sokolowski, E. Umbach, M. Böhringer, R. Berndt, W.-D. Schneider, Surf. Sci. **405**, 1 (1998).
31. V. Wagner, Phys. Stat. Sol. (a) **188**, 1297 (2001).
32. T. Muck, J. Geurts, M. Schneider, E. Umbach, Appl. Surf. Science **212-213** 520 (2003).

Chemistry and Morphological Properties of Metal Interfaces to Organic Semiconductors

D.R.T. Zahn, G. Salvan, G. Gavrila, and B.A. Paez

Institut für Physik, Technische Universität Chemnitz, 09107 Chemnitz, Germany

Summary. Metal contacts to organic semiconductors play a decisive role for the performance of organic based devices such as organic light emitting diodes, organic field effect transistors or organic solar cells. In particular when a metal is deposited onto an organic substrate, severe disruption of the interface may occur as a result of chemical reactions of the metal with organic molecules and/or diffusion of the metal into the organic layer. Here, the interaction of metals of different reactivity (Ag, In, Mg) with a variety of perylene derivatives as model molecules is probed employing *in situ* Raman spectroscopy. The results reveal that this technique allows an extreme interface sensitivity to be achieved via surface enhanced Raman scattering. The degree of reactivity and indiffusion can be derived from the analysis of the evolution in scattering by internal vibrational modes of the molecules and phonon-like external modes of the molecular crystal.

1 Introduction

Organic based devices are promising for new generation electronics, by providing low cost, lightweight and good performance [1, 2]. The effect of the interface formation between a metal contact and the organic active layer on the device performance can be dramatic. Therefore issues like interface chemistry, metal diffusion and metal morphology are currently intensively investigated [3, 4, 5]. In the present work the vibrational spectra of two perylene derivatives, 3,4,9,10-perylene tetra-carboxylic dianhydride (PTCDA) and N, N' dimethyl-3,4,9,10-perylene tetracarboximide (DiMe-PTCDI) are studied by Raman spectroscopy upon deposition of Ag, In and Mg. Only few investigations of metal-on-organic interfaces are reported to date and those are mainly performed by means of electron spectroscopies. In particular, photoemission spectroscopy (PES) studies revealed that Ag deposition onto PTCDA leads to the formation of an abrupt, non-reactive interface [6]. Even though a thorough understanding of the vibrational properties of PTCDA molecules has been achieved from a combination of theoretical calculations and Raman spectroscopy measurements on crystals [7, 8, 9], most of the work devoted to metal/PTCDA interfaces, however, refers to the vibrational properties of molecular thin films deposited on either low index single crystal or rough metal substrates. Representative for the case of single crystal metal substrates are the studies of PTCDA on Ag(111) and Ag(110) by means of high

B. Kramer (Ed.): Adv. in Solid State Phys. **45**, 313–324 (2005)

resolution electron energy loss spectroscopy (HREELS) and electron energy loss spectroscopy (EELS) [10, 11]. Regarding the In interface formation with the perylene derivatives there still is a controversy: PES studies indicate a strong reaction between In and PTCDA via the C=O bonds [12], whereas more recent near edge x-ray absorption fine structure revealed the presence of a fractional charge transfer between the PTCDA or DiMe-PTCDI molecules and In [13]. The deposition of some metals like Ag and In onto organic layers of PTCDA or DiMe-PTCDI gives rise to an enhancement of the normally Raman-active internal vibrational mode intensities, accompanied by the activation of normally infrared-active modes [14]. This effect is known as surface enhanced Raman scattering (SERS) [15, 16, 17]. The enhancement of normally Raman-active modes is a consequence of the enhancement of the electric field of the incoming and scattered radiation in the vicinity of a rough metal film. This enhancement affects molecules located up to even 10 nm away from the metal surface [16, 17]. The enhancement factors are essentially determined by the electronic properties of the metal and by the morphology of the metal film. The break-down of the selection rules is a characteristic of molecules in the very near vicinity of a metal surface and can be induced by several mechanisms: structural deformation of the molecule, charge transfer from the molecule into the metal vice versa, or formation of new chemical bonds. Thus the spectral changes induced by SERS can be used to extract information about the chemical reactions at the interface, as well as the morphology of the metal film.

2 Experimental

The organic layers were grown onto sulfur-passivated GaAs(100) substrates [14] by organic molecular beam deposition (OMBD) in ultra-high vacuum (base pressure=$2{\cdot}10^{-10}$ mbar). The molecules obtained from Syntec GmbH Wolfen were pre-purified by double sublimation prior to deposition by thermal evaporation. The evaporation rate for both organic materials was 0.3 nm/min as determined by a quartz microbalance calibrated using atomic force microscopy measurements. The evaporation rate was 1.6 nm/min for Ag and 2 nm/min for In and Mg. In order to record Raman signals in two spectral regions at different stages of metal deposition, the growth process was interrupted for the accumulation periods. All depositions took place at room temperature. For *in situ* Raman measurements, the UHV system is optically aligned with a triple monochromator Raman spectrometer (Dilor XY) equipped with a CCD camera for multichannel detection [14]. The samples were excited with the 488 nm (2.54 eV) Ar^+ laser line that lies in the first absorption maximum of both organic molecules and thus ensures resonance conditions for the Raman process. Two spectral regions were recorded upon step-wise metal deposition. The region between 25 cm^{-1} and 650 cm^{-1} contains the external modes as well as a mode corresponding to the breathing

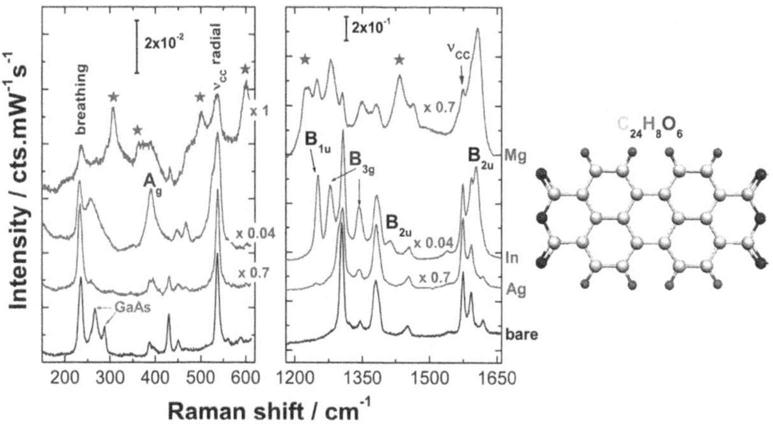

Fig. 1. Raman spectra of In (5nm), Ag (4.5 nm) and Mg (5 nm) coverages on 15 nm thick PTCDA films, compared with the spectrum of the bare PTCDA film in the spectral region of the internal breathing mode (*left*) and in the spectral region of C-H deformation and C=C stretching modes (*right*)

of the whole molecule and C-C stretching modes [18]. The frequency region between 1200 cm^{-1} and 1650 cm^{-1} contains internal modes with dominant C-H and C-C character.

3 Interaction of Metals with Perylene Derivatives

In Fig. 1 the Raman spectra of 15 nm films of PTCDA are shown for metal coverages of 5nm In, 4.5 nm Ag, and 5 nm Mg. The spectra in the low frequency windows are normalized to the height of the molecular breathing mode at 233 cm^{-1}. The normalization in the high frequency region is performed with respect to the C=C stretch mode (1572 cm^{-1}).

The deposition of Ag and In onto PTCDA leads to the appearance of the B_{1u} band at 1243 cm^{-1} and the increase in the relative intensity of the B_{3g} mode at 1338 cm^{-1}. These bands are strongly enhanced compared to the A_g modes during the Ag deposition onto a monolayer of PTCDA on S-GaAs [3], therefore they are considered to be a signature of the "first layer effect", *i.e.* of the molecules having direct contact with Ag [3]. The band at 1338 cm^{-1} was identified in reference [19] to be a B_{3g} band based on its frequency and intensity in the crystal spectra, while a band 1292 cm^{-1} is likely to be a shifted variant of the C-H deformation A_g mode at 1303 cm^{-1} in the single crystal [20]. The other bands correspond to modes that normally show infrared activity. These modes are also activated in the spectra of PTCDA monolayer covered with indium, only with higher intensities relative to those of the normally Raman active modes[3]. In reference [3] the observed break-down of the Raman-infrared selection rules was proposed to originate from a weak charge transfer between the molecules and the metal surface mediated by molecular vibrations. The Raman spectra of the (5 nm) Mg / (15 nm) PTCDA system also exhibit the break-down of selection rules, with the occurrence of the modes observed in the other two metal/organic heterostructures (see Fig.1). In addition, several modes with significant intensity appear at: 307 cm^{-1}, 502 cm^{-1}, 598 cm^{-1}, 696 cm^{-1}, 1088 $cm^{-1}$and at 1225 cm^{-1}, and 1433 cm^{-1}. The assignment of these modes cannot yet be done unambiguously. The frequency of the mode at 598 cm^{-1} is very close to the calculated value of 592 cm^{-1} for a B_{3g} mode of an isolated PTCDA molecule [18]. Frequency calculations performed with the same basis set and density functional method in Gaussian'98 as in reference [18] but for a modified PTCDA molecule, in which the central O atom in the anhydride group is removed, deliver several frequencies that may be candidates for the assignment of the experimentally observed modes: 308 cm^{-1}, 500 cm^{-1}, 581 cm^{-1}, 702 cm^{-1}, 1090 cm^{-1}. Raman active modes in MgO microcrystals were observed at 595 cm^{-1}, 719 cm^{-1} and 1096 cm^{-1} [21]. Thus the modes observed in the present work at 598 cm^{-1}, 696 cm^{-1} and 1088 cm^{-1}may also indicate the formation of MgO as a result of the interaction between Mg and PTCDA. Calculations for other reaction scenarios are currently being performed. No modes of PTCDA or the modified molecule are found in the vicinity of 1225 cm^{-1}. Whatever the final assignment of the new modes is, they are not activated in the molecules in contact with either Ag or with In. Therefore it can be concluded that the model of weak charge transfer is not sufficient to describe the interaction involved at the Mg/PTCDA interface. Indeed, recent PES measurements performed during the Mg deposition onto PTCDA show that the C1s and O1s core levels undergo dramatic changes which can only be explained by the breaking of the C-O bonds in the PTCDA molecule with formation of MgO [22]. In Fig.2 the spectra of 15 nm DiMe-PTCDI films for metal coverages of 5nm In, 4,5 nm Ag and 5 nm Mg are shown. The spectra in the low frequency windows are normalized to the height of the breathing mode at 221 cm^{-1}. The normalization in the high frequency region is performed with respect to the C-C stretch modes (1570 cm^{-1}). In the case of DiMe-PTCDI all the investigated metals, *i.e.* Ag, In and Mg, lead to the break-down of selection rules with the occurrence of normally infrared active modes at 1246 cm^{-1} and 1606 cm^{-1}. The breathing mode at 221 cm^{-1} survives with increasing metal coverage. Thus a chemical reaction between these metals and the O atoms of DiMe-PTCDI molecules can be ruled out [3]. Interestingly, the features potentially assigned to MgO phonons do not appear in the spectra even for higher coverages of Mg. It can thus be concluded that the imide-methyl group in the DiMe-PTCDI is less reactive compared to the O atoms in the anhydride group of PTCDA.

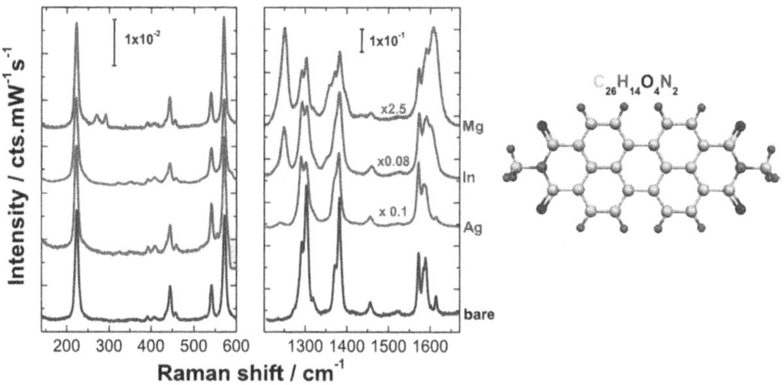

Fig. 2. Raman spectra of In (5nm), Ag (4.5 nm) and Mg (6 nm) coverages on 15 nm thick DiMe-PTCDI films, compared with the spectrum of the bare DiMe-PTCDI film

4 Morphology of the Metal Film

Besides the occurrence of internal modes related to molecules in direct contact with the metal the totally symmetric modes are also enhanced (as shown by the normalization factors in Fig. 1 and Fig. 2) in the spectra of Ag and In on 15 nm thick PTCDA and DiMe-PTCDI films. The latter effect originates from the coupling between the incident and scattered radiation with localized and/or collective plasmon resonances in the rough metal film. Therefore the intensity of the totally symmetric A_g modes is very sensitive to the morphology of the metal film. For a quantitative determination of the enhancement factors curve fitting of each set of spectra recorded during silver, indium and magnesium deposition onto PTCDA and DiMe-PTCDI was performed using Lorentzian peaks. The dependence of relative area on metal coverage is plotted in Fig. 3 for a representative totally symmetric mode and for a normally infrared active mode of each organic material. The relative intensities were calculated by dividing the intensities in the spectra at a given coverage to the initial intensities in the spectrum where the mode occurs for the first time. For example, the reference spectrum for the totally symmetric Raman band is that of the pure organic film, while the reference spectrum for normally infrared active band is that taken after the first metal deposition. The intensities of the A_g modes initially increase upon Ag and In deposition reflecting an increase in number and size of metal clusters as their plasmon energy approaches the energy of the laser electromagnetic field. When Mg is deposited onto PTCDA the intensities initially decrease, reflecting a reduction in number of Raman active PTCDA molecules. This is in good agreement with the conclusion drawn in the previous section regarding the disruption of PTCDA molecular structure upon reaction with Mg. Above 2.8 nm Mg nominal coverage, however, the A_g Raman modes start to be enhanced, in-

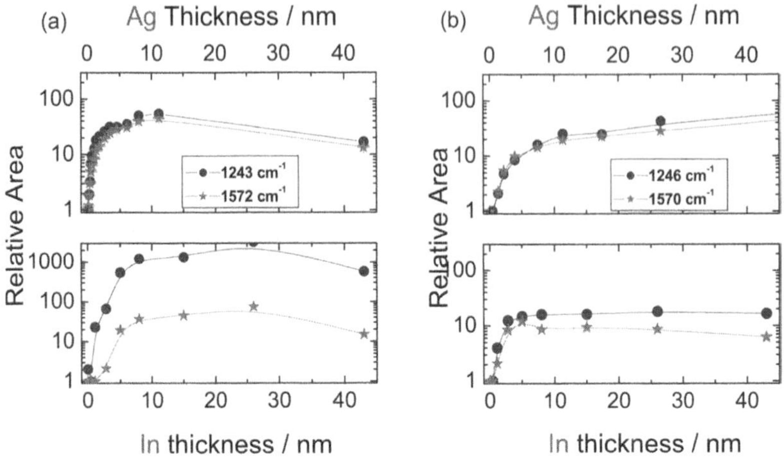

Fig. 3. Enhancement factors of the Bu mode (1243 cm^{-1} in PTCDA and 1246 cm^{-1} in DiMe-PTCDI) and of the C-C stretch Ag mode (1572 cm^{-1} in PTCDA and 1570 cm^{-1} in DiMe-PTCDI) for PTCDA (*left*) and DiMe-PTCDI (*right*) as a function of the metal coverage

dicating the formation of metallic clusters. Interestingly, the enhancement of the DiMe-PTCDI modes occurs only above 20 nm nominal Mg thickness. The large difference in nominal thickness for which the metallic character of Mg clusters is formed on PTCDA and DiMe-PTCDI is probably related to the different morphology of the underlying organic layer. The DiMe-PTCDI films have very large empty spaces between the organic islands, while the PTCDA films are much more compact [23].

The maximum enhancement of PTCDA modes for the Ag/PTCDA (15 nm) system is observed around 11 nm nominal Ag coverage (Fig. 3) and corresponds to the optimum cluster size for the dipolar plasmon resonance. The In film thicknesses yielding the maximum enhancement for PTCDA and DiMe-PTCDI films are 26 nm and 5 nm, respectively. Further increase in the metal thickness leads to increasing size of the metal clusters associated with decreasing strength of the plasmon coupling with the incident and scattered radiation. Furthermore, the absorption in the metal film also plays an important role in decreasing the Raman signal for higher nominal coverages, when the clusters start to percolate. The signal from PTCDA and DiMe-PTCDI internal modes remains visible even for a metal coverage of 43 nm, with higher intensity compared to the pure organic film. For Ag deposition onto DiMe-PTCDI no saturation of the signal intensity was observed up to a coverage of 263 nm. Considering that I_0 is the intensity of the light incident on the sample, $d$is the nominal thickness of the metal coverage and δ is the light penetration depth in the metal than the light intensity I scattered by the sample can be described by:

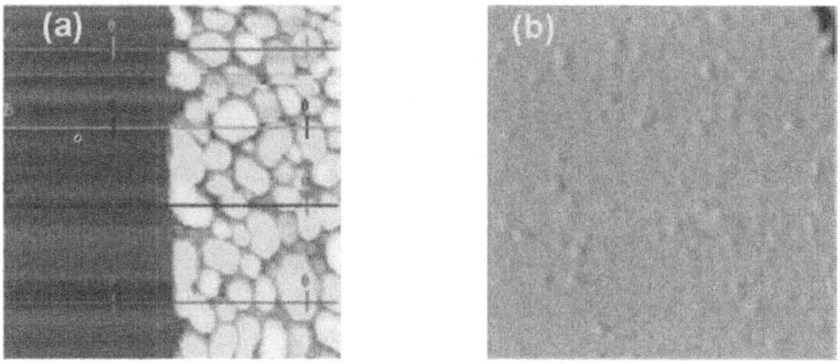

5μm × 5 μm   10μm × 10 μm

Fig. 4. AFM topographic images of a 30 nm thick In film on PTCDA (a) (right part of the image is covered with In) and of a 113 nm thick Mg film on PTCDA (b)

$$I \propto I_0 \cdot e^{-2 \cdot \frac{d}{\delta}}$$

A summary of the values obtained from the fitting of the experimental decay of the enhancement factors for the totally symmetric C=C stretching mode in all investigated heterostructures is given in table 1. The obtained values are much larger compared to the penetration depth of 488 nm light into smooth closed metal films. This is a clear indication that In and Ag films grown on PTCDA and DiMe-PTCDI are not closed and have a high degree of roughness. The apparent light penetration depth in Mg films grown on PTCDA and DiMe-PTCDI estimated from the decrease in intensity of the C=C stretching mode has comparable values with the penetration depth in a closed smooth Mg film. This indicates that the Mg film is smoother and that the efficiency of the 488 nm radiation in exciting dipolar resonances is lower for Mg. The AFM topographic images in Fig. 4 confirm the higher roughness of In compared to that of Mg films.

Table 1. Skin depth of smooth metallic films, apparent penetration depth of 488 nm light in In, Ag and Mg films grown on DiMe-PTCDI and PTCDA

	In	Ag	Mg
$\delta_{\lambda=488nm}$ smooth metal film	8 nm	2.5 nm	14 nm
$\delta_{\lambda=488nm}$(DiMe-PTCDI)	49 nm	50 nm	15 nm
$\delta_{\lambda=488nm}$(PTCDA)	98 nm	-	24 nm

5 Structural Properties of the Metal/Organic Interfaces

To recall, the A_g modes are enhanced via the long range electromagnetic effect, while the activation of B_u modes is characteristic for the molecules in

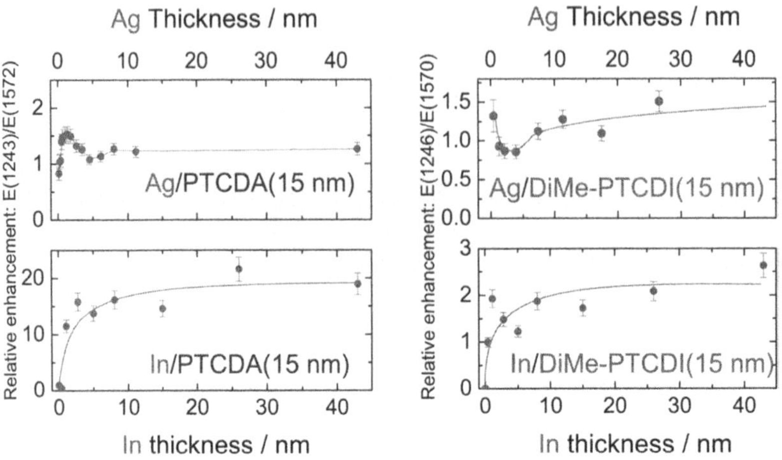

Fig. 5. Ratio between enhancement factor of the Bu mode (1243 cm^{-1} in PTCDA and 1246 cm^{-1} in DiMe-PTCDI) and of the C-C stretch Ag mode (1572 cm^{-1} in PTCDA and 1570 cm^{-1} in DiMe-PTCDI) as a function of the metal coverage

intimate contact with the metal. Therefore the intensity of B_u modes relative to that of A_g modes will be considered in the following to extract the metal diffusion depth into the organic films. In the case of Ag/PTCDA and Ag/DiMe-PTCDI the intensity of B_u modes is relatively low, indicating that only few molecules have intimate contact with Ag. This leads to the conclusion that Ag atoms diffuse very little into PTCDA layers. On the other hand, the B_u bands are stronger compared to the A_g modes in the spectra of In/PTCDA. This suggests that a large number of PTCDA molecules have intimate contact with the metal indicating a strong diffusion of In into PTCDA layers. In/DiMe-PTCDI represent an intermediate case between Ag/PTCDA and In/PTCDA.

The ratio between the area of the B_u mode at 1243 cm^{-1}(1246 cm^{-1}) and that of the A_g mode at 1572 cm^{-1}(1570 cm^{-1}) in PTCDA(DiMe-PTCDI) is shown as a function of metal thickness in Fig. 5. In the case of Ag/DiMe-PTCDI the maximum value is observed for the first Ag deposition, *i.e.* 0.4 nm Ag, whereas for PTCDA it increases up to a 1.4 nm nominal coverage of Ag. For In deposition onto both organics this ratio shows a saturation tendency only above 15 nm nominal In coverage, but its value is lower for In/DiMe-PTCDI. It is proposed that a maximum in the above defined ratio can be directly related to the metal diffusion length in the organic film. Thus the Ag atoms arriving at the organic film surface diffuse into the PTCDA or DiMe-PTCDI islands up to a nominal Ag coverage of 1.4 nm and 0.4 nm, respectively.

The conclusions regarding diffusion of the metal atoms into the organic polycrystalline layers drawn from the enhancement factors of the internal

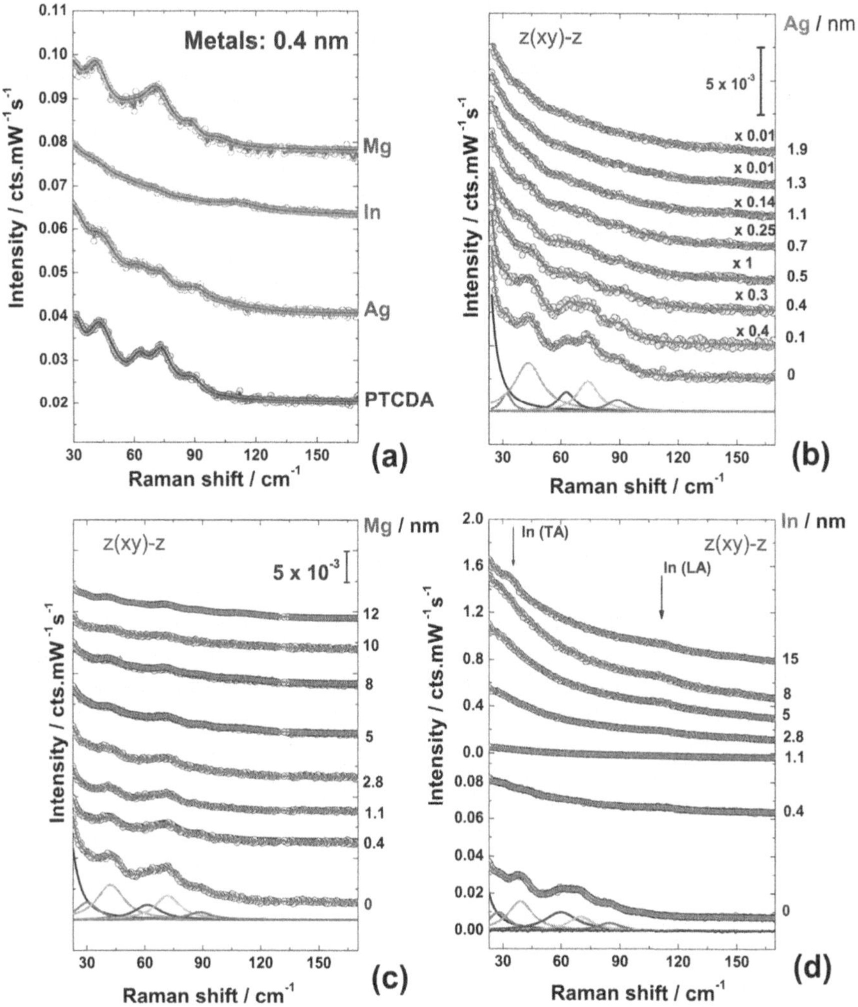

Fig. 6. (a) Raman spectra of 15 nm thick PTCDA films covered with Ag, In and Mg in the region of the external modes. Raman monitoring in the external mode region upon metal deposition: (b) Ag, (c) Mg, (d) In. The experimental spectra are shown by open symbols and the fitted spectra by red lines. The Lorentzian functions used for curve fitting of the Raman spectrum of the pure PTCDA film are shown by lines in the lower part of the figures. The spectra are normalized for Ag/PTCDA for a better observation of the phonons

modes are further confirmed by the spectral changes in the region of external modes of the organic layers. The external modes of PTCDA disappear almost completely already after depositing 0.4 nm In, while they only get broader and decrease in intensity after Ag and Mg deposition (Fig. 6). In Fig. 7 the spectra of the external modes are shown upon step-wise metal deposition onto 15 nm of PTCDA. For a quantitative evaluation the spectra of Ag/PTCDA and Mg/PTCDA were fitted using Lorentzian functions. The evolution of the FWHM as a function of Ag and Mg thickness is plotted in Fig. 7 for the external mode at 41 cm^{-1}. This mode is fairly well separated from its neighbours and hence the fitting parameters of the corresponding Lorentzian function are less correlated. As the metal thickness increases, the FWHM of the external modes increases faster in Ag/PTCDA compared to Mg/PTCDA (Fig. 7). For Mg the external modes are still visible at 12 nm coverage, whereas they are almost completely smeared out at 1.3 nm of Ag. This is a clear indication that the crystalline structure of the organic layers is less affected by the Mg deposition compared to Ag. However, it should be noted that curve fitting of the spectra in the case of Ag/PTCDA is complicated by the significant increase in the low frequency background (which was already subtracted in Fig. 6). The background evolution reflects an increasing degree of roughness, which is consistent with an increasing number of metallic clusters that diffusely scatter the light. A strong increase in the low frequency background is also observed for the case of In deposition onto PTCDA, while it hardly affects the spectra of Mg/PTCDA, supporting that the roughening due to Mg is lower compared to that of the Ag and In.

While the external molecular modes disappear already in the first deposition stages for In/PTCDA, two new modes develop at 33 cm^{-1} and 112 cm^{-1} above In coverage of 2.8 nm. They may correspond to the transverse acoustic and longitudinal acoustic phonons located at 34 cm^{-1} and 114 cm^{-1}, respectively, in bulk indium [24]. This observation corroborated by the concomitant increase in the low frequency background indicates the formation of metallic In clusters. Moreover, the enhancement of the internal modes also increases dramatically above 2.8 nm In, supporting the conclusion of metallic cluster formation.

6 Summary

Ag, In and Mg deposition onto 15 nm thick PTCDA and DiMe-PTCDI films on S-GaAs(100):2x1 was characterized *in situ* by Raman spectroscopy. The break-down of selection rules in the spectra of Ag/organic and In/organic heterostructures originates from a dynamical fractional charge transfer process modulated by molecular vibrations. In the spectra recorded during the Mg deposition onto PTCDA bands assigned to MgO develop, indicating a reaction of Mg with PTCDA molecules that causes the loss of the central O atom of the anhydride groups. Upon Ag and In deposition the totally symmetric

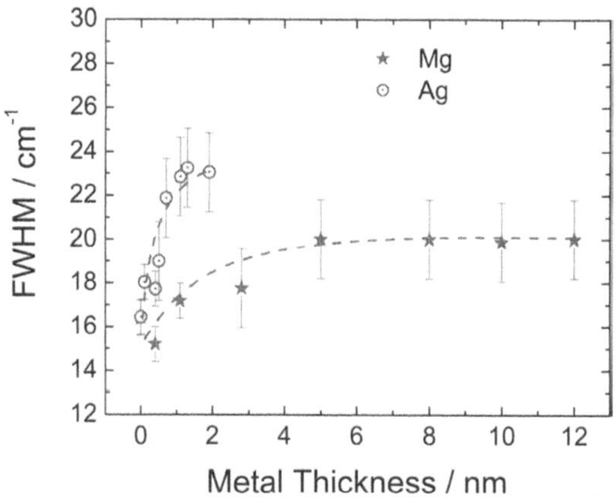

Fig. 7. Evolution of the FWHM of the external mode at 41 cm^{-1} as a function of the metal coverage relative to the initial values before the metal deposition: for Ag (a) and Mg (b). The dashed lines are a guide for the eyes

modes are initially strongly enhanced. Subsequently the signal is attenuated exponentially with an exponent that is much smaller than the penetration depth of the incident radiation in a smooth closed metal film, reflecting a high level of roughness of the metal overlayer. The intensity of the normally infrared active modes relative to the Raman active modes provides information on the metal diffusion depth in the organic films. Complementary information on the metal diffusion depth into the organic layers is provided by the attenuation rate of the external mode intensities as a function of the metal coverage. While Mg and Ag form abrupt interfaces, In strongly diffuses into the organic layers. For Mg, however, additional features are observed in the Raman spectra compared to those induced by Ag and In clearly indicating the reactive nature of the interface between Mg and PTCDA.

Acknowledgements

This work was supported by the EU funded Human Potential Research Training Network DIODE (Contract No.: HPRN-CT-1999-00164) and by the Deutsche Forschungsgemeinschaft within the OFET SSP (Project No. SPP 1121/ Za146/4-2).

References

1. S. Möller, C. Perlov, W. Jackson, C, Taussig, S. R. Forrest, Nature **426**, 166 (2003).
2. P. R. L. Malenfant, C. D. Dimitrakopoulos, J. D. Gelorme, L. L. Kosbar, T. O. Graham, A. Curioni , and W. Andreoni, Appl. Phys. Lett. **80**, 2517 (2002).

3. D. R. T. Zahn, G. Salvan, B. A. Paez, R. Scholz, J. Vac. Sci. Tech. A **22**, 1482 (2004).
4. S. R. Forrest, M. L. Kaplan, P. H. Schmidt, J. Appl. Phys. **56**, 543, (1984).
5. X. Crispin, V. Geskin, A. Crispin, J. Comil, R. Lazzaroni, W. R. Salaneck, J. L. Bredas, J. Am. Chem. Soc. **124**, 8131, (2002).
6. Y. Hiroshe, A. Kahn, V. Aristov, P. Soukiassian, V. Bulovic, S. R. Forrest, Phys. Rev. B **54** 13748 (1996).
7. R. Scholz, A.Yu. Kobitski, T.U. Kampen, M. Schreiber, D.R.T. Zahn, G. Jungnickel, M. Elstner, M. Sternberg, and Th. Frauenheim, Phys. Rev. B **61**, 13659 (2000).
8. A. Yu Kobitski, R. Scholz, D. R. T. Zahn, J. Molec. Struct. (Theochem.) **625**, 39 (2003).
9. D. A. Tenne, S. Park, A. Das, T. U. Kampen, D. R. T. Zahn, Phys. Rev. B **61**, 15464 (2000).
10. F. S. Tautz, S. Sloboshanin, J. A. Schaefer, R. Scholz, V. Shklover, M. Sokolowski, E. Umbach, Phys. Rev. B **61**, 16933 (2000).
11. F. S. Tautz, M. Eremtchenko, J. A. Schaefer, M. Sokolowski, V. Shklover, E. Umbach, Phys. Rev. B. **65**, 125405 (2000).
12. S. Kera, H. Setoyama, M.Onoue, K. Okudaira, Y. Harada, N. Ueno, Phys. Rev. B. **63**, 115204 (2001).
13. S. Park, PhD-Thesis Chemnitz http://archiv.tu-chemnitz.de/pub/2002/0004/index.html, (2002).
14. G. Salvan, D. A. Tenne, T.U. Kampen, R. Scholz, G. Jungnickel, Th. Frauenheim, D. R. T. Zahn, Appl. Surf. Sci. **179**, 113 (2001).
15. F. Brouers, S. Blacher, A. N. Lagarkov, A. K. Sarychev, P. Gadenne and V. M. Shalaev. Phys. Rev. B **55**, 13234 (1997).
16. J. A. Creighton, in: *Spectroscopy of Surfaces*, Chapter 2, *The selection rules for Surface-Enhanced Raman Spectroscopy*, edited by R. J. Clark and R. E. Hester (John Wiley and Sons Ltd., London 1988).
17. A. Otto, I. Mrozek, H. Graborn, W. Akemann, J. Phys.: Condens. Matter. **4**, 1134 (1992).
18. A. Yu Kobitski, R. Scholz, D. R. T. Zahn, J. Molec. Struct. (Theochem.) **625**, 39 (2004).
19. A. Yu. Kobitski, G. Salvan, R. Scholz, D. Tenne, T. U. Kampen, H. P. Wagner, D. R. T. Zahn, Appl. Surf. Sci. **190**, 386 (2002).
20. G. Salvan, D. R. T. Zahn, Europhys. Lett. **67**, 827 (2004).
21. H.K. Böckelmann and R.G. Schlecht, Phys. Rev. B **10**, 5225 (1974).
22. G. Gavrila, D. R. T. Zahn et al. in preparation.
23. M. Friedrich, G. Gavrila, C. Himcinschi, T. U. Kampen, A. Yu. Kobitski, H. Mendez, A. Paraian, S. Park, G. Salvan, D. R. T. Zahn, J. Phys. Cond. Mater. **15**, S2699 (2003).
24. K. Fleischer, S. Chandola, N. Esser, W. Richter, J. F. McGilp, Phys. Rev. B **67**, 235318 (2003).

Part V

Materials and Applications

The Role of Ion Irradiation Induced Viscous Flow in Thin Film Structuring: Hard Matter Going Soft

S. G. Mayr

I. Physikalisches Institut, Georg-August-Universitt Göttingen,
Friedrich-Hund-Platz 1, 37077 Göttingen, Germany
smayr@gwdg.de

Summary. Ion beams provide unique opportunities for modifying thin film structural and mechanical properties on length scales ranging from several nanometers to macroscopic dimensions. In a variety of materials, including polycrystalline and amorphous metals and semiconductors, plastic effects can dominate, which originates from processes in the small volumes of materials affected by a single ion impact. The underlying atomic-scale kinetics as well as its applications for self-organized structuring are investigated using a combination of experiments and molecular–dynamics (MD) computer simulations: We report experiments, that illustrate the impact of radiation induced viscous flow on morphology and stresses, such as interface smoothing, stress relaxation, radiation induced sintering or self–organized pattern formation. Using MD simulations we study the underlying atomic–scale mechanisms for the two energy regimes, where local melting along the ion track is either dominant or not. We show, that macroscopic plastic flow does not require melting, indicating, that defect–like entities are sufficient to mediate plastic flow.

1 Introduction

Since the seminal works of Seitz and Koehler [18] almost 50 years ago, it has been known that ion bombardment does not only introduce defects into a solid within a collision cascade, but can also lead to local regions of elevated temperature, viz. thermal spikes. Both instances, defects and thermal spikes, can lead to materials modification: Defects mostly constitute highly mobile entities of strongly elevated energy, which reside even locally out thermodynamic equilibrium. Relaxation towards an energetically more favorable configuration thus can involve materials transport over rather long distances by these "freely migrating defects" [16], and can even counteract the macroscopic equilibrium formation. In contrast, thermal spikes are local lattice excitations, which preferentially occur in dense matter towards the end of collision cascades with a lifetime, which is *per definition* long in comparison to a lattice vibration, i.e. the reciprocal Debye frequency. Within this local thermodynamic equilibrium, thermal excitations of materials processes can occur, which relax local configurations towards their corresponding local

B. Kramer (Ed.): Adv. in Solid State Phys. **45**, 327–338 (2005)

equilibrium configurations at that specific temperature. Naturally the thermodynamic accessible processes within a thermal spike only include those with a sufficiently low activation energy to be relevant during the spike lifetime (some picoseconds), among them melting and plastic flow. In fact, for heavy ion bombardment of metallic PdSi glasses with ion energies in the GeV range, where electronic stopping is highly dominant, plastic deformation effects based on cylindrical liquid regions along the ion track and the ion momentum have first been reported by Klaumunzer and Schumacher [7] and interpreted by Trinkaus [19]. Despite this rather long history, ion beams continue to be an exciting tool for materials modification [2], where cutting-edge investigations often employ ions to introduce external, non-thermal driving forces - in addition to their thermodynamic counterparts - to introduce self-organized structure, pattern and stress formation at surfaces and interfaces within the framework of driven systems (e.g. [8, 3, 4, 10, 11]). Within these studies, which employ energy levels ranging from some tens of eV up to MeV, the identification of the ion induced materials processes, which underlie structure formation, is highly relevant. The present work exactly focuses on this point, by employing critical experiments, molecular dynamics computer simulations as well as continuum modeling of materials processes, which are induced by bombardment with ion energies in a range, where nuclear stopping is dominant.

2 Phenomenology of Radiation Induced Viscous Flow in Metallic Glasses

As a model system to investigate the impact of MeV ion irradiation on surface morphologies we employ amorphous $Zr_{65}Al_{7.5}Cu_{27.5}$ thin films, which are prepared by condensation from three individually rate-controlled electron beam evaporators onto SiO_2 substrates in an ultra high vacuum chamber. As a main advantage, amorphous thin films do not exhibit crystallographic anisotropies or other lattice constrains, which can complicate and masquerade the identification of materials processes in crystalline systems due to dynamic and mechanical anisotropies. This allows for a most complete understanding of the basic kinetics, which underlies structure formation.

2.1 Amorphous Film Growth from Vapor

Independent of the details of the materials systems, glassy metal film growth from vapor proves to be basically dominated by three competing kinetic ingredients, which all act on the surface at temperatures sufficiently below the glass temperature, as reported before [12]: deposition noise, a shadowing instability and surface diffusion. The deposition noise is omnipresent in any deposition process due to the randomness of particle deposition onto the

Fig. 1. Surface topographs, as measured by AFM, for a 480nm amorphous $Zr_{65}Al_{7.5}Cu_{27.5}$ film, irradiated with different doses of 1.8MeV Kr^+ [10]

substrate. The shadowing instability is based on a preferential aggregation of particles onto protrusions due to interatomic interactions and size effects, while surface diffusion is driven by gradients in the surface chemical potential due to curvature modulations. The combination of these processes gives rise to a linear growth instability, which is reflected by a rough, columnar-like surface morphology with a characteristic structure size [12], which corresponds to the most unstable wave number of the instability, as seen in Fig. 1a. Concurrently with rough film growth, strong tensile stresses are generated, due to a liquid-like zipping of the top domes of the surface corrugations [13], resulting in a tensile stress generation [1] of −0.58GPa.

2.2 Impact of Ion Irradiation on Morphology

When rough a-$Zr_{65}Al_{7.5}Cu_{27.5}$ films are irradiated with 1.8MeV Kr^+ ions, the surface morphology $h(\boldsymbol{x})$ ($\boldsymbol{x} = (x, y)$ in the substrate plane) smoothens out pretty dramatically, as shown in Fig. 1, resulting in almost atomically smooth films at a dose of $\approx 2 \times 10^{16}\text{cm}^{-2}$. Fig. 2a shows the corresponding evaluation of the root mean square (rms) roughness

$$\xi = \sqrt{\langle h^2 \rangle_{\boldsymbol{x}} - \langle h \rangle_{\boldsymbol{x}}^2} \tag{1}$$

[1] In our convention, compressive stresses are defined to be positive.

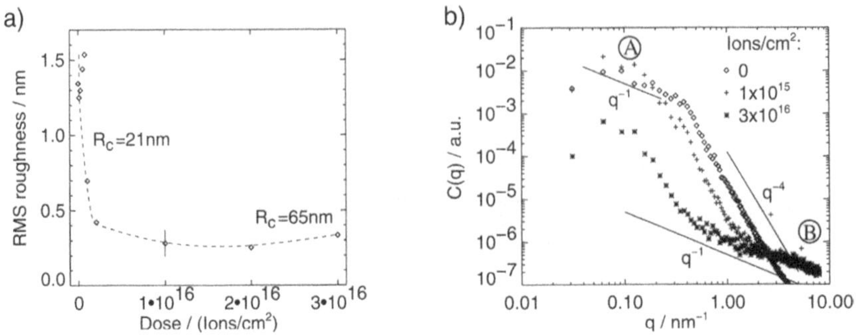

Fig. 2. (a) Evolution of surface roughness and lateral structure size, R_C, in dependence of the radiation dose. (b) Radially averaged spectral power densities for unirradiated and irradiated amorphous $Zr_{65}Al_{7.5}Cu_{27.5}$ films [10]

as a function of dose, as well as the spectral power density,

$$C(q) = \left\langle \left| \hat{h}\left(\boldsymbol{q}\right) \right|^2 \right\rangle_{|\boldsymbol{q}|=q} . \tag{2}$$

Here $\hat{h}\left(\boldsymbol{q}\right)$ denotes the Fourier transfomation of $h(\boldsymbol{x})$. The latter has the exciting property to converge towards a steady state, which is characterized by power law behavior, $C(q) \propto q^{-i}$, starting from high to low wave numbers [20]. Here, i is an integer, which is characteristic for the dimensionality of the dominant smoothening process, that is, $i = 1$ for viscous flow and $i = 4$ for curvature driven surface diffusion [5]. As seen in Fig. 2b, the *as grown* sample reaches a steady state dominated by curvature driven surface transport for high enough wave numbers. Upon ion bombardment, the slope of $C(q)$ on a double logarithmic scale changes almost instantaneously to $i \approx 1$, which is characteristic for radiation-induced viscous flow; after bombardment with sufficiently high doses ($3 \times 10^{16}\mathrm{cm}^{-2}$) viscous flow is observed for $q \gtrapprox 0.5\mathrm{nm}^{-1}$, or, equivalently, for length scales smaller than ≈ 12.6nm. This is interpreted as the maximum extent of the thermal spike, beyond which smoothing is impossible. Following the works of Mullins [14], the radiation induced fluidity, H (reciprocal of the radiation induced viscosity $\eta \cdot \dot{\Phi}$) is readily calculated as

$$H = \left(\gamma \cdot q \cdot \Phi_S\right)^{-1} , \tag{3}$$

where γ denotes the surface energy and Φ_S is the dose required for saturation at wave numbers of q and higher. If the dose is normalized to damage production (i.e. measured in displacements per atom (dpa)), $H = (1.1 \pm 1.0)$(Pa dpa)$^{-1}$ is obtained, estimating $\gamma \approx 1.4$N/m. Another interesting observation is, that for wave number of $q = 2\mathrm{nm}^{-1}$ and higher, ion irradiation cases an increase of the roughness towards the steady state, which is defined by q^{-1},

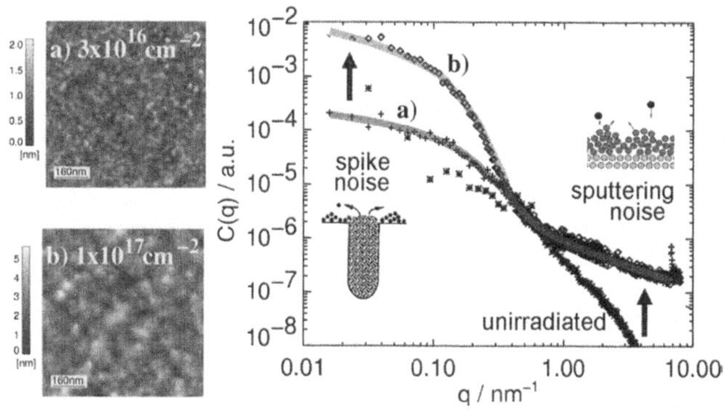

Fig. 3. At high doses, smooth films roughen. The roughness is created on two different length scales: the atomic scale ("sputtering noise") and the scale of the thermal spike ("spike noise")

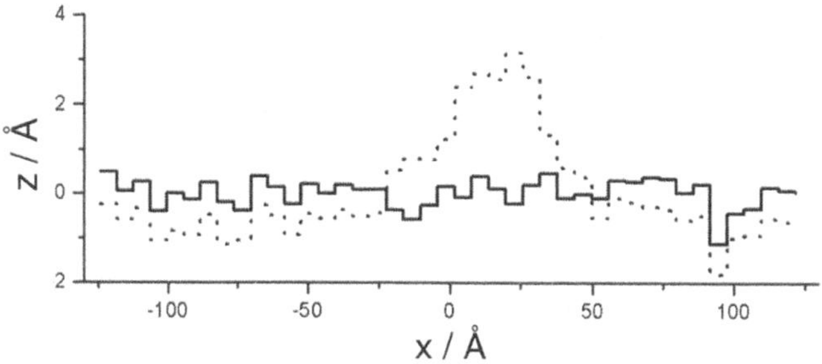

Fig. 4. Surface profile before (*solid*) and after (*dotted*) bombardment of an a-CuTi surface with a 40keV Xe^+ ion. Simulations performed in cooperation with Y. Zhong, University of Illinois at Urbana-Champaign

i.e. for length scales of $\approx$ 3nm and smaller. This is due to the competing dynamics of roughening by random erosion by sputtering ("sputtering noise") and viscous smoothening dynamics.

Smoothening does not continue infinitely, but ceases at $\Phi \approx 2 \cdot 10^{16}\text{cm}^{-2}$; higher doses cause roughening beyond that threshold. To identify the origin of roughness creation, the roughness changes, as decomposed in Fourier modes, are visualized in Fig. 3. It is readily seen, that the roughness increases on two different length scale: On the atomic scale, as well as on a scale of the size of the thermal spike and beyond. To trace down the underlying mechanism, molecular dynamics computer simulations on 40keV Xe irradiation of a-CuTi (using a potential based on Sabochick and Lam – see Section 3) are performed, and the change in surface morphology is monitored during the course of the irradiation event. From Fig. 4 it is clearly evident, that the melting and thermal expansion of thermal spikes in surface proximity leads

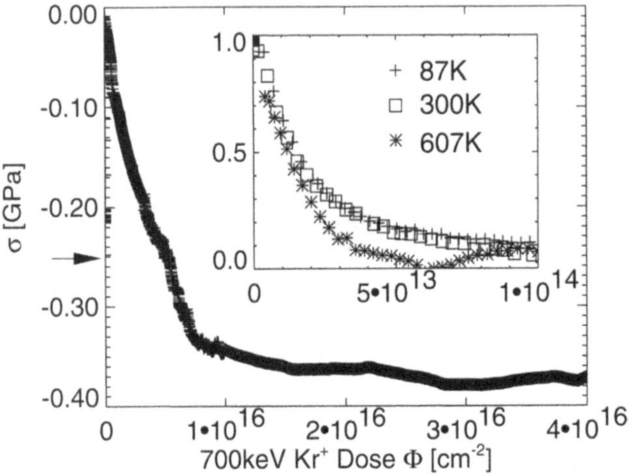

Fig. 5. Biaxial film stresses in $Zr_{65}Al_{7.5}Cu_{27.5}$ during bombardment with 700keV Kr^+ ions. The arrow indicates the initial stress level. For low doses the stresses relax with an exponential decay (*Inset*), which is measured temperature dependent and normalized to the initial stress. Only a weak temperature dependence is detectable [11]

to material flow form inside the sample onto the surface, leaving a "bump" of material behind – as indicated schematically in Fig. 3. The roughness created by numerous individual recoils is thus accumulated, leading to pronounced surface roughening.

2.3 Impact of Ion Irradiation on Mechanical Stresses

Before irradiation with 700keV Kr^+ ions, our 480nm thick a-$Zr_{65}Al_{7.5}Cu_{27.5}$ films are under tensile stresses (≈ -0.25GPa) due to the preparation conditions by co-condensation from vapor. During bombardment with low doses ($\lessapprox 1 \cdot 10^{14}\text{cm}^{-2}$), these stresses relax almost completely (Fig. 5). As seen in the inset of Fig. 5, stress relaxation follows about an exponential decay, which is greatly independent of the sample temperature. Assuming, that Newtonian viscous flow is the underlying mechanism, which relaxes the initial biaxial stress load, the relaxation behavior is readily calculated from $\frac{\sigma}{6 \cdot \eta} = B \cdot \dot{\sigma}$, where B denotes the biaxial modulus of the film. For radiation induced viscous flow, the natural time unit is the dose, i.e. the defect production, resulting in

$$\sigma = \sigma_0 \cdot \exp\left(B \cdot H \cdot \Phi / 6\right), \tag{4}$$

where H denotes the radiation induced fluidity, as defined above. The fluidities, as determined from exponential fits, result in $(3.1 \pm 0.5) \cdot 10^{-9}$,

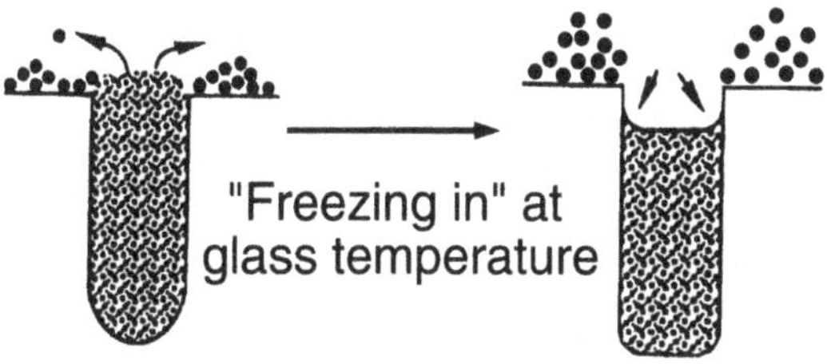

Fig. 6. Illustration of the mechanism of tensile stress generation in amorphous and nanocrystalline samples during ion irradiation

$(2.6 \pm 0.5) \cdot 10^{-9}$ and $(5.2 \pm 0.5) \cdot 10^{-9}$ $(\text{Pa dpa})^{-1}$ for 87, 300 and 607K, respectively. These values are greatly temperature independent and compare reasonably well with the value determined from the surface smoothing reaction $((1.1 \pm 1.0)(\text{Pa dpa})^{-1})$.

For high irradiation doses, tensile stresses (≈ -0.39GPa) are generated, which are interpreted within the picture of material flow from inside the solid onto the surface, as indicated by the MD simulations in Fig. 4 and illustrated in Fig. 6 [1, 11]: Thermal spikes in proximity to the surface lead to material flow onto the surface; during subsequent resolidification, the configuration is frozen–in at the corresponding glass temperature. Cooling down to the sample temperature leads to tensile stresses. Using this rather simple picture, the expected stress levels are readily estimated from $\sigma \approx B \cdot \nu \cdot \Delta T$, where B denotes the biaxial modulus, ν the coefficient of thermal expansion and ΔT the difference between glass and sample temperature. With $T_G \approx 640$K, an expected stress level of -0.48GPa is predicted, which compares reasonably well with the measurement.

3 Atomic-Scale Mechanism of Radiation Induced Viscous Flow

To identify the atomic-scale mechanisms, which underlie radiation induced viscous flow in metallic amorphous samples, classical molecular dynamics computer simulations [15] are performed on the model glass, CuTi, using the interatomic potential of Sabochick and Lam [17]. The short-range interactions of the potential were splined to the Ziegler-Biersack-Littmark [21] universal potential using a third degree polynomial, while the electron densities in the corresponding region were smoothly joined away, as described before [15]. Amorphous simulation cells are prepared by successive heating and quenching cycles to obtain a most relaxed amorphous simulation cell. Prior to simulated ion bombardment, the cells were loaded with a biaxial stress of 1GPa, while the z direction remained stress–free. Ion irradiation was simulated by assigning the desired recoil energy to randomly selected atoms of random type into a random direction. The relaxation of the stresses

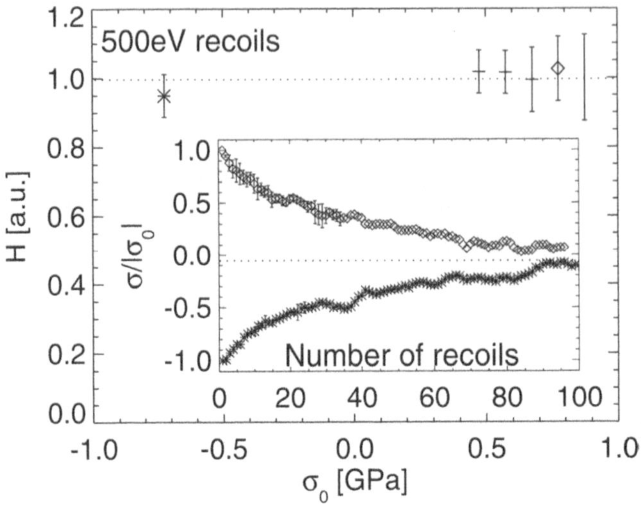

Fig. 7. The radiation induced fluidity H is calculated from the relaxation curves for different stresses and signs of the stresses, as show in the inset. The independence of the initial stress, σ_0, suggests Newtonian flow [9]

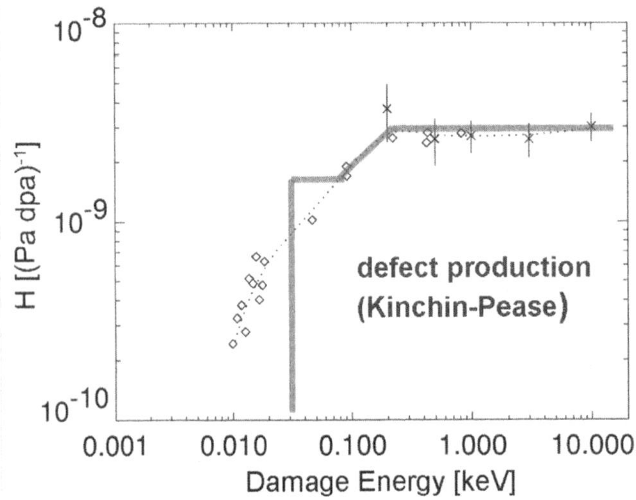

Fig. 8. The dependence of the RIF H on the recoil energy – and qualitative comparison with the Kinchin-Pease prediction for defect generation

was monitored in dependence of the sign of the original stress load and damage energy, following an exponential decay for energies of 10eV and higher - see Fig. 7. The defect production is again measured in dpa, as determined from the Kinchin-Pease expression [6], using the displacement energy, 10eV for the present potential. Similar to the experiments, the radiation induced fluidity, H, is readily calculated by an exponential fit, using values for B as determined directly from MD calculations. The fluidities as a function of damage energies are shown in Fig. 8 [9], which indicate a universal value of $\approx 3 \times 10^{-9}(\text{Pa dpa})^{-1}$ for energies higher than $\approx$ 200eV. In fact, the $H(E)$ curve of Fig. 8 closely resembles the Kinchin-Pease expression [6] of damage production, which suggests a close relation of defect production and radiation induced viscous flow. To shade some more light into this issue, two different types of defects, vacancies and interstitials, are gently inserted randomly into the amorphous, pre-stressed CuTi simulation cell, while, again, stress modification by defect insertion is monitored. It is found [9], that each individually created defect – no matter whether vacancy or interstitial – acts as local shear instability, which is capable of inducing stress relaxation in its neighbouring environment by local reorientations of the atomic configurations. For the present case of defect insertion, calculation of the corresponding defect production is trivial, which enables – again – direct calculation of the damage induced fluidity ($4 \cdot 10^{-9}(\text{Pa dpa})^{-1}$) [9], which agrees very well with the radiation induced fluidity (see Fig. 8 and section 2). This indicates, that radiation induced viscous flow in amorphous samples can be explained solely on the basis of defect production and does not require a liquid thermal spike. In fact, radiation induced viscous flow is observed in materials, such as semiconductors, and under low energy conditions, where the presence of melting within a thermal spike can surely be excluded. As investigated by detailled studies of the volumes affected by melting within thermal spikes, and defects, respectively, we came to the conclusion that the volume affected by heat spikes is always smaller than the volume affected by point defects. In fact, a liquid thermal spike — if present – is always embedded into a bigger volume [9], which is relaxed by a flow effect due to the presence of defects. Thus, within the present picture, the liquid-like heat spike simply does not count for the radiation induced fluidity.

4 Generalization to Crystalline Systems

The studies on keV to MeV heavy ion bombardment of a-$Zr_{65}Al_{7.5}Cu_{27.5}$ films suggest radiation induced viscous flow as a main mechanism, that profoundly influences surfaces, morphologies and stresses in amorphous metal films. A central issue for application of these techniques to tailor structure and stresses of functional thin films, is the question of transferibility of these results. Clearly, they are transferible to different glassy metallic systems, such as CuTi or ZrCu, which – according to our investigations – show qualitatively

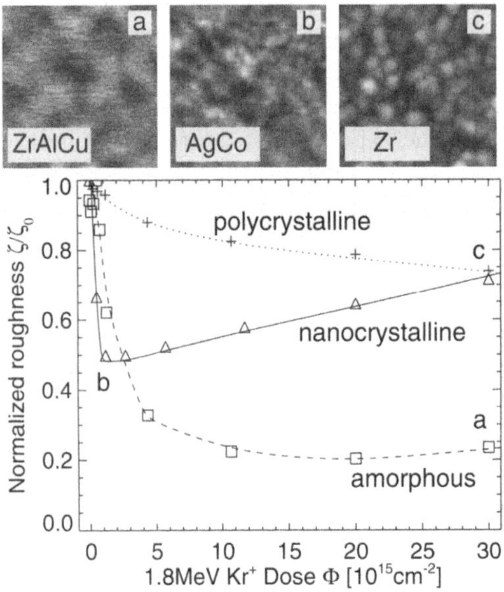

Fig. 9. Evolution of the surface morphologies of amorphous $Zr_{65}Al_{7.5}Cu_{27.5}$, nanocrystalline $Ag_{50}Co_{50}$ and columnar polycrystalline Zr films during 1.8MeV Kr^+ ion beam irradiation. Ion bombardement of amorphous films enables the creation of atomically smooth films. Nanocrystalline films initially smoothen, but roughen for higher doses due to growth of surface grains. Columnar polycrystalline films hardly change the surface morphology during irradiation [11]

identical behavior of their morphologies and mechanical stresses. To investigate, whether crystalline systems behave any different, the smoothening reaction of nanocrystalline AgCo films and columnar polycrystalline Zr films, which have been deposited by electron beam evaporation at liquid nitrogen and room temperature, respectively, is compared to the amorphous alloy in Fig. 9. In nanocrystalline AgCo the grain size is significantly smaller than the size of the thermal spikes. It is therefore not surprising, that the AgCo nanosystem behaves similar to an amorphous systems, as long as the grain size remains significantly smaller than the size of the thermal spike. In later stages, the Ag and Co grain grow by radiation induced grain growth, leading to an increase of surface roughness due to the surface grain morphology. Columnar polycrystal Zr films smoothen just slightly under the action of ion irradiation; here the roughness remains to be dominated by the top surface domes of the columnar microstructure. Two factors prevent the surface domes from significantly changing their shape: Firstly the columnar grain structure is hardly modified by energetic ions, which do not fully penetrate the film, as the grains remain firmly attached to the substrate. Secondly, the force equilibirum between grain boundary energy and surface energy determines

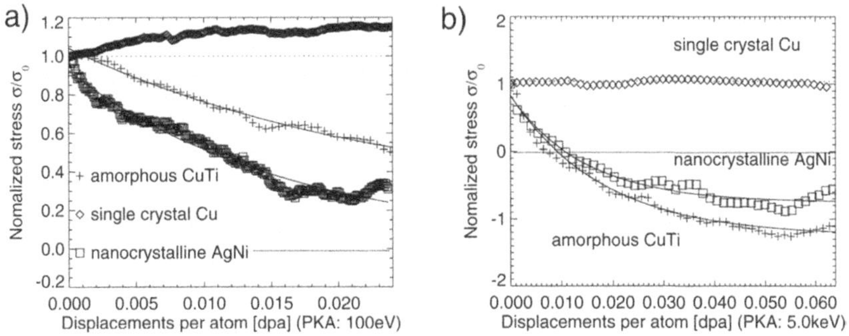

Fig. 10. MD simulation of the relaxations of the normalized biaxial film stresses σ/σ_0 by (a) 100eV and (b) 5 keV primary knockon atoms (PKA): Stress relaxation is only observable for amorphous and nanocrystalline films, and does not occur for single crystal Cu films. Instead compressive stress is built up for 100eV recoils in Cu due to defect generation. The exponential fits are employed for the determination of the (effective) radiation-induced viscosity [11]

the equilibrium angle in the triple point between two domes and the corresponding grain boundary, which prevents the surface from getting smoother.

The dependence of radiation induced stress relaxation on the thin film phase and microstructure was investigated experimentally and with MD simulations for a variety of systems. The general findings are shown exemplary in Fig. 10, which shows ion induced stress modifications by ion bombardment in an amorphous, nanocrystal and single crystal system for two different ion energies. We find that ion irradiation induced stress relaxation does only occur in systems, which are not a single crystal. This indicates the necessity for a non-conservation of lattice sites, or, equivalently, the requirement of defect sinks and sources for radiation induced viscous flow. This finding is consistent with our defect-based picture of radiation induced fluidity, as presented in section 3.

5 Conclusions

Ion irraditation induced viscous flow is a phenomenon, which can be utilized to relax and structure morphology and stresses in a broad class of materials, including amorphous, nanocrystalline and polycrystalline metals. The basic requirement is a non-conservation of lattice sites in the overall systems. This allows for a plastic deformation of the regions affected by individual cascades. Plastic deformation is triggered by the generation of point defects, which relax under the bias of any stress applied.

Acknowledgements

Financial support by the German DFG-SFB 602, TPB3, as well as a grant of computer time by the Gesellschaft für Wissenschaftliche Datenverarbeitung Göttingen (GWDG) are gratefully acknowledged.

References

1. R. S. Averback and T. Diaz de la Rubia. *Solid State Physics*, 51:281, 1998.
2. M. B. H. Breese. *MRS Bulletin, Volume 20.* Materials Research Society, Warrendale, 2000.
3. S. Facsko, T. Dekorsy, C. Koerdt, C. Trappe, H. Kurz, A. Vogt, and H. L. Hartnagel. Formation of Ordered Nanoscale Semiconductor Dots by Ion Sputtering. *Science*, 285(5433):1551, 1999.
4. F. Frost, A. Schindler, and F. Bigl. *Phys. Rev. Lett.*, 85:4116, 2000.
5. C. Herring. *J. Appl. Phys.*, 50, 1950.
6. G.H. Kinchin and R.S. Pease. *Rep. Prog. Phys.*, 18:1, 1955.
7. S. Klaumunzer and G. Schumacher. *Phys. Rev. Lett.*, 51:1987, 1983.
8. G. Martin and P. Bellon. *Solid State Physics*, 50:307, 1996.
9. S. G. Mayr, Y. Ashkenazy, K. Albe, and R. S. Averback. *Phys. Rev. Lett.*, 90:55505, 2003.
10. S. G. Mayr and R. S. Averback. *Phys. Rev. Lett.*, 87:196106, 2001.
11. S. G. Mayr and R. S. Averback. *Phys. Rev. B*, 68:214105, 2003.
12. S. G. Mayr, M. Moske, and K. Samwer. *Phys. Rev. B*, 60:16950, 1999.
13. S.G. Mayr and K. Samwer. Model for intrinsic stress formation in amorphous thin films. *Phys. Rev. Lett.*, 87:36105, 2001.
14. W. W. Mullins. *J. Appl. Phys.*, 30, 1959.
15. K. Nordlund and R. S. Averback. *Phys. Rev. B*, 56:2421, 1997.
16. L. E. Rehn, P. R. Okamoto, and R. S. Averback. *Phys. Rev. B*, 30:3073, 1984.
17. M. J. Sabochick and N. Q. Lam. *Phys. Rev. B*, 43:5243, 1991.
18. F. Seitz and J.S. Koehler. *Solid State Physics*, 2:305, 1956.
19. H. Trinkaus and A. I. Ryazanov. *Phys. Rev. Lett.*, 74:5072, 1995.
20. J. Villain. *J. Phys. I*, 1, 1991.
21. J. F. Ziegler, J. P. Biersack, and U. Littmark. *The Stopping and Range of Ions in Matter.* Pergamon Press, New York, 1985.

Atomic-Scale Properties of High-k Dielectrics: Ab Initio Study for Pr-Based Materials

Jarek Dąbrowski[1], Andrzej Fleszar[2], Gunther Lippert[1], Grzegorz Lupina[1], Anil Mane[1], Hans-Joachim Müssig[1], Thomas Schroeder[1], Roland Sorge[1], Hans Thieme[1], Christian Wenger[1], and Peter Zaumseil[1]

[1] IHP, Im Technologiepark 25, 15236 Frankfurt(Oder), Germany
[2] Universität Würzburg, Am Hubland, 97074 Würzburg, Germany

Summary. We discuss the atomic and electronic structures and energetics of native point defects and of impurities (Si, Ti, B, moisture) in $PrO_{1.5}$ and, to some extent, also in PrO_x and in $PrSiO_{3.5}$, as obtained from ab initio total energy calculations. We introduce the concept of Silicon-related Nitrogen-Coordinated Oxygen (SiNCO) which we then use to explain the origin of fixed charge in classical SiO_2 films thermally grown on Si substrates and in high-k dielectrics deposited on Si substrates.

1 Introduction

In the nearest future, a dielectric with a dielectric constant k several times higher than that of SiO_2 will be needed for the fabrication of CMOS (Complementary Metal-Oxide-Semiconductor) devices. The search for an optimal dielectric to substitute SiO_2 in transistors and capacitors used in CMOS circuits is not complete. Numerous metal oxides and silicates are investigated and various deposition and annealing techniques are developed. They try to utilize the effects attributed to alloying, incorporation of nitrogen, gettering of oxygen, etc. At the same time, the basic knowledge on the microscopic properties of these materials is poor. Ab initio calculations assist in the interpretation of experiments and in the choice of the direction of further work.

We present our fundamental understanding of point defects in Pr-based dielectrics ($PrO_{1.5}$ and $PrSiO_{3.5}$) in the context of their influence on the electrical properties of the Metal Oxide Semiconductor (MOS) structure. From this point of view, there are three major issues associated with the presence of point defects: fixed charges, Trap Assisted Tunneling (TAT) centers, and electrically active interface states. The paper addresses the first of these issues. We derive a model of Si-related fixed charge and argue that this model is valid also for typical high-k dielectrics and also for thermal SiO_2/Si films.

2 Theoretical Approach

The calculations were done by the ab initio pseudopotential plane wave code fhi96md [1, 2]. We applied the Local Density Approximation (LDA) for the

B. Kramer (Ed.): Adv. in Solid State Phys. **45**, 339–350 (2005)

exchange and correlation energy [3, 4] and nonlocal pseudopotentials in the Trouller-Martins scheme [5, 6] with 40 Ry cutoff for plane waves. Since numerous defect structure have been investigated in this work and the typical cells have no symmetry but many atoms, any low-symmetry special k-point sampling scheme would require prohibitively high numerical effort. The Brillouin zone was thus routinely sampled at the Γ point corresponding to the cell of the dimensions as close as possible to the dimensions of the $PrO_{1.5}$ cell (cube with the lattice constant of ~1 nm). Tests with more converged samplings indicate that although more exact calculations are needed to confirm the quantitative results presented here, the qualitative picture is valid.

Because of the open f-shell of Pr atoms, a key problem in calculations involving Pr is the construction of a reliable Pr pseudopotential [7, 8]. It turns out that in practice two different Pr pseudopotentials are needed: a pseudopotential with two core f electrons for $PrO_{1.5}$ (trivalent Pr(III), +3 ionic charge), and with only one f electron for PrO_x (tetravalent Pr(IV), +4 ionic charge). Thus, Pr^{+3} and Pr^{+4} are treated by us as distinct species. We calibrate the pseudopotential energy difference such that the experimental difference in formation enthalpies of $PrO_{1.5}$ and PrO_x is reproduced [9, 10]; similarly, the chemical potential of Pr metal is not calculated solely from first principles but obtained from the computed total energy of $PrO_{1.5}$ and the experimental formation enthalpy of this compound, whereby the chemical potential of O_2 is adjusted such that the experimental formation enthalpy of SiO_2 is reproduced by the calculation (the required correction is of the order of 0.3 eV). The fundamental bulk properties (lattice constant, bulk modulus) of $PrO_{1.5}$ and PrO_x obtained with the Pr pseudopotentials used here are in agreement with experimental data; the discrepancies are well within the range typical for LDA calculations. The drawback of the approach adapted here for Pr pseudopotentials is that the wavefunctions responsible for this electron transition level cannot be computed. On the other hand, since all f electrons are now contained in the core, we are not troubled by the fact that, by the electron counting rule, $PrO_{1.5}$ would be an f-band metal in LDA (but it is an insulator due to strong electron correlation effects in the f shell).

Defect formation energies in a compound (and, generally, impurity formation energies) depend on the chemical potentials of the components. In particular, for the native defects in Pr oxides we have, for X = (Pr, O):

$$\text{X interstitial:}\quad G_{\mathrm{f}}(\mathrm{X}_I) = G_f^{\mathrm{o}}(X_{\mathrm{I}}) - \mu(X), \tag{1}$$

$$\text{X vacancy:}\quad G_{\mathrm{f}}(\mathrm{X}_V) = G_f^{\mathrm{o}}(X_{\mathrm{I}}) + \mu(X), \tag{2}$$

$$\text{equilibrium with } \mathrm{PrO}_{1.5}\text{:}\quad \mu(Pr) + 1.5\mu(\mathrm{O}) = G_f^{\mathrm{o}}(\mathrm{PrO}_{1.5}), \tag{3}$$

where G_f^{o} are the standard (i.e., corresponding to room temperature and atmospheric pressure) free energies of formation. Since we compute total energies at zero K, we use for calibration and comparison with experiment the corresponding formation enthalpies rather than free energies.

The important regimes of the chemical potential of oxygen are:

$$\begin{aligned}
PrO_{1.5} \text{ in contact with Pr metal:}\quad & \mu(O) = G_f^o(PrO_{1.5})/1.5 \\
PrO_{1.5} \text{ in contact with } SiO_2/Si\text{:}\quad & \mu(O) = G_f^o(SiO_2)/2 \\
PrO_{1.5} \text{ in contact with } PrO_2\text{:}\quad & \mu(O) = 2(G_f^o(PrO_2) - G_f^o(PrO_{1.5})) \\
PrO_2 \text{ in contact with air:}\quad & \mu(O) = 0
\end{aligned}$$

Since point defects are usually charged, we must also consider the dependence of the defect formation energy on the electron chemical potential, that is, on the Fermi energy E_F:

$$\text{positive charge } n_+ > 0\text{:}\quad G_f(n_+, E_{X_F}) = G_f(n_+, 0) + n_+ E_{X_F} \qquad (4)$$

$$\text{negative charge } -n_- < 0\text{:}\quad G_f(n_-, E_{X_F}) = G_f(n_-, 0) - n_- E_{X_F} \qquad (5)$$

This means that the formation energy of charged defects in a dielectric in electrical contact with the Si substrate is determined by the position of the Fermi level in the substrate and by the valence band offsets between Si and the dielectric. Since the latter is affected by the electrical dipole moment at the dielectric/substrate interface, the chemical character and electrical quality of the interface may have a noticeable effect on the defect formation energies and, consequently, on the defect population in the dielectric film.

Here we assume the valence band offset between Si and $PrO_{1.5}$ to be about 1.8 eV. One should keep in mind that this is an estimated value.

3 Fixed Charges

A fixed charge means that there is a charged ion somewhere in the dielectric. This charge is fixed in the electrical sense, that is, the ion cannot be re-loaded by charge carriers. The ions which may move in external electric fields are still "fixed" charges from electrical point of view. Conceptually, fixed charge centers in a dielectric film behave as shallow dopants in semiconductors. The donated carriers are swept away from the film (e.g., into the substrate). Experimentally, we observed negative fixed charge in as-grown $PrO_{1.5}$ MBE films. The magnitude of this fixed charge is significantly reduced when the film is annealed in N_2 atmosphere (in rare cases, annealing leads to the appearance of positive charge). On the other hand, the fixed charge in Pr silicate films is always positive. Also TiO_2 admixture to $PrO_{1.5}$ produces a positive fixed charge; a similar effect is observed in Ti-added Pr silicates.

Fixed charges are problematic because the electrical dipole between the charge in the film and the compensating charge in the substrate acts as an additional potential applied across the dielectric. This potential shifts the CV characteristic of the MOS structure along the voltage axis, affecting in this way the electrical performance of the capacitor. Large shifts render the MOS useless for device application. Indeed, the action of a MOS Field Effect Transistor (MOSFET) is controlled by the capacitance of the MOS structure; shifts comparable in magnitude to the electrical potential difference needed to switch a classical SiO_2-based transistor between its ON and OFF states are unacceptable. Analogous shifts in the CV characteristics occur due to the charge transfer between the metal gate and the substrate in response to the work function difference between these two materials. Fixed charge should not be, however, viewed as a good means to fix these work function differences, nor work function engineering is a good solution to the fixed charge problem. This is because the presence of numerous fixed charges degrades the working speed of the MOSFET by degrading the mobility of carriers in the channel between MOSFET source and drain. The reason for this degradation is scattering on random electric fields associated with the fixed charges, particularly those close to the interface with the substrate. This effect is noticeable only if the electrical quality of the interface is as high as needed for technological applications: the areal density of the charged species should not exceed $10^{11}\,\mathrm{cm}^{-2}$ (in SiO_2 gate dielectrics, it is of the order of $10^{10}\,\mathrm{cm}^{-2}$).

In Section 4 we argue that native defects (interstitials, vacancies) are an important source of fixed charge. The example of $PrO_{1.5}$ shows that the sign of this charge depends on the ambient condition (e.g., on the presence of a capping layer during post-deposition annealing and on the affinity of this layer to oxygen). A comparison between $PrO_{1.5}$, PrO_x, and PrO_x with $1.5 < x < 2$ illustrates the dependence of this intrinsic fixed charge on the crystallographic details of the atomic structure of the dielectric and on electronic properties of the metal component. When the fixed charge is generated in an unprotected film (i.e., when the film has a chemical contact to O_2), we argue that the source of the positive fixed charge is similar in SiO_2 and high-k dielectrics and comes from a defect which we label SiNCO (Silicon-related Nitrogen-Coordinated Oxygen). This defect is associated with a dangling bond belonging to a silicon interstitial atom located next to a regular oxygen atom in the atomic network of the host material. The atomic structure of SiNCO is described and the SiNCO formation energy in $PrO_{1.5}$ is estimated in Section 5. Table 1 summarizes the major sources of fixed charge in $PrO_{1.5}$.

Table 1. Major sources of fixed charge in $PrO_{1.5}$ films on Si substrates for the oxygen chemical potential defined by various ambients. The defects NCO (Nitrogen Coordinated Oxygen) and SiNCO (Si-related NCO) are described in the text. Note that longer exposure to oxidizing ambients causes oxidation of $PrO_{1.5}$ to PrO_x with x> 1.5, and longer exposure to moisture converts it into $Pr(OH)_3$

Uncapped in dry O_2	Uncapped in air	Capped on SiO_2	Capped under Pr or Ti
$SiNCO^+$	$SiNCO^+$, OH_O^+	$SiNCO^+$, NCO^{+2}	O_V^{+2}, Pr_i^{+3}, Ti_i^{+4}
O_I^{-2}, SiO_{Pr}^-	OH_I^-	O_I^-	-

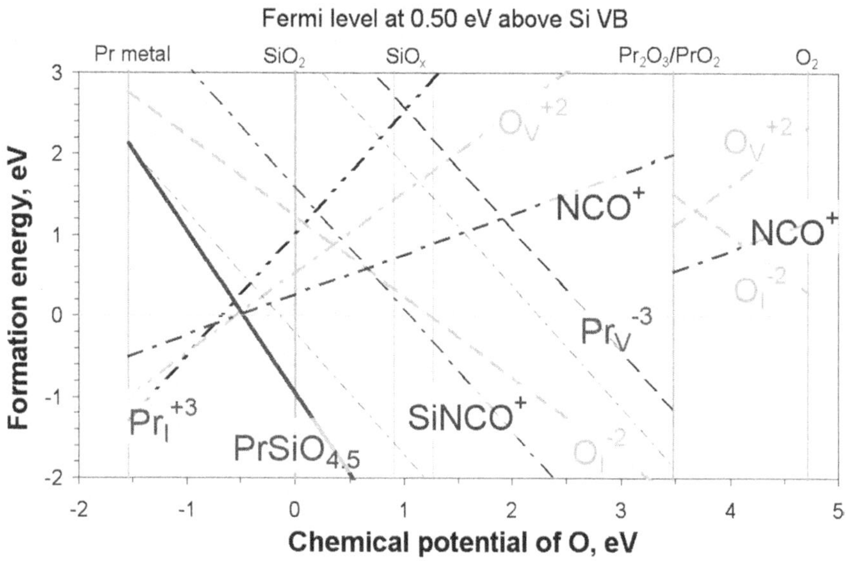

Fig. 1. Formation energies of various charged defects in Pr oxides as a function of the chemical potential of oxygen. The Fermi level corresponds to intrinsic Si. 1.8 eV is assumed for the valence band offset between Si and $PrO_{1.5}$, and 1.0 eV for the valence band offset between Si and PrO_x. For the SiNCO defect (Fig. 3), the upper and lower bound estimates and the average value are shown. The energy of the intrinsic NCO is estimated as a half of that of the oxygen vacancy. The calculated formation energy of G-type $PrSiO_{3.5}$ (per Si atom) is also displayed

4 Native Point Defects

As first we consider the classical native point defects, namely interstitials and vacancies on both sublattices. Since the point defects are created during processing (deposition, annealing), we show the formation energies obtained for the Fermi level aligned with its position in intrinsic Si. Indeed, slightly doped Si ($10^{16}\,cm^{-3}$) is nearly intrinsic already for temperatures around 300° C. The dependence of the defect formation energies on the chemical potential of oxygen is presented in Fig. 1. Consider, for example, the case of O_I^{-2}. As given by Eq. (1), the formation energy of O_I^{-2} decreases with increasing $\mu(O)$: it is as high as 2.8 eV for $\mu(O)$ corresponding to equilibrium between $PrO_{1.5}$, oxygen and Pr metal, and drops below -2 eV when $\mu(O)$ corresponds to equilibrium between $PrO_{1.5}$ and PrO_x. Thus, when the $PrO_{1.5}$ film remains in the contact with dry oxidizing ambient (chemical potential of oxygen in the range of the transition between $PrO_{1.5}$ and PrO_x, or higher), oxygen is inserted in the form of negatively charged interstitial atoms, O_I^{-2}. This insertion occurs with energy gain, meaning that the concentration of the charged defects increases until the energy loss due to electrostatic repulsion (and substantiated by the change of the band offset between Si and the oxide caused by the appearance of the dipole moment between these materials) compensates the energy gain due to the formation of O_I^{-2}. A rough estimate gives the upper limit for this concentration in the range of $10^{13}\,cm^{-2}$. High-temperature annealing in oxygen-poor N_2 should thus create a negative fixed charge in films deposited on Si. Self-compensation of these charges by conversion of some of the Pr atoms to Pr^{+4} takes place only when energy lost due to the electrostatic repulsion exceeds the energy gain due to the capture of electrons from the Fermi level in the semiconductor to oxygen acceptors[3]. Thus, the film is expected to remain negatively charged even if it is partially oxidized to PrO_x with x increased towards 1.75, i.e., even when the film becomes a mixture of the majority component $PrO_{1.5}$ (with Pr^{+3} ions) and the minority component PrO_x (with Pr^{+4} ions). Such a negative charge, typical for $PrO_{1.5}$/Si(001) MBE films grown by us, is also consistent with the fact that PrO_x crystals are typically p-type when x is below approximately 1.75 [11]). Only when the chemical potential of oxygen is as low as that determined by the exchange of O atoms with SiO_2 (e.g., when the film is annealed under a Si cap), a positive fixed charge can be created in the film by classical native point defects. This occurs mostly by oxygen vacancies O_V^{+2}. In amorphous material and along grain boundaries and dislocations an intrinsic NCO^+ (Nitrogen Coordinated Oxygen) defects related to O vacancies and discussed in more detail later may become the major intrinsic source of positive charges. Finally, when the film is brought into contact with Pr metal, positively charged Pr interstitial atoms can be injected into $PrO_{1.5}$. In this metal-rich extreme there is no significant equilibrium concentration of native defects carrying negative charges.

Let us now return to the oxide-rich regime. According to Fig. 1, Pr vacancies Pr_V^{-3} may also be responsible for fixed charge. However, since the Frenkel pair formation on the Pr lattice is relatively high (above 5 eV), kinetic constraints may hinder the appearance of these defects in relevant concentrations unless the vacancies are created directly during the deposition process.

Finally, we briefly consider other Pr-based dielectrics. As seen in the right-hand side of Fig. 1, the formation energy of oxygen interstitials O_I^{-2} is significantly lower in PrO_x than in $PrO_{1.5}$; in contrast to that, the formation energy of positively charged oxygen vacancies O_V^{+2} is significantly higher in PrO_x than in $PrO_{1.5}$. The first effect is due to the fact that oxygen interstitials in $PrO_{1.5}$ occupy these sites in the lattice from which oxygen is removed when the stoichiometry of the oxide changes from PrO_x to $PrO_{1.5}$. In PrO_x these low-energy sites are already filled with oxygen and the interstitial atoms have

---

[3] More precisely, the excess electrons captured at the O_I^{-2} acceptor may arrive either locally from Pr(III) neighbors or from the outside of the dielectric. In $PrO_{1.5}$ films grown on Si the latter source is more favorable. As the oxygen content x in PrO_x increases towards 1.75, the f shell of Pr(III) experiences more and more repulsion from the electrostatic charge of O atoms and becomes a more and more competitive source of the electrons.

to be placed in less convenient locations where they experience a remarkable compressive stress from the crystalline neighborhood. The second effect (the decrease of the O_V^{+2} energy in PrO_x) is apparently associated with the tendency of Pr oxides to form intermediate PrO_x phases which are structurally equivalent to PrO_x with an (ordered) array of oxygen vacancies [12].

The important difference between native point defects in $PrO_{1.5}$ and in $PrSiO_{3.5}$ is that the oxygen vacancy is not a charged defect in the latter material. This is because O atoms in $PrSiO_{3.5}$ have either only Si neighbors or both Si and Pr neighbors. The O vacancy created by removal of an O atom from between two Si atoms results in a formation of an electrically neutral SiSi bond. The removal of an O atom from a site where it had, besides Pr neighbors, also a Si neighbor results in the formation of a Si dangling bond. Although this dangling bond does capture an electron, this electron comes locally from the metal atoms: the negative charge localized now at the Si dangling bond is exactly the same charge that was collected from the metal atoms by the removed O atom. Another important difference between $PrO_{1.5}$ and $PrSiO_{3.5}$ is that the formation energy of O_I^{-2} is significantly higher $PrSiO_{3.5}$ and becomes negative only close to the O_2 extreme of the chemical potential of oxygen. The reason for this is similar as it was in the case of PrO_x: in contrast to $PrO_{1.5}$, there is no interstitial site in $PrSiO_{3.5}$ that would be naturally suited for oxygen to fill.

It becomes therefore clear that in order to explain the positive fixed charge in Pr silicate films and in certain as-deposited $PrO_{1.5}$ films prepared without a protective capping, defects other than intrinsic point defects must be invoked.

5 Si-related Nitrogen-Coordinated Oxygen

The origin of intrinsic fixed charges has been unclear even in SiO_2. Here we argue that the positive fixed charge in SiO_2 comes from a triple-coordinated oxygen atom that is associated with a Si dangling bond arising from an incompletely oxidized Si atom injected into the oxide during the process of thermal oxidation. We will then adapt a similar model to explain the appearance of fixed charges in $PrO_{1.5}$ and in transition metal oxides and rare earth metal oxides in general.

In silicon oxynitride, each Si atom is bonded to four oxygen or nitrogen atoms, each O atom is bonded to two Si atoms, and each N atom is bonded to three Si atoms. This saturates all the valencies; the oxynitride is an insulator.

Similarly, a nitrogen atom can be incorporated into the amorphous SiO_2 network without generating any localized states when it becomes bonded to sp3 orbitals of three Si neighbors. Each Si atom remains fourfold and each O atom remains twofold coordinated, but the impurity N atom is threefold coordinated. One can view this configuration as a Si dangling bond saturated by a N atom substituting a nearby O site.

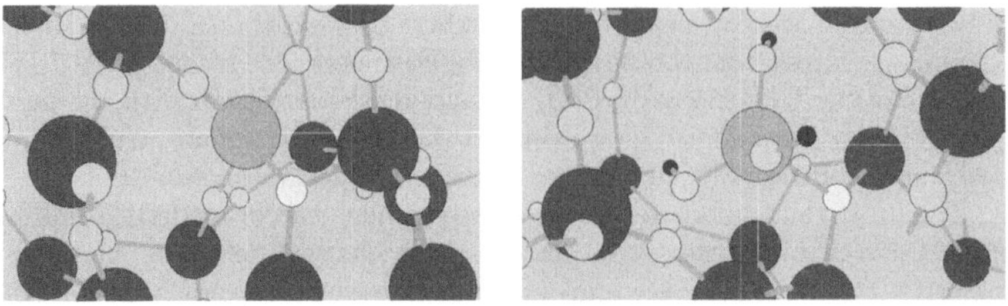

Fig. 2. *Left*: Si-related Nitrogen-Coordinated Oxygen (SiNCO) defect in amorphous SiO_2. *Right*: SiNCO modeled by a $Si(OH)_3$ molecule attached to an O atom in the SiO_2network. SiNCO is a shallow donor and is a plausible source of positive fixed charge generated during thermal oxidation of Si. Si atoms are red, O atoms are yellow, the NCO atom is white, the incompletely oxidized Si atom inserted into the SiO_2 network is pink; H atoms in the $Si(OH)_3$ molecule are black

Imagine that this threefold-coordinated, group-V nitrogen atom is now replaced by a group-VI oxygen atom. The latter has one electron more than needed to saturate the valencies of Si, and therefore the additional electron is donated to the conduction band (CB) or bonded only weakly on the defect. Since the CB bottom of SiO_2 is 3 eV above the CB bottom of the Si substrate, the electron is then transferred to the substrate and a fixed charge is formed in the SiO_2 film. We will name this defect a Silicon-related Nitrogen Coordinated Oxygen (SiNCO) defect. We have verified that SiNCO (Fig. 2) indeed behaves as described above. Although we did not make an attempt to estimate the formation energy of this defect, we note that this energy should be noticeably smaller than the formation energy of a Si dangling bond in SiO_2.

We now return to the case of $PrO_{1.5}$. We computed that the presence of oxygen in the ambient (as during post-deposition annealing) promotes dissolution of Si atoms from the substrate into the $PrO_{1.5}$ film. In particular, energy is gained when a Pr atom is replaced by a Si atom taken from the substrate and subsequently oxidized to Pr oxide. In an otherwise perfect lattice, such a substitutional Si_{Pr} has a dangling Si bond, which introduces electron transition states in the the upper part of the Si band gap. In an amorphous network or at a grain boundary, this dangling bond may arrange itself next to an oxygen atom from the oxide, thus forming a SiNCO similar to that in SiO_2. We will now estimate the formation energy of such a defect.

Building a model of such a configuration in order to verify the conjunction that such a defect would act as a fixed charge would be a tedious task. Instead, we have adopted a model similar to that used to simplify the SiNCO configuration in SiO_2 (Fig.2, right). We have placed an interstitial SiH_3 molecule in two configurations in crystalline $PrO_{1.5}$: in the perfect crystal and in a $PrO_{1.5}$ void created in an otherwise perfect $PrO_{1.5}$ crystal (Fig. 3). In both cases, the defect behaves as expected: it delivers a positive fixed charge. However, the formation energies E_{Si-NCO} (with respect to Si-Si bond in Disilane,

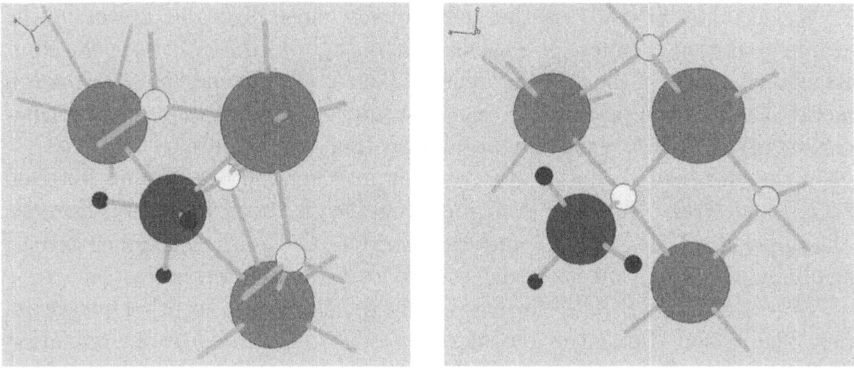

Fig. 3. Interstitial SiH_3 molecule as a model for NCO in $PrO_{1.5}$. *Left*: SiNCO in a perfect crystal; the Si atom is forced to a close neighborhood with lattice Pr atoms. *Right*: SiNCO in a void created by removal of a $PrO_{1.5}$ moiety from the crystal; the Si atom has enough free space. Pr atoms are blue, the Si atom is red, O atoms are yellow, the NCO atom is white, H atoms in the SiH_3 molecule are black

Si_2H_6) is very different: when computed for the Fermi level corresponding to that of intrinsic Si, it amounts to 2.4 eV in the first configuration, and to -1.2 eV in the second configuration. The difference comes apparently from the high compressive stress in the first configuration: the Si atom is forced to a site close to three Pr atoms (note that in spite of that, a regular bond with the oxygen atom is formed). In contrast to that, the second configuration allows the Si atom to find a place reasonably distant from the Pr neighbors without compromising the Si-O bond length. This leads to a significant lowering of the formation energy in the second configuration. We will treat these two formation energies as the upper and lower bound estimate of the bond energy between the dangling bond of Si and an NCO atom in $PrO_{1.5}$.

In order to estimate the SiNCO formation energy we still need an estimate of the energy of a regular bond between Si atom and an O atom in $PrO_{1.5}$, such as in the case of a substitutional Si_{Pr}. It is not straightforward to use the formation energy of Si_{Pr}, as it contains a dangling bond [13]. Therefore, we took another approach: we inserted a SiO_2 moiety into $PrO_{1.5}$. This moiety finds a configuration in which the Si atom is bonded to a tetraeder of four O atoms. The formation energy $4E_{\mathrm{Si-OPr}}$ of this defect with respect to SiO_2 is about 1.3 eV and corresponds to the formation of four regular Si-O bonds in $PrO_{1.5}$. Since the SiNCO defect has three such bonds and one SiNCO bond, we are now in a position to estimate its formation energy:

$$E_f(\mathrm{SiNCO}) = E_{\mathrm{Si-NCO}} + 3E_{\mathrm{Si-OPr}} \tag{6}$$

The result has been plotted as a function of the chemical potential of oxygen in Fig. 1. The two thin red dash-dot line correspond to the upper and lower bounds, while the line $SiNCO^+$ corresponds to their arithmetic average.

We see that $SiNCO^+$ is the energetically most favorable defect in the regime of oxygen chemical potentials already slightly (0.5 eV) above the equilibrium with SiO_2, if the average value is taken as the estimate of its formation energy. Even in the pessimistic case (the upper bound estimate) the formation energy of SiNCO is only slightly higher than the formation energy of O_i^{-2} when the chemical potential of oxygen approaches the equilibrium between $PrO_{1.5}$ and PrO_x. What is even more significant, both formation energies (the upper limit for $SiNCO^+$ and the energy of O_i^{-1}) are clearly negative in this limit, meaning that these defects are formed spontaneously.

We expect that SiNCO formation energy is low also in other metal oxides. The reason is that the low energy of SiNCO is caused by the oxidation of the Si atom. Even when the SiNCO formation energy is estimated from quite unfavorable atomic configurations (SiO_2 interstitial in $PrO_{1.5}$ and SiH_3 interstitial in perfect $PrO_{1.5}$ lattice) the fact that Si–O bonds are formed overweights the geometrical constraints associated with the particular defect site. Since the presence of metal neighbors to these O atoms affects the strength of this bond only moderately (as proven by the fact that amorphous silicates stable up to several hundred degrees C can be obtained not only with "silifilic" rare-earth oxides but also with "silifobic" materials such as HfO_2 and TiO_2), the appearance of SiNCO as a major positive fixed charge source in, e.g., HfO_2/Si(001), seems plausible. More to the point, HfO_2 is purposely grown on Si in amorphous form, and this is the amorphous host that is the natural environment for the SiNCO atomic configuration.

6 Impurities as Fixed Charges

Moisture is definitely the factor that has to be examined in the context of fixed charge formation [13]. As other rare earth oxides, $PrO_{1.5}$ readily absorbs water, to the extent that it is easily converted to a hydroxide. When a water molecule is dissolved in $PrO_{1.5}$, it dissociates into OH_I^- interstitial and H^+. The latter becomes attached to a lattice oxygen atom, forming a defect which may be termed a substitional OH group, OH_O^+. These defects do not introduce any localized states in the gap of $PrO_{1.5}$ and, since they have the opposite charge, a dissolved H_2O molecule cannot act as a fixed charge. Nevertheless, it is possible that the charge balance is affected by defect reactions in the film. For example, if the oxygen atom from dissociated H_2O is used to oxidize silicon in the substrate to (fully relaxed) SiO_2, the OH_I^- becomes converted to OH_O^+. Two positive fixed charges are thus created by each H_2O molecule taking part in such a reaction. We calculated that this oxidation reaction is energetically favorable by about $\simeq$0.3 eV per H atom in p-type Si. This means that, in principle, one cannot exclude that such processes take place in ultrathin $PrO_{1.5}$/Si films exposed to moisture. The calculated formation energy of OH_I^- in $PrO_{1.5}$ is also negative within a broad range of O potential in the oxidizing regime [4], meaning that water from air would be a source of negative fixed charge in $PrO_{1.5}$. The formation energy of OH_O^+ is sufficiently small only when the chemical potential of oxygen approaches the UHV range. Although the formation energy of OH_O^+ is negative in equilibrium with SiO_2, this does not seem to have a direct relevance to the fixed charge formation. Positively charged defects might be formed in this way if, for example, a water-contaminated oxide is sealed with a Si layer and then annealed.

Boron is the traditional acceptor used in MOSFET channels and in polysilicon gates. It is not a direct source of fixed charge in Pr oxides. Although boron atoms strongly segregate from Si to $PrO_{1.5}$, they substitute Pr in the lattice. As B_{Pr}, B atoms are isovalent impurities. They are electrically neutral and introduce no localized states, at least not in the hazardous energy region within approximately one eV to the band gap of Si. Nevertheless, boron segregation may be responsible for fixed charge generation is by kick-out of Pr interstitials, Pr_I^{+3} by B interstitials, B_I^+. The kick-out is energetically favorable by 0.8 eV when the Fermi level is aligned with that of intrinsic Si. This means that each B atom that makes it to the oxide produces one Pr_I^{+3} ion. If the annealing of an uncapped layer takes place in an atmosphere containing enough oxygen to oxidize these ions to $PrO_{1.5}$ (given Fig. 1, this is true also for all "inert" ambients), this effect is irrelevant. However, if the annealing takes place under a capping layer (as it is likely to be the case during technological processing), the Pr interstitial atoms may remain unoxidized in the film, producing a fixed charge that is particularly problematic because it may move across the dielectric in the electric field and render the device unreliable by causing a hysteresis in its CV characteristics.

Hazardous positively charged interstitial metal atoms may be also injected from such metal films as **titanium**. The detailed discussion of the interaction between Ti and Pr oxides is out of the scope of this paper, but we note that this injection becomes energetically unfavorable (by at least 2 eV) when the Ti source is oxidized to TiO_2. Nevertheless, nanosize inclusions of TiO_2 in $PrO_{1.5}$ and in $PrSiO_{3.5}$ may trap positive charge. Similarly to the charge trapped on Si_{Pr}, this positive charge is re-loadable and can be neutralized by electrons. This means that such inclusions act as Trap Assisted Tunneling centers, contributing to the leakage current flowing across the dielectric.

7 Summary and Conclusions

Given the results of ab initio total energy calculations and electrical data obtained for Pr-based dielectric films of various compositions and produced in various ways on Si(001) substrates, we have analyzed the influence of native

[4] We assume here that the chemical potential of H corresponds to that in H_2O remaing in thermodynamic equilibrium with H, that is, that the sum of $2\mu(H)$ and $\mu(O)$ yields the formation energy of water.

point defects and certain impurities (Si, B, Ti, moisture) on the formation of fixed charges in the dielectrics. Depending on the environment which determines the chemical potential of oxygen in the film, the fixed charge may come from positively charged metal interstitials, positively charged O vacancies, or negatively charged O interstitials; moreover, negatively charged metal vacancies may lead to a fixed charge when these defects are created during growth. We argued that the Si substrate is a source of Si impurities which give rise to positive fixed charge not only in $PrO_{1.5}$ but also in other high-k metal oxides, including HfO_2, and in thermal SiO_2/Si films. We have presented a conceptual model of Silicon-related Nitrogen-Coordinated Oxygen (SiNCO) atomic configuration of the defect responsible for this fixed charge.

Acknowledgements

The calculations have been done on IBM Regatta in von Neumann Institute for Computing, Jülich, Germany (project hfo06).

References

1. M. Bockstedte, A. Kley, J. Neugebauer, and M. Scheffler, Comp. Phys. Comm. **107**, 187 (1997).
2. A. Fissel, J. Dąbrowski, and H. J. Osten, J. Appl. Phys **91**, 8986 (2002).
3. D.M Ceperley and B.J. Alder, Phys. Rev. Lett. **45**, 567 (1980).
4. J. P. Perdew and A. Zunger, Phys. Rev. B **23**, 5048 (1981).
5. D. R. Haman, Phys. Rev. B **40**, 2980 (1989).
6. G. B. Bachelet, D. R. Hamann, and M. A. Schlüter, Phys. Rev. B **26**, 4199 (1982).
7. H.-J. Müssig, H. J. Osten, E. Bugiel, J. Dąbrowski, A. Fissel, T. Guminskaya, K. Ignatovich, J. P. Liu, B. P. Zaumseil, and V. Zavodinsky, Proceedings of the 2001 IEEE Integrated Reliability Workshop, p.1 (2001).
8. J. Dąbrowski, V. Zavodinsky, and A. Fleszar, Microel. Reliability **41**, 1093 (2001).
9. H. Bergman, *Gmelin Handbuch der Anorganischen Chemie, Seltenerdelemente, Teil C1* (Springer Verlag, Berlin 1974).
10. David R. Lide (ed), *Handbook of Chemistry and Physics, 73rd Edition* (CRC Press, 1993-1994).
11. G. V. Subba Rao, S. Ramdas, P. N. Mehrotra, and C. N. R. Rao, J. Solid State Chem. **2**, 377 (1970).
12. H. Inkaba and K. Naito, J. Solid State Chem. **50**, 100 (1983).
13. G. Lippert, J. Dąbrowski, V. Melnik, R. Sorge, Ch. Wenger, P. Zaumseil, and H.-J. Müssig, Appl. Phys. Lett. **86**, 042902 (2005).

Size Effects in Ba(Pb)TiO_3 Nanopowders by EPR and NMR

Emre Erdem[1], Rolf Böttcher[1], Hans Jürgen Gläsel[2], Eberhard Hartmann[2], Gert Klotzsche[1], and Dieter Michel[1]

[1] Institute for Experimental Physics II, University of Leipzig, Linnéstrasse 5, 04103 Leipzig, Germany
[2] Leibniz Institute of Surface Modification (IOM e.V.), Permoserstrasse 15, 04318 Leipzig, Germany

Summary. The aim of the present paper is to study the influence of the particle size on the properties of $BaTiO_3$ and $PbTiO_3$ materials by means of a NMR and EPR spectroscopy. Mn^{2+} EPR and ^{137}Ba NMR measurements were performed on $BaTiO_3$ particles with average diameters between 15 nm and 255 nm. Cr^{3+} EPR spectra were measured on $PbTiO_3$ fine powders with mean particle diameters ranging from 6 nm to 210 nm. Typical changes in the fine structure and the quadrupole coupling tensors are observed in the ferroelectric tetragonal phase which allow information about the structure model for the nanoparticles and the critical particle size.

1 Introduction

Surface and size effects in ferroelectrics have become a subject of increasing importance due to the rapid progress in the fabrication of the new-generation ferroelectric random access memories (FRAM) and other ferroelectric thin film devices [1]. Barium titanate ceramics and related materials are especially interesting because of their high permittivity values. The properties of these materials have been extensively investigated [2]. For barium titanate, the transition from the cubic non-polarized phase to the tetragonal ferroelectric phase occurs 407 K. The tetragonal phase is stable above 278 K. The phase transitions in $BaTiO_3$ are commonly considered as displacive type transitions [3] with an overdamped transverse-optical soft mode [4, 5]. In several papers a possible coexistence of displacive (underdamped soft mode) and order-disorder components (strongly overdamped or relaxor dynamics) in the mechanism of phase transition has been discussed [6, 7]. The debate about the nature of the phase transition still continues [8]. The displacive type phase transition for $PbTiO_3$ from the cubic paraelectric phase into the ferroelectric phase occurs at 766 K. A high spontaneous polarization P_S (60 $\mu C/cm^2$ for $PbTiO_3$, 20 $\mu C/cm^2$ for $BaTiO_3$) was measured at room temperature [9].

Previous Ba and Ti NMR studies on $BaTiO_3$ [10-12] were made on single crystals over a wide range of temperatures. The quadrupole coupling constant is found to be proportional to the square of P_S in the tetragonal phase [11, 13]. Mn^{2+} and Cr^{3+} ions in the $BaTiO_3$ and $PbTiO_3$ host lattice which

B. Kramer (Ed.): Adv. in Solid State Phys. **45**, 351–361 (2005)

substitute for Ti ions, form a very suitable local probe for studying size effects in nanocrystalline powders by means of EPR spectroscopy. A survey to the respective measurements on bulk samples is found in the refs. [14-16]. The aim of the present paper is to study the influence of the particle size on the properties of $BaTiO_3$ and $PbTiO_3$ materials. The investigations are based on an advanced sample preparation route, admitting reproducible adjustment of mean particle size and doping by paramagnetic probe ions [17, 18]. The measurements are performed on ultrafine particles of $BaTiO_3$ with average diameters d between 15 nm and 255 nm and of $PbTiO_3$ with d values from 6 nm to 210 nm in comparison to polycrystalline samples and to single crystals from which the polycrystalline material was produced.

2 Experimental

2.1 Sample Preparation and Characterization

Our preparation route of nanopowders [17, 18] enabled an adjustment of efficient particle sizes in the range from a few nm up to μm. Since doping by paramagnetic ions can affect the grain size of $Ba(Pb)TiO_3$ samples to a significant extent we especially aim in our route at a homogeneous doping by paramagnetic ions with very low concentrations. The size dependence on the reaction temperature has been carefully controlled by means of XRD. The product (yield 97 %) was characterized with the aid of elemental analysis and spectroscopic methods.

2.2 NMR and EPR Measurements

The ^{137}Ba NMR spectra were obtained using a Bruker MSL 500 spectrometer (55.579 MHz, nominal field of 11.747 T) and a Bruker AVANCE 750 spectrometer (83.349 MHz, 17.6157 T) using a probe with a 7 mm coil. For the broad lines, the spectra were obtained by means of a two pulse echo sequence $\pi/2$ - τ - π - τ' - acquisition in combination with a 16 step phase cycle. For the narrow lines in the cubic phase, the spectra were obtained from single pulse Bloch decays. The $\pi/2$ pulse duration ranged between 2.6 and 5.5 μs. EPR measurements were carried out in the X band with a Bruker ESP 380 spectrometer. Temperature dependent EPR measurements were carried out using a Varian E 112 spectrometer with a variable temperature accessory. In the temperature range from 300 K to 830 K, a stability better than 1.0 K was achieved.

3 Results and Discussion

3.1 EPR Studies on $BaTiO_3$ Particles

For the interpretation of the Mn^{2+} EPR spectra (electron spin quantum number $S = 5/2$), a spin-Hamiltonian was used which includes the electron Zeeman term, the second-order fine structure (FS) term, and the hyperfine structure term (HFS). The axial FS term $D[S_z^2 - \frac{1}{3}S(S+1)]$ contains the major part of structural information. FS parameter D is temperature dependent and drops to zero at the tetragonal-to-cubic phase transition. In the spectra of the nanopowder samples, only the central FS transition $M_S=-\frac{1}{2} \Leftrightarrow M_S=\frac{1}{2}$ could be detected. In order to deduce the spin-Hamiltonian parameters D and the distribution widths ΔD from the EPR powder spectra, a simulation program was applied [19]. This allows to evaluate the line shape according to

$$I(B) \propto \sum_i \sum_{M_I=-\frac{5}{2}}^{\frac{5}{2}} \int_{-\infty}^{+\infty} dD \int_0^{\pi/2} \sin\Theta \, d\Theta \, P(D) W(\Theta) \left| \frac{dB_{RES}(\frac{1}{2}, M_I, i)}{d(h\nu)} \right| \cdot F \left[\frac{dB_{RES}(\frac{1}{2}, M_I, i) - B}{\Delta B_{pp}} \right] . \quad (1)$$

In Eq. (1), $W(\Theta)$ denotes the angular dependent transition probability, i is the summation index of the allowed and forbidden transitions, Θ is the angle between the external magnetic field B and the symmetry axis of the local crystal field (here: $c \parallel z$). $F[(B_{RES} - B)/\Delta B_{pp}]$ is the line shape function which was assumed to be Gaussian, and $P(D)$ describes a statistical distribution of the FS parameter D which was also assumed to be Gaussian. For the interpretation of the spin-Hamiltonian parameter D, the superposition model [14, 20] was applied. In this approximation the parameter D is proportional to P_S^2 and may be correlated to the lattice parameters a and c,

$$D \propto \left(\frac{c}{a} - 1\right) \propto P_S^2 . \quad (2)$$

In order to investigate the influence of the mean particle size on the EPR spectra, we have measured Mn^{2+} EPR spectra of doped $BaTiO_3$ micro-powders, prepared as a reference sample. The powder pattern of the central FS transition in Fig. 1a is a superposition of tetragonal and cubic spectra (marked by asterisks). The six intense lines, observed both for microcrystalline and nanopowders, are due to the ^{55}Mn HFS interaction. From the simulation of the axial EPR spectrum the parameter D is estimated, yielding for the micropowder sample at room temperature a value of about 24.5 mT. When going from micro- to nanopowder samples the resolution of Mn^{2+} EPR spectra of $BaTiO_3$ becomes quite poor (Fig. 1b). At first, for the 155 nm-sized sample the outer derivative features are broadened and an asymmetric line occurs in between them.

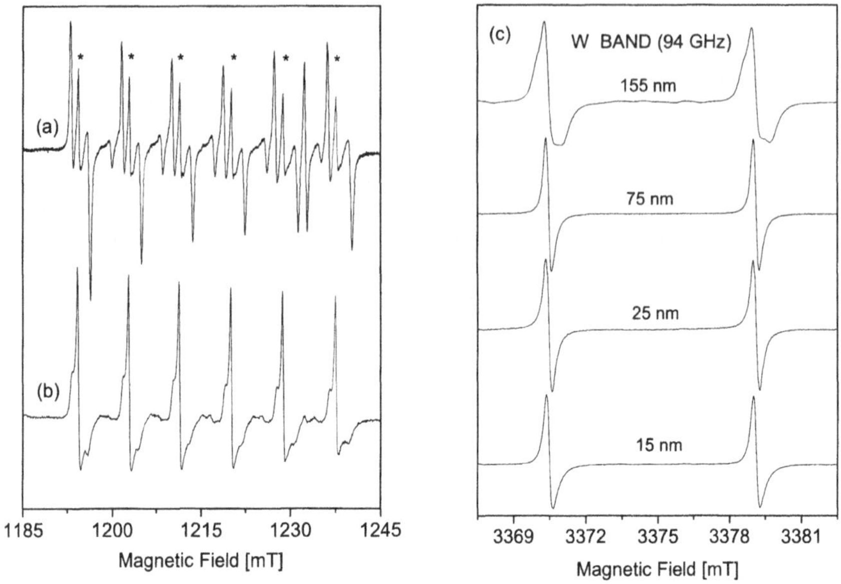

Fig. 1. Mn^{2+} EPR spectra of microcrystalline (a) and nanocrystalline $BaTiO_3$ samples (b) measured in the Q band. (c) Mn^{2+} EPR central transition lines ($M_S = -\frac{1}{2} \Leftrightarrow M_S = \frac{1}{2}, M_I = -\frac{3}{2}, -\frac{5}{2}$) measured in the W band at room temperature for powder samples of various mean sizes

In the EPR spectra of powders with smaller grain size ($<$ 100 nm), the two outer peaks are no longer observed and the central FS transition consists of several allowed and forbidden HFS lines with asymmetric shapes (see Fig. 1c). The best-fit value of the g-factor is 2.002 $\pm$0.001. The HFS constant ($A = 8.64 \pm 0.02$ mT) does not vary with the particle size and is typical for Mn^{2+} ions incorporated in $BaTiO_3$ single crystals [14].

The mean FS parameters D and the widths ΔD of the distribution reveal a pronounced grain-size dependence. An optimal simulation of the spectra of powder samples with intermediate grain size (e.g. 155 nm) is attained when two axial spectra are superimposed. One of the two D values still lies quite close to the FS parameters of single crystals [14] and microceramics [15]. The other FS parameter set opens a series in which the mean FS parameter D decreases and the distribution width ΔD grows with decreasing particle size. For greater grain size (when $D > \Delta D/2$) a cubic spectral component has to be taken explicitly into account. For finer nanopowder samples such a broad FS parameter distribution ($D < \Delta D/2$) was found that an adequate cubic fraction (around $D = 0$) is always contained. The inevitable assumption of a pronounced grain size distribution in the simulation of the experimental data throws some new light on a presumable coexistence of tetragonal and cubic phases within a particular $BaTiO_3$ particle as it has been deduced preferentially from the XRD results [21]. For small particles the scattering of the D values is clearly larger than the distortions due to the unit cell deformation in the tetragonal unit cell. The ΔD values measured for the 25 nm and 15 nm grains, indicate subregions with strong local distortions which have nothing to do with the limited tetragonal distortions normally occurring in the ferroelectric phase. This finding strongly supports our picture of a small particle as an aggregate of inner regular core and heavily distorted outer layer mediated by a tetragonality gradient as will be outlined below.

3.2 NMR Studies

In the ^{137}Ba NMR measurements on $BaTiO_3$ powder samples, only the central line transitions could be detected in the various phases. The line shape for the central transition NMR lines is dominated by the quadrupole coupling tensor (proportional to the electric field gradient tensor, EFG) in second-order perturbation theory. This assumption has been checked by comparing the results of NMR line shape measurements on microcrystalline samples with those ones derived from the rotation patterns measured for single crystals and by measuring at two different d. c. magnetic fields ($B_0 = 11.747$ and 17.6157 T). The central transition ^{137}Ba NMR line shapes in the various phases are shown in Fig. 2. Relatively sharp ^{137}Ba NMR lines (with Gaussian line shape and widths at half-heights of 150 Hz) were observed for the polycrystalline material in the cubic phase.

The lines are broadened abruptly on cooling below the phase transition into the tetragonal phase. The quadrupole perturbed NMR line shape for microcrystalline samples was simulated by means of the programs WinNMR and WinFit. For very small particles an own simulation program was developed. In the tetragonal phase, the quadrupole coupling constant $C_Q = |e^2qQ/h|$ may be directly derived from the distance between the edge singularities in the powder pattern for the second-order broadened central transition. The values are in a complete agreement with the results from single crystal measurements in the refs. [12, 13]. The quantity eQ denotes the nuclear quadrupole moment, $eq = V_{zz}$ is the electric field gradient along the crystallographic c (z) axis which is parallel to the direction of the P_S. The proportionality between the square η^2 of the order parameter, $\eta = P_S(T)/P_S(0)$, and the coupling constant C_Q [11] is confirmed in Fig. 3.

The typical powder pattern of the ^{137}Ba NMR spectra in the tetragonal phase (see Fig. 4) for samples with mean particle sizes of 155, 75, and 25 nm may be explained by a superposition of two tetragonal contributions. For each of which the line shape calculation has been carried out in a similar way as described in Eq. (1) for the EPR spectra. To simulate the line shape $I(\nu)$,

$$I(\nu) \propto \sum_{k=1,2} p_k \int_{-\infty}^{+\infty} dC_Q \int_0^{\pi/2} \sin\Theta \, d\Theta \, P(C_{Q,k}) \cdot F\left[\frac{\nu^{(2)}(\Theta) - \nu}{\Delta\nu_{pp}}\right], \quad (3)$$

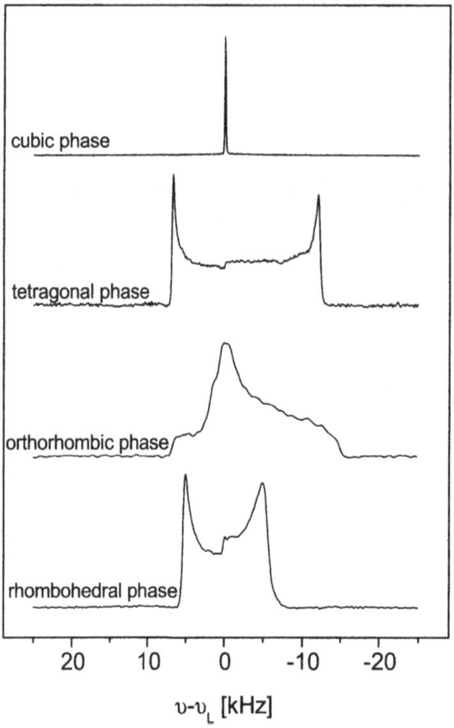

Fig. 2. The ^{137}Ba NMR spectra of microcrystalline $BaTiO_3$ powder, prepared from single crystals. Resonance frequency of 55.6 MHz, d. c. magnetic field of 11.7 T. Measuring temperatures: rhombohedral phase at 160 K, orthorhombic phase at 260 K, tetragonal phase at 300 K, and cubic phase at 416 K

the frequencies $\nu^{(2)}(\Theta)$ of the central transitions (relative to the Larmor frequency ν_L) were averaged over the polar angles Θ by taking into account of Gaussian distribution of the quadrupole coupling constant C_Q with mean values $< C_{Q,k} >$ and widths $\Delta C_{Q,k}$. The intensity fraction of the lines (k = 1,2) is denoted by p_k, $\Delta\nu_{pp}$ is a line shape parameter similarly chosen as in the line shape simulation for the microcrystalline powder where no distribution of the coupling constant is necessary. One line shape component for the nanoparticles shows the typical line shape for the tetragonal phase ($< C_{Q,k} >$ = 2.6 MHz) and a relatively small distribution of the quadrupolar constant ($\Delta C_{Q,2}$ = 0.5 MHz). This situation leads to the conclusion that one part of the sample possesses a tetragonal symmetry with small distortions. The other component obviously belongs to regions where the mean quadrupole coupling constant is smaller (between 2.2 MHz and 1.6 MHz, depending on the particle size) and where a relatively broad distribution occurs.

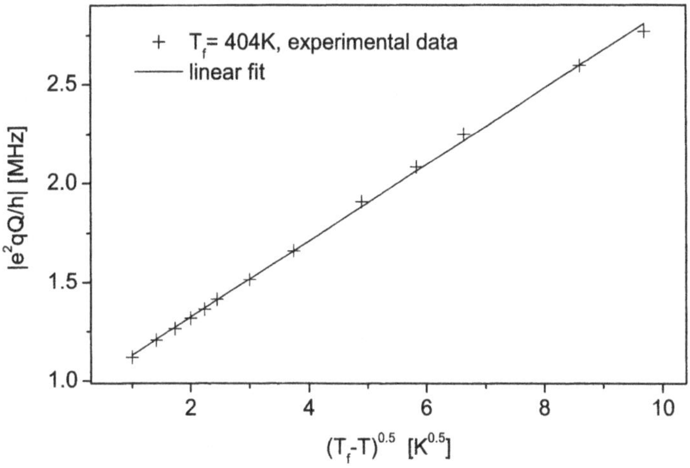

Fig. 3. Experimental values for the quadrupole coupling constant, $C_Q = |e^2qQ/h|$, in the tetragonal phase as a function of $(T_f - T)^{1/2}$ (with $T_f = 404$ K). The linear plot confirms the proportionality between C_Q and the square of the order parameter

3.3 EPR Studies on $PbTiO_3$ Particles

Cr^{3+} $PbTiO_3$ nanopowders with mean particle diameters from 6 nm to 210 nm were used. Mean particle sizes and the size distributions were derived from XRD measurements by an analysis of the (200) and (002) Bragg reflection shapes (Fig. 5).

Using a simulation program *X-SOPHE* [22] the fine structure (FS) parameter D of the axial spin-Hamiltonian of the Cr^{3+} centers and, particularly, the distributions of the FS parameters D and E have been determined from multi-frequency EPR measurements. Typical X-band Cr^{3+} EPR spectra are shown in Fig. 6. Spectra taken from a sample of relatively large mean particle size ($d = 210$ nm from XRD) reveal in their center still a multi-line spectrum of minor importance which may be attributed to Cr^{5+} sites ($S = 1/2$) [23].

In our context the Cr^{3+} ($S = 3/2$) spectrum is of much greater relevance. The multi-frequency EPR measurements reveal three Cr^{3+} centers where only the C1 center is consistent with the spectra parameter derived from single crystal Cr^{3+} EPR measurements [24]. The spectrum for the C1-site merges into the cubic Cr^{3+} spectrum at temperatures above T_C. With lowering the particle size, the FS parameter D and the tetragonality $(c/a - 1)$ decrease.

In Fig. 7 normalised quantities D, $(c/a - 1)$, T_C are shown for various particle sizes. The normalised values are referred to the respective values measured for large particle sizes. The normalised Curie temperature T_C was both derived from EPR spectra and from dielectric measurements [25]. The normalised tetragonality values were obtained from the XRD data. All quantities may be described by the semi-empirical expression [26]

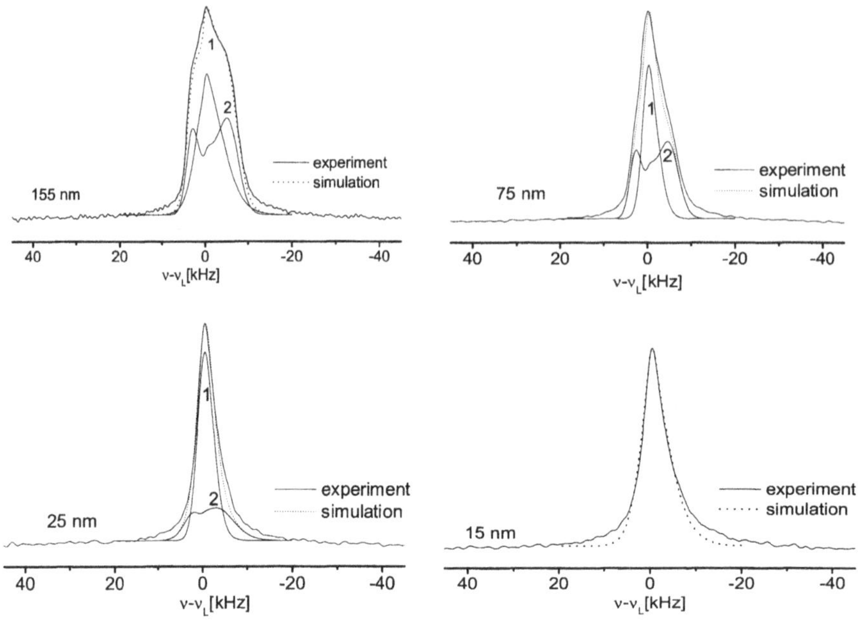

Fig. 4. Central transition ^{137}Ba NMR spectra in $BaTiO_3$ powder sample with an average diameter d_m. The spectra were measured at 300 K (in the tetragonal phase) and at a d.c. magnetic field of B = 17.6157 T. Apart from the scaling proportional to 1/B, the same line shapes were measured at B = 11.744 T. The lines 1 and 2 denote the simulated NMR powder spectra. The sum of the simulated spectra is denoted by dots. The intensity ratio of the two simulated spectra is denoted by $p_{1,2} = p_{line1} : p_{line2}$. Top left: $d_m = 155$ nm; $p_{1,2} = 1 : 1$; Line *1*: $< C_Q > = 2.2$ MHz, $\Delta C_Q = 1$ MHz. Line *2*: $< C_Q > = 2.65$ MHz, $\Delta C_Q = 0.5$ MHz. Top right: $d_m = 75$ nm; $p_{1,2} = 1 : 1.2$, Line *1*: $< C_Q > = 1.6$ MHz, $\Delta C_Q = 0.7$ MHz. Line *2*: $< C_Q > = 2.55$ MHz, $\Delta C_Q = 0.5$ MHz. Bottom left: $d_m = 25$ nm; $p_{1,2} = 2.2 : 1$, Line *1*: $< C_Q > = 1.6$ MHz, $\Delta C_Q = 0.7$ MHz. Line *2*: $< C_Q > = 2.6$ MHz, $\Delta C_Q = 0.8$ MHz. Bottom right: $d_m = 15$ nm; one line simulated with $< C_Q > = 2.2$ MHz, $\Delta C_Q = 1.1$ MHz

$$y(d) = y_\infty - \frac{A}{d - d_{cr}} \tag{4}$$

from which a critical size of 6 nm is derived by extrapolation. The extrapolated value is approximately in agreement with the theoretical value 4.2 nm of Zhong et al. [27] calculated by means of the phenomenological Landau theory.

4 Conclusion

The simulation of the EPR and NMR spectra is nicely consistent with the following structural model for ultrafine $BaTiO_3$ particles. A weakly distorted

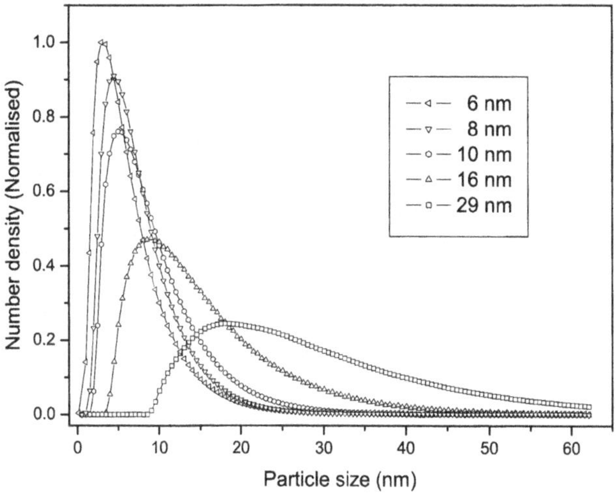

Fig. 5. Size distribution spectra (normalized number density) of $PbTiO_3$ ultrafine nanopowders

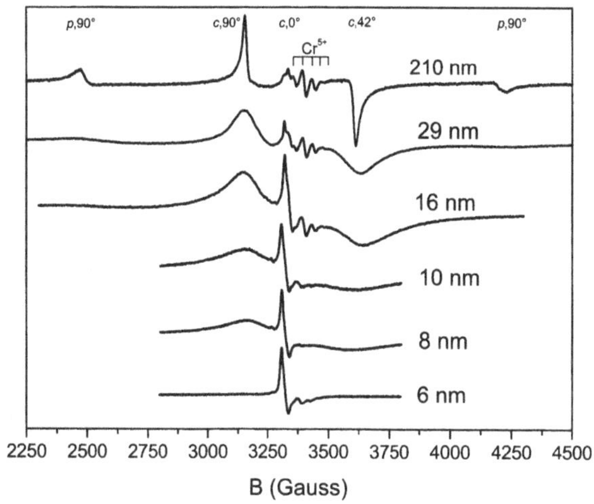

Fig. 6. Cr^{3+} EPR spectra of $PbTiO_3$ nanopowders in dependence on the mean particle size taken at room temperature

tetragonal core is surrounded by a highly distorted shell where the local symmetry can be only characterized by a strong distribution of the internal electrical fields. The fraction of the distorted regions increase when the particle diameters are reduced from 155 nm to 25 nm. At 25 nm still a tetragonal ordered part occurs. For the sample with the smallest diameters (15 nm) the strongly distorted component dominates and it was not possible to detect the tetragonal core. This supports the conclusion that an ordered tetragonal phase disappears when the mean diameters are close and smaller than this

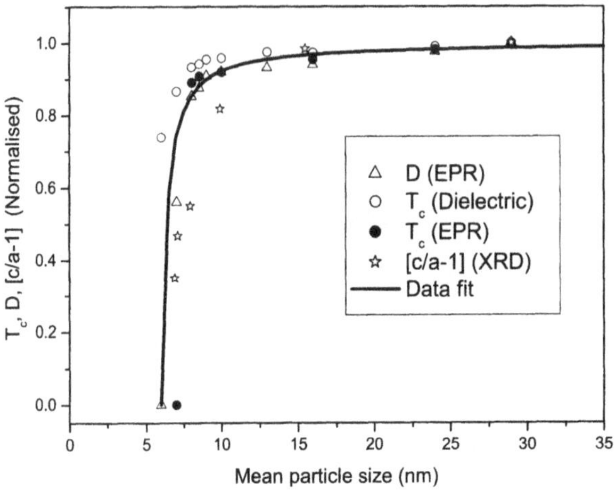

Fig. 7. Correlation between the measured data for the FS parameter D, tetragonality $(c/a-1)$ and the Curie temperature T_C for nanocrystalline $PbTiO_3$ samples with various mean particle diameters. Normalised values are shown which were referred to the respective values measured for large particle sizes

size. Our picture of a small particle as an aggregate of inner regular core and heavily distorted outer layer mediated by a tetragonality gradient is consistent with that by Anliker et al. [28]. The main qualitative conclusions may be also compared with those of Frey ([29], critical value of 9 nm), of Tanaka ([30], 20 nm as a practical critical size for crystallites and an extrapolated value of about 10 nm for the critical size at which the T_C drops to zero). Critical diameters of 20 nm for size driven ferroelectric phase transition in $BaTiO_3$ nanograins were calculated by Glinchuk et al. [31]. The order of magnitude for the critical diameter derived in our work for $BaTiO_3$ particles is consistent with the results achieved for $PbTiO_3$. EPR measurements in combination with the results of XRD and dielectric studies lead to a critical particle size of 6 nm.

Acknowledgements

The authors are indebted to Prof. G. Völkel for his comments and suggestions and to Dr. S. M. Osman for his help in the NMR measurements. The support of the Deutsche Forschungsgemeinschaft (DFG) is gratefully acknowledged.

References

1. J. Dreßen, A. Hoffknecht, Phys. Journal **1**, 39 (2003).
2. E. K. Akdogan, M. R. Leonard, A. Safari, *Size Effects in Ferroelectric Ceramics*, in: *Handbook of Low and High Dielectric Constant Materials and their Applications, Materials and Processing*, vol. II (Academic Press, London 1999).

3. S. M. Shapiro, G. Shirane, T. Riste, J. D. Axe, Phys. Rev. B **6**, 4332 (1973).
4. J. Harada, J. D. Axe, G. Shirane, Phys. Rev. B **4**, 155 (1971).
5. H. Vogt, J. A. Sanjurjo, G. Rossbroich, Phys. Rev. B **26**, 5904 (1982).
6. K. A. Müller, W. Berlinger, Phys. Rev. B **34**, 613 (1986).
7. A. D. Bruce, R. A. Cowley, *Structural Phase Transitions* (Taylor and Francis, London 1981).
8. B. Zalar, V. V. Laguta, R. Blinc, Phys. Rev. Lett. **90**, 37601 (2003).
9. M. E. Lines, A. M. Glass, *Principles and applications of ferroelectrics and related materials* (Clarendon Press, Oxford 1977).
10. T. J. Bastow, H. J. Whitfield, Sol. Stat. Commun. **117**, 483 (2001).
11. O. Kanert, H. Schulz, J. Albers, Sol. Stat. Commun. **91**, 465 (1994).
12. A. Taye, G. Klotzsche, D. Michel, S. M. Osman, R. Böttcher, J. Phys. Condens. Matter **11**, 871 (1999).
13. T. J. Bastow, J. Phys. Condens. Matter **1**, 4985 (1989).
14. K. A. Müller: *Structural Phase Transitions II* (Springer Verlag, Berlin 1991).
15. B. Milsch, phys. stat. sol. (a) **133**, 455 (1992).
16. R. N. Schwartz, B. A. Wechsler, Phys. Rev. B **48**, 7057 (1993).
17. H. J. Gläsel, E. Hartmann, D. Hirsch, R. Böttcher, C. Klimm, D. Michel, H. C. Semmelhack, J. Hormes, H. Rumpf, J. Mater. Sci. **34**, 2319 (1999).
18. E. Erdem, R. Böttcher, H. C. Semmelhack, H. J. Gläsel, E. Hartmann, D. Hirsch, J. Mater. Sci. **38**, 3211 (2003).
19. R. Böttcher, C. Klimm, D. Michel, H. C. Semmelhack, G. Völkel, H. J. Gläsel, E. Hartmann, Phys. Rev. B **62**, 2085 (2000).
20. D. J. Newman, W. Urban, Adv. Phys. **24**, 793 (1975).
21. T. Takeuchi, K. Ado, T. Asai, H. Kageyama, Y. Saito, C. Masquelier, O. Nakamura, J. Amer. Cer. Soc. **77**, 1665 (1994).
22. D. Wang, G. R. Hanson, J. Magn. Res. A **117**, 1 (1995).
23. R. Böttcher, W. Brunner, B. Milsch, G. Völkel, W. Windsch, S. T. Kirillov, Chem. Phys. Lett. **129**, 546 (1986).
24. V. V. Laguta, T. V. Antimirova, M. D. Glinchuk, I. P. Bykov, J. Rosa, M. Zaritskii, L. Jastrabik, J. Phys. Condens. Matter **9**, 10041 (1997).
25. E. Erdem, R. Böttcher, H. J. Gläsel, E. Hartmann, Magn. Res. Chem. (2005, in print).
26. K. Ishikawa, K. Yoshikawa, N. Okada, Phys. Rev. B **37**, 5852 (1988).
27. W. L. Zhong, Y. G. Wang, P. L. Zhang, B. D. Qu, Phys. Rev. B **50**, 698 (1994).
28. M. Anliker, H. R. Brugger, W. Kanzig, Helv. Phys. Acta **27**, 99 (1954).
29. M. H. Frey, D. A. Payne, Phys. Rev. B **54**, 3158 (1996).
30. M. Tanaka, Y. Makino, Ferroelectrics Lett. Sect. **24**, 13 (1998).
31. M. D. Glinchuk, P. I. Bykov, J. Phys. Condens. Matter **16**, 6779 (2004).

AlGaN/GaN Electrolyte-Gate Field-Effect Transistors as Transducers for Bioelectronic Devices

Georg Steinhoff[1], Barbara Baur[1], Hans-Georg von Ribbeck[1], Günter Wrobel[2], Sven Ingebrandt[2], Andreas Offenhäusser[2], Martin Stutzmann[1], and Martin Eickhoff[1]

[1] Walter Schottky Institut, Technische Universität München, Am Coulombwall 3, 85748 Garching, Germany
[2] Institut für Schichten und Grenzflächen ISG-2, Forschungszentrum Jülich, Leo-Brandt-Straße, 52428 Jülich, Germany

Summary. AlGaN/GaN electrolyte gate field-effect transistors are assessed as a transducer device for biosensor applications. Their low-frequency noise power spectral density exhibits a $1/f$ characteristic with a dimensionless Hooge-parameter of 5×10^{-3}. The equivalent gate-input noise under operation conditions has a peak-to-peak amplitude of 15 µV, one order of magnitude smaller than for common silicon-based devices. The untreated GaN device surface showed a high linear pH response of 56 - 57 mV/pH, whereas the sensitivity to other ions is comparably low. Arrays of these transistors have been used for monitoring of action potentials from a confluent layer of rat heart muscle cells cultivated directly on the non-metallized gate surfaces and signals with an amplitude of 1.2 mV were recorded.

1 Introduction

Biosensors based on the combination of biological recognition mechanisms with planar electronic transducer devices are of increasing interest for biomedical and biotechnological applications. In these devices proteins, enzymes or even whole cells or cell networks are immobilized or cultured directly on the device surface as biological receptor units. In enzyme-based biosensors, the respective analyte is catalytically converted by enzymes immobilized at the transducer surface. The accompanying change in concentration of the ionic byproducts is detected by the transducer, an ion-sensitive field-effect transistor (ISFET) or electrode [1, 2, 3].

In whole-cell biosensors, either the electrical cell activity or cell metabolism are monitored. By recording the extracellular acidification rate, changes in the physiological states of cultured cells to chemical and physical stimuli can be detected [4, 5, 6, 7]. Whole-cell biosensors benefit from the high sensitivity and selectivity of specific receptors in the cellular membrane and signal amplification in second messenger pathways of the cells. These cell-sensor hybrids are suitable for a variety of applications such as drug discovery and screening in pharmacology, detection of toxins and environmental monitor-

B. Kramer (Ed.): Adv. in Solid State Phys. **45**, 363–374 (2005)

ing [8, 9]. Both biosensor concepts require a noninvasive multisite transducer device suitable for long-term measurements under physiological conditions. This excludes the recording of cell action potentials by conventional invasive patch-clamp techniques [10] or the utilization of pH- or voltage-sensitive dyes [11], which are often toxic on illumination. Extracellular monitoring of electrical cell activity can be performed by micro-electrode arrays [12, 13, 14] or field-effect transistor (FET) device arrays, as it was first reported for muscle fibers and neuronal slices by Bergveld [15]. The electrolyte-oxide FETs which are generally used as transducers for this purpose as well as the ISFETs or capacitive sensors for determination of H^+-concentration are usually realized in silicon technology [16, 17, 19, 4, 5, 6]. The main drawbacks of these devices are their long-term drift in electrolytes due to the electrochemical instability of the SiO_2-surface during exposure to electrolye solutions [20]. Furthermore they exhibit a high noise level due to mobility fluctuations caused by trapping and detrapping of carriers to and from trap states at the interface to the surface oxide layer [21]. Whereas the device drift and stability can be significantly improved by the deposition of additional metal-oxide (Al_2O_3, Ta_2O_5) or silicon nitride (Si_3N_4) layers on the exposed sensor surface, the reduction of device noise requires complex device concepts like buried-channel FETs [22] or floating-gate devices [23, 24]. An alternative approach is the utilization of wide bandgap materials such as silicon carbide, diamond or group III-nitrides for the realization of electrolyte-gate FETs (EGFETs). Apart from their capabilities in high temperature, high power and high frequency electronics, these materials exhibit excellent properties for sensor applications in harsh environments [25, 26, 27, 28]. Group III-nitrides are chemically stable under physiological conditions as well as non-toxic to living cells [29] and AlGaN/GaN FETs have recently been demonstrated to exhibit promising properties for sensor applications in electrolytic environments [29, 30]. In this contribution, we summarize the properties of AlGaN/GaN heterostructure EGFET arrays as transducer devices in field-effect biosensors. Basic device properties such as surface ion sensitivity, transistor characteristics and low-frequency noise are discussed and the extracellular recording of action potentials from a cardiac myocyte syncytium cultivated on the transistor surface is reported.

2 Device Structure and Processing

EGFET-arrays consisting of 16 individual transistors were processed on AlGaN/GaN heterostructures grown by metalorganic vapor phase epitaxy on sapphire substrates [31]. The sample structures are shown in Fig.1.

The heterostructures consisted of 1.5 µm GaN buffer and a 24 - 26 nm undoped AlGaN barrier, with a 3 nm undoped GaN-cap layer. The Al-content in the AlGaN barriers was 20 - 30%, determined by high resolution X-ray diffraction measurements. Room temperature sheet carrier concentrations and

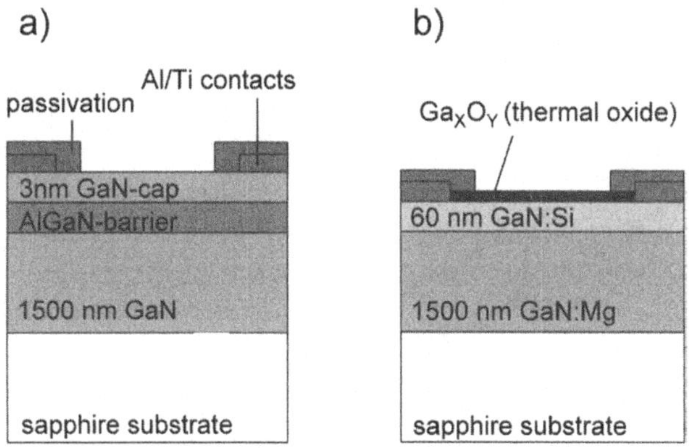

Fig. 1. Schematic layout of (a) the GaN/AlGaN/GaN and (b) the thermally oxidized GaN:Si/GaN:Mg heterostructure

mobilities obtained from Hall measurements were between 1×10^{13} cm^{-2} and 1.2×10^{13} cm^{-2} and ranged from 1100 cm^2/Vs to 1240 cm^2/Vs, respectively. For comparison of the ion-induced surface potential changes of as-deposited and thermally oxidized surfaces, additional transistor structures based on GaN:Si/GaN:Mg double layers have been investigated. These transistors consisted of a 60 nm thick Si-doped channel on a 1.5 μm GaN buffer layer grown on sapphire substrates by plasma-induced molecular beam epitaxy. The GaN buffer layer was compensated by Mg-doping to achieve a high resistivity compared to the thin Si-doped channel. The carrier densities at room temperature obtained from C -V measurements were 3×10^{18} cm^{-3} in the Si-doped channel and 5×10^{15}cm^{-3} in the Mg-compensated GaN buffer layer. Thermal oxidation in dry oxygen at 700°C for 2 h lead to the formation of a thin Ga_xO_y layer, as observed by X-ray photoelectron spectroscopy (XPS) [32]. Single transistor elements with a channel width of 35 μm and a source-drain spacing of 35 μm were patterned by Ar ion-beam etching. The gate length was defined by photolithographically opening the epoxy-based passivation layer. The EGFET arrays were mounted on a standard 28 DIL ceramic chip carrier, wire bonded and partially encapsulated to form an electronic culture dish.

3 Ion-Sensitivity of GaN Surfaces

For the operation in electrolyte solutions the chemical response of the gate surface to changes in the ionic composition of the ambient electrolyte is of major importance. As H^+ is the main ion with respect to cell metabolism monitoring [4, 5, 6, 7], we have investigated the sensitivity of both, as grown and thermally oxidized GaN surfaces towards pH changes. The measurements

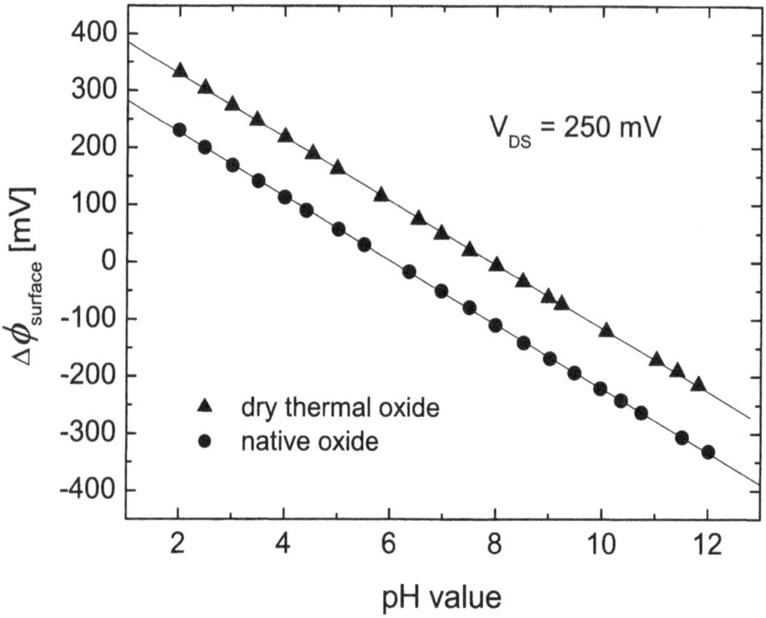

Fig. 2. Variation of the surface/electrolyte potential as a function of pH. Datasets have been shifted by a constant potential offset

were performed at a constant temperature of 23°C with a three-electrode setup. The pH of the solution was measured by a pH glass electrode and adjusted by titration with diluted NaOH and HCl with a background of 100 mM NaCl/10 mM HEPES to adjust the overall ionic strength. During pH changes the channel current I_{DS} was kept constant by adjusting the gate potential by ΔV_{GS} via the Ag/AgCl reference electrode and the potentiostat used in this setup. For these measurement conditions, the pH-dependent change in the GaN surface potential $\Delta\phi_{surface}$ is balanced by the gate potential variation ΔV_{GS}. In Fig. 2 the change in surface potential as a function of the electrolyte pH for the two investigated surfaces is shown.

A linear dependence of 56.6 mV/pH for the thermally oxidized surface and of 56.0 mV/pH for the as grown GaN surface was obtained in the investigated range between pH 2 and pH 12. These values are close to the theoretical Nernstian limit of 58.7 mV/pH for the response at this temperature.

XPS analysis of as-deposited GaN surfaces via the surface sensitive Ga2p and the O1s core level spectra revealed the almost immediate formation of a thin surface oxide after exposure to atmosphere as similarly reported by other groups [33]. Therefore, we attribute the observed pH response to the metal oxidic Ga_xO_y surface according to the site-binding model by Yates et al. [34], which was first applied to ISFETs by Siu and Cobbold [35], and further developed by Bousse [36] and Fung [37]. Following this model, amphoteric hydroxyl groups are formed at oxidic surfaces in contact with aqueous solutions. These hydroxyl groups are either neutral, protonized or deprotonized, depending on the H^+-concentration and the equilibrium constants K_a and K_b for the relevant dissociation reactions:

$$GaOH \overset{K_a}{\rightleftharpoons} GaO^- + H^+ \;, \tag{1a}$$

$$GaOH \overset{K_b}{\rightleftharpoons} GaOH_2^+ + H^+ \;. \tag{1b}$$

This leads to a pH-dependent net surface charge and an additional voltage drop at the solid/liquid interface. Bayer et al. calculated the pH sensitivity of Ga_xO_y surfaces and the best agreement with our experimental results was found for dissociation constants of $K_a \sim 10^{-8}$ and $K_b \sim 10^{-6}$ and a point of zero charge of $pH_{pzc} = 6.8$ [38]. For comparison values of $K_a \sim 10^{-10}$ and $K_b \sim 10^{-6}$ can be found for Al_2O_3 in the literature [39].

Furthermore, the surface sensitity towards Na^+, K^+, Ca^{2+}, which are the main ions exchanged during cell action potentials, and Cl^- ions was investigated. For this purpose, the Ag/AgCl reference electrode was replaced by a double chamber reference electrode with a 1M tris(hydroxymethyl)amoniomethane (TRIS) solution as an outer electrolyte. The electrolyte solution was a 50 mM HEPES buffer at pH 7. Ion concentrations were measured by ion-sensitive glass or polymer membrane electrodes. Compared to the pH sensitivity, significantly lower responses were measured towards the other investigated ions. The results are summarized in table 1.

Table 1. Change of surface potential $\Delta\phi_{surface}$ with change in ion concentration

Ion	$\Delta\phi_{surface}/dec.$
H^+	56 – 57mV
K^+	< 2mV
Na^+	< 5mV
Ca^{2+}	< 2mV
Cl^-	< 2mV

4 Electronic Characteristics of AlGaN/GaN EGFETs

For electrical measurements, the potential V_{GS} of the electrolytic gate with respect to the grounded source contact was adjusted via an Ag/AgCl-wire electrode. Fig. 3 shows the transfer- and transconductance characteristics of a single AlGaN/GaN-EGFET element.

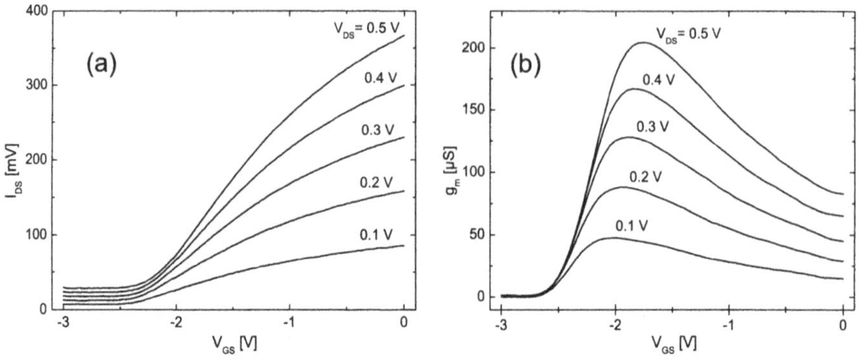

Fig. 3. (a) Transfer- and (b) transconductance characteristics of a single AlGaN/GaN EGFET element

Under usual operating conditions (drain-source voltage V_{DS} = 0.5 V), the devices exhibited a maximum transconductance g_m of 0.2 mS at V_{GS} = -1.8 V. The leakage current I_{GS} through the electrolytic gate was negligible (<2 nA).

The sensitivity of the transducer is limited by its signal-to-noise ratio and determines the resolution of the whole biosensor. As cell action potentials and other biological signals usually occur at low frequencies, the low frequency noise (1Hz < f < 1 kHz) of the transistor arrays was analyzed for different gate-source voltages at room temperature. A low-pass filter at 3 kHz was used to reduce the noise level during signal recording. From the observed drain-source current noise S_{IDS} the equivalent gate-input noise S_{VGS} was calculated by $S_{VGS} = S_{IDS}/g_m^2$.

The typical noise power density spectrum displayed in Fig. 4a shows a $1/f$ frequency dependence. The exponent γ varies with V_{GS} between $\gamma = 0.9 \ldots 1.3$. The corresponding dimensionless Hooge parameter α [40] was calculated to $\gamma = 5 \times 10^{-3}$ at V_{GS} = 0 V from the relative source-drain current spectral density S_{IDS}/I_{DS}^2. The noise power spectral density exhibits a minimum for V_{GS} = -1.8 V, as shown in Fig. 4b for different frequencies. By further optimization of the device structure, reduction of the low frequency noise by at least one order of magnitude is possible [42, 43].

The high signal-to-noise ratio of the AlGaN/GaN EGFET devices is demonstrated in Fig. 5, which presents the change in equivalent gate-source voltage calculated from the change in the source-drain current upon application of a 100 µV test pulse to the electrolytic gate for a time interval of 100 ms. The signal is clearly resolved with a S/N-ratio of $> 5:1$. The background noise under these conditions presents a peak-to-peak amplitude of 15 µV, which is one order of magnitude lower than reported for comparable silicon-based devices [17, 41].

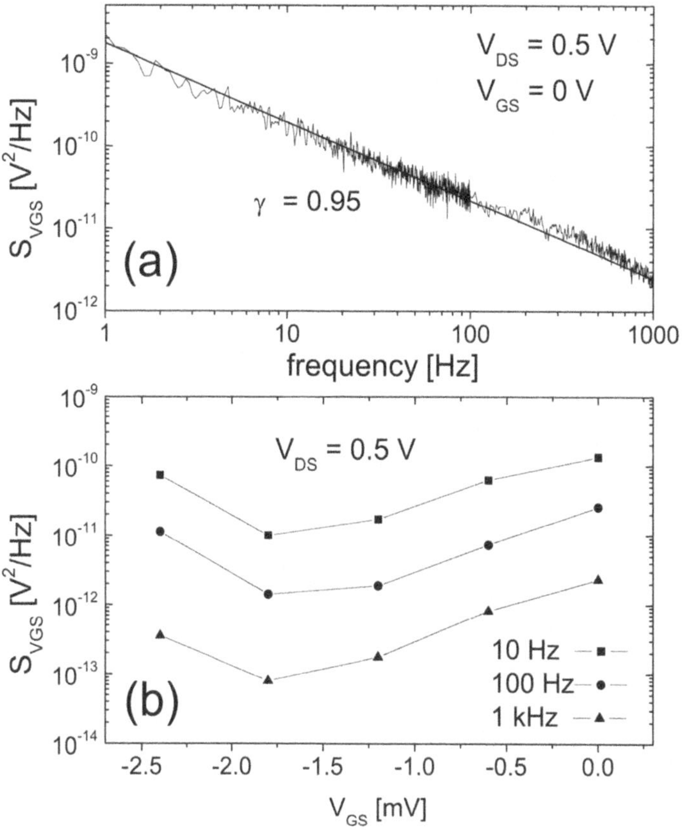

Fig. 4. (a) Noise power density spectrum of a single AlGaN/GaN EGFET element at $V_{GS} = 0V$. (b) Variation of the equivalent gate-voltage noise power spectral density as a function of V_{GS} for different frequencies

5 Recording of Cell Action Potentials

Cardiac myocyte cells of embryonic Wistar rats (Charles River GmbH, Sulzbach, Germany, E 19) were prepared as described in previously published protocols [44]. The AlGaN/GaN chips were cleaned with 70% (v/v) ethanol (p.a.) and coated for 30 min with 20 µl of 12.5 µg/ml fibronectin (Sigma-Aldrich) in Hanks' balanced salt solution (HBSS) at 37°C. The surfaces were rinsed with phosphate-buffered saline (PBS) buffer solution (137 mM NaCl, 2.7 mM KCl, 4.3 mM Na_2HPO_4, 1.4 mM NaH_2PO_4, pH 7.3). The cells were then plated onto the chips at densities of 3000 - 5000 cells per mm^2 exposed chip area (effective chip surface: 6.2 mm^2). The culture was incubated at 37°C and 5% CO_2 in humidified atmosphere. The Ham's F10 medium (Sigma-Aldrich) containing 5% (v/v) foetal bovine serum (FBS), 1% (v/v) penicillin-streptomycin (10000 units/ml penicillin and 10 mg/ml

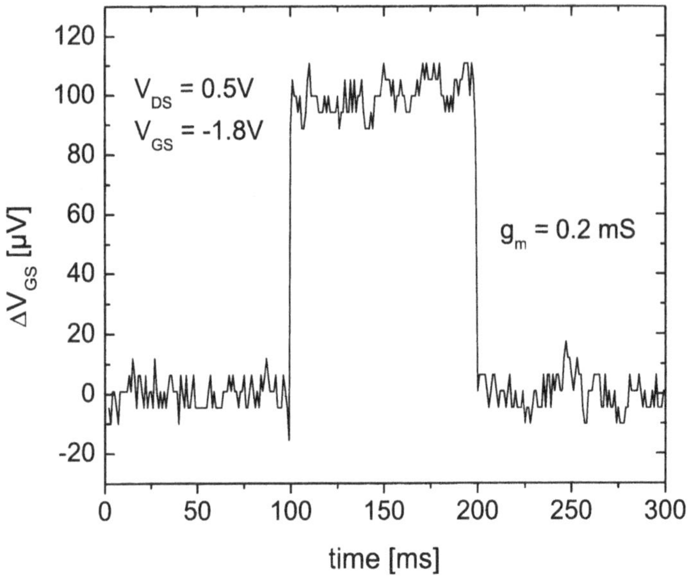

Fig. 5. Transient response of a single AlGaN/GaN EGFET element to a 100 μV pulse applied for 100 ms via the electrolyte gate

streptomycin, Sigma-Aldrich) and 0.5% (v/v) ITS (Sigma-Aldrich), adjusted to pH 7.2 was replaced several times. Under these conditions, after 5 - 6 days in culture, a confluent monolayer of cells (syncytium) which exhibited spontaneous rhythmic activity developed on the sensor surface and action potentials were recorded.

For this purpose, the culture medium was replaced by extracellular solution (145 mM NaCl, 3 mM KCl, 2 mM $MgCl_2$, 3 mM $CaCl_2$, 10 mM HEPES, 8 mM glucose, adjusted to pH 7.3 with NaOH). Figure 6 shows cardiac myocyte cells on the gates of the AlGaN/GaN transistors.

The cells were beating spontaneously with a stable frequency and the electrical activity of the cells was recorded by means of the voltage V_J in the junction area between cell and transistor gate. The monitored transistor signals were 100 ms - 150 ms in duration, firing at a stable frequency for several minutes (Fig. 7). Signals with amplitudes of 1.2 mV were clearly detected because of the low background noise of the AlGaN/GaN transistors [45].

Action potentials of fully developed cultured cardiac cells are mainly generated by ionic fluxes of sodium, calcium and potassium across the cellular membrane. The initial fast-rising action potential is triggered by an influx of sodium ions. An inward directed calcium flux carries the plateau phase of the action potential and the following potassium efflux repolarizes the cell. A first approximation to describe the shape of action potentials recorded with planar sensors is a point-contact model [17, 46, 47], which considers mainly two contributions to the electrical signal.

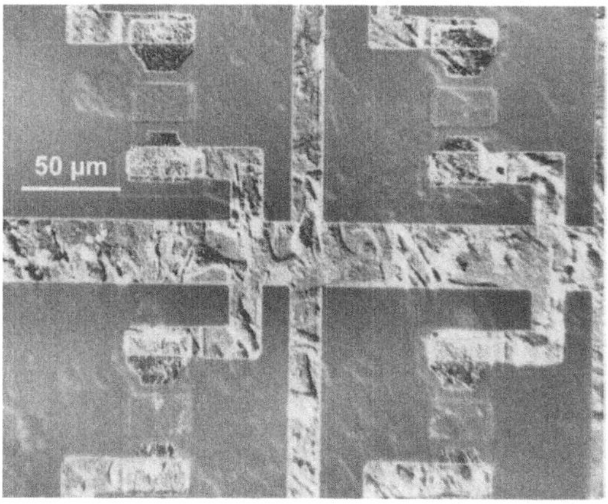

Fig. 6. Syncytium of cardiac myocyte cells cultivated on the surface of a AlGaN/GaN EGFET array

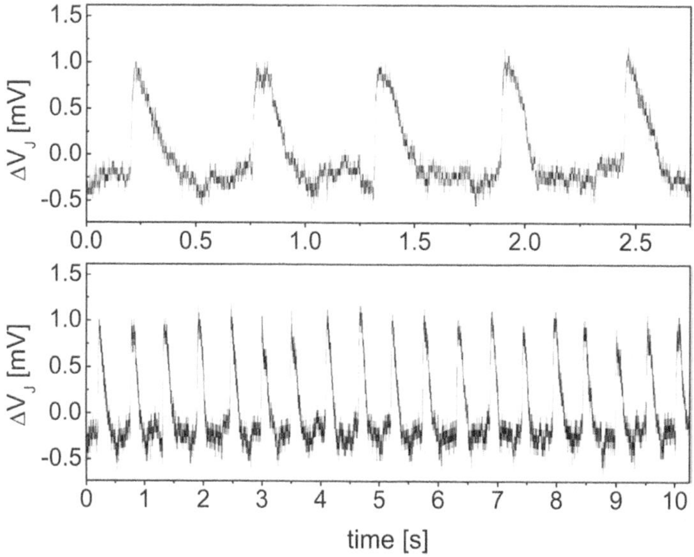

Fig. 7. Extracellular potential of spontaneously beating rat heart muscle cells recorded with an AlGaN/GaN EGFET. The cells were cultivated on the device surface [45]

First, changes of the intracellular potential lead to capacitive currents across the attached part of the cellular membrane into the contact point. Secondly, the ionic currents through ion channels in the attached part of the membrane have to be considered. Both currents are forced to flow along a small cleft formed by the cellular membrane and the sensor surface acting as an electrical resistor. This leads to a voltage change at the sensor surface and both contributions have to be summed for the interpretation of the recorded signal shape [18]. The signal shapes recorded by the AlGaN/GaN transistors (Fig. 7) seem to consist mainly out of the potassium signal part [18, 47]. The exact origin of this behavior shape is not yet clarified.

6 Summary

In conclusion, we have summarized the application of AlGaN/GaN EGFETs as a transducer device for biosensors. The devices show stable operation under physiological conditions and due to the low noise a much higher signal resolution than currently used Si-based devices. Native as well as thermally oxidized GaN surfaces showed a high and linear pH sensitivity of 56 - 57 mV/pH which is close to the theoretical Nernstian limit whereas their response to other ions is negligible. The recording of spontaneous extracellular action potentials of rat heart muscle cells with arrays of these devices has been demonstrated.

Acknowledgements

This work was supported by the Deutsche Forschungsgemeinschaft DFG (SFB 563) and by the European Community within the strategic research project GaNano (Contract No. NMP4-CT-2003-505641). The authors thank R. Helpenstein for the preparation of the cardiac myocyte cells and Armin Dadgar and Alois Krost from the Otto-von-Guericke Universitä Magdeburg for supply of AlGaN/GaN heterostructures.

References

1. S. Caras, J. Janata, Anal. Chem. **52**, 1935 (1980).
2. A. Phoghossian, M. J. Schöning, P. Schroth, A. Simonis, H. Lüth, Sens. Actuators B **78**, 519 (2001).
3. D. G. Pijanowska, M. Dawgul, W. Torbicz, Sensors **3**, 160 (2003).
4. J. W. Parce, J. C. Owicki, K. M. Kercso, G. B. Sigal, H. G. Wada, V. C. Muir, L. J. Bousse, K. L. Ross, B. I. Sikic, H. M. McConnell, Science **246**, 243 (1989).
5. H. M. McConnell, J. C. Owicki, J. W. Parce, D. L. Miller, G. T. Baxter, H. G. Wada, S. Pitchford, Science **257**, 1906 (1992).
6. J. C. Owicki, J. W. Parce, Biosensors Bioelectron. **7**, 255 (1992).

7. W. H. Baumann, M. Lehmann, A. Schwinde, R. Ehert, M. Brischwein, B. Wolf, Sens. Actuators B **55**, 77 (1999).
8. G. W. Gross, B. K. Rhoades, H. M. E. Azzazy, M.-C. Wu, Biosensors Bioelectron. **10**, 553 (1995).
9. C.-K. Yeung, S. Ingebrandt, M. Krause, A. Offenhäusser, W. Knoll, J. Pharmacol. Toxicol. Meth. **45**, 207 (2001).
10. E. Neher, B. Sakmann, Nature **260**, 799 (1976).
11. A. Grinvald, Ann. Rev. Neurosci. **8**, 263 (1985).
12. C. A. Thomas, P. A. Springer, G. E. Loeb, Y. Berwald-Netter, L. M. Okun, Exp. Cell Res. **74**, 61 (1992).
13. G. W. Gross, W. Y. Wen, J. W. Lin, J. Neurosci. Methods **15**, 243 (1985).
14. F. Heer, W. Franks, A. Blau, S. Taschini, C. Ziegler, A. Hierlemann, H. Baltes, Biosensors Bioelectron. **20**, 358 (2004).
15. P. Bergveld, J. Wiersma, H. Meertens, IEEE Trans. Biomed. Eng. **23**, 136 (1976).
16. P. Fromherz, A. Offenhäusser, T. Vetter, J. Weis, Science **252**, 1290 (1991).
17. A. Offenhäusser, C. Sprössler, M. Matsuzawa, W. Knoll, Bioelectron. **12**, 819 (1997).
18. S. Ingebrandt, C.-K. Yeung, M. Krause, and A. Offenhäusser, Biosensors Bioelectron. **16**, 565 (2001).
19. A. Poghoossian, M. Thust, M.J. Schöning, M. Müller-Veggian, P. Kordos, H. Lüth, Sens. Actuators B **68**, 260 (2000).
20. H. Abe, M. Esashi, T. Matsuo, IEEE Trans. Electr. Dev. **ED-26**(12), 1939 (1979).
21. C. Jakobsen, I. Bloom, Y. Nemirovsky, Solid-St. Electron. **42**, 1807 (1998).
22. M. Völker, P. Fromherz, small **1**, 1 (2005).
23. A. Offenhäusser, J. Ruhe, W. Knoll, J. Vac. Sci. Technol. A **13**, 2606 (1995).
24. A. Cohen, M. E. Spira, S. Yitshaik, G. Borghs, O. Shwartzglass, J. Shappir, Biosensors Bioelectron. **19**, 1703 (2004).
25. A. Lloyd Spetz, A. Baranzahi, I. Lundström, Phys. Stat. Sol (a) **162**, 493 (1997).
26. E. Kohn, P. Gluche, M. Adamschik, Diam. Relat. Mater. **8**, 934 (1998).
27. J. Schalwig, G. Müller, M. Eickhoff, O. Ambacher, M. Stutzmann, Sens. Actuators B **87**, 425 (2002).
28. M. Eickhoff, J. Schalwig, G. Steinhoff, O. Weidemann, L. Görgens, R. Neuberger, M. Hermann, B. Baur, G. Müller, O. Ambacher, M. Stutzmann, Phys. Stat. Sol. (c) **0**, 1908 (2003).
29. G. Steinhoff, O. Purrucker, M. Tanaka, M. Stutzmann, M. Eickhoff, Adv. Funct. Mater. **13**, 841 (2003).
30. G. Steinhoff, M. Hermann, W. Schaff, L. F. Eastman, M. Stutzmann, M. Eickhoff, Appl. Phys. Lett. **83**, 178 (2003).
31. R. Dimitrov, M. Murphy, J. Smart, W. Schaff, J. R. Shealy, L. F. Eastman, O. Ambacher, M. Stutzmann, J. Appl. Phys. **87**, 3375 (2000).
32. M. Eickhoff, R. Neuberger, G. Steinhoff, O. Ambacher, G. Müller, M. Stutzmann, Phys. Stat. Sol (b) **228**(2), 519 (2001).
33. M. Kocan, A. Rizzi, H. Lüth, S. Keller, U. Mishra, Phys. Stat. Sol. (b) **234**(3), 773 (2002).
34. D. E . Yates, S. Levine, T. W. Healy, J. Chem. Soc. Farady Trans. I **70**, 1807 (1974).

35. W. M. Siu, R. S. C. Cobbold, IEEE Trans. Electr. Dev. **ED-26**(11), 1805 (1979).
36. L. Bousse, N. F. De Rooij, P. Bergveld, IEEE Trans. Electr. Dev. **ED-30**(10), 1263 (1983).
37. C. D. Fung, P. W. Cheung, W. H. Ko, IEEE Trans. Electr. Dev. **ED-33**(1), 8 (1986).
38. M. Bayer, C. Uhl, P. Vogl, J. Appl. Phys **97**, 033703 (2005).
39. R. E. G. van Hal, J. C. T. Eijkel, P. Bergveld, Sens. Actuators B **24**, 201 (1995).
40. F. N. Hooge, IEEE Trans. Electr. Dev. **ED-41**(11), 1926 (1994).
41. S. Ingebrandt, C. K. Yeung, W. Staab, T. Zetterer, A. Offenhäusser, Biosensors Bioelectron. **18**, 429 (2003).
42. S. Rumyantsev, M. E. Levinshtein, R. Gaska, M. S. Shur, J. W. Yang, M. A. Khan, J. Appl. Phys. Lett. **87**, 1849 (2000).
43. S. A. Vitusevich, S. V. Danylyuk, N. Klein, M. V. Petrychuk, A. E. Belyaev, J. Appl. Phys. **95**, 5625 (2004).
44. M. C. T. Denyer, M. Riehle, J. Hayashi, M. Scholl, C. Sproessler, S. T. Britland, A. Offenhäusser, W. Knoll, In Vitro Cell. Dev. Biol.-Animal **35**, 352 (1999).
45. G. Steinhoff, B. Baur, G. Wrobel, S. Ingebrandt, A. Offenhäusser, A. Dadgar, A. Krost, M. Stutzmann, M. Eickhoff, Appl. Phys. Lett. **86**, 033901 (2005).
46. W. G. Regehr, J. Pine, C. S. Cohan, M. D. Mischke, D. W. Tank, J. Neurosci. Methods **30**, 91 (1989).
47. C. Sprössler, M. Denyer, S. Britland, W. Knoll, A. Offenhäusser, Phys. Rev. E **60**, 2171 (1999).

Advanced MOSFET Gate Dielectrics for High-Performance Microprocessors: Materials Selection and Analytical Challenges

E. Zschech[1], H.J. Engelmann[1], S. Ohsiek[1], B. Tracy[2], E. Adem[2], S. Robie[2], J. Bernard[2], and D. Schmeisser[3]

[1] AMD Saxony LLC & Co. KG, Dresden, Germany
[2] Spansion Inc., Sunnyvale/CA, USA
[3] Applied Physics-Sensor Technologies, Brandenburg University of Technology, Cottbus, Germany

1 Introduction

For semiconductor industry, the challenges to future process technology and advanced materials are outlined in the International Technology Roadmap for Semiconductors (ITRS) [1]. Microelectronic and further nanoelectronic products of the next technology generations will need advanced materials particularly for device structures since the traditional shrinking of feature sizes is reaching fundamental physical limits. Currently, leading-edge microprocessor (MPU) technology is pushing material innovations in the gate stack of metal-oxide-semiconductor field effect transistors (MOSFETs).

The high frequency performance of MPUs, in order to obtain intrinsic MOSFET gate delays in the range of picoseconds (or terahertz performance levels) well before the end of the decade, is a major driving force for solutions that respond to the challenges outlined in the ITRS [1]. Consequently, one key transistor performance metric is $C_{gate}V_{dd}/I_d$ – the product of the gate capacitance and the drain voltage divided by the drive current. This value measures the individual transistor's switching time. As the gate length shrinks, improved performance arises from smaller capacitance and voltage, and increased drive current. Actually, the gate capacitance is responsible for switching the transistor. For MOSFETs with reduced gate length, the amount of charge being switched is smaller, and consequently, less capacitance is needed. However, reducing the gate dielectric thickness causes high leakage current, one of the main concerns for future CMOS technology nodes. Since the dominant transport mechanism through SiO_2 films with a thickness of less than 2 nm is direct tunneling of electrons or holes, the leakage current increases exponentially with decreasing thickness. Considering the metric mentioned above, a continuous transistor performance increase will be only possible if improvements from gate length scaling, equivalent oxide thickness (EOT) for gate dielectrics, threshold voltage and carrier mobility in the channel will be combined ([2], see also Table 1). Material approaches which have the potential to increase the transistor performance significantly

B. Kramer (Ed.): Adv. in Solid State Phys. **45**, 375–388 (2005)

Table 1. Technology requirements for mainstream logic devices [2]

Front-end technology for high-performance logic devices					
Tech. nodes	90 nm	65 nm	45 nm	32 nm	22 nm
Gate stack	SiO_2/poly	SiO_2/poly	High-κ/metal Chemical oxide starting layer	High-κ/metal Direct epitaxy? $\kappa \sim 30$	
Channel	Strained Si planar	Strained Si planar	3D Multi-gate?	– Strained Si – High mobility semiconductors (Ge, III-V)	
Physical gate length (nm)	37	25	18	13	9
EOT (nm)	1.2	0.8	0.7	0.6	0.5
Drive current NMOS (mA/µm)	1.2	1.5	1.9	2.1	2.4

are high-k gate dielectric materials for gate leakage reduction and strained silicon (or high mobility semiconductors) for carrier mobility increase in the transistor channel.

High-k gate dielectric materials with significantly higher permittivity than that of SiO_2 and with a sufficient band gap will be necessary to improve the transistor performance and to maintain its reliability. It is expected that advanced high-k dielectric/metal electrode stacks will be integrated into the manufacturing process of MOSFETs for low power (low operating power and low standby power) MPUs and for high-performance MPUs beginning with the 65 nm or 45 nm CMOS technology node [3]. This need is particularly acute for low-power logic since it requires a very low gate leakage current. Using high-k gate dielectrics with large physical thickness (about 3 nm), it will be possible to scale down the EOT to less than 1 nm maintaining the gate leakage current at acceptable levels below 1 A/cm^2[4]. Unfortunately, transistors with advanced high-k ggate dielectrics suffer from mobility degradation for reasons which are currently not fully understood. One approach to overcome this problem temporarily is to deposit at first a very thin SiO_2 layer of 2–5 Å thickness and subsequently the high-k material [4]. This layer stack does not reduce the channel mobility, however the EOT is limited by the SiO_2 layer and it limits further scaling to 32 and 22 nm CMOS technology nodes where EOTs as low as 5 Å are required (see Table 1, [2]). For such low EOT values, ultimately clean interfaces are needed.

For gate dielectric applications in Si-based transistor devices, it is necessary to grow either amorphous or epitaxial oxides directly on the single-crystalline Si, particularly on (001)-oriented Si which has been the favorite substrate orientation in semiconductor industry for many years. Epitaxial growth provides the general possibility of exactly designed interfaces. It has been shown that Y_2O_3 [5, 6] and rare earth oxides with cubic symmetry such as Pr_2O_3 [7] can be grown crystalline on Si (001). Acceptable epitaxy quality has been demonstrated for several oxides, however the limited wafer throughput of molecular beam epitaxy (MBE) has prevented the use of this approach in high volume manufacturing. An alternative, successfully applied thin film deposition method is atomic layer deposition (ALD) which has the capability to produce thin, smooth, conformal and pure films with defined stoichiometry and well controllable thickness at low growth temperatures (100 – 400 °C) [8, 9, 10]. Scarel et al. have illustrated how ALD allows further engineering of film properties (roughness, crystallinity, density, impurity content, interfacial layer, electrical properties) and structures (films on different substrates, multilayers and nano-scale defined structures) [11].

In this paper, several steps in the evolution process - starting with silicon oxynitride up to real high-k metal oxide gate dielectrics - as well as the need of leading-edge analytical and metrology techniques, including data analysis procedures, are discussed. Particularly, the application of spectroscopic and high-resolution imaging techniques is demonstrated – as prerequisites for the optimization of thin films that are necessary to improve the product performance and to maintain the product reliability of high-performance MPUs.

2 High-*k* Materials Selection, Materials Properties, Integration Challenges and Analytical Needs

Silicon oxide (SiO_2) has a dielectric constant of about 4 with a unique combination of properties: stable, easy to grow and to control, and compatible to silicon. Thermally grown, amorphous SiO_2 has been used for decades to isolate the transistor gate from the channel for Si-based MOSFETs. However, with thicknesses close to 1 nm, physical limits have been reached for leading edge logic products. Since the capacitance of the gate dielectric layer depends on its thickness and on its dielectric constant, the electrical thickness can be reduced by increasing the permittivity of the gate dielectric material even for a constant physical thickness. That means, the transistor performance can be further increased by metal oxide gate dielectrics with a high dielectric constant (high-k materials). The particular need for MOSFETs with low gate leakage current is driving the implementation of high-k gate dielectric materials. However, until real high-k dielectrics will be introduced into the CMOS process, silicon oxynitrides (SiO_xN_y) will being used as an interim solution. For the 90 nm CMOS technology node, and probably also for the 65 nm CMOS technology node, manufacturers are staying with SiO_xN_y. This material provides a first step into the high-k materials direction, however, its permittivity offers only little benefit (increase of the k value less than a factor of 2).

High-k dielectric thin films material selection is needed before deposition, e. g. using ALD, considering several criteria that have to be fulfilled simultaneously [9]: (1) permittivity, (2) band structure and the associated band offsets for carrier transport, (3) thermodynamic stability considerations at the gate stack interfaces, (4) interface quality, (5) film morphology, (6) gate electrode compatibility, (7) process compatibility, and (8) reliability. The formation of a thin but optimal interface layer in contact with the substrate is probably the most critical one [12]. To ensure a high carrier mobility in the channel and to avoid a significant metal interdiffusion to the channel region, it is important to have no metal oxide or silicide phase present at the channel interface, i. e., the interface layer has to be thick enough. On the other hand, the thickness of a SiO_2 or SiO_2-containing interface layer has to be compromised to minimize the capacitance gain from any high-k layers in the gate stack.

Several high-k materials have been studied extensively, such as Al_2O_3, HfO_2, ZrO_2, TiO_2, and rare earth oxides. Al_2O_3 has been considered mainly because of its high optical band gap, however, its dielectric constant is only in a medium range. There is a great competition between suitable transition metal oxides (mainly HfO_2 and ZrO_2) and the whole series of rare earth oxides (from La to Lu). The latter exhibit the potential of fine tuning some structural, thermal, mechanical, vibrational, electric, and electronic properties due to the progressive filling of the f shell [13], and the variation of ionic radius and electronegativity [14] with the atomic number of the rare earth element (see Table 1, [2]). Among the research groups involved in the ALD deposition of oxides, many of them are investigating high-k transition metal oxide films, but only a few are exploring the potentials of rare earth oxide films [10,15]. The main difficulty in the ALD of rare earth oxides is the identification of suitable precursors.

The introduction of new high-k materials and respective equipment like ALD tools into the manufacturing process will require to solve several integration issues. New material integration, including interface engineering, and material characterization rather than by the materials properties itself are major challenges. There are several factors which will make it inherently more difficult to control film thickness and composition uniformity, considering also a possible additional interface layer, and to control the interface quality: For the SiO_2 process, metrology solutions for thickness measurements have been well establish for decades. It has been sufficient to measure the oxide thickness on blanket wafers, e. g. applying ellipsometry. For high-k materials, a new and critical metrology parameter – not present before high-k adoption – is the film or film stack composition. That means, both film thickness and composition of high-k gate dielectrics and of possible interface layers have to be determined, two values that might be correlated in measuring signals. Even gradual changes in the chemical composition of the gate dielectrics have to be detected to guarantee process stability. In contrast to

SiO_2, the high-k material/Si interface quality has to be controlled in addition. That means, materials characterization of thin ($< 5\,nm$) films and of interfaces is a key requirement to analytics and metrology, for which solutions have to be developed.

For unpatterned wafers, X-ray photoelectron spectroscopy (XPS) has been established as an potential out-of-fab analytical technique and – for special applications like N concentration monitoring in SiO_xN_y films – as in-line metrology technique of choice. At patterned structures, transmission electron microscopy (TEM) at cross-sectioned samples is the preferred technique. The gate dielectric thickness determination is discussed in [16] extensively. In addition, the N concentration – and maybe the distribution across the gate dielectric – has to be determined for SiO_xN_y layers, and the metal concentration for metal oxide layers.

3 Silicon Oxynitride

Silicon oxynitride layers (SiO_xN_y) are presently used as replacements for SiO_2 [16]. SiO_xN_y thin films provide less impurity penetration through the dielectric, are more reliable and have a slightly higher dielectric constant than SiO_2 [17]. These films can be produced by thermal oxynitridation of Si with NO or N_2O, or by thermal nitridation of SiO_2 films in NO, N_2O, N_2 or NH_3. They can also be grown by chemical vapor deposition (CVD) or ALD, or N can be incorporated into SiO_2 by low energy ion implantation. Depending on the growth process and the process parameters, nitrogen can pile up at or near the Si/SiO_2 interface, be uniformly distributed through the film, or be enriched at the interface to the poly-Si gate electrode [17]. The N distribution affects the electrical properties of the gate dielectric layer significantly, and therefore, it has to be analyzed. Analytical techniques are required that can measure N in the 5 – 15 at% range with sub-nm depth resolution – both for unpatterned films and at transistor structures [18].

For blanket wafers, SiO_xN_y layers can be analyzed using surface analytical techniques like X-ray photoelectron spectroscopy (XPS) or time-of-flight secondary ion mass spectroscopy (TOFSIMS) [17]. In the XPS technique, only photoelectrons which escape from the sample without energy loss contribute to the observed signal whereas electrons which have lost energy in inelastic collisions contribute to the background. The distance travelled by an electron between two inelastic collisions is called inelastic mean free path (IMFP). In practice, the attenuation length which also takes elastic scattering into account is used to describe the electron transport in a sample.

Plotting the N/O ratio as a function of the emission angle of the photoelectrons provides qualitative depth information, i. e., on the location of the nitrogen within a thin SiO_xN_y film. Fig. 1 shows an example for three thin SiO_xN_y layers prepared with different recipes. Samples 1 and 2 show a nitrogen enrichment towards the interface whereas in sample 3 the nitrogen

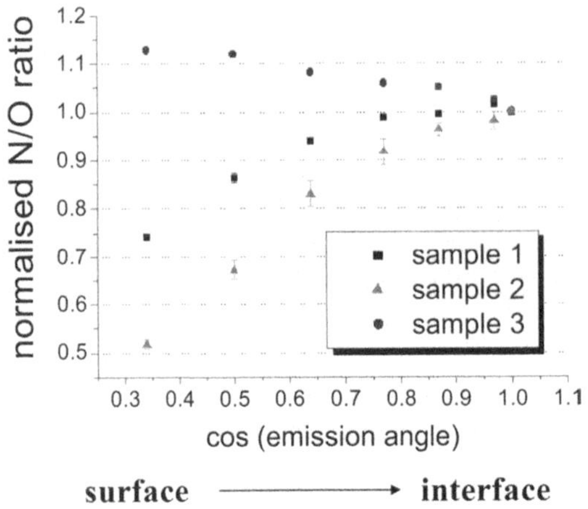

Fig. 1. ARXPS data for 3 SiO_xN_y samples, prepared with different recipies

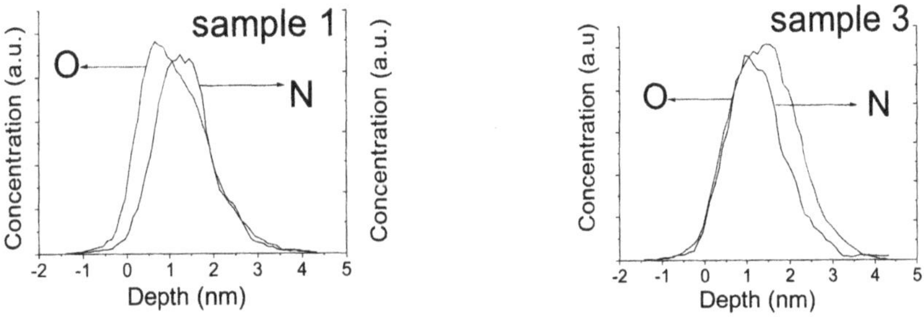

Fig. 2. HR-ERD data for 2 SiO_xN_y samples

is enriched towards the sample surface. These findings are in good agreement with absolute data from high resolution elastic recoil detection (ERD) (see Fig. 2).

In addition to the location of the nitrogen, XPS provides information about the chemical bonding in SiO_xN_y films. Despite a similar location of the nitrogen within the films, the binding states can vary considerably. The spectra of samples 1 and 2 acquired at an emission angle of 50 ° are compared in Fig. 3. Sample 1, a silicon oxynitride, is characterized by a N 1s signal at a binding energy of 398.3 eV. The Si 2p peak contains contributions of the Si substrate, silicon sub-oxides, silicon oxide and silicon oxynitride. In addition to the silicon oxynitride component, sample 2 clearly shows the presence of a nitride species. This becomes obvious from the shoulder in the Si 2p signal in which the nitride component appears at a binding energy of 101.8 eV. The N 1s signal also contains contributions of a nitride and an oxynitride component [19] at binding energies of 397.4 and 398.0 eV, respectively.

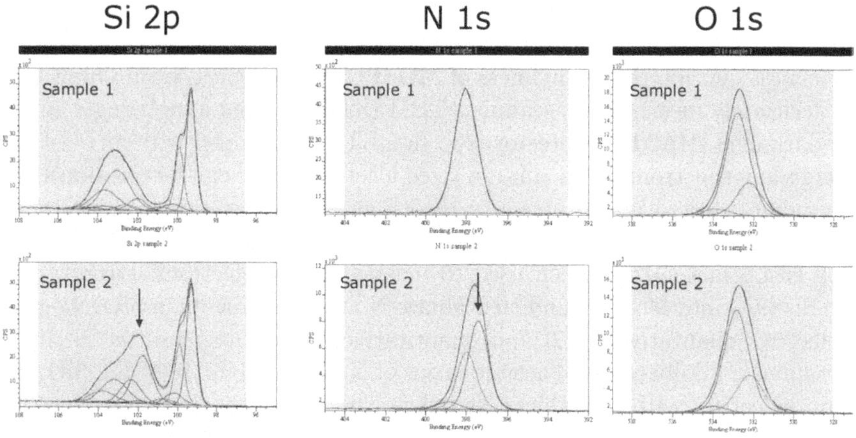

Fig. 3. Chemical binding information for 2 SiO_xN_y samples

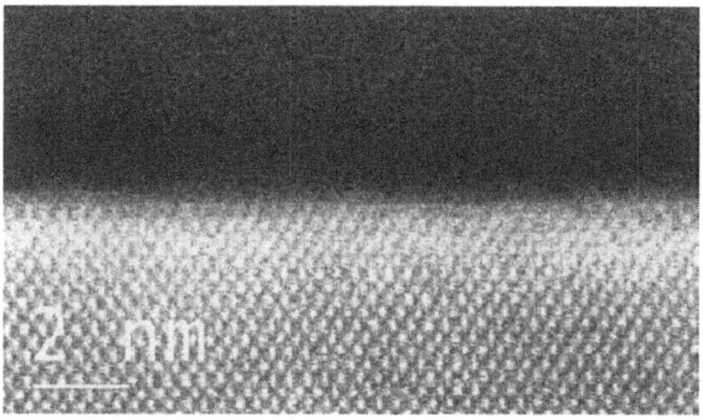

Fig. 4. STEM-HAADF profile of a SiO_xN_y gate dielectric layer

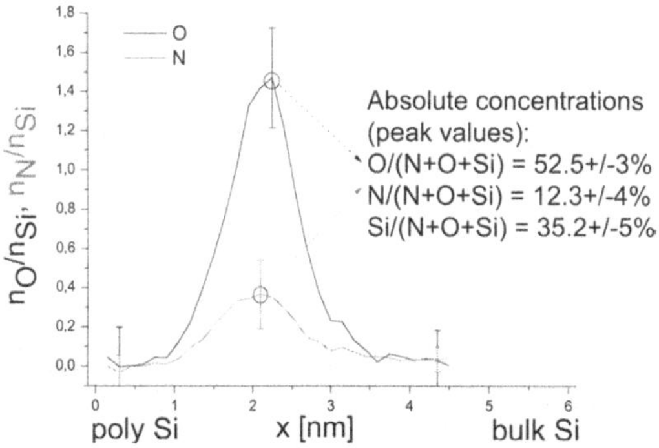

Fig. 5. Atomic ratio profiles across a SiO_xN_y gate dielectric layer [18]

The examination of SiO_xN_y films in MOSFET structures (current physical gate length about 50 nm) requires the high spatial resolution of TEM. Thickness and interface roughness of MOSFET gate dielectric thin films can be accurately measured by scanning TEM (STEM) using a high angle annular dark filed (HAADF) detector [20]. In a STEM equipped with an electron energy loss spectrometer, a sub-nm sized electron probe can be positioned at any point selected in the image to record an electron energy loss spectrum (EELS). Fig. 4 shows a STEM-HAADF image of a SiO_xN_y gate dielectrics. The EELS method has been used to investigate the electronic structure of the Si/SiO_2 interface [21] and to evaluate N and O gradients in SiO_xN_y gate dielectrics qualitatively [22] and quantitatively [23]. Stegmann et al. have measured spatially resolved atomic ratios of N, O and Si in MOSFET SiO_xN_y gate dielectrics with film thickness below 2 nm. Quantitative values for the layer composition were determined by fitting a set of reference spectra to the experimental EELS data [23]. Fig. 5 shows the atomic ratio profiles, n_N/n_{Si} and n_O/n_{Si}, across the gate dielectric layer for a SiO_xN_y layer depositied with plasma-enhanced CVD. Deposition parameters and electrical transistor properties can thus be correlated with the N and O distributions across the oxynitride gate dielectric of transistors [18].

4 Aluminum Oxide

Among the group-III oxides, alumina (Al_2O_3) is a very stable and robust dielectric material. Regarding its usefulness as an alternative gate dielectrics, it has many favourable properties, including a high band gap (8.7 eV) and a thermodynamic stability on Si up to high temperatures. It is amorphous for typical manufacturing and operations conditions. The drawback is its relatively low k-value of about 8–10. One approach published in literature is to determine the band gap from the loss region below the O1s peak in the XPS spectrum [24]. The energy difference between the O 1 s peak center and the start of the increase in spectral background due to excitation of valence band electrons to the conduction band is a rough measure for the band gap, however this method is limited by the inherent width of the O1s signal and seems to underestimate the band gap for some materials.

The XPS sputter depth profiles of as-deposited and of chemically oxidized Al_2O_3 are shown in Figs. 6 and 7. The clearly visible SiO_x binding after chemical oxidation indicates a thin SiO_2 interlayer beneath the Al_2O_3 film, which results in a changed etch behaviour of the layer.

Thin Al_2O_3 layers were studied using X-ray reflectometry (XRR) and grazing incidence X-ray diffraction (GIXRD). A 5-layer XRR model was applied to simulate the density gradients found in these films. GIXRD peak integration was applied to the (440) diffraction peak of γ-Al_2O_3. Fig. 8 shows how film thickness, density and XRD peak intensity vary for an Al_2O_3 film as a function of anneal temperature. This figure shows that the onset of film

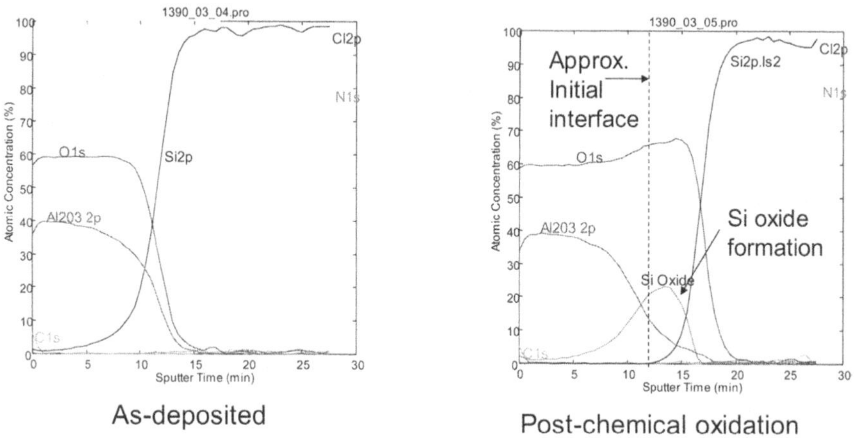

Fig. 6. XPS sputter depth profiles of as-deposited and chemically oxidized Al_2O_3

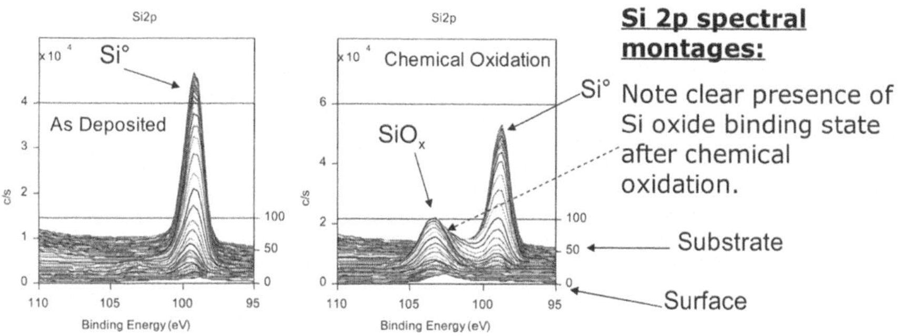

Fig. 7. XPS Si2p sputter depth profiles of as-deposited and chemically oxidized Al_2O_3

shrinkage and densification occurs in tandem with conversion of the Al_2O_3 from the amorphous state to a crystalline phase related to γ-Al_2O_3 at an anneal temperature near 850 °C. This result demonstrates that X-ray techniques are particularly useful for analyzing the changes to high-k materials during processing.

5 Metal Oxides

HfO_2 offers good thermal stability and a dielectric constant between 15 and 28 [25]. However, this material has not been introduced into high volume production since some issue like transient charging, degradation of carrier mobility and unstable threshold voltage characteristics have not been solved.

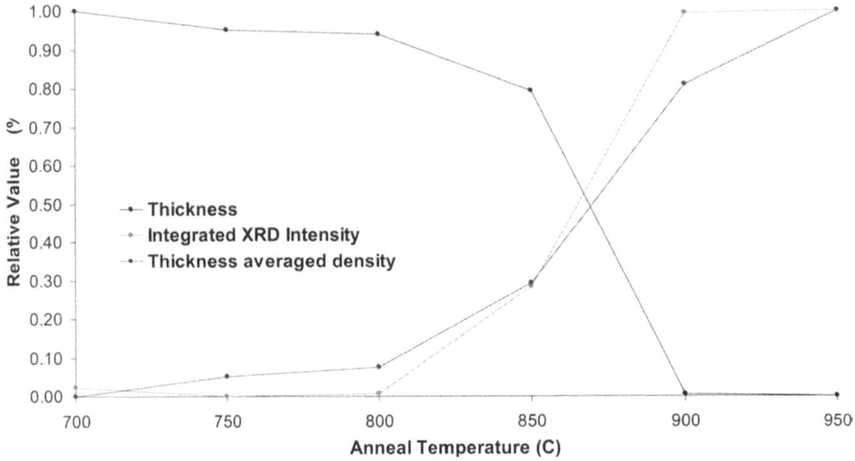

Fig. 8. Effect of anneal temperature on Al_2O_3 thickness, density and crystallinity

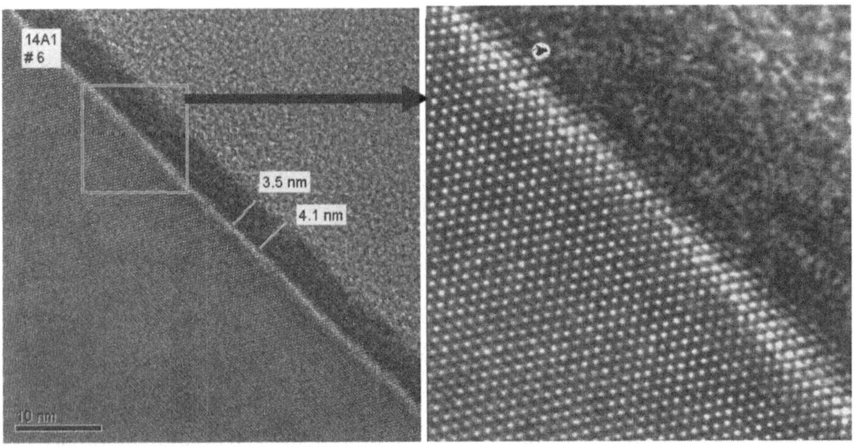

Fig. 9. HAADF-STEM image for annealed HfON on Si substrate

For HfO_2, the thickness of the interlayer – e. g. interfacial SiO_2 or a Hf silicate - between Si and HfO_2 seems to control the rate of traps that charge and decharge rapidly, which can cause charge accumulation, and consequently, carrier mobility degradation as well as threshold voltage instability. Interface engineering, e. g. by nitrogenintroduction and HfON formation, can improve stability. The interlayer near the Si substrate has to be optimized, additional studies are necessary, however, the effect of N on the electronic band structure has to be considered too. Fig. 9 shows a HAADF-STEM image for an annealed HfON gate dielectrics film. The HAADF-STEM image provides the thickness of the dielectric layer and some first hints to interlay-

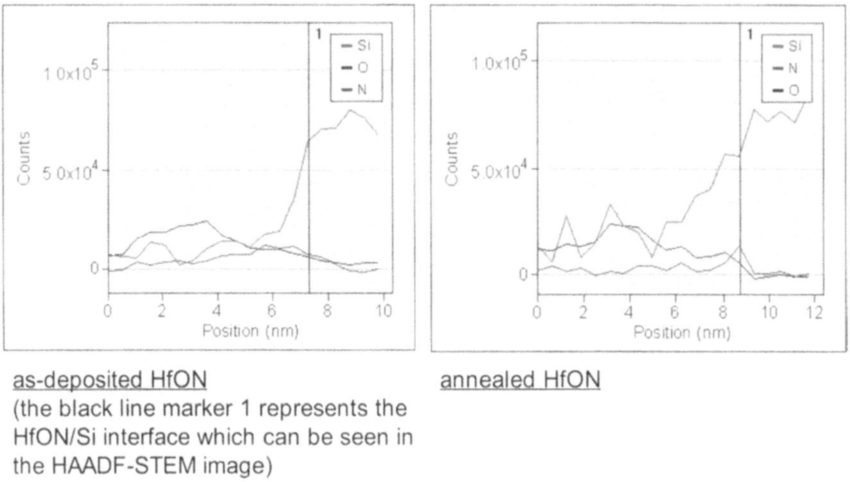

Fig. 10. EELS line scans for as-deposited and annealed HfON on Si substrate

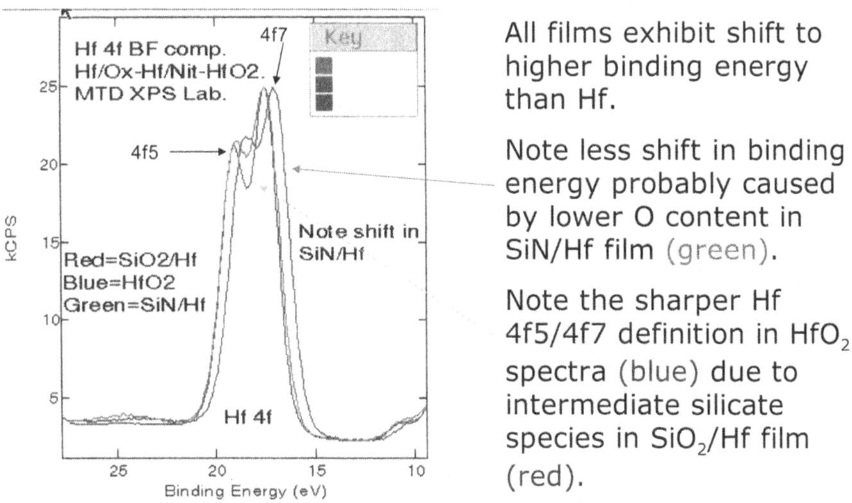

Fig. 11. Hf 4f5/4f7 XPS spectra of various Hf-based gate dielectrics: SiO_2/Hf, HfO_2, SiN/Hf

ers. The respective EELS line scans provide the Si, O and N element profiles, in Fig. 10 presented for the as-deposited and the annealed states of a HfON thin film on a Si substrate. XPS and X-ray absorption spectroscopy (XAS) are techniques which allow to proof the formation of silicates. The comparison of the EELS line scans shows a broad N distribution and only few Si in the as-deposited dielectric film, and a more significant Si signal in the

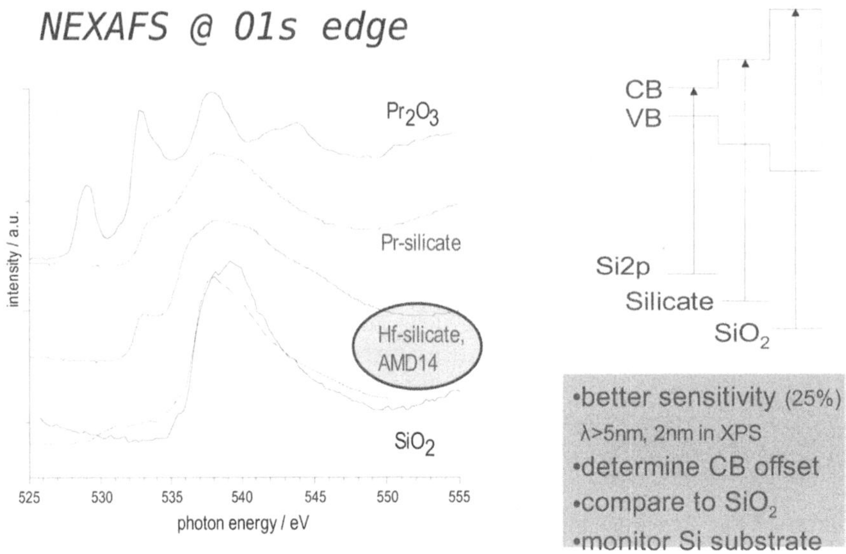

Fig. 12. O-K NEXAFS spectra to demonstrate silicate formation in annealed HfON on Si substrate

film (which is an indication for a Hf silicate formation) and a pronounced N amount at the substrate/film interface after annealing. Fig. 11 demonstrates that a careful quantitative analysis of the Hf 4f5/4f7 XPS spectrum proofs the formation of a silicate. The Near-edge X-ray absorption fine structure (NEXAFS) at the O-K (1 s) absorption edge shows an edge onset between the energies for Si and SiO_2 which again indicates the silicate formation in the annealed Hf(Si)ON film (see Fig. 12). The energy for the O-K absorption edge is nearly at the same energy as for Pr silicate, and it differs significantly from the fine structures usually observed in the O-K NEXAFS of the bulk rare earth oxides [7].

Another subject is the interface between HfO_2 and the poly-silicon gate electrode. Hf-Si bonds at the interface seem to be responsible for Fermi-level pinning. Charge defects accumulate in the oxide near the interface and cause poly-Si charge depletion. The created dipoles change the transistor flatband voltage and shift the threshold voltage. Metal gate electrodes seem to be easier to be integrated with metal oxide gate dielectrics, and are expected to be introduced soon too.

6 Summary and Outlook

The ideal gate dielectric stack should have an interface layer comprised of several atomic layers of Si-O or Si-O-N bonds, which results in the well-known high-quality Si/SiO_2 interface, and a high-k metal oxide layer on top.

The use of medium-k materials like SiO_xN_y will be an interim solution as long as the metal oxide gate dielectrics will not be available for high volume manufacturing.

The integration of high-k dielectric materials into carbon nanotube field effect transistors (CNTFET) will benefit from the good compatibility of carbon with metal oxide. A nanotube encapsulation with metal oxide thin films will not cause a reaction interlayer since CNTs have no dangling bonds. Since carbon does not react with the metal oxide, no interlayer or interface states are formed that decrease the EOT [26]. With high-k dielectrics like ZrO_2 and HfO_2, sub-threshold slopes ranging between 130 mV/decade and 70 mV/decade were demonstrated. This is competitive with the sub-threshold slopes of silicon MOSFETs, and it is close to the theoretical limit of 60 mV/decade at room temperature [27, 28].

The traditional down-scaling of MOSFETs leads to performance limitations that have to be overcome by the introduction of new materials in the gate stack, which result in new and exciting challenges to solid-state physicists and materials scientists. With other words: More than in the past, the future micro- and nanoelectronic products will rely on the introduction of new materials. These new materials for future high-performance transistors will result in both manufacturing process and metrology challenges.

Solid-state physicists and materials scientists will play a key role in the introduction of new high-k gate dielectric materials, and consequently, they will determine the direction of the transistor stack for the next technology nodes, together with engineers in technology development and integration departments. The development and introduction of the ALD process for gate dielectric material deposition creates new demands for thin film analytics and metrology techniques that can be used for process characterization and monitoring of very thin films. The introduction of medium-k and high-k materials (SiO_xN_y and metal oxides) for gate dielectrics creates not only additional demands for thin film metrology, but demands for new kinds of film metrology that were not available or not required for the SiO_2 process.

Acknowledgements

The authors wish to thank Heiko Stegmann, now with Carl Zeiss NTS Oberkochen, for developing a strategy for quantitative EELS analysis of gate dielectrics and Günter Dollinger and his group at the Munich Tandem Accelerator for providing the HR-ERD data.

References

[1] International Technology Roadmap for Semiconductors (ITRS): Semiconductor International Association (SIA), 2003 Edition, http://public.itrs.net
[2] A. Dimoulas, in "Materials for Information Technology" (Eds. E. Zschech, C. Whelan, T. Mikolajick), Springer, London 2005 (in press)

[3] R. Chau, International Workshop on Gate Insulators, Tokyo, Nov 6-8 (2003)
[4] W. Tsai, L. Ragnarsson, P.J. Chen et al., 2003 Symposium on VLSI Technology Digest of Technical Papers, Kyoto, p. 21 (2003)
[5] Y. Nishikawa, N. Fukushima, N. Yasuda, K. Nakayama, S. Ikegawa, Jpn. J. Appl. Phys. 41, 2480 (2002)
[6] A. Dimoulas, G. Vellianitis, A. Travlos, V. Ioannou-Sougleridis, A.G. Nassiopoulou, J. Appl. Phys. 92, 426 (2002)
[7] D. Schmeißer, H.-J. Müssig, J. Phys. of Condensed Matter 16, 153 (2004)
[8] M. Ritala and M. Leskelä, in "Handbook of Thin Film Materials" (Ed. H. S. Nalwa), Vol. 1, Academic Press, San Diego/CA 2002, p. 103
[9] G. D. Wilk, R. M. Wallace, and J. M. Anthony, J. Appl. Phys. 89, 5243 (2001)
[10] L. Niinistö, J. Päiväsaari, J. Niinistö, M. Putkonen and M. Nieminen, Phys. Stat. Sol. (a) 201, 1433 (2004)
[11] G. Scarel, M. Fanciulli, in "Materials for Information Technology" (Eds. E. Zschech, C. Whelan, T. Mikolajick), Springer, London 2005 (in press)
[12] G. Lucovsky, Y. Wu, H. Niimi, V. Misra, J. C. Phillips, Appl. Phys. Lett. 74, 2005 (1999)
[13] G. V. Samsonov, I. Ya. Gil'man, Poroshkovaya Metallurgiya 13, 73 (1974)
[14] S. Jeon, H. Hwang, J. Appl. Phys. 93, 6393 (2003)
[15] G. Scarel, E. Bonera, C. Wiemer et al., Appl. Phys. Lett. 85, 630 (2004)
[16] M. L. Green, E. P. Gusev, R. Degraeve, E. L. Garfunkel, Appl. Phys. Lett, 90, 2057 (2001)
[17] E. Gusev, H. Lu, E. Garfunkel, T. Gustafsson, M. Green, IBM J. Res. Dev. 43, 265 (1999)
[18] H. Stegmann, E. Zschech, Appl. Phys. Lett. 83, 5017 (2003)
[19] P. Hoffmann, D. Schmeißer, G. Roters, Z. Nenyei, Thin Solid Films 428, 216 (2003)
[20] D. Muller, Characterization and Metrology for ULSI Technology, 2000 Int. Conf., AIP Conf. Proc. Vol. 550 (American Institute of Physics, Woodbury, NY, 2001), 500
[21] D. Muller, T. Schorsch, S. Moccio et al., Nature (London) 399, 758 (1999)
[22] E. Principe, A. Hegedus, C. Kisielowski et al., Proc. of the AVS, 48th Int. Symp., San Francisco/CA (American Vacuum Society, New York, 2001)
[23] H. Stegmann, E. Zschech, Characterization and Metrology for ULSI Technology, 2005 Int. Conf., AIP Conf. Proc. (American Institute of Physics, Woodbury, NY, 2005), (in press)
[24] H. Bender, T. Conard, O. Richard et al., E-MRS Spring Meeting, Strasbourg 2003
[25] M. Houssa, M. M. Heyns, in "High-k dielectrics" (Ed. M. Houssa), Institute of Physics, Bristol, 2004, p. 10
[26] R. Seidel, Carbon Nanotube Devices, Thesis, TU Dresden (2004)
[27] S. J. Wind, J. Appenzeller, R. Martel et al., Appl. Phys. Lett. 80, 3817 (2002)
[28] A. Javey, H. Kim, M. Brink et al., 2003 Int. Symp. on VLSI Technology, Systems and Applications, 42 (2003)

Neutron Tomography as Tool for Applied Research and Technical Inspection

Eberhard H. Lehmann, Peter Vontobel, and René Hassanein

Paul Scherrer Institut, 5232 Villigen PSI, Switzerland

Summary. Similar to the imaging with X-ray, neutron imaging systems have been shifted more and more from film exposure towards direct digital detection in the last decade. This approach enables the option for tomography because the image content is measured and understood as dataset now. With the help of mathematical algorithms it becomes possible to derive the necessary information about the full three-dimensional volume of an object from a set of projections from different viewing angles.

Neutron tomography (NT) with best performance requires a strong and well collimated neutron beam. Such beam lines for either thermal or cold neutrons are available at reactor based sources (FRM-2, HMI, ILL) or at spallation neutron sources (PSI). The beam geometry for neutrons is in the best case a parallel one, delivering a sharp 1:1 transmission image of the object under investigation. Therefore, the limited detector resolution is responsible for the tomography resolution generally.

In practical reality, objects with outer dimensions up to 40 cm can be inspected with a nominal resolution of up to 0.1 mm. This compares well with similar X-ray tomography systems used in non-destructive testing (NDT) tasks.

However, the application fields in neutron tomography are different, given by the different interaction probability with matter. Neutrons have the advantage to penetrate most metals (especially heavy ones) very efficiently and to detect hydrogenous compounds very sensitively. Therefore, neutron tomography has application fields in moisture detection, test of adhesive connections, structure analysis for soil and geologic samples and the non-invasive investigation of objects from our cultural heritage.

Based on some practical applications, the paper describes the present state of the art of this technique and gives a prospect for further improvement.

1 Introduction

The non-invasive investigation of inner structures of materials by transmitting radiation became possible first by the invention of X-ray by C. Röntgen at the end of the 19th century. Since this time, X-ray studies have been routine procedures in medical diagnostics where the soft tissue becomes transparent and the heavier bones deliver higher contrast. The most common medical method is the production of transmission (shadow) images of the body parts on an X-ray sensitive film as an integral of all layers in beam direction. Such

B. Kramer (Ed.): Adv. in Solid State Phys. **45**, 389–405 (2005)

investigations are common too in non-destructive testing (NDT) procedures of industrial components when failures or indications for destructive defects should be inspected. In such technical examinations, higher doses can be applied than to human bodies and other X-ray energies (up to 450 kV than 120 kV high-voltage) than common in hospitals.

If the third dimension (in beam direction in the transmission studies) should be investigated too, mathematical procedures have to be applied and the CT (computed tomography) is the right tool. The mathematical principles for such an approach were already derived in 1917 by Radon [1] but a technical concept for a CT device was realized much later for practical medical problems by Hounsfield in 1972. The results of tomography investigations is not only one single image of the object (as in radiography) but a numerical data set representing the whole volume, where outer views, dedicated slices but also segmented parts can be derived.

This article describes the present situation of neutron tomography (NT) as tool for alternative non-destructive testing as it is now established at different neutron sources. Because the mathematical principles have already been described in detail in books and papers [2]–[4], we will only sketch the used routines at PSI but focus on the application of this method to practical problems and samples and describe the occurring problems and limitations with some examples. Based on own experience, the authors will explain and demonstrate where NT can play an important role, with advantages compared to other techniques. Selected cases will demonstrate the performance of the existing tools.

2 Neutron Interaction with Matter in Comparison to X-ray

As described before, the X-ray techniques have been available as standard tools since many years and are often simplifying called "computerized tomography" or CT. However, when neutrons or other kinds of transmitting radiation are applied as shown in Fig. 1, the principles for data acquisition, and later data treatment and visualization are very similar.

From the source of radiation a beam is guided through the collimator to the object and the shadow image is registered in the two-dimensional detector placed perpendicular to the beam direction. The interaction of the radiation with the object delivers the needed contrast to enable the visualization of material distributions as its shadow image on the detector screen.

It is well known that X-ray collision with matter happens via an interaction with the electrons in the atomic shell as Compton scattering, photo effect and pair production. Therefore the number of electrons per atoms is a clear measure for the interaction probability, that means heavier elements delivers much higher contrast than light ones. On the other hand, the X-ray

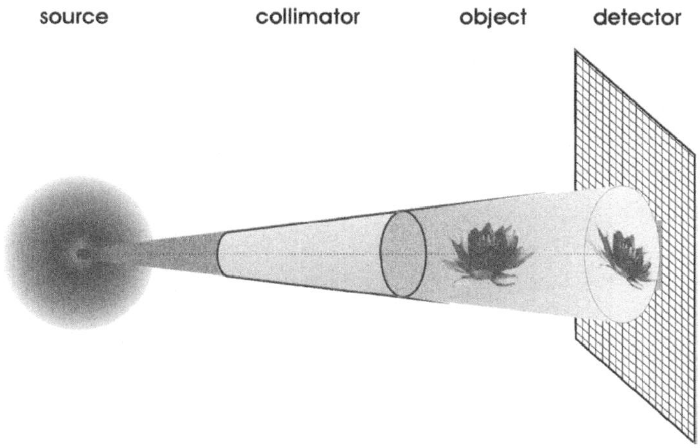

Fig. 1. Principle setup for radiation transmission imaging as basis for tomographic investigations of macroscopic material samples. In praxis, a good radiation shielding around the assembly is also necessary. The obtained image represents a two-dimensional data matrix of transmission values, corresponding to the attenuation in the object

interaction with matter is depending on the initial photon energy. This is important for the samples contrast on the one side, but the detector response on the other side. The energy dependence of attenuation coefficients are shown in Fig. 2 for some relevant materials. The typical photon energy range is from few keV (especially synchrotron radiation) up to 450 keV (for X-ray tubes) and some MeV for gamma sources.

Similar conditions are true for neutrons too. However, the interaction mechanism with the atom is completely different: neutrons interact with the nucleus and do not recognize the electrons. Two interaction mechanisms are mainly relevant for imaging purposes – neutron absorption and neutron scattering. In first order, all neutrons missing in the direct line behind the sample are considered as lost. This is in principle true for all neutron capture reactions. Scattered neutrons are often more difficult to handle when they contribute to the resulting image behind the sample in the form of a diffuse sky shine (see below Sect. 5).

In the practical neutron tomography thermal neutrons (E_n around 25 meV) play the most important role. There are two major reasons in respect to this neutron spectrum: reactor based sources (and this is still the majority) need for the chain reaction in the core the moderation to the thermal energy; the largest community dealing with the neutron research – the neutron scattering people – require this energy range because of the most efficient interaction with matter. In the recent years, the neutron energy could be shifted by cold moderators (T about 25 K) towards lower values (E_n around 3 meV). Due to the higher interaction probability with matter compared to

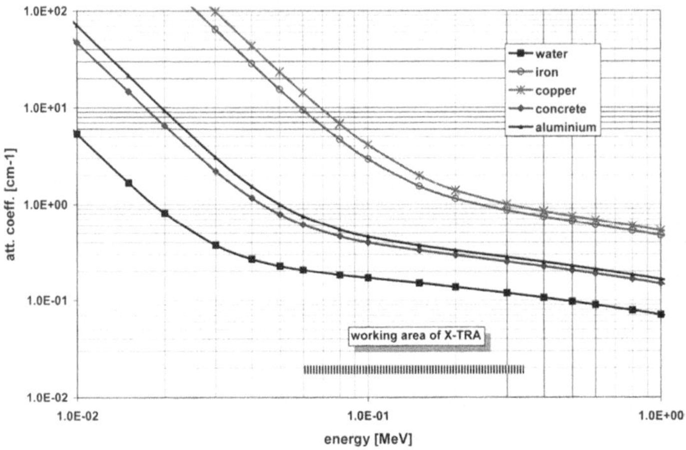

Fig. 2. Attenuation coefficients for several common materials for X-ray in the energy range between 10 keV and 1 MeV; the spectral range of a X-ray tube with 320 kV high voltage is marked too. For comparison, the values for thermal neutrons are: water – 3.2 cm^{-1}, Al – 0.1 cm^{-1}, Fe – 1.2 cm^{-1}, concrete – 0.35 cm^{-1}

thermal neutrons both the material contrast and the detection efficiency is increasing with cold neutrons. A variation of the neutron energy as possible for X-ray by changing the high-voltage is therefore only possible for neutrons in a limited range.

In respect to individual material interaction there is no clear systematic as in the case of X-rays. Four regions in the periodic table of elements are important, where large differences exist between interaction probabilities of thermal neutrons and X-rays:

- Most of the light elements (H, Li, B, C) have stronger neutron interaction than X-ray
- Heavy elements (Pb, Bi, U, W) have only low interaction with thermal neutrons compared to X-ray
- There are some especially strong neutron absorbers (Gd, In, Cd) with extremely high cross-sections
- Most of the technically relevant metals are more transparent for neutrons than for X-ray.

Due to these differences in the attenuation behavior of the individual materials it becomes very clear that both kinds of radiation are complementary to each other even if the tomography methods and procedures are very similar in principle. Therefore, the fields of applications must differ – X-rays are ideal e.g. for the investigation of large hydrogenous assemblies as the human body, where neutrons must fail completely due to the strong scattering at hydrogen. The advantages of neutron tomography (NT) are described in more detail below.

The interaction probability defines also the dimensions of the samples under investigation in respect to the amount of the transmitted radiation. As long as the Lamberts-Beer's law of attenuation is valid, the product of the attenuation coefficient Σ and the sample thickness d should follow the condition:

$$(\Sigma \cdot d)_{\max} \leq \ln\left(\frac{S-B}{N}\right) \tag{1}$$

(S(signal)/N(noise) describes the valid dynamic range, B the background level.)

Because Σ is given by the investigated material, the largest observable material layer d in beam direction is also very material dependent. This is true in the same way for the lowest detectable material thickness in the following relation:

$$(\Sigma \cdot d)_{\min} \geq \ln\left(\frac{S}{S-2 \cdot N}\right) \tag{2}$$

Deviations from the exponential law by the energy dependence of the initial and collided beam and by multiple scattering in the sample material will be discussed in more detail below.

3 Set-Up of a Neutron Tomography Device

The principle of NT consists in the observation of the object by different projections, which are obtained by a stepwise rotation around the vertical axis of the object under investigation. Because neutron sources are huge installation, the relative motion is only possible by the sample, but never by the source itself.

This experimental procedure is followed by the treatment of the n projection data $T_n(x, y)$ with the reconstruction algorithm with the results of the three-dimensional matrix $\Sigma(i, j, k)$ of attenuation coefficients for the complete volume under investigation. The segmentation into defined numbers of volume elements (voxel) requires a detection system with a fixed number of pixels and the pixel size defines the spatial resolution of the whole system. The details of the tomography reconstruction algorithm are described in Sect. 4 and in [5]. Some aspects of the used methods are discussed below.

The beam geometry can also define the performance of the tomography set-up and the required reconstruction method. For neutron sources, the intensity is a limiting factor in most cases. Therefore it is impossible to achieve the case of a strong point source where the resulting cone beam can be used for a magnifying projection of the object. The only reasonable approach is a parallel beam with a 1:1 shadow image. The advantage of this set-up is the simplified reconstruction algorithm, the requirement of an observation angle of only 180° and reasonable high beam intensity.

Such a setup is limited thus in the resolution as given from the detection system. The size of the object, when within the limitations in (1), is defined either by the beam diameter or the detector area.

The NT system at SINQ (NEUTRA station at the thermal beam port 32) is built flexibly in a manner that three positions in the beam can be used under different conditions in respect to beam collimation (L/D between 250 and 550), the intensity (from $3.6 \cdot 10^6$ to $2.5 \cdot 10^7\,\mathrm{cm}^{-2}\mathrm{s}^{-1}$) and the beam diameter (between 15 and 40 cm diameter).

The detector for NT purposes has to fulfill the following conditions:

- It has to deliver a two-dimensional array of digital values in equidistant manner per pixel. Typical numbers of pixels per line are in the order of 256 to 4000.
- Each projection has to be obtained under identical conditions in respect to position and applied intensity (the later can be calibrated in the following numerical operations).
- The dynamic range should be as large as possible to enable the inspection of also asymmetric objects. The S/N level should be as large as possible to enable high data quality.

From the presently available digital detectors for imaging purposes (CCD + scintillator, imaging plates, CMOS, a-Si flat panel + scintillator) the imaging plates are not usable for NT even if they provide excellent conditions in radiography. However, it is impossible to replace IP's many times under identical positioning with the required precision in the beam and the scanner as well (about 100 μm).

The most common setup is a highly sensitive CCD-camera system which detects the emitted light from a neutron sensitive scintillator [6]. To avoid damage by direct beam exposure, the light from the scintillator is mirror reflected by 90 °. With suited lenses, the field-of-view (FOV) and the pixel size in the images can be varied. The PSI setup covers the range from 4 to 40 cm, which delivers a pixel size (with a 4 Mega pixel chip) between 0.20 and 0.02 mm. These are only nominal values. The real limitations in spatial resolution have other origin as described and discussed below.

A second system has been already used for NT – the amorphous silicon flat panel [7]. In this case, the scintillator and the light sensitive area are placed into the direct beam with zero light loss via the optical path. This gains in higher detection efficiency and (coupled with faster read-out) much higher frame rate. The disadvantage is in the lower dynamic range of the system and the higher noise level. It can favorably be used for the tomography of time dependent processes.

First tests for NT have been made recently at NEUTRA with a modified MEDIPIX-2 chip [8]. The potential of this device is promising because the readout of each pixel is via an individual amplifier on the chip. In this manner, the inherent resolution of the pixel array can be exploited and the noise level can be limited.

A summary of the properties of the different setups is given in Table 1 for conditions at the NEUTRA station at PSI.

Table 1. Performance of different detection systems suited for tomography with neutrons under the conditions at the NEUTRA station at SINQ (PSI)

tomography system	*CCD+scintillator*	*a-Si+scintillator*	*MEDIPIX-2*
active area in the beam [cm]	4×4 to 30×30	10×15 to 20×25	1.4×1.4
pixel size, nominal [mm]	0.02 to 0.15	0.127	0.05
pixel number	1024 to 2048		256 to 256
dynamic range	16 Bit	12 Bit	13 Bit
signal-to-noise ratio	120	25	180
exposure time per frame [s]	2 to 60	0.067	150
read-out time per frame [s]	2	included in the frame	0.02
frame rate		0.5 to 30	

The complete NT system is sketched in Fig. 3, where several computers are needed for data acquisition, the control of the turning table and all motions of the object, the volume reconstruction and the final data visualization. Software programs in LabView [9] are used for the first two tasks, whereas the reconstruction tools implemented by P. Vontobel are based on IDL [10] routines. As most powerful for the data visualization the VG Studio [11] programs have been found.

4 Principles and Features of the PSI Reconstruction Tools

Neutron tomography relies on the same principles and uses the same mathematical reconstruction methods as used in X-ray tomography. For an overview see e.g. Natterer (1999) [3]. Neutron tomography setups mainly use "parallel" beam geometry. Therefore the reconstruction problem can be solved for individual slices perpendicular to the rotation axis. Let $\Sigma(x, y)$ be the distribution of the linear neutron attenuation coefficient in such a plane two-dimensional cross section of an object. A thin neutron beam is attenuated along a line of response (LOR) extending from the source to the detector according to the exponential law of radiation attenuation given in (3).

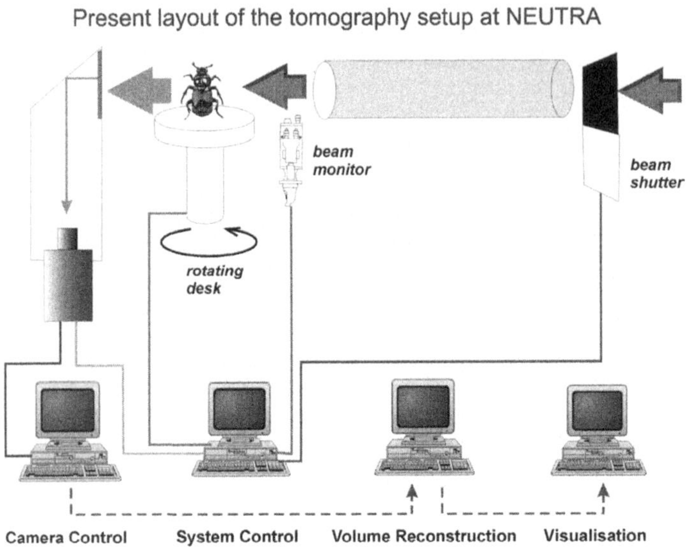

Fig. 3. Principle layout of the tomography system as realized at the NEUTRA station of the spallation neutron source SINQ (PSI)

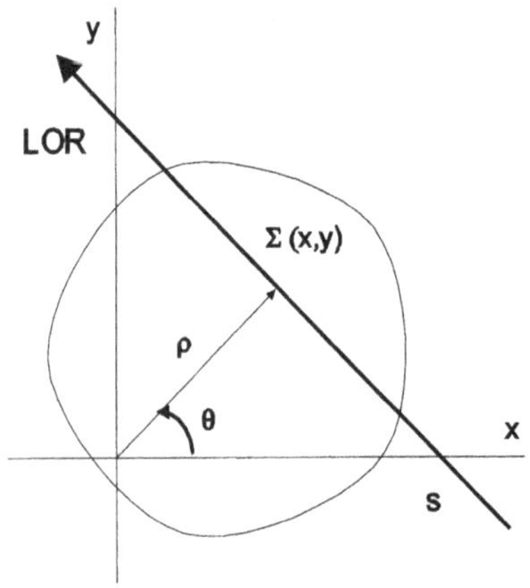

Fig. 4. Cross section of an object with unknown $\Sigma(x, y)$ distribution

Measuring the neutron beam intensity before and past the object allows to determine the projection $P(\rho, \theta)$, which is the line integral of the unknown attenuation coefficient distribution along a line of response. $P(\rho, \theta)$ is called the Radon transform of $\Sigma(x, y)$. The problem to compute $\Sigma(x, y)$ from $P(\rho, \theta)$ has been solved by Radon (1917) [1] and is therefore known as inverse Radon transform. A short form of one of the most used inversion schemes is given in (3) and is called filtered back projection, $\hat{P}$ being the filtered projections.

$$I(\rho,\theta) = I_0 \cdot \exp\left(-\int_{-\infty}^{\infty} \Sigma(x,y)\,\mathrm{d}s\right) \tag{3}$$

$$P(\rho,\theta) = \ln\left(\frac{I_0}{I(\rho,\theta)}\right) = \int_{-\infty}^{\infty} \Sigma(x,y)\,\mathrm{d}s \tag{4}$$

$$\Sigma(x,y) = B(x,y) = \int_0^{\pi} \hat{P}(x\cos\theta + y\sin\theta,\theta)\,\mathrm{d}\theta \tag{5}$$

Another direct inversion algorithm is based on the Fourier slice theorem and can be implemented using fast Fourier transform (FFT) [3]. The data acquired in a neutron tomography setup is always sampled, i.e. there are finite size detector elements and the object is rotated in discrete angular steps around 180 ° in parallel beam geometry. For this scanning geometry it can be shown [2], that for optimal resolution the number of angular projections N should be related to the number of detector elements M by $N/M = \pi/2$. For modern detector hardware used in X-ray or neutron tomography typical values of M are in the order of a few 100 to some 1000 detector elements, which would lead to a large number of angular projections required. For practical reasons the used number of projections in neutron tomography is rather in the order $N/M = 0.5 \div 1$.

After tomography data acquisition, reconstruction of slices is done in 3 steps. In a first step the radiographic images are normalized using flat field and dark current images and speckle noise caused by γ-ray hits on the CCD chip is removed. In a second step the sinogram is created and the center of rotation is determined, ring artefacts due to insufficient alignment of the object or setup are filtered out. In the third stage the sinogram data is reconstructed either using a filtered back projection algorithm or a Fourier transform based inversion scheme. Because the objects investigated by neutron tomography vary considerably in size and composition, the experimental setups as well as the used software tools have to be adapted to the individual investigations. The software tools used for neutron tomography allow the visualization of the pre-processing and reconstruction steps. Thereby threshold settings can be checked and tailored to the individual requirements. This can be nicely implemented in programming languages providing a broad choice of image processing modules and visualization capabilities like e.g. IDL [10], LabView [9], etc.

Due to the large amount of projection data which has to be processed, computing performance issues have to be solved. This is done either by taking advantage of the inherent parallelism of the parallel beam scanning geometry. The reconstruction of individual slices can be distributed to several PC's like in the LabView based program Octopus [12]. Another possibility is the use of a highly efficient Fourier transform based reconstruction algorithm like the fast filtered back transform implemented by G.H. Campbell and R.B. Marr [4]. It is based on a gridding technique proposed by J.D. O'Sullivan [13]. This algorithm, called "Gridrec" and implemented in an external C module, makes use of highly optimized Fast Fourier Transforms from the Intel® Math

Kernel library. It is the default reconstruction algorithm in an IDL based reconstruction package provided by M. Rivers [14], which has been adapted to neutron tomography by P. Vontobel at the Paul Scherrer Institute [5].

The hardware presently used for reconstruction (and visualization) is a 2.3 GHz PC with Dual XEON processors and a 3.5 GBytes RAM memory, which enable the reconstruction of one horizontal layer within 1 to 2 seconds, depending on the size per image.

5 Neutron Scattering Artefacts and Their Correction with the Help of Monte-Carlo Simulations

If the neutron beam transmits the sample, most of the neutrons are scattered by the atomic nuclei, only a few neutrons are absorbed. The scattered neutrons can reach the scintillator and appear in the radiographies. Hence the transmission of the neutron beam through the sample is overestimated by the measurement. The attenuation and the sample density are underestimated, if the scattering contribution would be neglected. In the tomographic reconstruction evolves a dip in the attenuation values from the scattering contribution comparable to the effect of beam hardening in X-ray tomography.

Also in neutron radiographies beam hardening can be observed, but compared to the scattering it is only a secondary effect. Only for strong neutron absorbers (Gd, Cd, B-10, ...) beam hardening effects can dominate. For most materials (except hydrogen) beam hardening is only important for samples, which are anyway too thick to be evaluated quantitatively. Figure 5 shows the comparison of the effective attenuation coefficient due to beam hardening and the expected attenuation coefficient for thin samples. The curves are obtained by simulations.

While beam hardening is a local effect and can be corrected pixel wise by reference curves, the neutron scattering is a neighborhood effect and is corrected by a de-convolution. One method of scattering correction is based on Point Scattered Functions (PScF) [15]. It is an iterative pre-processing algorithm, which is applied on every single radiography. In this approach the scattering distribution of neutrons from a directed point source through a layer of the sample material is simulated by the MCNPX Monte Carlo code [16]. The energy spectrum of the neutrons, the sample material and thickness and the detector type and distance are input parameters for this simulation (see Fig. 6a). The resulting scattering distribution on the detector is called the PScF. A library of PScF is pre-computed for different combinations of the mentioned parameters.

The obtained discrete PScF are approximated by continuous functions so that they can be interpolated for the arbitrary parameters, which occur in a real experiment. The approximation is based on isotropic scattering with the scattering intensity and the detector distance as free parameters. Moreover

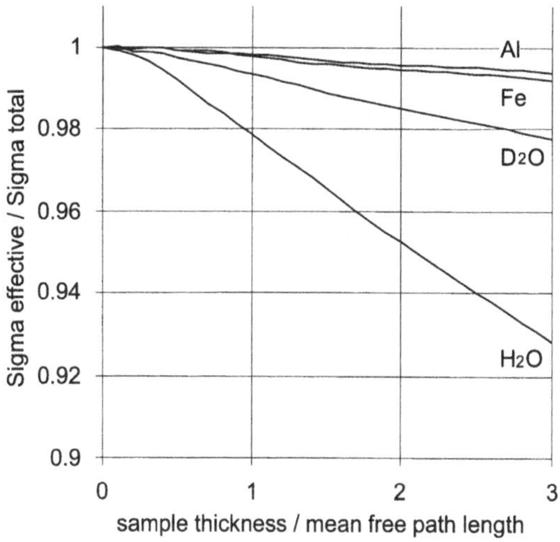

Fig. 5. Effect of beam hardening on the effective attenuation coefficient for the materials H_2O, D_2O, Fe, Al

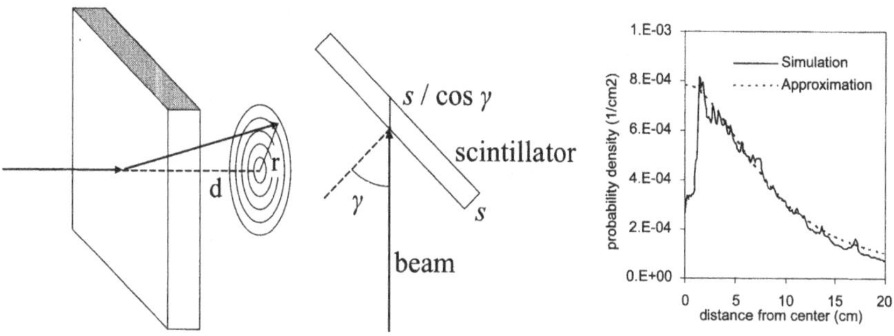

Fig. 6. (**a**) Simulation setup with the beam, sample layer and the radially symmetric detector. (**b**) The longer path length of scattered neutrons as reason for the angular sensitivity of thin detectors. (**c**) The PScF of 2 mm water in 10 cm distance from the detector, simulated (*solid line*) and approximated (*dotted line*)

the angular sensitivity of thin detectors is important and has to be taken into account.

Because a scattered neutron, that hits the detector plane not perpendicularly, has a longer path length through the detector, the detection probability is higher therefore.

$$\mathrm{PScF}(r) = S_\mathrm{A} \cdot \frac{d_\mathrm{A}}{4\pi(d_\mathrm{A}^2 + r^2)^{3/2}} \cdot \frac{1 - \exp(-\Sigma_\mathrm{D} \cdot s_\mathrm{D} \cdot \sqrt{d_\mathrm{A}^2 + r^2}/d_\mathrm{A})}{1 - \exp(-\Sigma_\mathrm{D} \cdot s_\mathrm{D})} \qquad (6)$$

r is the distance from a detector point to the projection of the source point on the detector plane. S_A and d_A are the fitting parameters with the physi-

Fig. 7. Slice through the tomographic reconstruction of the (**a**) uncorrected and (**b**) scattering corrected radiographies. (**c**) The diagram displays the profiles along the white lines

cal meaning of the approximated scattering intensity and the approximated distance between the sample layer and the detector.

They depend on the mentioned simulation parameters. s_D and Σ_D are the detector thickness and the macroscopic cross section of the detector reaction respectively. A comparison of the simulated and approximated PScF is shown in Fig. 6c. The peaks and also the dip at $r \approx 0$ in the simulation curve are no physical effects but arise from inaccuracies by the MCNPX interpolation of the scattering libraries [17].

The iterative correction algorithm performs the following steps:

1. The uncorrected transmission image (containing the transmitted and scattered neutrons) is used to find a first estimation of the sample thickness from the reference curves of the main scatterer in the material.
2. For every pixel the set of PScF for the main scatterer is interpolated to the estimated sample thickness and the experimental sample-detector distance. The sum of all these PScF yields the total estimated scattering distribution.
3. The corrected transmission image is then the difference of the uncorrected image and the scattering distribution.
4. This correction can be used to repeat the steps 1–3 iteratively, resulting in the corrected sample thickness. As a rule the algorithm converges after three iterations.

The scattering correction was applied e.g. for the tomography of a sand column, that was drained with heavy water. The visual comparison in Fig. 7 shows already that the corrected tomography is more homogeneous than the uncorrected one. And also the dip in the profiles of the attenuation values becomes smaller. Unfortunately, the corrected profile is noisier than the uncorrected one. This additional noise arises because the transmission values move closer to zero by the scattering correction. Therefore the same absolute noise results in a greater relative noise in the transmission image. But the attenuation values themselves become greater by the correction. In the above example (Fig. 7) the quantification of the tomography yields around 100 % higher densities than without correction. The total noise in an open beam area of the radiography is in the region of 1 %. With that, quantification is reasonable in the range of 0.05–3 mean free path lengths, cf. (1) and (2). Beyond this limits the noise becomes dominant.

6 Examples for Practical Applications from Science and Industry

Because in neutron imaging a parallel beam is the only reasonable option, the resulting 1:1 images have resolution and a field-of-view (FOV) as given by the detector. Furthermore, the samples for the tomography investigations should fit into this FOV and the beam size as well in order to avoid incomplete projection data.

Neutron tomography has been used in the recent years as standard tool for both studies with scientific background and for material testing purposes with pragmatic requests from different customers. The method can favorably be used when the material thickness to transmit is too thick for X-ray or if the sensitivity to detect small material amounts (especially hydrogenous containing) is small.

An example for a bulky assembly is given in Fig. 8. It is an exhaust manifold of an old-timer (where no construction plans exist) made of gray cast iron and has a total wall thickness in the order of few centimeters. The aim of the study was the non-destructive analysis of the object in order to get a CAD drawing for replica casting processes. This technology is called "Reverse Engineering", where the tomography plays the important role to generate the 3D volume data of the object.

The non-invasive investigation of thin adhesive layers between metallic structures is another interesting and relevant problem, which can be solved very reliable with neutron tomography means. The three examples in Fig. 9 demonstrate how inhomogeneous the glue is often distributed inside the assembly where the risk for a failure would occur. The layers of 0.1 mm thickness and below can easily be inspected even in thicker metallic enclosure.

The investigation of objects from cultural heritage should be done non-destructively per definition to enable a transfer to later human generation. Neutron tomography as such a tool is of high importance e.g. when the transmission by other kinds of radiation would fail. This was the case in a batch of Roman bronze sculptures from the Swiss National Museum in Zurich containing high amounts of lead. It was possible to investigate both outer structures, corrosion effects and the inner surface, as shown in Fig. 10.

Scientific work is mainly concentrated onto samples and processes where the distribution of moisture is of importance. The matrix for that can be soil, wood, stone or organic material as shown e.g. in Fig. 11 for the 3D study of the behavior of carnations during growing under well defined conditions.

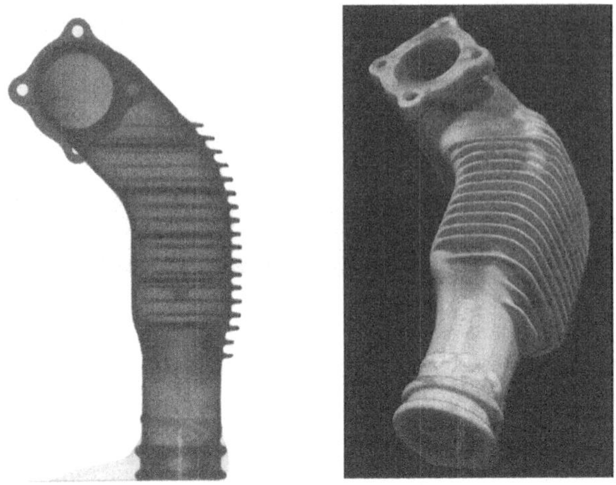

Fig. 8. (**a**) Transmission radiography and (**b**) a tomographic view of an exhaust manifold with about 42 cm length and 10 cm diameter. With the help of the volume data the process for Reverse Engineering could be initiated

Fig. 9. (**a**) Adhesive connections between copper tubes, (**b, c**) car components. The dark areas describe the distribution of the glues inside the metallic structure. Depending on the needed performance, the connections might be right or non-acceptable, respectively. This decision can be made without destruction of the connection

The options for neutron tomography studies are manifold and can only be sketched here. Potential users should have in mind that it is powerful in the macroscopic scale from about 1 cm^3 to sample size up to 30 cm, depending on the material.

7 Options for Further Improvements

Neutron tomography is limited in certain respect: spatial resolution, observation time, data handling and visualization performance. There are options in a certain amount to overcome the one or other drawback in the future.

Fig. 10. The "Mars-Amor from Zurich", a bronze sculpture attributed to Roman origin with a height of ca. 20 cm is unable to be inspected by X-ray methods due to its high lead content. Neutron tomography was the right tool to derive the outer shape, but also virtual cuts to investigate the inner parts of this cast part

Fig. 11. The three-dimensional analysis of carnations during growing under certain climate conditions was possible due to the high contrast of neutrons for organic structures and moisture in particular. The experiment was done inside a climate chamber made of aluminum, which is transparent for the neutron beam

- The present limitation in the spatial resolution is by the detector, the neutron sensitive scintillator screen. With a thickness of some hundred microns, the resolution is in the same order. Thinner scintillators require more beam intensity and a more efficient light collection to get the same result. The PSI approach is to use cold neutrons at another beam line ICON, enabling higher detection efficiency, the use of optimized optics and a CCD with 2048 × 2048 pixels. The micro-tomography project has the goal of 10 micron pixel size nominal, but 20–30 microns in reality.
- The computation and data handling needs will increase accordingly with the third order of the pixel number. The requirement to reconstruct and

to visualize this large data volumes needs high end computer performance and 64-bit operational systems for the related software tools.

- Although already tested at some examples, the speedy tomography with a-Si flat panels can be exploited much more for the study of time-dependent study phenomena in three dimensions. The data amount obtained is very huge in this case too.
- The final option will be to apply energy selection of the arriving neutrons by either turbine type velocity selectors or a dedicated installation at a pulsed spallation source. Both the increased inherent contrast and the option to use the Bragg edges to distinguish materials will increase the application range for neutron tomography in the future.

8 Conclusions

Neutron tomography is established now as tool for non-invasive investigations of samples and processes on the macroscopic scale from 1 to 30 cm. It complements similar techniques with X-ray and can advantageously be applied in cases when light elements should be detected and thick metallic layers have to be penetrated. Further improvements by using cold neutrons or energy selective methods together with higher spatial resolution and faster image acquisition and data processing will contribute to establish neutron tomography as a unique technique for special interesting applications.

Acknowledgements

The authors want to express their thanks to M. Jaggi and M. Estermann for their contributions to analytical work in many tomography tasks and to K. Hunger for their work on the archaeological samples, mentioned in the paper.

References

1. J. Radon, *Über die Bestimmung von Funktionen durch ihre Integralwerte längs gewisser Mannigfaltigkeiten*, Berichte Sächsische Akademie der Wissenschaften, Math.-Phys, Kl, **69**, 262 (Leipzig 1917).
2. F. Natterer, F. Wübbeling, *Mathematical Methods in Image Reconstruction*, SIAM monographs on mathematical modeling and computation, (Philadelphia 2001).
3. F. Natterer, *Numerical Methods in Tomography*, Acta Numerica, submitted; http://arachne.uni-muenster.de:8000/num/Preprints/
4. B. A. Dowd, G. H. Campbell, R. B. Marr et al, *Developments in synchrotron X-ray computed microtomography at the National Synchrotron Light Source.* SPIE Conf. on Developments in X-Ray Tomography II, Denver, Colorado, Jul. 1999.

5. P. Vontobel, E. Lehmann, G. Frei, *Performance characteristics of the tomography setup at the PSI NEUTRA thermal neutron radiography facility*, Proceedings of Computed Tomography and Image Processing for Industrial Radiology, June 23–25, 2003, Berlin, Germany.
6. H. Pleinert, E. Lehmann, S. Körner, Nucl. Instr. & Meth. A **399**, 382 (1997) .
7. M. Estermann et al., *The performance of an amorphous Si flat panel for neutron imaging under the conditions at NEUTRA*, PSI-Annual Report (2003), Vol. III, p. 188, ISSN 1423-7326.
8. J. Jakubek et al., *High resolution tomography with MEDIPIX*, PSI-Annual Report (2003), Vol. III, in press.
9. LabView, National Instruments Corporation, Austin, Texas, USA.
10. IDL, Interactive Data Language, RSI, Research Systems Inc, Boulder, Colorado, USA.
11. http://www.volumegraphics.com/
12. M. Dierick, B. Masschaele and L. Van Hoorebeke, Meas. Sci. Technol. **15**, 1366 (2004).
13. J. D. O'Sullivan, IEEE Trans. On Medical Imaging, **4**, 4, 200 (1985).
14. M. Rivers, IDL Software for tomography: http://cars9.uchicago.edu/software/idl/tomography.html
15. R. Hassanein, E. Lehmann, P. Vontobel, *Methods of scattering corrections for quantitative neutron radiography*, to be published in Nuclear Instruments and Methods in Physics Research A (2005).
16. L. S. Waters (Ed.), MCNPX user's manual, version 2.4.0, LA-CP-02-408, Los Alamos National Laboratory, Sep 2002.
17. D. E. Cullen et al., Thermal Scattering Law Data: Implementation and Testing using the Monte Carlo neutron transport codes COG, MCNP and TART, U.S. Department of Energy, Lawrence Livermoore National Laboratory, May 17, 2003.

Absorption- and Phase-Based Imaging Signals for Neutron Tomography

W. Treimer[1,2], N. Kardjilov[2], U. Feye-Treimer[1], A. Hilger[1,2], I. Manke[2,3], and M. Strobl[1,2]

[1] FB II, University of Applied Sciences (TFH) Berlin, 13353 Berlin, Germany
[2] SF3, Hahn-Meitner-Institut Berlin, 14109 Berlin, Germany
[3] Faculty 3, Technical University Berlin, 10623 Berlin, Germany

Summary. Thermal and cold neutrons are a perfect tool to investigate massive and large (several cm^3) samples by means of computerized tomography (CT). The main interaction to visualize structures in samples is absorption, similar to x-rays in medicine or material research. Based on absorption new techniques such as energy dependent CT or real time CT have gained importance within the last years and were used to show e.g. the movement of pistons in engines. Absorption, however, is not the only interaction that can be used for computerized tomography. Other interactions such as interference effects, so-called phase based interactions, were exploited for CT to visualize details in samples that could not be detected by absorption. Phase contrast imaging was first applied to x-ray CT, but soon it was also applied to neutron CT. Based on the same interaction, refraction and ultra small angle scattering of neutrons were studied and successfully applied to neutron tomographic investigations of samples. These new imaging signals (refraction and ultra small angle scattering) require entirely different techniques, different experimental set ups (high resolution double crystal diffractometer) and special data evaluation, to detect these signals and separate them from absorption and noise. Using absorption and phase-based signals the length scale of tomography with neutrons is more than five orders of magnitude, which covers a large field of non-destructive testing in material research and fundamental physics.

1 Introduction

Within the last ten years x-ray and neutron tomography have experienced an increasing attention due to a variety of new applications in material science and in the case of neutrons, tomography is one of the topic research field of the new generation of neutron sources. The interactions used for imaging are very limited: absorption (emission), magnetic resonance and phase are up to now the only signals. They were invented many years ago and successfully applied to a number of problems in medicine and material science. The fundamental physical problem of tomography is how to use interactions of radiation or matter waves with structures in samples and recover a 2D (or 3D) image of them. Principally there are two images one can recover, a real image and a Fourier transformed one, depending on the interaction with the structure. In scattering physics (e.g. elastic and small angle scattering) usually the Fourier Transform of a structure is measured, in classical

B. Kramer (Ed.): Adv. in Solid State Phys. **45**, 407–420 (2005)

tomography up to now only real images have been reconstructed. In the latter case the structures under investigation are much larger than the wave length of the radiation and any interaction with them (absorption or emission) can be seen as an incoherent signal. However, every interaction contains "coherent" information which can be recovered by proper experimental and mathematical methods. If, for example, a neutron (or x-ray) beam passes a region wherein the index of refraction changes, then the absorption may be the same, but the momentum changes (elastically). This change is not determined in "classical" tomography because the momentum transfer is extremely small ($\sim 10^{-4}$ nm^{-1}) and it was seen as a minor, negligible effect. We will show, that just these "small" effects contain sufficient information for image reconstruction. But first these effects have to be determined and then mathematically correctly recovered (restored). Strictly spoken, to distinguish between coherent and incoherent interactions one has to measure them, and that simultaneously, to distinguish what is the coherent (phase-based) part and what is the incoherent (absorption-based) part in a signal and which part can be used for image reconstruction. A ray that e.g. experiences absorption and refraction yields two kinds of information, an incoherent one (decrease of intensity) and a coherent one (change of the momentum of the ray). From a pure absorption measurement, as done in the classical CT, one cannot calculate the momentum transfer from it, therefore one ignores and neglects the coherent information. In the case of the determination of the momentum transfer the coherent part of information is measured and the incoherent part can be determined as well within this measurement. This holds also for other interactions, such as depolarization or small angle scattering.

2 Theoretical Background

If monoenergetic radiation (neutrons, x-rays, etc.) penetrates a layer of material having the mass thickness d and the density ρ the ratio of incident and transmitted intensity is given by

$$\frac{I}{I_0} = e^{-\frac{\mu}{\rho}d} \tag{1}$$

$\mu/\rho =$ mass attenuation coefficient. The mass thickness d here is defined as the mass per unit area and is obtained by $d = \rho t$, with $t=$ thickness of the layer. In the case of neutrons $\mu = \mu(E)$ depends on the energy of the radiation and on the nucleus (isotope) and the absorption cannot be calculated from quantum mechanics as can be done in the case of x-rays. This is due to the different kind of interactions of neutrons and x-rays: neutrons interact with the nucleus of the atom, x-rays with the electrons surrounding the nucleus. X-ray absorption depends on the energy of the photons as well, but it depends on the number of electrons of the atom. To distinguish different structures in an object certain differences of intensities must be registered, known as the contrast C. The contrast (in general) of a part in an image is defined as the ratio

$$C = \frac{I_{max} - I_{min}}{I_{max} + I_{min}} \tag{2}$$

where the particular intensities I_{max} and I_{min} are determined in adjacent pixels of a 2D detector or film.

Attenuation involves both, absorption and scattering. Therefore attenuation of a beam by an inhomogeneous sample means absorption and/or phase (refraction or small angle) contrast. One can show that in spite of small absorption (contrast $< 5\%$) structures can be visualized by e.g. phase contrast. In the case of absorption one has an entirely incoherent interaction of radiation with matter, in the case of phase contrast coherent interaction of the incident and transmitted beam has to be taken into account.

From optic lectures it is well known that two (or more) points of a wave emerging from a point source can be seen as to be coherent to each other if there is a constant phase relationship between them or if there is a given function, that describes the evolution of the phase in space or time. It is further known, that neutrons can be seen as particles as well as waves because a neutron has after de Broglie a wave length $\lambda = h/p$, (h = Planck' constant, $p = mv$). Therefore the distinction between incoherent signals and coherent ones simplifies the discussion about different interactions in neutron radiography and tomography. The fundamental entity for all interactions (coherent, incoherent) is the index of refraction n, given by

$$n = 1 - \delta + \mathrm{i}\beta \tag{3}$$

with [1]

$$n = \frac{K}{k} \sim 1 - \lambda^2 \frac{N}{2\pi} \sqrt{b_c^2 - \left(\frac{\sigma_t}{2\lambda}\right)^2} + \mathrm{i}\frac{\sigma_t N \lambda}{4\pi} \tag{4}$$

Here, K is the wavevector inside, k the wavevector outside of the material. λ the vacuum wavelength, N atomic density, b_c coherent scattering length of the material (for neutrons), and $\sigma_t = \sigma_a + \sigma_{incoh}$ is the sum of the absorption and incoherent cross section. The material is seen as a region having the mean potential $\bar{V}$ which is connected with n as

$$n = \sqrt{1 - \frac{\bar{V}}{E}} \tag{5}$$

The real part δ of equ.(3) describes the phase-based interaction, if $\sigma_t = \sigma_a + \sigma_{incoh}$, $\sigma_t/2\lambda$ is much smaller than 1, i.e. if δ becomes

$$\delta = \lambda^2 \frac{N b_c}{2} \pi \tag{6}$$

The imaginary part β is then given by

$$\beta = \frac{\sigma_t N \lambda}{4\pi} \tag{7}$$

and it is responsible for the absorption part of the index of refraction. It is apparent, that δ corresponds with the wave-nature, β with the particle-like nature of the neutron.

In the case of (neutron) tomography a sample, which is defined in a bounded region, is scanned by pencil rays that traverse the sample, where $\mu(x,y)$ characterizes the position dependent absorption of the sample by (mono-energetic) neutrons . (In the case of tomography with a double crystal diffractometer (DCD) it describes the attenuation of the neutron beam, because additionally to absorption scattering effects alter the momentum direction of neutrons that are not reflected by the analyzer). If $\mu = \mu(x,y)$ denotes the 2-dimensionally function that describes the index of absorption within a defined slice in the (x,y) plane, $\mu(x,y)$ can be represented by the following path-integral

$$\mu(x,y) = -\frac{4\pi}{\lambda} \int_{path} \beta(s) ds \tag{8}$$

where s (path) means the way along a straight line over $\mu(x,y)$. The intensity ratio I/I_0 is then given by

$$\frac{I}{I_0} = exp\left(- \int_{path} \mu(x,y) ds \right) \tag{9}$$

and a point of a so-called projection $P_\theta(t)$ (see Fig.1) is determined by

$$P_\theta(t) = \ln \frac{I_0}{I} = \int_{path} \mu(x,y) ds \tag{10}$$

One projection is a total set of such line integrals covering the whole region (sample). The mathematical solution to reconstruct a two dimensional function from its projections is well known [2]. With the so called method of filtered back projection (FBP) [3], [4] it is possible to reconstruct a slice of an object which was scanned over 180° by calculating the two dimensional Fourier transform $M(u,v)$ of the image function $\mu(x,y)$ from the one dimensional Fourier transform $S(\omega,\alpha)$ of the projection $P(s,\alpha)$. The back transformation yields the two-dimensional function $\mu(x,y)$. If $S_\theta(\omega)$ is the Fourier transform of $P_\theta(t)$

$$S_\theta(\omega) = \int P_\theta(t) e^{-i2\pi t\omega} dt \tag{11}$$

the function $\mu(x,y)$ can be reconstructed with

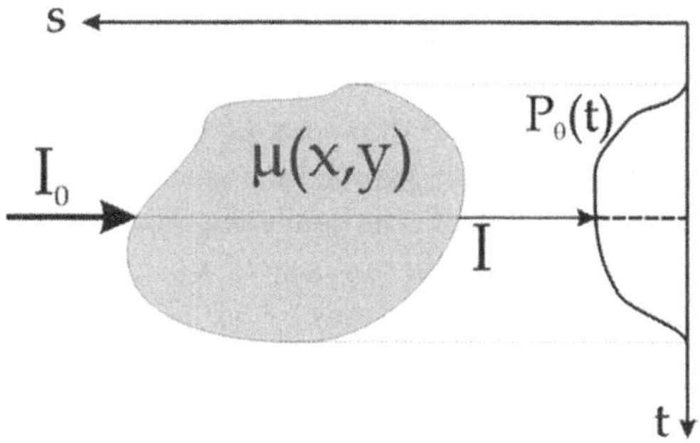

Fig. 1. A pencil ray traverses an object that is characterized by its 2D function $\mu = \mu(x, y)$.

$$\mu(x,y) = \int_0^{\pi} \int_{-\infty}^{\infty} S(\omega,\theta)\, |\omega|\, e^{i2\pi(x\cos\theta + y\sin\theta)} d\omega d\theta \tag{12}$$

using the convolution theorem one can calculate $\mu(x, y)$ as

$$\mu(x,y) \sim P_\theta(t) \otimes FT^{-1}\{|\omega|\} \tag{13}$$

In the case of absorption contrast, $P_\theta(t)$ is the logarithmic ratio of the incident and transmitted intensity based on the particular absorption processes in the sample. It is measured, as indicated above, by scanning the sample with pencil rays for sample orientations θ with $0 < \theta < 180°$. Absorption as an imaging signal can be used in different manners, such as energy-dependent radiography and tomography, using the wavelength-dependent Bragg-edges of a cold neutron beam, real time radiography and tomography, CT with monochromatic neutrons.

Problems arise if structures do not contribute significantly to the absorption signal. The question is how to get an image from these structures. Other interactions than absorption, such as phase based interactions have to be considered.

3 Phase Contrast

The idea of phase contrast imaging is the following: Consider a (neutron) wave which is coherent in a region given by the first Fresnel zone. The size of the first Fresnel zone (FZ) A_{FZ} is

$$A_{FZ} = \frac{Rr}{R+r} \pi\lambda \frac{1}{\sin(\theta)} \tag{14}$$

with R distance point source - object, r distance object - detector and θ the angle between the ray propagation vector and the plane of observation (usually $\theta = 90°$). This size A_{FZ} determines the coherent source for any interference effects. For standard geometries (e.g. $R \sim 500$ cm, $r \sim 1$ cm, $\lambda \sim 0.5$ nm) A_{FZ} is app. 1.6x10^{-7} cm^2 and the diameter of the coherent source $\sim$ 4 µm. So one has to create a "point-like" source by a small pinhole, that determines together with the divergence of the beam the size of the coherent area d_{coh} of a real neutron beam. The coherent size of a source d_{coh} can also be estimated by simple geometric considerations as [5]

$$d_{coh} = \lambda \frac{1}{2} \frac{R}{D} \tag{15}$$

d_{coh} is the size of the monochromatic source, R/D the common known "L/D - ratio", D distance object - detector. A short calculation demonstrates, that D must be kept very small, of the order some hundred microns or even less, to achieve a reasonable size of the coherent "source area".

In the case of phase contrast the coherent part of the wave propagates through an object and experiences position dependent phase shifts, i.e. the propagating wave front is modulated by different indices of refractions in the sample. The modulation occurs quite rapidly in the so-called near field region ($z < d^2/\lambda$) , z being the distance object-detector, d is the diameter of the object and it broadens with increasing z [6]. The interference pattern caused e.g. by an edge is washed out in the case of too large divergence and/or source size. To observe the strongest phase contrast, one has to operate in the near field region, where $z < d^2/\lambda$, where the modulation of the amplitude is best. It can be shown that the optimal contrast occurs at $z = d^2/2\lambda$. This was (partially) proved by the investigations of the contrast to realize experimentally phase images of steel needles at different detector distances and L/D [7], [8]. One approach to realize phase contrast was performed by neutron radiography some years ago [9]. The achieved contrast, however, was not as brilliant as it should be. This was due to a number of experimental conditions, mainly due to the small intensity (several hours exposure time). First experiments based on phase effects (refraction, diffraction enhanced intensity DEI) were already performed in 1989 [10].

One can calculate the contrast, knowing that the Fourier-Transform of the structure is measured. The phase development close behind the sample yields strong intensity oscillations which can be measured at different distances from the object. To calculate this behavior one can also use simple ray-tracing, adding up all amplitudes emerging from the structure which causes the phase gradients that contribute to a measuring point. Fig.2 shows the result of such calculations for a phase boundary consisting of Al and Si which have very similar indices of refraction. Therefore for equal thickness no intensity variation occurs (both 50 µm). In the calculation it was assumed, that both slabs were illuminated by a plane wave. This explains the small oscillation that appear app. more than 200 µm off the boundary at 600 µm.

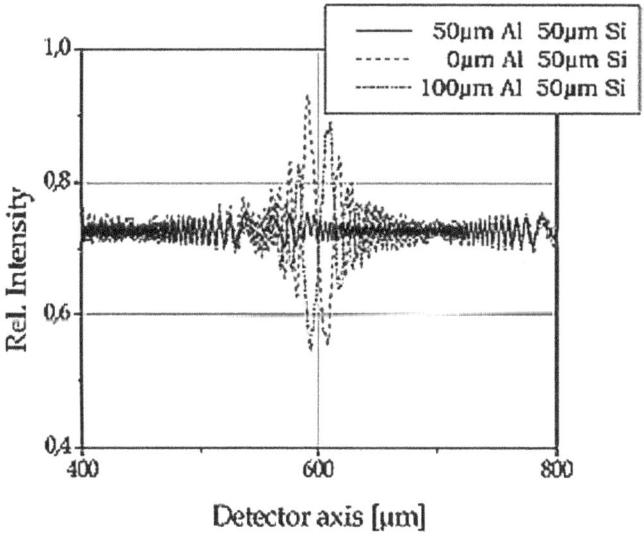

Fig. 2. Phase boundary at 600 μm between two slabs (Al and Si) having nearly the same index of refraction, but different thickness.

4 Refraction Contrast

Another approach to image phase contrast was done by a different way. One has to keep in mind, that due to interaction with matter the coherent partial waves form a wave front that changes its direction for $z > d^2/\lambda$ [11]. If the interface between two media is not perpendicular to the flight direction of the neutron, refraction occurs. This is for tomography usually the case, where the incident flight direction θ changes from 0 to 180°. The change of direction is due to the (extreme small) interaction potential of a nucleus with neutrons, which is of the order of μrad (for glancing angles) and therefore not observed in conventional tomography. Considering this interaction more closely, one has to distinguish between two cases, the one where the incident coherent wave-front is larger, and the other where it is smaller than the interface of the object. In the case of refraction the coherent wave front is much smaller then the interface. A neutron (wave) is then many times deflected (zigzag path) by the structures in the sample and leaves the sample with a different flight direction compared to the one in front of the sample [12]. The refraction along a certain path can be described by a position dependent phase shift function $\triangle\varphi(x, y)$ as

$$\triangle\varphi(x, y) = -k \int_{path} \delta(x, y) ds \tag{16}$$

Note, that the $\triangle$ is not the Laplace operator. The entities one has to determine are the refraction angles $\Theta = \Theta(\theta, t)$, where θ is orientation of the object related to the incident beam. One measures only remarkable Θ if the interface is nearly parallel to the incident beam, so Θ can be approximated by

$$\Theta(\theta, t) = \frac{1}{k} \nabla_{\perp} \left(\triangle\varphi(x, y) \right) \tag{17}$$

A point of a projection $P_\theta(t)$ can then be written as

$$\begin{aligned} P_\theta(t) &= \int_0^t \Theta\left(x, y(x, \theta, t')\right) dt' \\ &= \int_{path} \delta\left(x, y(x, \theta, t) ds\right) dx dy \\ &= \int_{-\infty}^{\infty} \int_{-\infty}^{\infty} \delta_D(x \cos\theta + y \sin\theta - t) \delta(x, y) \end{aligned} \tag{18}$$

(δ_D = Dirac function).

The other case to consider is, if neutrons traverse structures that have sizes of the order or less of the coherent wave-front. In this cases, small angle scattering arises (or large angle scattering if diffraction is involved), which decreases the contrast in a conventional tomogram.

5 Small Angle Contrast

The other case to consider is, if neutrons traverse structures that have sizes of the order or less of the coherent wave-front. In this case, small angle scattering (SAS) arises (or large angle scattering if diffraction is involved), which decreases the contrast in a conventional tomogram. Small angle scattering is an established method in x-ray and neutron physics and has a wide field of applications. Therefore it should be of high interest to investigate the possibilities to use it for tomography. In the case of SAS absorption should not be the dominating interaction, however, it is present in each measurement. In a SAS measurement one determines the Fourier transform of the shape function of a structure which can be characterized by different parameters, one of them is the radius of gyration. A coherent wave front traverses the structure and is deformed in phase by the shape of the structure. Far off the structure, a coherent superposition of all partial waves leads to an interference function that characterizes the structure. The size of the scattering angles is inverse proportionally to the size of the structure, i.e. the larger the structure, the smaller the scattering angles and vice versa. The problem to use SAS for tomography was to transform the scattering angles into data, that can be used as conventional tomography data, i.e. as sums of line integrals registered within a tomography experiment. Similar to the considerations in the case of refraction, one measures in the case of SAS a broadening of the

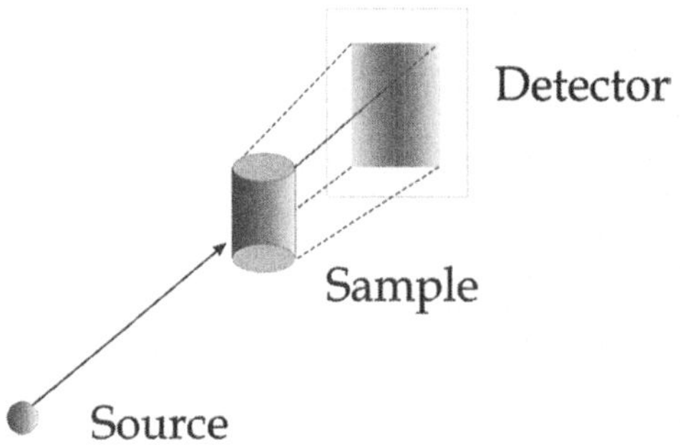

Fig. 3. Set up for classical tomography: source-sample ~ 5 m - 10 m, sample-detector ~ 10 cm - 50 cm, the "source" is usually a (pin)hole ~ 0.5 mm - 10 mm

incident beam (wave) that is characteristic for the structure. The broadening B is given as [13]

$$B = \sqrt{\int_{path} \frac{\sigma(x,y)N(x,y)}{R(x,y)^2} ds} \tag{19}$$

where $\sigma(x,y)$ is the scattering cross section, N the density of the structures and R is a parameter that describes the mean size of the structure. If B is bounded and finite, then also B^2, so if $f_B(x,y)$ can be set to

$$f_B(x,y) = \frac{\sigma(x,y)N(x,y)}{R(x,y)^2} ds \tag{20}$$

we can calculate the Radon transform as

$$P_\theta(t) = B^2(\theta,t) = \int_{path} f_B(x, y(x,\theta,t))ds \tag{21}$$

So we can use the broadening B as an imaging signal for tomography.

6 Some Experiments

"Classical" tomographies were performed with "standard - set ups", that means a small hole, being the neutron source and some meters downstream the sample, that can be rotated around a vertical axis. Close behind the sample is the detector unit, consisting of a neutron converter and a CCD camera (Fig.3)

Usually one takes the total spectrum (thermal, cold) for a tomogram, involving some difficulties that are close connected with beam hardening und scattering effects (as unwanted image background), activation, etc. It can be shown, that the image contrast with cold neutrons is better than using the

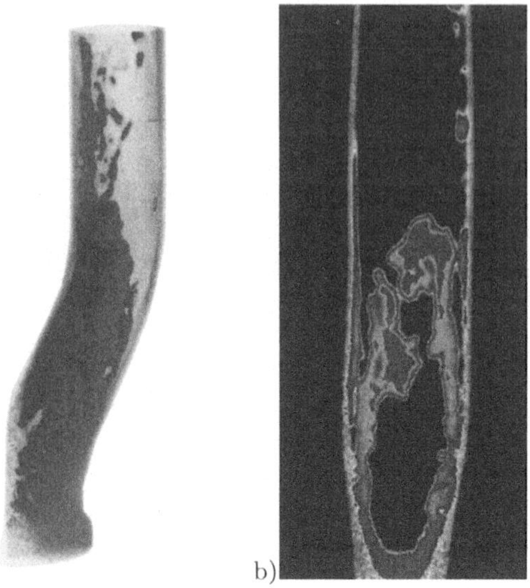

a) b)

Fig. 4. a) Vent tube with deposits of a mixture of oil and carbon. b) A cut through the tube showing the distribution of the deposit. Outer diameter was app. 15 mm.

thermal spectrum, and therefore it might be possible to use thinner converter foils for imaging [15]. Converters are one of the limiting factors in neutron tomography, so any reduction of the thickness of a converter improves the spatial resolution in neutron tomography and therefore increases the quality of reconstructions.

The use of cold and monochromatic neutrons (λ app. 0.5 nm) is sometimes more efficient, esp. if samples must not be activated. Here only a small wavelength band is used, so the available intensity is much less than in the case taking the integral neutron flux. Fig.4 shows one example of a 3D n-CT using monochromatic neutrons.

Very impressing are the results of energy-dependent (wave length dependent) tomography. Here one uses the fact that some materials have the same or similar absorption coefficients for different energies (or wave lengths). If one calculates the intensity ratio for two different energies as

$$\frac{\frac{I_1}{I_{01}}}{\frac{I_2}{I_{02}}} = \frac{I_1 I_{02}}{I_2 I_{01}} = \frac{e^{-\Sigma(\lambda_1)d}}{e^{-\Sigma(\lambda_2)d}} = e^{[\Sigma(\lambda_2)-\Sigma(\lambda_1)]d} \tag{22}$$

, where I_i/I_{i0} is ratio transmitted/incident neutron intensity, $\Sigma(\lambda_i)$ is the attenuation (absorption) coefficient for λ_i at the energy i, one can suppress certain parts of the material and so enhance other components of the sample. Strong attenuating parts in a samples can be made "semi-transparent" and weak absorbing components can be enhanced. Fig.5 shows the result of an experiment performed at the PSI (Swiss) [16].

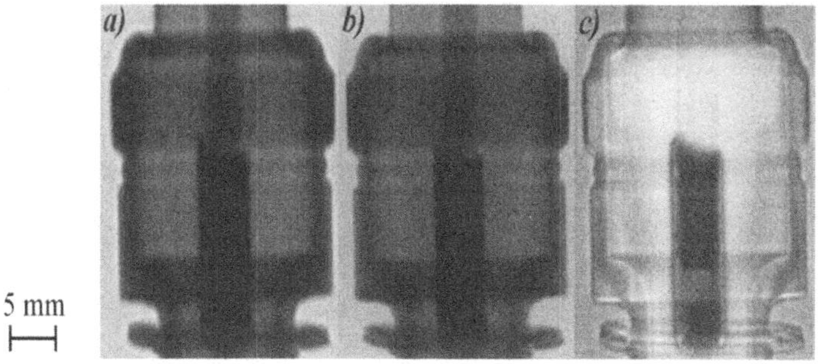

Fig. 5. Energy dispersive radiograms of a spark plug at 0.69 nm (a) and 0.32 nm (b). The division of both images yields(c) [8].

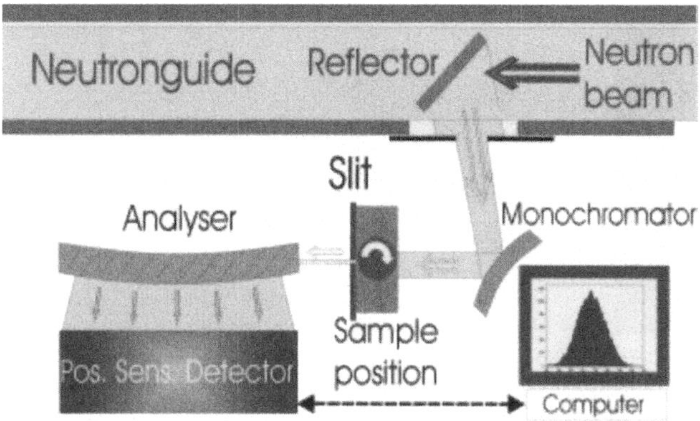

Fig. 6. Special double crystal diffractometer for USANS and refraction and USANS tomography.

Another option for absorption contrast is real time radiography and tomography. Having a strong neutron flux ($> 10^6$ n/cm^{2}s) it is possible to visualize rapid processes in e.g. automotive industry, combustion processes, in hydrodynamics, water flows (with D_2O), stress and strain developments, but also in biology and other fields of science. This option uses the advantage either to trigger periodic processes or to have enough image contrast due to strong absorbing parts in the sample. In both cases the incident neutrons should be preferably cold due to the better contrast behavior.

The experimental results of refraction (also 3D) and USANS tomography are based on the assumption to visualize extreme weakly absorbing structures and distributions of nano particles dissolved in a liquid. The experiments are described in detail in [11], [13], [14] and were performed with a double crystal diffractometer at the BER II reactor of the Hahn-Meitner-Institute (Berlin). This is an instrument, which can measure extreme small angles, less than 0.1 µrad. It works with two highly perfect Si crystals which act as monochromator and analyzer. Neutrons are Bragg reflected by the monochromator to

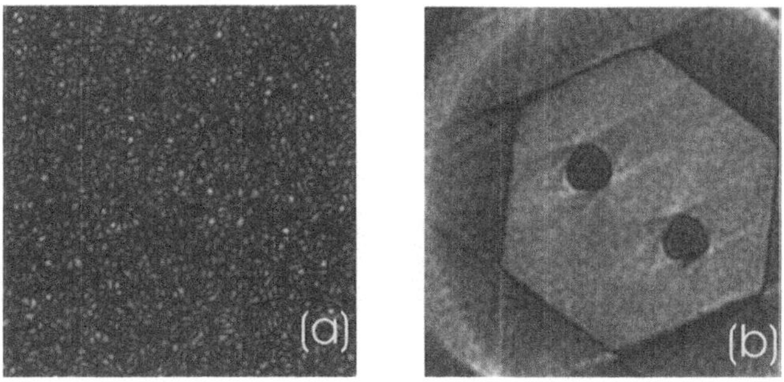

Fig. 7. Absorption (a) and refraction (b) contrast: Al piece, diameter $\sim$ 6 mm, holes 2 mm, monochromatic neutrons, $\lambda = 0.467(1)$ nm

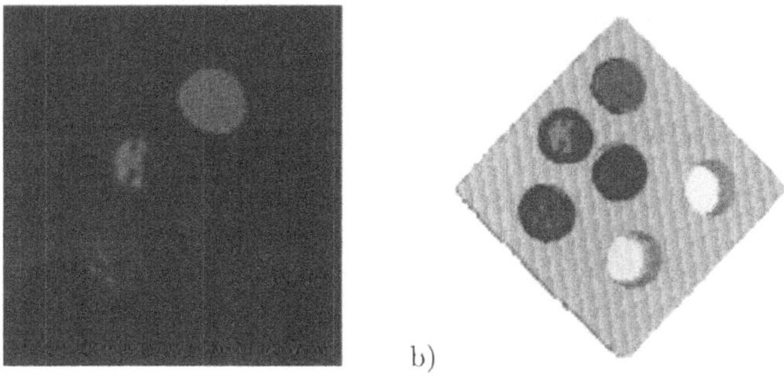

a) b)

Fig. 8. Al box with six holes, four of them filled with D_2O and β-carotene, from top to bottom: 12%, 5.8%, 3.8%, 0% + 2 empty holes. (a) USANS-data only, (b) USANS + refraction tomography data. [13]

the analyzer, which itself reflects them only if the reflecting atomic planes of both are parallel within fraction of μrad. A special version of this DCD is shown in Fig. 6

The sample is put between the monochromator and analyzer and scanned by moving it perpendicular to the neutron beam. The position sensitive detector registers the full rocking curve from which one can deduce the data for absorption- (integral intensity), for refraction- (shift of the rocking curve) and for USANS - (broadening of the rocking curve) tomography. Fig.7 shows some results for refraction tomography.

Also 3D-imaging could be realized, however, with a much less spatial resolution of app. 500 μm [12]. The most astonishing result was the detection of clusters of app. 150 nm large particles (β-carotene), dissolved in D_2O ($5\% - 12\%$). In this experiment both signals, refraction and USANS, could be combined to a single reconstruction showing the different concentration of these nano particles (Fig.8).

Finally it should be mentioned, that tomography with polarized neutrons will open a new wide field for tomographic investigations of magnetic mate-

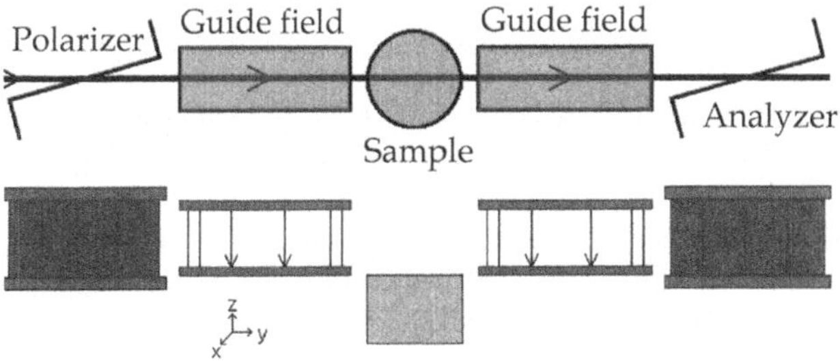

Fig. 9. Experimental set up for CT with polarized neutrons

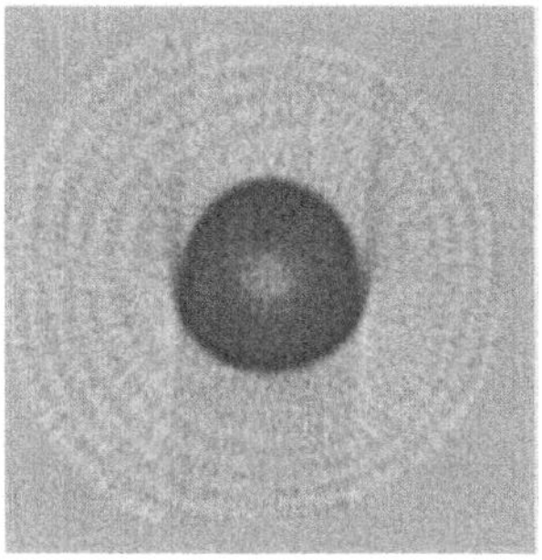

Fig. 10. 2D reconstruction of a stray field of a small permanent magnet, first CT with polarized neutrons.

rials. That this works was realized already many years ago [17], [18]. Fig.9 shows the set up to polarize a neutron beam and analyze its spin state after it has traversed a magnetic field. The spins of the polarized neutrons experience the magnetic stray field of the sample and start to Larmor precession, being depolarized when they reach the analyzer. This causes changes of the beam intensities, that have to be transformed into data, which can be used in classical CT reconstruction algorithms. Due to the limited beam time and because of the symmetry of the stray field of the magnet only one projection was measured and used for reconstruction. Fig.10 shows the reconstruction of the magnetic stray field, the first CT with polarized neutrons.

References

1. M.L. Goldberger and Seitz F. (1947) Phys. Rev. 71, 294
2. J. Radon, Berichte Sächsische Akademie der Wissenschaften 29 (1917) 262
3. G.T. Herman, Topics in Applied Physics, vol. 32, Springer, Berlin, 1979
4. G.T. Herman, Image Reconstruction from Projections, Computer Science and Applied Mathematics, Akademic Press, New York, 1980
5. X. Wu, H. Liu Med. Phys. 30 (8), August (2003)

6. A. Pogany. D. Gao, S. W. Wilkins Rev. Sci. Instrum. 68 (7), July 1997
7. N. Kardjilov, E. Lehmann, E. Steichele, P. Vontobel Nucl. Instr. & Methods in Physics Research A 527 (2004) 519-530
8. N. Kardjilov, S. Baechler, M. Basturk, M. Dierick, J. Jolie, E. Lehmann, T. Materna, B. Schillinger, P. Vontobel, Nuclear Instruments and Methods in Physics Research A 501 (2003), 536-546
9. B. E. Allman, P. J. McMahon, K. A. Nugent, D. Paganin, D. L. Jacobson, M. Arif, and S. A. Werner, Nature (London) 408, 158 (2000).
10. K. M Podurets, V. A. Somenkov, and S. Sh. Shil'shtein, Sov. Phys. Tech. Phys. 34, 654 (1989)
11. W. Treimer, M. Strobl, A. Hilger, C. Seifert, U. Feye-Treimer Appl. Phys. Lett., 83, 2 (2003)
12. M. Strobl, W. Treimer, A. Hilger, Nucl . Instr. & Meth. B 222, 653 - 658 (2004)
13. M. Strobl, W. Treimer, A. Hilger, Appl. Phys. Lett., Vol 85,3, 488-490 (2004)
14. W. Treimer, A. Hilger, N. Kardjilov, M. Strobl, Nucl. Instr. & Meth. A 542, 367 - 375 (2005)
15. N. Kardjilov, Special features of tomography with cold neutrons, presented on DPG 2005
16. N. Kardjilov, S. Baechler, M. Basturk, M. Dierick, J. Jolie, E. Lehmann, T. Materna, B. Schillinger, P. Vontobel, Nucl. Instr. & Methods A, 501,536-546 (2003)
17. W. Treimer, C. Ernst experiments were performed at the E8 instrument at the HMI Berlin
18. W. Treimer, A. Hilger, N. Kardjilov, M. Strobl, Nucl. Instr. & Meth. A 542 , 367 - 375 (2005)

Index

3,4,9,10-perylene-tetracarboxylic-dianhydride, 301

Aarts, Dirk G.A.L., 15
Adem, E., 375
adenine, 45
advanced material, 375
$Ag_{50}Co_{50}$, 327
AlGaN/GaN, 363
amorphous
  metal, 327
  semiconductor, 327
Anderson phase, 109
Andreev reflection, 213, 214
anisotropy ratio, 239
application
  biomedical, 363
  biotechnological, 363
atom
  bosonic, 109
  fermionic, 109
atomic structure, 339

B, 339
Böttcher, Rolf, 351
Bakin, A., 275
band structure, 197
Bardeen-Cooper-Schrieffer
  generalized theory, 214
$BaTiO_3$, 351
Baur, Barbara, 363
Belzig, Wolfgang, 123
Bernard, J., 375
BeTe(100), 301
BeTe/ZnSe, 301
Bethe ansatz, 185, 186, 188
$Bi_2Sr_2CaCu_2O_{8+\delta}$ (Bi-2212), 227
Bichler, Max, 173
bioelectronic device, 363
biological recognition mechanism, 363
biomolecular layer, 45
biosensor, 363
Blonder-Tinkham-Klapwijk
  excess current, 222
Bose-Einstein condensate, 123
Brückner, J., 275
break-junction technique, 215, 217
Bretinger, Helmut, 149
Brownian motion, 15
Bunde, Armin, 59

C(100), 161
capillary waves, 15
Carnot, Sadi, 3
CCD camera, 389
$CeCoIn_5$, 213, 219, 220
central charge, 189
charge transfer event, 123
charge-ordering fluctuation, 227
classical superconductor, 213
Clausius, Rudolf, 3
CMOS, 339
colloid-polymer, 15
complex system, 83, 95
conformal field theory, 185, 189, 194
conventional superconductivity, 214
correlated electron system, 135
correlations
  two-point, 15
counting statistic, 123
critical exponent, 185
critical field, 239
cuprate, 227
Curie temperature, 197
current statistic, 123
cytosine, 45

D'yakonov, 197
Dąbrowski, Jarek, 339
Dahm, Thomas, 239
Decker, M., 275
decoherence time, 173
defect, 327
Del Sole, Rodolfo, 161
delocalization transition, 185
density fluctuation statistic, 123
Density Functional Theory, 161
density functional theory, 45
deterministic dynamical evolution, 83
device, 375
dielectric, 339
diffractometer, 407
diffusion, 29
diffusion coefficient, 59
disorder, 109
DMFT
  linearized, 135
DNA, 45
dynamical mean-field theory (DMFT), 109, 135

Ebeling, Werner, 3
ecology, 95
Eickhoff, Martin, 363
Einstein, Albert, 3, 15
electrolyte gate, 363
electronic structure, 339
ellipsometry, 45
Engelmann, H. J., 375
ensemble
  multi-flavor, 109
EPR
  Cr^{3+}, 351
  spectroscopy, 351
Erb, Andreas, 227
Erdem, Emre, 351
Essler, Fabian H. L., 185
Ester, Patrick, 173
evolution
  directed, 95
excess current, 222
exciton, 275

fermionic quantum gas, 123
ferroelectric, 351
ferroelectric film, 149
Feye-Treimer, U., 407
field-effect transistor, 363
flavor degeneracy, 109
Fleszar, Andrzej, 339
fluctuating force, 83
fluctuation, 83, 95
Frahm, Holger, 185
Friedrich, R., 83
FRM-2, 389
Fujita, Toshitsu, 227

(Ga, Mn)As, 197
GaAs, 197
GaN, 363
Gavrila, G., 313
genus, 75
Geoghegan, Mark, 29
Ginzburg Landau theory, 239
Gläsel, Hans Jürgen, 351
Goll, Gernot, 213
guanine, 45

H_2O, 339
Hackl, Rudi, 227
Hartinger, Christine, 227
Hartmann, Eberhard, 351
Hartree-Fock theory, 135
Hassanein, René , 389
Hauschild, R., 275
heavy-fermion compounds, 219
heavy-fermion superconductors, 213, 219
Helmholtz, Hermann, 3
high-performance microprocessor, 375
Hilger, A., 407
HMI, 389
Hofstetter, Walter, 109
Hooge-parameter, 363
Hubbard model, 135

II-VI semiconductor, 301
ILL, 389
InAs(110), 161
InGaAs, 173
Ingebrandt, Sven, 363
interdigitation, 29
interface property, 301
inverse opal, 149
ion beam analysis, 29

Jensen, Henrik Jeldtoft, 95

Kärger, Jörg, 59
Kalt, H., 275
Kardjilov, N., 407
Kikugawa, Naoki, 227
Kirchner, Ch., 275
Kleinhans, D., 83
Kling, R., 275
Klingshirn, C., 275
Klotzsche, Gert, 351
Knudsen regime, 59
Konjhodzic, Denan, 149

$La_{2-x}Sr_xCuO_4$ (LSCO), 227
Lambert diffusion, 59
Landau quantization, 185
Langevin equation, 83
Lehmann, Eberhard H. , 389
Lekkerkerker, Henk N.W., 15
Levy walk, 59
Li, Hongliang, 149
Lippert, Gunther, 339
localization, 227
Lupina, Grzegorz, 339

Müssig, Hans-Joachim, 339
magnetic relaxation, 95
Mane, Anil, 339
Manke, I., 407
Marlow, Frank, 149
Marsili, Margherita, 161
Mayer, Robert, 3
Mayr, S. G., 327
Mecke, Klaus R., 15
Membrane Dynamics, 71
mesoscopic system, 123
metal interface, 313
metal-insulator transition, 135
metallic glass, 327
MgB_2, 213, 215, 216
MgB_2, 239
Michel, Dieter, 351
microprocessor, 375
microscopy
  scanning confocal, 15
microwave conductivity, 239
Migdal Eliashberg equation, 239
moisture, 339
molecular
  collision, 59
  diffusion, 59
molecular crystal, 313
monolayer, 301
MOSFET, 375
Mott phase, 109
Mott transition, 135
multi-band superconductivity, 213, 215
Murakami, Shuichi, 197

nanoelectronic, 375
nanoparticle, 351
Nb_3Sn, 213, 216, 217
Nernst, Walther, 3
network model, 185, 186
neutron, 407
  cold, 407
  imaging system, 389
  source, 407
  thermal, 407
  tomography, 389, 407
neutron reflectometry, 29
NMR
  ^{137}Ba, 351
  Mn^{2+}, 351
  spectroscopy, 351
noise
  dynamical, 83
nonlinear absorption, 173
nonlinear dynamics, 83

Offenhäusser, Andreas, 363
Ohsiek, S., 375
opal, 149
opalescence, 15
optical lattice, 109
optical spectroscopy, 45
optical spectrum, 161
order-parameter symmetry, 215
organic
  field effect transistor, 313
  light emitting diode, 313
  solar cell, 313
  substrate, 313
organic field effect transistor, 45
organic molecule, 301
organic semiconductor, 313
oxide superconductors, 213, 219

Paez, B. A., 313
pairing
  singlet, 219
  triplet, 219
Palummo, Maurizia, 161
$PbTiO_3$, 351
Peinke, J., 83
percolation, 186, 195
Perel, 197
perturbation theory, 161
phase contrast, 407
photocurrent spectroscopy, 173
photonic crystal, 149
plateau transition, 185, 195
point-contact spectroscopy, 213, 214
Poisson statistic, 95
polariton, 275
polymer
  brush, 29
  diffusion, 29
  network, 29
porous media, 59
Potthoff, Michael, 135
Pr, 339
Priller, H., 275
PrO_x, 339
$PrO_{1.5}$, 339
probability density, 95
$PrSiO_{3.5}$, 339
PTCDA, 301
Pulci, Olivia, 161
pulsed laser field, 173

quadrupole coupling, 351
quake, 95
quantum
  correlations, 123
  dot, 123
  many-body phenomena, 109
  noise, 123
  simulator, 109
quantum Hall effect, 185
quantum spin chain, 185
qubit, 173

Röntgen, C., 389
Rabi oscillation, 173
Raman scattering, 227
Raman spectroscopy, 301, 313
random variable, 83
record dynamics, 95
reflectance anisotropy spectroscopy, 45, 161
refraction, 407
refractive index, 149
Reuss, F., 275
Robie, S., 375
roughness, 59
Russ, Stefanie, 59
ruthenate, 219

Saleur, Hubert, 185
Salvan, G., 313
Sauer, R., 275
Schmeisser, D., 375
Schmidt, Matthias, 15
Schroeder, Thomas, 339
scintillator, 389
self-energy-functional, 135
semiconductor quantum dot, 173
shot noise, 123
Si(111), 45
sigma model, 185
Silaghi, Simona D., 45
silica film, 149
SiNCO, 339
SiO_2, 327, 339
SiO_x, 339
small angle scattering, 407
sol gel, 149
Sorge, Roland, 339
spin, 185
spin glass, 95
spin Hall effect, 197
spin injection, 197
Spin Quantum Hall Effect, 185
spin quantum Hall effect, 186
spin-orbit coupling, 197
spintronic, 197
Sr_2RuO_4, 213, 222
Steinhoff, Georg, 363
Strobl, M., 407
structure analysis, 389
Stufler, Stefan, 173
Stutzmann, Martin, 363
substrate
  Si, 339
superconducting contact, 123

superconductivity, 109, 227
  conventional, 214, 215
  high-temperature, 109
  two-gap, 239
  unconventional, 213
superconductor, 95
  heavy-fermion, 219
  ruthenate, 219
  unconventional, 214, 219
superconductor-superconductor contacts, 217
superspin chain, 185, 186
surface
  metallic, 301
  optical property, 161
  phase, 135
  reconstruction, 301
  state, 135
swelling, 29
symmetry breaking, 109

tangled nature, 95
Tassini, Leonardo, 227
thermodynamics, 3
Thieme, Hans, 339
Thonke, K., 275
thymine, 45
Ti, 339
time reversal symmetry, 95
time series, 83
TiO_2, 149, 339
tomography, 389, 407
Tracy, B., 375
trajectory, 83
transducer, 363
transition temperature, 239
Treimer, W., 407
two-dimensional electron gas, 185
two-level system, 173

ultracold atom, 109
unconventional superconductor, 214, 219
universality, 186, 193, 195
UPS, 301

variational principle, 135
Venturini, Francesca, 227
viscous flow, 327
von Ribbeck, Hans-Georg, 363
Vontobel, Peter , 389

Waag, A., 275
Wagner, Veit, 301
waveguide, 149
Wehmann, H., 275
Wenger, Christian, 339
Wrobel, Günter, 363
WZW model, 186, 190, 193, 195

X-ray, 389
  tomography, 407
XPS, 301

Yang-Baxter equation, 187
$YBa_2Cu_3O_{6+x}$ (Y-123), 227

Zahn, Dietrich R. T., 45, 313
Zaumseil, Peter, 339
Zeller, J., 275
zero-bias conductance anomaly, 221
ZnO, 275
$Zr_{65}Al_{7.5}Cu_{27.5}$, 327
Zrenner, Artur, 173
Zschech, E., 375
Zschiegner, Stephan, 59

Printing: Krips bv, Meppel
Binding: Stürtz, Würzburg

The manufacturer's authorised representative in the EU is Springer Nature Customer Service Centre GmbH, Europaplatz 3, 69115 Heidelberg, Germany. If you have any concerns regarding our products, please contact ProductSafety@springernature.com

Printed and bound by CPI Group (UK) Ltd, Croydon, CR0 4YY
25/09/2026
02222839-0008